Monographs in Theoretical Computer Science
An EATCS Series

More information about this series at http://www.springer.com/series/776

Monographs in Theoretical Computer Science
An EATCS Series

Editors: J. Hromkovič · M. Nielsen
Founding Editors: W. Brauer · G. Rozenberg · A. Salomaa

On behalf of the European Association
for Theoretical Computer Science (EATCS)

More information about this series at http://www.springer.com/series/776

Stanisław Gawiejnowicz

Models and Algorithms of Time-Dependent Scheduling

Second Edition

 Springer

Stanisław Gawiejnowicz
Faculty of Mathematics and Computer Science
Adam Mickiewicz University in Poznań
Poznań, Poland

ISSN 1431-2654   ISSN 2193-2069   (electronic)
Monographs in Theoretical Computer Science. An EATCS Series
ISBN 978-3-662-59364-6   ISBN 978-3-662-59362-2   (eBook)
https://doi.org/10.1007/978-3-662-59362-2

This Springer imprint is published by the registered company Springer-Verlag GmbH, DE part of
Springer Nature.
The registered company address is: Heidelberger Platz 3, 14197 Berlin, Germany

To Agnieszka and Mirosława

Preface

This book is a revised and updated edition of monograph [2]. During the eleven years that elapsed from its publication, time-dependent scheduling significantly developed. This fact is confirmed by the volume of the time-dependent scheduling literature, counting today about three hundred and fifty references published mainly in SCI journals. Time-dependent scheduling problems also began to be discussed in monographs devoted to such domains as multi-agent scheduling [1] or scheduling with rate-modifying activities [3].

The new edition differs substantially from the previous one. First, in order to increase the overall readability of the monograph, its organization has been changed: now it is composed of twenty one chapters organized into six parts and an appendix. Second, the book has been enriched by the results published after 2008, which considerably increased the volume of the book. Third, pseudo-code of the discussed algorithms is presented in a more readable layout using **begin .. end** commands and vertical lines which show the internal structure of the algorithms. Fourth, in order to give the reader a deeper insight into the presented material, a number of new examples have been added. Finally, some misprints and inaccuracies have been corrected.

The previous edition of the book focused on computational complexity of time-dependent scheduling problems. In this edition, not forgetting about complexity issues, we concentrate on models of time-dependent job processing times and algorithms for solving scheduling problems with such job processing times. This is reflected in a slightly different title of the present book.

The first part of the book consists of three chapters and includes the mathematical background of time-dependent scheduling. The aim of this part is to give the reader an introductory view of the presented topics. Therefore, only fundamental notions and concepts are discussed. In Chap. 1, the mathematical notation, basic definitions and auxiliary results used in this book are given. In Chap. 2, essential concepts related to decision problems and algorithms are recalled. The part is completed by Chap. 3, where the definitions and fundamental results of the theory of $\mathcal{NP}$-completeness are presented.

The second part of the book includes the basic definitions, notions and models of the whole scheduling theory. This part is composed of three chapters. In Chap. 4, the basics of the classical scheduling theory are given, where job processing times are numbers. In Chap. 5, a similar material concerning the modern scheduling theory is presented, where job processing times are position- or resource-dependent functions. Chapter 6, closing the second part, concerns time-dependent scheduling, where job processing times are functions of the starting times of the jobs, and discusses terminology, notation and applications of that domain.

Each chapter of these two parts is completed with bibliographic notes including a list of selected references in which the reader may find a more detailed presentation of these topics.

The next two parts include detailed discussion of polynomially solvable and $\mathcal{NP}$-hard time-dependent scheduling problems, respectively.

The third part of the book consists of three chapters and is devoted to polynomially solvable time-dependent scheduling problems. Chaps. 7, 8 and 9 are devoted to single, parallel and dedicated machine time-dependent scheduling problems which can be solved in polynomial time.

The fourth part of the book, organized in a similar way as the previous one, focuses on intractable time-dependent scheduling problems. In Chaps. 10, 11 and 12 are discussed $\mathcal{NP}$-hard single, parallel and dedicated machine time-dependent scheduling problems.

For most of the results presented in Chaps. 7–12 there is given either the pseudo-code of a polynomial algorithm or a sketch of an $\mathcal{NP}$-hardness proof. Each chapter is completed with a list of references.

The fifth part of the book is dedicated to algorithms for intractable time-dependent scheduling problems. This part consists of five chapters. Chapter 13 presents exact algorithms for $\mathcal{NP}$-hard single, parallel and dedicated machine time-dependent scheduling problems. The main emphasis is given to branch-and-bound algorithms which are the most popular exact algorithms. The next four chapters concern different classes of approximation and heuristic algorithms. In Chap. 14, approximation algorithms and approximation schemes for single, parallel and dedicated machine time-dependent scheduling problems are discussed. In Chap. 15, two greedy algorithms for a single machine time-dependent scheduling problem are presented. The fifth part is completed with Chaps. 16 and 17, where we present constructive heuristic and meta-heuristic algorithms for time-dependent scheduling problems, respectively. Each of the chapters ends with a list of references.

The last, sixth part of the book includes advanced topics in time-dependent scheduling. This part consists of four chapters. In Chap. 18, results on time-dependent scheduling with precedence constraints are presented. In Chap. 19, matrix methods and their application to time-dependent scheduling are studied. In Chap. 20, bi-criteria time-dependent scheduling problems are considered. The sixth part of the book is completed with Chap. 21, where new topics in time-dependent scheduling are addressed, such as scheduling on machines with limited availability, scheduling with mixed job processing times and two-agent scheduling problems. As previously, each chapter of this part is completed with a list of references.

The sixth part is followed by an appendix, where several most challenging open time-dependent scheduling problems are discussed. For each of these problems, a summary of principal results and a list of references are given.

The book is intended for researchers in the scheduling theory, Ph. D. students and everybody interested in recent advances in computer science. Though some parts of the book are addressed to more advanced readers, it can also be used by graduate students, since the prerequisities for reading the book are only the standard courses in discrete mathematics and calculus, fundamentals of the theory of algorithms and basic knowledge of any high-level programming language.

The book can be used as a reference for different courses. The first two parts of the book can serve as a basis for an introductory course in modern scheduling theory. The material from the next three parts can be used in a one-semester course on time-dependent scheduling. The last part and appendix may be a starting point for a research seminar in time-dependent scheduling.

I wish sincerely to thank my co-authors: Alessandro Agnetis (University of Siena, Italy), Jean-Charles Billaut (University of Tours, France), Ming-Huang Chiang (National Taiwan University, Taipei, Taiwan), Alexander Kononov (Sobolev Institute of Mathematics, Novosibirsk, Russia), Wiesław Kurc (Adam Mickiewicz University Poznań, Poland), Tsung-Chyan Lai (Harbin Engineering University, Harbin, P. R. China) Wen-Chiung Lee (Feng Chia University, Taichung, Taiwan), Bertrand Miao-Tsong Lin (National Chiao Tung University, Hsinchu, Taiwan), Dario Pacciarelli (Roma Tre University, Rome, Italy), Lidia Pankowska (Adam Mickiewicz University Poznań, Poland), Ameur Soukhal (University of Tours, France), Cezary Suwalski (Adam Mickiewicz University Poznań, Poland) and Chin-Chia Wu (Feng Chia University, Taichung, Taiwan) for joint work on different modern scheduling problems. A number of the results presented in the book were obtained jointly with these researchers.

I also wish to thank the following people for remarks on some of time-dependent scheduling problems discussed in the book: Jin-Yi Cai (University of Wisconsin-Madison, Madison, USA), Nir Halman (Hebrew University of Jerusalem, Jerusalem, Israel), Mikhail Y. Kovalyov (National Academy of Sciences of Belarus, Minsk, Belarus), Gur Mosheiov (Hebrew University of Jerusalem, Jerusalem, Israel), Dvir Shabtay (Ben Gurion University of the Negev, Beer Sheeva, Israel), and Vitaly A. Strusevich (University of Greenwich, London, United Kingdom).

I thank Ronan Nugent (Springer, Heidelberg) for constructive cooperation while working on this book.

I direct special thanks to Mrs. Krystyna Ciesielska, M.A., M.Sc., who helped me to improve the English of this book.

Last but not the least, I thank my wife Mirosława and my daughter Agnieszka for their love and support.

References

1. A. Agnetis, J-C. Billaut, S. Gawiejnowicz, D. Pacciarelli and A. Soukhal, *Multiagent Scheduling: Models and Algorithms*, Berlin-Heidelberg: Springer 2014.
2. S. Gawiejnowicz, *Time-Dependent Scheduling*, Berlin-Heidelberg: Springer 2008.
3. V. A. Strusevich and K. Rustogi, *Scheduling with Time-Changing Effects and Rate-Modifying Activities*. Berlin-Heidelberg: Springer 2017.

Poznań, September 2019   *Stanisław Gawiejnowicz*

Preface to the first edition

The book presented to the reader is devoted to time-dependent scheduling. Scheduling problems, in general, consist in the allocation of resources over time in order to perform a set of jobs. Any allocation that meets all requirements concerning the jobs and resources is called a feasible schedule. The quality of a schedule is measured by a criterion function. The aim of scheduling is to find, among all feasible schedules, a schedule that optimizes the criterion function. A solution to an arbitrary scheduling problem consists in giving a polynomial-time algorithm generating either an optimal schedule or a schedule which is close to the optimal one, if the given scheduling problem has been proved to be computationally intractable. The scheduling problems are subject of interest of the scheduling theory, originated in mid-fifties of the twentieth century. The theory has been developing dynamically and new research areas constantly come into existence. The subject of this book, time-dependent scheduling, is one of such areas.

In time-dependent scheduling, the processing time of a job is variable and depends on the starting time of the job. This crucial assumption allows us to apply the scheduling theory to a broader spectrum of problems. For example, in the framework of the time-dependent scheduling theory we may consider the problems of repayment of multiple loans, fire fighting and maintenance assignments. In this book, we will discuss algorithms and complexity issues concerning various time-dependent scheduling problems.

Time-dependent scheduling is a relatively new subject. Although the first paper from the area appeared in late 1970s, most results have been published in the last 10 years. So far, time-dependent scheduling has not gained much attention in books devoted to the scheduling theory. This book, summarizing the results of almost 15 years of the author's research into time-dependent scheduling, hopefully fills this gap.

The book is composed of fourteen chapters, organized into four parts.

The first part of the book consists of five chapters and includes the mathematical background used in subsequent chapters. The aim of the part is to give the reader an introductory view of presented topics. Therefore, only fundamental notions and concepts are discussed. In Chap. 1, the mathematical notation, the basic definitions and results used in this book are given. In Chap. 2, the essential concepts related to decision problems and algorithms are recalled. Chapter 3 includes the definitions and fundamental results of the theory of $\mathcal{NP}$-completeness. The part is completed by Chaps. 4 and 5, where the basics of the scheduling theory and time-dependent scheduling theory are given, respectively. Each chapter of this part is completed with bibliographic notes including a list of selected references in which the reader may find a more comprehensive presentation of the particular topics.

The second part of the book includes a detailed survey of the time complexity of time-dependent scheduling problems. This part is composed of three chapters. In Chap. 6, single-machine time-dependent scheduling problems are discussed. Chapters 7 and 8 cover results concerning time-dependent scheduling on parallel and dedicated machines, respectively.

The third part of the book is devoted to suboptimal algorithms for $\mathcal{NP}$-hard time-dependent scheduling problems. This part starts with Chap. 9, which presents constructive approximation and heuristic algorithms. Chap. 10 introduces two greedy algorithms which exploit the properties of the so-called signatures of sequences of job deterioration rates. Finally, local search heuristics for time-dependent scheduling problems are discussed in Chap. 11.

The fourth part of the book includes selected advanced topics in time-dependent scheduling. This part begins with Chap. 12, in which applications of matrix methods to time-dependent scheduling problems are discussed. Chapter 13 is devoted to scheduling proportionally and linearly deteriorating jobs under precedence constraints. In Chap. 14, closing the book, time-dependent scheduling problems with two criteria are studied.

Each chapter of these two parts ends with concluding remarks. Chapters of the fourth part include also comments on selected open problems.

The book is intended for researchers into the scheduling theory, Ph. D. students and everybody interested in recent advances in computer science.

The prerequisities for reading the book are the standard courses in discrete mathematics and calculus, fundamentals of the theory of algorithms and basic knowledge of any high-level programming language. Hence, this book can also be used by students of graduate studies.

The second part of the book can serve as a basis for an introductory course in time-dependent scheduling. The material from the next two parts can be used as a starting point for a research seminar in time-dependent scheduling.

The research presented in the book has been partially supported by grant N519 18889 33 of the Ministry of Science and Higher Education of Poland. While working on the book, I was also supported in different ways by different people. It is my pleasure to list here the names of the people to whom I am mostly indebted for help.

I heartily thank Dr. Alexander Kononov (Sobolev Institute of Mathematics, Novosibirsk, Russia), Dr. Wiesław Kurc and Dr. Lidia Pankowska (both from Adam Mickiewicz University, Poznań, Poland) for many stipulating discussions on different aspects of time-dependent scheduling. Many of the results presented in the book are effects of my joint work with these researchers.

I sincerely thank Professor Jacek Błażewicz (Poznań University of Technology, Poznań, Poland) for his continuous encouragement and support during many years of my research into time-dependent scheduling.

I also thank the following people for help in obtaining references concerning time-dependent scheduling: Gerd Finke (Leibniz-IMAG, Grenoble, France), Yi-Chih Hsieh (National Chengchi University, Hsinchu, Taiwan), Shi-Er Ju (Zhongshan University, Guangzhou, China), Li-Ying Kang (Shanghai University, Shanghai, P. R. China), Mikhail Y. Kovalyov (National Academy of Sciences of Belarus, Minsk, Belarus), Wiesław Kubiak (Memorial University of Newfounland, St. John's, Canada), Bertrand Miao-Tsong Lin (National

Chiao Tung University, Hsinchu, Taiwan), Yakov M. Shafransky (National Academy of Sciences of Belarus, Minsk, Belarus), Prabha Sharma (Indian Institute of Technology Kanpur, Kanpur, India), Vitaly A. Strusevich (University of Greenwich, London, United Kingdom), Yoichi Uetake (Adam Mickiewicz University, Poznań, Poland), Ji-Bo Wang (Shenyang Institute of Aeronautical Engineering, Shenyang, China), Gerhard J. Woeginger (Eindhoven University of Technology, Eindhoven, The Netherlands), Dar-Li Yang (National Formosa University, Yun-Lin, Taiwan).

I direct special thanks to Mrs. Krystyna Ciesielska, M.A., M.Sc., who helped me to improve the English of this book.

Last but not the least, I thank my wife Mirosława and my daughter Agnieszka for their love, patience and support during the many months of my work on the book, when I was not able to be with them.

Poznań, April 2008   *Stanisław Gawiejnowicz*

Contents

Part II SCHEDULING MODELS

Part IV NP-HARD PROBLEMS

Part VI ADVANCED TOPICS

Part A APPENDIX

FUNDAMENTALS

1

Preliminaries

The scheduling theory uses notions and methods from different disciplines of mathematics. Therefore, any systematic presentation of an arbitrary domain in the theory needs some mathematical background. The first part of the book introduces this background.

This part of the book is composed of three chapters. In Chap. 1, we present the mathematical notation, basic definitions and auxiliary results. The essential concepts related to decision problems and algorithms are recalled in Chap. 2. The definitions and the most important results of the theory of $\mathcal{NP}$-completeness are presented in Chap. 3.

Chapter 1 is composed of two sections. In Sect. 1.1, we introduce the notation and terminology used in this book. In Sect. 1.2, we give some mathematical preliminaries, used in subsequent chapters. The chapter is completed with bibliographic notes and a list of references.

1.1 Mathematical notation and inference rules

We assume that the reader is familiar with basic mathematical notions. Therefore, we explain here only the notation that will be used throughout this book.

1.1.1 Sets and vectors

We will write $a \in A$ ($a \notin A$) if a is (is not) an element of a set A. If $a_1 \in A$, $a_2 \in A, \ldots, a_n \in A$, we will simply write $a_1, a_2, \ldots, a_n \in A$.

If an element a is (is not) equal to an element b, we will write $a = b$ ($a \neq b$). If $a = b$ by definition, we will write $a := b$. In a similar way, we will denote the equality (inequality) of numbers, sets, sequences, etc.

The set composed only of elements $a_1, a_2, \ldots, a_n$ will be denoted by $\{a_1, a_2, \ldots, a_n\}$. The maximal (minimal) element in set $\{a_1, a_2, \ldots, a_n\}$ will be denoted by $\max\{a_1, a_2, \ldots, a_n\}$ ($\min\{a_1, a_2, \ldots, a_n\}$).

© Springer-Verlag GmbH Germany, part of Springer Nature 2020
S. Gawiejnowicz, *Models and Algorithms of Time-Dependent Scheduling*,
Monographs in Theoretical Computer Science. An EATCS Series,
https://doi.org/10.1007/978-3-662-59362-2_1

If set A is a subset of set B, i.e., every element of set A is an element of set B, we will write $A \subseteq B$. If A is a strict subset of B, i.e., $A \subseteq B$ and $A \neq B$, we will write $A \subset B$. The empty set will be denoted by $\emptyset$.

The number of elements of set A will be denoted by $|A|$. The power set of set A, i.e., the set of all subsets of A, will be denoted by 2^A.

For any sets A and B, the union, intersection and difference of A and B will be denoted by $A \cup B$, $A \cap B$ and $A \setminus B$, respectively.

The Cartesian product of sets A and B will be denoted by $A \times B$. The Cartesian product of $n \geqslant 2$ copies of a set A will be denoted by A^n.

A partial order, i.e., a reflexive, antisymmetric and transitive binary relation, will be denoted by $\prec$. If $x \prec y$ or $x = y$, we will write $x \preceq y$.

The set-theoretic sum (product) of all elements of a set A will be denoted by $\bigcup_{a_i \in A} a_i$ ($\bigcap_{a_i \in A} a_i$). The union (intersection) of a family of sets A_k, $k \in K$, will be denoted by $\bigcup_{k \in K} A_k$ ($\bigcap_{k \in K} A_k$).

The sets of all natural, integer, rational and real numbers will be denoted by $\mathbb{N}$, $\mathbb{Z}$, $\mathbb{Q}$ and $\mathbb{R}$, respectively. The subsets of positive elements of sets $\mathbb{Z}$, $\mathbb{Q}$ and $\mathbb{R}$ will be denoted by $\mathbb{Z}_+$, $\mathbb{Q}_+$ and $\mathbb{R}_+$, respectively. The subset of $\mathbb{N}$ composed of the numbers which are not greater than a fixed $n \in \mathbb{N}$ will be denoted by I_n or $\{1, 2, \ldots, n\}$.

Given a set A and a property p_A, we will write $B = \{a \in A : p_A \text{ holds for } a\}$ to denote that B is the set of all elements of set A for which property p_A holds. For example, a closed interval $\langle a, b \rangle$ for $a, b \in \mathbb{R}$, $a \leqslant b$, will be defined as the set $\langle a, b \rangle := \{x \in \mathbb{R} : a \leqslant x \leqslant b\}$.

A $(n \geqslant 1)$-dimensional vector space over $\mathbb{R}$, its positive orthant and the interior of the orthant will be denoted by $\mathbb{R}^n$, $\mathbb{R}^n_+$ and $\mathrm{int}\mathbb{R}^n_+$, respectively.

A row (column) vector $x \in \mathbb{R}^n$ composed of numbers $x_1, x_2, \ldots, x_n$ will be denoted by $x = [x_1, x_2, \ldots, x_n]$ ($x = [x_1, x_2, \ldots, x_n]^\top$). A norm (the l_p norm) of vector x will be denoted by $\|x\|$ ($\|x\|_p$). The scalar product of vectors x and y will be denoted by $x \circ y$.

1.1.2 Sequences

A sequence composed of numbers $x_1, x_2, \ldots, x_n$ will be denoted by $(x_j)_{j=1}^n$ or $(x_1, x_2, \ldots, x_n)$. If the range of indices of elements of sequence $(x_j)_{j=1}^n$ is fixed, the sequence will be denoted by (x_j). In a similar way we will denote sequences of sequences, e.g., the sequence of pairs $(x_1, y_1), (x_2, y_2), \ldots, (x_n, y_n)$ will be denoted by $((x_j, y_j))_{j=1}^n$, $((x_1, y_1), (x_2, y_2), \ldots, (x_n, y_n))$ or $((x_j, y_j))$.

A sequence (z_k) that is a concatenation of sequences (x_i) and (y_j) will be denoted by $(x_i | y_j)$. If A and B are sets of numbers, the sequence composed of elements of A followed by elements of B will be denoted by $(A|B)$.

A sequence (x_j) in which elements are arranged in non-decreasing (non-increasing) order will be denoted by $(x_j \nearrow)$ ($(x_j \searrow)$). An empty sequence will be denoted by (ϕ).

The algebraic sum (product) of numbers $x_k, x_{k+1}, \ldots, x_m$ for $k, m \in \mathbb{N}$, will be denoted by $\sum_{i=k}^m x_i$ ($\prod_{i=k}^m x_i$). If the indices of components of the sum (product) belong to a set J, then the sum (product) will be denoted by $\sum_{j \in J} x_j$ ($\prod_{j \in J} x_j$). If $k > m$ or $J = \emptyset$, then $\sum_{i=k}^{m} x_i = \sum_{j \in J} x_j := 0$ and $\prod_{i=k}^{m} x_i = \prod_{j \in J} x_j := 1$.

1.1.3 Functions

A function f from a set X to a set Y will be denoted by $f : X \to Y$. The value of function $f : X \to Y$ for some $x \in X$ will be denoted by $f(x)$.

If a function f is a monotonically increasing (decreasing) function, we will write $f \nearrow$ ($f \searrow$).

For a given set X, the function f such that $f(x) = 1$ if $x \in X$ and $f(x) = 0$ if $x \notin X$ will be denoted $\mathbf{1}_X$.

The absolute value, the binary logarithm and the natural logarithm of $x \in \mathbb{R}$ will be denoted by $|x|$, $\log x$ and $\ln x$, respectively. The largest (smallest) integer number not greater (not less) than $x \in \mathbb{R}$ will be denoted by $\lfloor x \rfloor$ ($\lceil x \rceil$).

Given two functions, $f : \mathbb{N} \to \mathbb{R}_+$ and $g : \mathbb{N} \to \mathbb{R}_+$, we will say that function $f(n)$ *is of order* $O(g(n))$, in short $f(n) = O(g(n))$, if there exist constants $c > 0$ and $n_0 \geqslant 0$ such that for all $n \geqslant n_0$ the inequality $f(n) \leqslant cg(n)$ holds. We will write $f(n) = \Omega(g(n))$, if $g(n) = O(f(n))$. If both $f(n) = O(g(n))$ and $f(n) = \Omega(g(n))$, we will write $f(n) = \Theta(g(n))$.

Permutations of elements of set I_n, i.e., bijective functions from set I_n onto itself, will be denoted by small Greek characters. For example, permutation σ with components $\sigma_1, \sigma_2, \ldots, \sigma_n$, where $\sigma_i \in I_n$ for $1 \leqslant i \leqslant n$ and $\sigma_i \neq \sigma_j$ for $i \neq j$, will be denoted by $\sigma = (\sigma_1, \sigma_2, \ldots, \sigma_n)$. In some cases, permutations will also be denoted by small Greek characters with a superscript. For example, σ' and σ'' will refer to two distinct permutations of elements of set I_n. Partial permutations defined on I_n, i.e., bijective functions between two subsets of set I_n, will be denoted by small Greek characters with a superscript in brackets. For example, $\sigma^{(a)} = (\sigma_1^{(a)}, \sigma_2^{(a)}, \ldots, \sigma_k^{(a)})$ is a partial permutation of elements of set I_n. The set of all permutations (partial permutations) of set I_n will be denoted by $\mathfrak{S}_n$ ($\hat{\mathfrak{S}}_n$).

The sequence $(x_j)_{j=1}^{n}$ composed of numbers $x_1, x_2, \ldots, x_n$, ordered according to permutation $\sigma \in \mathfrak{S}_n$, will be denoted by $x_\sigma = (x_{\sigma_1}, x_{\sigma_2}, \ldots, x_{\sigma_n})$.

Because of the nature of the problems considered in this book, we will assume, unless stated otherwise, that all objects (e.g. sets, sequences, etc.) are finite.

1.1.4 Logical notation and inference rules

A negation, conjunction and disjunction will be denoted by $\neg$, $\wedge$ and $\vee$, respectively. The implication of formulae p and q will be denoted by $p \Rightarrow q$. The equivalence of formulae p and q will be denoted by $p \Leftrightarrow q$ or $p \equiv q$. The existential and general quantifiers will be denoted by $\exists$ and $\forall$, respectively.

In the proofs presented in this book, we will use a few proof techniques. The most often applied proof technique is the *pairwise job (element) interchange argument*: we consider two schedules (sequences) that differ only in the order of two jobs (elements) and we show which schedule (sequence) is the better one. A number of proofs are made *by contradiction*: we assume that a schedule (a sequence) is optimal and we show that this assumption leads to a contradiction. Finally, some proofs are made by *mathematical induction*.

The use of the rules of inference applied in the proofs will be limited mainly to *De Morgan's rules* $(\neg(p \wedge q) \equiv (\neg p \vee \neg q), \neg(p \vee q) \equiv (\neg p \wedge \neg q))$, *material equivalence* $((p \Leftrightarrow q) \equiv ((p \Rightarrow q) \wedge (q \Rightarrow p)))$ and *transposition* $((p \Rightarrow q) \equiv (\neg q \Rightarrow \neg p))$ rules.

1.1.5 Other notation

Lemmas, theorems and properties will be numbered consecutively in each chapter. In a similar way, we will number definitions, examples, figures and tables. Examples will be ended by the symbol '♦'. Algorithms will be numbered separately in each chapter.

Most results will be followed either by a full proof or by the sketch of a proof. In several cases, no proof (sketch) will be given and the reader will be referred to the literature. The proofs, sketches and references to the sources of proofs will be ended by symbols '■', '□' and '◇', respectively.

1.2 Basic definitions and results

In this section, we include the definitions and results that are used in proofs presented in this book.

1.2.1 Elementary lemmas

Lemma 1.1. (Elementary inequalities)
(a) *If $y_1, y_2, \ldots, y_n \in \mathbb{R}$, then $\max\{y_1, y_2, \ldots, y_n\} \geqslant \frac{1}{n} \sum_{j=1}^{n} y_j$.*
(b) *If $y_1, y_2, \ldots, y_n \in \mathbb{R}$, then $\frac{1}{n} \sum_{j=1}^{n} y_j \geqslant \sqrt[n]{\prod_{j=1}^{n} y_j}$.*
(c) *If $a, x \in \mathbb{R}$, $x \geqslant -1$, $x \neq 0$ and $0 < a < 1$, then $(1+x)^a < 1 + ax$.*

Proof. (a) This is the *arithmetic-mean inequality*; see Bullen et al. [9, Chap. 2, Sect. 1, Theorem 2].

(b) This is a special case of the *geometric-arithmetic mean inequality*; see Bullen et al. [9, Chap. 2, Sect. 2, Theorem 1].

(c) This is *Bernoulli's inequality*; see Bullen et al. [9, Chap. 1, Sect. 3, Theorem 1].   ◇

Lemma 1.2. (Minimizing or maximizing a sum of products)
(a) *If $x_1, x_2, \ldots, x_n, y_1, y_2, \ldots, y_n \in \mathbb{R}$, then the sum $\sum_{i=1}^{n} x_{\sigma_i} \prod_{j=i+1}^{n} y_{\sigma_j}$ is minimized (maximized) when it is calculated over the permutation $\sigma \in \mathfrak{S}_n$ in which indices are ordered by non-decreasing (non-increasing) values of the ratios $\frac{x_i}{y_i - 1}$.*
(b) *If $(x_1, x_2, \ldots, x_n)$ and $(y_1, y_2, \ldots, y_n)$ are two sequences of real numbers, then the sum $\sum_{j=1}^{n} x_j y_j$ is minimized if the sequence $(x_1, x_2, \ldots, x_n)$ is ordered non-decreasingly and the sequence $(y_1, y_2, \ldots, y_n)$ is ordered non-increasingly or vice versa, and it is maximized if the sequences are ordered in the same way.*

Proof. (a) By the pairwise element interchange argument; see Kelly [18, Theorems 1-2], Rau [23, Theorem 1], and Strusevich and Rustogi [26, Chap. 2].

(b) By the pairwise element interchange argument; see Hardy et al. [16, p. 269].   ◇

Definition 1.3. (*V*-shaped and Λ-shaped sequences)
(a) *A sequence $(x_1, x_2, \ldots, x_n)$ is said to be V-shaped (has a V-shape), if there exists an index k, $1 \leqslant k \leqslant n$, such that for $1 \leqslant j \leqslant k$ the sequence is non-increasing and for $k \leqslant j \leqslant n$ the sequence is non-decreasing.*
(b) *A sequence $(x_1, x_2, \ldots, x_n)$ is said to be Λ-shaped (has a Λ-shape) if the sequence $(-x_1, -x_2, \ldots, -x_n)$ is V-shaped.*

In other words, sequence $(x_j)_{j=1}^{n}$ is *V*-shaped (Λ-shaped) if the elements which are placed before the smallest (largest) x_j, $1 \leqslant j \leqslant n$, are arranged in non-increasing (non-decreasing) order, and those which are placed after the smallest (largest) x_j are in non-decreasing (non-increasing) order.

The *V*-shaped and Λ-shaped sequences will be also called *V-sequences* and Λ-*sequences*, respectively. Moreover, if index k of the minimal (maximal) element in a *V*-sequence (Λ-sequence) satisfies the inequality $1 < k < n$, we will say that this sequence is *strongly V-shaped (Λ-shaped)*.

Remark 1.4. The V-shapeness of a schedule is an important property of some scheduling problems with variable job processing times; see, e.g., Strusevich and Rustogi [26, Chap. 5] and Sect. A.1 for details.

Definition 1.5. (The partial order relation $\prec$)
Let $(u, v), (r, s) \in \mathbb{R}^2$. The partial order relation $\prec$ is defined as follows:

$$(u, v) \prec (r, s) \text{ if } (u, v) \leqslant (r, s) \text{ coordinatewise and } (u, v) \neq (r, s). \quad (1.1)$$

Lemma 1.6. *The relation $(u, v) \prec (0, 0)$ does not hold when either $(u > 0$ or $v > 0)$ or $(u = 0$ and $v = 0)$.*

Proof. By Definition 1.5, $(u, v) \prec (0, 0)$ if $(u, v) \leqslant (0, 0)$ coordinatewise and $(u, v) \neq (0, 0)$. By negation of the conjuction, the result follows.   ∎

1.2.2 Graph theory definitions

Definition 1.7. (A graph and a digraph)
(a) *A graph (undirected graph) is an ordered pair $G = (N, E)$, where $N \neq \emptyset$ is a finite set of nodes and $E \subseteq \{\{n_1, n_2\} \in 2^N : n_1 \neq n_2\}$ is a set of edges.*
(b) *A digraph (directed graph) is an ordered pair $G = (V, A)$, where $V \neq \emptyset$ is a finite set of vertices and $A \subseteq \{(v_1, v_2) \in V^2 : v_1 \neq v_2\}$ is a set of arcs.*

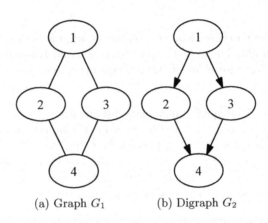

(a) Graph G_1 (b) Digraph G_2

Fig. 1.1: Graph vs. digraph

Example 1.8. Consider graph G_1 and digraph G_2 given in Figure 1.1.

In the graph $G_1 = (N, E)$, presented in Figure 1.1a, the set of nodes $N = \{1, 2, 3, 4\}$ and the set of edges $E = \{\{1, 2\}, \{1, 3\}, \{2, 4\}, \{3, 4\}\}$.

In the digraph $G_2 = (V, A)$, presented in Figure 1.1b, the set of vertices $V = \{1, 2, 3, 4\}$ and the set of arcs $A = \{(1, 2), (1, 3), (2, 4), (3, 4)\}$.   ♦

In this book, we will consider mainly directed graphs. Therefore, all further definitions and remarks will refer to digraphs, unless stated otherwise.

Definition 1.9. (Basic definitions concerning digraphs)
(a) *A digraph $G' = (V', A')$ is called a* subdigraph *of a digraph $G = (V, A)$ if $V' \subseteq V$ and $(x, y) \in A'$ implies $(x, y) \in A$.*
(b) *A directed path in a digraph $G = (V, A)$ is a sequence $(v_1, v_2, \ldots, v_m)$ of distinct vertices from V such that $(v_k, v_{k+1}) \in A$ for each $k = 1, 2, \ldots, m-1$. The number m is called the* length *of the path.*
(c) *A vertex $x \in V$ is called a* predecessor (successor) *of a vertex $y \in V$ if in a digraph $G = (V, A)$ there is a directed path from x to y (from y to x). If the path has unit length, then x is called a* direct predecessor (successor) *of y.*

(d) *A vertex* $x \in V$ *that has no direct predecessor (successor) is called an* initial *(a* terminal) *vertex in a digraph* $G = (V, A)$. *A vertex* $x \in V$ *that is neither initial nor terminal is called an* internal *vertex in the digraph.*
(e) *A digraph* $G = (V, A)$ *is* connected *if for every* $x, y \in V$ *there exists in* G *a directed path starting with* x *and ending with* y; *otherwise, it is* disconnected.

For a given graph (digraph) G and $v \in N$ ($v \in V$), the set of all predecessors and successors of v will be denoted by $Pred(v)$ and $Succ(v)$, respectively.

Definition 1.10. (Parallel and series composition of digraphs)
Let $G_1 = (V_1, A_1)$ *and* $G_2 = (V_2, A_2)$ *be two digraps such that* $V_1 \cap V_2 = \emptyset$ *and let* $Term(G_1) \subseteq V_1$ *and* $Init(G_2) \subseteq V_2$ *denote the set of terminal vertices of* G_1 *and the set of initial vertices of* G_2, *respectively. Then*
(a) *digraph* G_P *is said to be a* parallel composition *of digraphs* G_1 *and* G_2 *if* $G_P = (V_1 \cup V_2, A_1 \cup A_2)$;
(b) *digraph* G_S *is said to be a* series composition *of digraphs* G_1 *and* G_2 *if* $G_S = (V_1 \cup V_2, A_1 \cup A_2 \cup (Term(G_1) \times Init(G_2)))$.

In other words, digraph G_P is a disjoint union of digraphs G_1 and G_2, while digraph G_S is a composition of G_1, G_2 and the arcs from all terminal vertices in G_1 are connected to all initial vertices in G_2.

Definition 1.11. (Special classes of digraphs)
(a) *A* chain $(v_1, v_2, \ldots, v_k)$ *is a digraph* $G = (V, A)$ *with* $V = \{v_i : 1 \leqslant i \leqslant k\}$ *and* $A = \{(v_i, v_{i+1}) : 1 \leqslant i \leqslant k - 1\}$.
(b) *An* in-tree (out-tree) *is a digraph which is connected, has a single terminal (initial) vertex called the* root *of this in-tree (out-tree) and in which any other vertex has exactly one direct successor (predecessor). The initial (terminal) vertices of an in-tree (out-tree) are called* leaves.
(c) *A digraph* $G = (V, A)$ *is a* series-parallel *digraph (sp-digraph, in short) if either* $|V| = 1$ *or* G *is obtained by application of parallel or series composition to two series-parallel digraphs* $G_1 = (V_1, A_1)$ *and* $G_2 = (V_2, A_2)$, $V_1 \cap V_2 = \emptyset$.

Remark 1.12. A special type of a tree is a *2-3 tree*, i.e., a balanced tree in which each internal node (vertex) has two or three successors. In 2-3 trees the operations of insertion (deletion) of a node (vertex) and the operation of searching through the tree can be implemented in $O(\log k)$ time, where k is the number of nodes (vertices) in the tree (see, e.g., Aho et al. [6, Chap. 2]).

Example 1.13. Consider the four digraphs depicted in Figure 1.2. The chain given in Figure 1.2a is a digraph $G_1 = (V_1, A_1)$ in which $V_1 = \{1, 2\}$ and $A_1 = \{(1, 2)\}$. The in-tree given in Figure 1.2b is a digraph $G_2 = (V_2, A_2)$ in which $V_2 = \{1, 2, 3\}$ and $A_2 = \{(2, 1), (3, 1)\}$. The out-tree given in Figure 1.2c is a digraph $G_3 = (V_3, A_3)$ in which $V_3 = \{1, 2, 3\}$ and $A_3 = \{(1, 2), (1, 3)\}$.
   The sp-digraph depicted in Figure 1.2d is a digraph $G_4 = (V_4, A_4)$ in which $V_4 = \{1, 2, 3, 4\}$ and $A_4 = \{(1, 3), (1, 4), (2, 3), (2, 4)\}$. The sp-digraph G_4 is

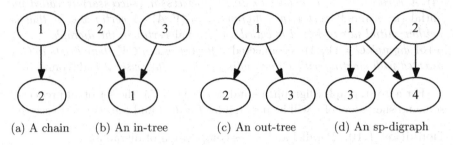

(a) A chain   (b) An in-tree   (c) An out-tree   (d) An sp-digraph

Fig. 1.2: Examples of digraphs from Definition 1.11

a series composition of sp-digraphs $G_5 = (V_5, A_5)$ and $G_6 = (V_6, A_6)$, where $V_5 = \{1, 2\}$, $V_6 = \{3, 4\}$ and $A_5 = A_6 = \emptyset$. Notice that the sp-digraph G_5, in turn, is a parallel composition of single-vertex sp-digraphs $G_5' = (V_5', A_5')$ and $G_5'' = (V_5'', A_5'')$, where $V_5' = \{1\}$, $V_5'' = \{2\}$ and $A_5' = A_5'' = \emptyset$. Similarly, the sp-digraph G_6 is a parallel composition of single-vertex sp-digraphs $G_6' = (V_6', A_6')$ and $G_6'' = (V_6'', A_6'')$, where $V_6' = \{3\}$, $V_6'' = \{4\}$ and $A_6' = A_6'' = \emptyset$. ◆

Remark 1.14. From Definition 1.10 it follows that every series-parallel digraph $G = (V, E)$ can be represented in a natural way by a binary *decomposition tree* $T(G)$. Each leaf of the tree represents a vertex in G and each internal node is a series (parallel) composition of its successors. Hence we can construct G, starting from the root of the decomposition tree $T(G)$, by successive compositions of the nodes of the tree. For a given series-parallel digraph, its decomposition tree can be constructed in $O(|V| + |E|)$ time (see Valdes et al. [28]). The decomposition tree of the sp-digraph from Figure 1.2d is given in Figure 1.3.

Remark 1.15. Throughout the book, the internal nodes of a decomposition tree that correspond to the parallel composition and series composition will be labelled by P and S, respectively.

1.2.3 Mean value theorems

Theorem 1.16. (Mean value theorems)
(a) *If functions $f : \langle a, b \rangle \to \mathbb{R}$ and $g : \langle a, b \rangle \to \mathbb{R}$ are differentiable on the interval (a, b) and continuous on the interval $\langle a, b \rangle$, then there exists at least one point $c \in (a, b)$ such that $\frac{f'(c)}{g'(c)} = \frac{f(b)-f(a)}{g(b)-g(a)}$.*
(b) *If function $f : \langle a, b \rangle \to \mathbb{R}$ is differentiable on the interval (a, b) and continuous on the interval $\langle a, b \rangle$, then there exists at least one point $c \in (a, b)$ such that $f'(c) = \frac{f(b)-f(a)}{b-a}$.*

Proof. (a) This is the *generalized mean-value theorem*; see Rudin [25, Chapter 5, Theorem 5.9].   ◇

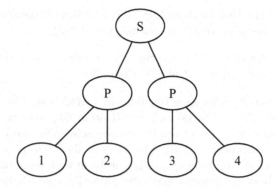

Fig. 1.3: The decomposition tree of the sp-digraph from Figure 1.2d

(b) This is the *mean-value theorem*. Applying Theorem 1.16 (a) for $g(x) = x$, we obtain the result. ∎

Remark 1.17. The counterparts of mean-value theorems for functions defined in vector spaces are given, e.g., by Maurin [20, Chapter VII].

Definition 1.18. (Hölder's vector norm l_p)
A norm on a vector space X is the function $\|\cdot\| : X \to \mathbb{R}$ such that for all $x, y \in X$ and any $a \in \mathbb{R}$ the following conditions are satisfied:
(a) $\|x + y\| \leqslant \|x\| + \|y\|$,
(b) $\|ax\| = |a|\|x\|$,
(c) $\|x\| = 0 \Leftrightarrow x = 0$.
The value $\|x\|$ is called a norm of vector $x \in X$.

Definition 1.19. (Hölder's vector norm l_p)
Given an arbitrary $p \geqslant 1$, the l_p-norm of vector $x \in \mathbb{R}^n$ is defined as follows:

$$\|x\|_p = \begin{cases} \left(\sum_{i=1}^{n} |x_i|^p \right)^{\frac{1}{p}} & , 1 \leqslant p < +\infty, \\ \max_{1 \leqslant i \leqslant n} \{|x_i|\} & , p = +\infty. \end{cases}$$

1.2.4 Priority-generating functions

Definition 1.20. (A priority-generating function)
Let $\pi' = (\pi^{(1)}, \pi^{(a)}, \pi^{(b)}, \pi^{(2)})$, $\pi'' = (\pi^{(1)}, \pi^{(b)}, \pi^{(a)}, \pi^{(2)}) \in \mathfrak{S}_n$, where $\pi^{(1)}, \pi^{(a)}, \pi^{(b)}, \pi^{(2)} \in \hat{\mathfrak{S}}_n$.
(a) *A function $\mathcal{F} : \mathfrak{S}_n \to \mathbb{R}$ is called a* priority-generating function *if there exists a function $\omega : \hat{\mathfrak{S}}_n \to \mathbb{R}$ (called* priority function*) such that either the implication $\omega(\pi^{(a)}) > \omega(\pi^{(b)}) \Rightarrow \mathcal{F}(\pi') \leqslant \mathcal{F}(\pi'')$ or the implication $\omega(\pi^{(a)}) = \omega(\pi^{(b)}) \Rightarrow \mathcal{F}(\pi') = \mathcal{F}(\pi'')$ holds.*
(b) *If $\pi^{(a)}, \pi^{(b)} \in \hat{\mathfrak{S}}_1$, then a priority-generating function is called* 1-priority-generating function.

Remark 1.21. Notice that by Definition 1.20 every priority-generating function is a 1-priority-generating function (but not vice versa).

Remark 1.22. Definition 1.20 concerns the priority-generating function of a single variable (see Tanaev et al. [27, Chap. 3]).

Remark 1.23. Priority-generating functions of a single variable are considered by Gordon et al. [13] and Strusevich and Rustogi [26] who explore the relationship between priority functions for various scheduling problems (see [13, Theorems 1–2], [26, Theorem 3.4]). The authors also identify several cases in which a criterion function for a scheduling problem with some time-dependent job processing times is a priority-generating function (see [13, Sects. 7–9], [26, Chaps. 8–9, 11]).

Remark 1.24. Priority-generating functions of many variables are considered by Janiak et al. [17].

Theorem 1.25. (Tanaev et al. [27]) *If $\mathcal{F} : \mathfrak{S}_n \to \mathbb{R}$ is a 1-priority generating function and $\omega : \mathfrak{S}_n \to \mathbb{R}$ is a priority function corresponding to $\mathcal{F}$, then the permutation in which the elements are arranged in non-increasing order of their priorities minimizes $\mathcal{F}$ over $\mathfrak{S}_n$.*

Proof. See Tanaev et al. [27, Chap. 3, Theorem 7.1].   ◇

1.2.5 Bi-criteria optimization definitions

Let X denote the set of feasible solutions of a bi-criterion optimization problem and let $f : X \to \mathbb{R}^2$, $f = (f_1, f_2)$ be the minimized criterion function, where $f_i : X \to \mathbb{R}$ are single-valued criteria for $i = 1, 2$.

Definition 1.26. (Pareto optimal solutions)
(a) *A solution $x^\star \in X$ is said to be* Pareto optimal *if there is no $x \in X$ such that $f(x) \prec f(x^\star)$.*
(b) *A solution $x^\star \in X$ is said to be* weakly Pareto optimal *if there is no $x \in X$ such that $f_i(x) < f_i(x^\star)$ for $i = 1, 2$.*

The set of all Pareto (weakly Pareto) optimal solutions in X will be denoted by X_{Par} ($X_{\mathrm{w-Par}}$). A solution $x^\star$ is Pareto (weakly Pareto) optimal if $x^\star \in X_{\mathrm{Par}}$ ($x^\star \in X_{\mathrm{w-Par}}$).

The images of sets X_{Par} and $X_{\mathrm{w-Par}}$ under function $f = (f_1, f_2)$, $f(X_{\mathrm{Par}})$ and $f(X_{\mathrm{w-Par}})$ will be denoted by Y_{eff} and $Y_{\mathrm{w-eff}}$, respectively. (Notice that $X_{\mathrm{Par}} \subset X_{\mathrm{w-Par}}$ and $Y_{\mathrm{eff}} \subset X_{\mathrm{w-eff}}$.)

We illustrate the notions of Pareto and weak Pareto optimality with the following example.

Example 1.27. (Ehrgott [11]) Consider a set X and a function $f = (f_1, f_2)$, where $X := \{(x_1, x_2) \in \mathbb{R}^2 : 0 < x_1 < 1 \wedge 0 \leqslant x_2 \leqslant 1\}$, $f_1 := x_1$ and $f_2 := x_2$. Then $Y_{\text{eff}} = \emptyset$ and $Y_{\text{w}-\text{eff}} = \{(x_1, x_2) \in X : 0 < x_1 < 1, x_2 = 0\}$.

If we define $X := \{(x_1, x_2) \in \mathbb{R}^2 : 0 \leqslant x_i \leqslant 1$ for $i = 1, 2\}$, then for the function f as above we have $Y_{\text{eff}} = \{(0, 0)\}$ and $Y_{\text{w}-\text{eff}} = \{(x_1, x_2) \in X : x_1 = 0 \vee x_2 = 0\}$. ♦

Lemma 1.28. (Scalar optimality vs. Pareto optimality)
If $x^\star$ is an optimal solution with respect to the scalar criterion $\omega \circ f$ for a certain $f = (f_1, f_2)$ and $\omega = (\omega_1, \omega_2)$, then
(a) if $\omega \in \mathbb{R}^2$, then $x^\star \in X_{\text{w}-\text{Par}}$,
(b) if $\omega \in \text{int}\mathbb{R}^2$, then $x^\star \in X_{\text{Par}}$.

Proof. (a), (b) See Ehrgott [11, Proposition 3.7].   ◇

Definition 1.29. (A convex function)
A function f is convex on the interval $\langle a, b \rangle$ if for any $x_1, x_2 \in \langle a, b \rangle$ and any $\lambda \in \langle 0, 1 \rangle$ the inequality $f(\lambda x_1 + (1 - \lambda)x_2) \leqslant \lambda f(x_1) + (1 - \lambda)f(x_2)$ holds.

In other words, a convex function is such a continuous function that the value at any point within every interval in its domain does not exceed the value of a convex combination of its values at the ends of the interval.

Remark 1.30. If in Definition 1.29 the symbol '$\leqslant$' is replaced by '$<$', then the function f is *strictly* convex.

Remark 1.31. A function f is (strictly) *concave* if $-f$ is (strictly) convex.

Definition 1.32. (A convex combination and convex set)
(a) A convex combination of elements x_1 and x_2 is an element $y := \lambda x_1 + (1 - \lambda)x_2$, where $\lambda \in \langle 0, 1 \rangle$ is a given number.
(b) A set $X \subseteq \mathbb{R}^n$ is convex if for any $x_1, x_2 \in X$ and any $\lambda \in \langle 0, 1 \rangle$ the convex combination $\lambda x_1 + (1 - \lambda)x_2 \in X$.

In other words, a set X is convex if the line segment joining any pair of points of X lies entirely in X.

Remark 1.33. Basic facts concerning convex functions and convex sets are given, e.g., by Rockafellar [24] and Walk [29, Chap. 1].

Lemma 1.34. (Pareto optimality vs. scalar optimality)
If X is a convex set and f_1, f_2 are convex functions, then if $x^\star \in X_{\text{w}-\text{Par}}$, there exists $\omega \in \text{int}\mathbb{R}^2$ such that $x^\star$ is an optimal solution with respect to the scalar criterion $\omega \circ f$.

Proof. See Ehrgott [11, Proposition 3.8].   ◇

With this lemma, we end the presentation of notation, definitions and auxiliary results used throughout the book. In subsequent chapters, we will introduce basic definitions and results concerning algorithms (Chap. 2) and $\mathcal{NP}$-complete problems (Chap. 3).

Bibliographic notes

Mathematical notation and inference rules

A comprehensive presentation of basic mathematical notions and mathematical notation may be found in Rasiowa [4]. Inference rules and proof techniques are discussed in Copi [1], Eccles [2], Lakins [3] and Velleman [5].

Basic definitions and results

Bullen et al. [9], Hardy et al. [16] and Mitrinović et al. [21] give a wide range of various inequalities.

Berge [7], Harary [15] and Wilson [30] present the graph theory from different perspectives. Brandstädt et al. [8] study the properties of different classes of graphs. Applications of graphs in computer science and engineering are discussed by Deo [10].

Maurin [20] and Rudin [25] give a concise presentation of calculus and mathematical analysis.

Priority-generating functions are discussed by Gordon et al. [13], Strusevich and Rustogi [26] and Tanaev et al. [27, Chap. 3]. The extension of these functions to the multiple criteria case is presented by Janiak et al. [17].

The properties of sp-(di)graphs and applications of these (di)graphs in the scheduling theory are discussed, e.g., by Gordon [12], Gordon et al. [14], Lawler [19], Möhring [22] and Valdes et al. [28].

Ehrgott [11] presents a comprehensive introduction to Pareto optimality.

References

Mathematical notation and inference rules

1. I. M. Copi, *Introduction to Logic*, 7th ed. New York-London: MacMillan 1986.
2. P. J. Eccles, *An Introduction to Mathematical Reasoning: Numbers, Sets and Functions*. Cambridge: Cambridge University Press, 2007.
3. T. J. Lakins, *The Tools of Mathematical Reasoning*. Providence: American Mathematical Society 2016.
4. H. Rasiowa, *Introduction to Contemporary Mathematics*. Amsterdam: North-Holland 1973.
5. D. J. Velleman, *How to Prove It: A Structured Approach*, 2nd. ed. Cambridge: Cambridge University Press 2006.

Basic definitions and results

6. A. V. Aho, J. E. Hopcroft and J. D. Ullman, *The Design and Analysis of Computer Algorithms*. Reading: Addison-Wesley 1974.
7. C. Berge, *Graphs and Hypergraphs*. Amsterdam: North-Holland 1973.

8. A. Brandstädt, L. Van Bang and J. P. Spinrad, *Graph Clases: A Survey.* Philadelphia: SIAM 1999.

9. P. S. Bullen, D. S. Mitrinović and P. M. Vasić, *Mean and Their Inequalities.* Dordrecht: Reidel 1988.

10. N. Deo, *Graph Theory with Applications to Engineering and Computer Science.* Englewood Cliffs: Prentice-Hall 1974.

11. M. Ehrgott, *Multicriteria Optimization.* Lecture Notes in Economics and Mathematical Systems **491**. Berlin-Heidelberg: Springer 2000.

12. V. S. Gordon, Some properties of series-parallel graphs. *Vestsi Akademii Navuk BSSR, Seria Fizika-Matematychnykh Navuk* **1** (1981), no. 1, 18–23 (in Russian).

13. V. S. Gordon, C. N. Potts, V. A. Strusevich and J. D. Whitehead, Single machine scheduling models with deterioration and learning: handling precedence constraints via priority generation. *Journal of Scheduling* **11** (2008), no. 5, 357–370.

14. V. S. Gordon, J-M. Proth and V. A. Strusevich, Single machine scheduling and due-date assignment under series-parallel precedence constraints. *Central European Journal of Operations Research* **13** (2005), no. 1, 15–35.

15. F. Harary, *Graph Theory.* Reading, MA: Addison-Wesley 1969.

16. G. H. Hardy, J. Littlewood and G. Polya, *Inequalities.* Cambridge: Oxford Press 1934.

17. A. Janiak, Y. M. Shafransky and A. V. Tuzikov, Sequencing with ordered criteria, precedence and group technology constraints. *Informatica* **12** (2001), no. 1, 61–88.

18. F. P. Kelly, A remark on search and sequencing problems. *Mathematics of Operations Research* **7** (1982), no. 1, 154–157.

19. E. L. Lawler, Sequencing jobs to minimize total weighted completion time subject to precedence constraints. *Annals of Discrete Mathematics* **2** (1978), 75–90.

20. K. Maurin, *Analysis.* Dordrecht: Reidel 1976.

21. D. S. Mitrinović, J. E. Pečarić and A. M. Fink, *Classical and New Inequalities in Analysis.* Dordrecht-Boston: Kluwer 1993.

22. R. H. Möhring, Computationally tractable classes of ordered sets. In: I. Rival (ed.), *Algorithms and Order.* Dordrecht: Kluwer 1989, pp. 105–193.

23. G. Rau, Minimizing a function of permutations of n integers. *Operations Research* **19** (1971), no. 1, 237–240.

24. R. T. Rockafellar, *Convex Analysis.* Princeton: Princeton University Press 2015.

25. W. Rudin, *Principles of Mathematical Analysis*, 3rd ed. New York: McGraw-Hill 1976.

26. V. A. Strusevich and K. Rustogi, *Scheduling with Times-Changing Effects and Rate-Modifying Activities.* Berlin-Heidelberg: Springer 2017.

27. V. S. Tanaev, V. S. Gordon and Y. M. Shafransky, *Scheduling Theory. Single-stage Systems.* Dordrecht: Kluwer 1994.

28. J. Valdes, R. E. Tarjan and E. L. Lawler, The recognition of series parallel digraphs. *SIAM Journal on Computing* **11** (1982), no. 3, 298–311.

29. M. Walk, *Theory of Duality in Mathematical Programming.* Berlin: Akademie-Verlag 1989.

30. R. J. Wilson, *Introduction to Graph Theory*, 2nd ed. London: Longman Group Ltd. 1979.

2
_____

Problems and algorithms

Scheduling problems considered in this book are formulated as either decision or optimization problems. These problems, in turn, are solvable by algorithms of different types. Therefore, in this chapter we present the basic concepts related to decision and optimization problems and to algorithms.

Chapter 2 is composed of five sections. In Sect. 2.1, we recall the notions of decision and optimization problems. In Sect. 2.2, we present the basic concepts related to algorithms. In Sects. 2.3, 2.4 and 2.5 we discuss the main types of exact, approximation and heuristic algorithms, respectively. The chapter is completed with bibliographic notes and a list of references.

2.1 Decision and optimization problems

A _problem_ is a general question to be answered. The question concerns a certain mathematical object and it is expressed in terms of a number of _parameters_, whose values are left unspecified. A problem is formulated by giving a general description of all its parameters and a statement concerning properties that the mathematical object must have. This object is called a _solution_ to the problem.

Example 2.1. (_a_) An example of a problem is the following one: given two integers which are not both equal to zero what is the greatest positive integer that divides each of the integers?
(_b_) The solution to the above problem is _the greatest common divisor of two integers_ and the problem itself is called the _problem of finding the greatest common divisor of two integers._ ◆

In this book we will consider decision and optimization problems. A _decision problem_ is a problem of existence of a solution that has properties specified in the formulation of the problem. An _optimization problem_ is a problem in which a solution which optimizes (i.e., minimizes or maximizes) a certain _objective (criterion) function_ is searched for.

© Springer-Verlag GmbH Germany, part of Springer Nature 2020
S. Gawiejnowicz, _Models and Algorithms of Time-Dependent Scheduling_,
Monographs in Theoretical Computer Science. An EATCS Series,
https://doi.org/10.1007/978-3-662-59362-2_2

Example 2.2. An example of a decision problem is the SUBSET PRODUCT problem: given a set of integer numbers and a *threshold* integer value, does there exist a subset of this set such that the product of all elements of this subset is equal to the threshold? ◆

Example 2.3. The problem formulated in Example 2.1 (a) is an example of an optimization problem. ◆

Remark 2.4. Formulations of decision problems considered in this book are given in Sect. 3.2.

An optimization problem may have a number of solutions which may be optimal or suboptimal. A solution to an optimization problem is *optimal* if the value of the minimized (maximized) objective function for this solution is minimal (maximal). A solution is *suboptimal* if the value of objective function for this solution is only close to an optimal value. Optimal and suboptimal solutions will be also called *feasible* solutions.

Remark 2.5. A *combinatorial optimization problem* is a special case of an optimization problem in which the set of solutions is finite (see Sect. 17.1 for more details). Scheduling problems considered in this book are examples of combinatorial optimization problems.

The optimization and decision problems are closely related to each other. For example, suppose that in an optimization problem, an objective function is minimized. Then, in a decision counterpart of this problem, we ask if there exists a solution for which the value of the objective function is not greater than a given threshold value. (If the objective function is maximized, then we ask for a solution for which the value of the objective function is not lower than a given threshold.) Therefore, though scheduling problems considered in this book are formulated mainly as combinatorial optimization problems, they can be considered either in the optimization or in the decision version.

2.1.1 Encoding schemes

By assigning a specific value to each parameter of a problem, we define an *instance* of the problem. (The set of all instances of a given problem P will be denoted by D_P.) The instance, encoded in a certain format, is the input of any procedure used for finding a solution to this instance. The rules which describe the coding format constitute an *encoding scheme*. The encoding scheme encodes any instance into a sequence of symbols from a certain finite alphabet. The length of this sequence is called the *length of input* for this instance. (The number of symbols needed for encoding a number k and the length of input for an instance I in an encoding scheme e will be denoted by $|k|_e$ and $|I|_e$, respectively.)

Remark 2.6. There exist many different encoding schemes. Each encoding scheme specifies the rules of encoding numbers, sets, graphs and other mathematical objects. The most important rules concern the encoding of numbers. The rules should ensure that the encoding scheme is concise, i.e., it does not cause an exponential growth of the length of input, and that numbers are represented in any positional numeral system.

The simplest scheme is the *unary encoding scheme u*, in which a number n is represented by n 1's, i.e., $|n|_u = n$. The scheme, however, is not concise.

Throughout this book, we will use the *binary encoding scheme b*. In this scheme, a number n is represented in the binary system, i.e., $|n|_b = \lceil \log n \rceil = O(\log n)$. This scheme is concise and unlike the unary encoding scheme, it does not cause an exponential growth of the length of input.

Remark 2.7. The n number may be an integer or a rational number. If the n is an integer, then it is represented by a single binary encoded string of characters as above. A rational number n is represented by two such strings.

2.1.2 Undecidable and decidable problems

A problem is solved if a solution to any instance of the problem may be found. However, the time needed for finding the solution may or may not be finite. Hence, from the point of view of the computation time, all problems are divided into two classes of problems: undecidable and decidable ones.

A problem is said to be *undecidable (unsolvable)*, if there is no finite procedure that solves the problem.

Example 2.8. (*a*) An example of an undecidable problem is the *halting problem*: given a Turing machine and an input, does the machine halt for this input?
(*b*) Another example of an undecidable problem is *Hilbert's tenth problem*: given an arbitrary *Diophantine equation* (i.e., an equation in which only integer solutions are allowed), does there exist a solution of this equation?   ◆

Remark 2.9. Though from the undecidability of a problem there follows a negative answer to the question of existence of a finite procedure solving any instance of the problem, it does not exclude the possibility of finding a solution procedure for a particular type of instance of the problem. For example, there are procedures that determine if a Turing machine halts for a given type of an input and procedures that solve particular types of Diophantine equations.

A problem is said to be *decidable (solvable)* if there exists a finite procedure which solves the problem. This means that for any instance of the problem, a solution to the instance can be found in a finite time.

In this book, we will consider only decidable scheduling problems.

2.2 Basic concepts related to algorithms

A finite procedure which finds a solution to an arbitrary instance of a problem is called an *algorithm*. (We say that the algorithm solves the problem.) Generally speaking, an algorithm consists of an *input*, a sequence of *steps* and an *output*. The input describes a specific instance. The sequence of steps must be performed by the algorithm to find a solution to the instance. Each step, in turn, can be decomposed into a finite number of *elementary operations*. Examples of elementary operations are *arithmetical operations* (addition, subtraction, multiplication, division), *logical operations* (negation, conjuction, disjunction), the *assignment statement* (assigning a value to a variable) and *control operations* (conditional jump, call of a function or a procedure). *Complex operations* which can appear in an algorithm (conditional loops, iterative loops, functions and procedures) are sequences of elementary operations. The algorithm, for a given input, performs the steps and produces an output, which is the solution to the input.

Remark 2.10. The solution produced by an algorithm at some step will be called a *partial solution*, contrary to the *complete solution* obtained after the completion of the algorithm.

2.2.1 Time and space complexity of algorithms

For a given problem, there may exist various algorithms which have different efficiency. We will now introduce two basic measures of efficiency of an algorithm: time and space.

The efficiency of an algorithm may be measured by the number of elementary operations which must be performed by the algorithm to find a solution to any instance of a given problem. The number of operations is a function of the input of this algorithm: if the input is longer, the algorithm will perform more operations. The *time complexity* of this algorithm is the function which maps each length of input into the maximal number of elementary operations needed for finding a solution to any instance of that length.

The efficiency of an algorithm may also be measured by the total space needed for the execution of the algorithm. The *space complexity* of this algorithm is the function which maps each length of input into the maximal amount of computer memory cells needed for the execution of the algorithm for any instance of that length.

In this book, we will mainly consider the time complexity of algorithms.

2.2.2 Pseudo-code of algorithms

Throughout this book, algorithms will be presented in a pseudo-code similar to Pascal. The formulation of an algorithm in the pseudo-code will start with a header with the name of the algorithm, followed by a description of its input and output. The remaining part of the pseudo-code will be divided into sections which will correspond to particular steps of the algorithm.

As an example of application of the pseudo-code, we present an algorithm for the problem formulated in Example. 2.1 (see Algorithm 2.1).

Algorithm 2.1. for computing the greatest common divisor

1 **Input** : integers a, b
2 **Output:** the largest positive integer c such that c divides a and c
    divides b
  ▷ Step 1
3 **while** $(a \neq b)$ **do**
4 $\quad$ **if** $(a > b)$ **then**
5 $\quad\quad$ $a \leftarrow a - b;$
$\quad$ **else**
6 $\quad\quad$ $b \leftarrow b - a;$
$\quad$ **end**
end
7 **return** a.

Remark 2.11. Algorithm 2.1 is commonly called *Euclid's algorithm.*

Remark 2.12. Euclid's algorithm is example of a weakly polynomial algorithm (cf. Definition 2.16).

In the pseudo-code, we use standard Pascal statements such as conditional jump **if .. then .. else**, iterative loop **for .. do**, conditional loops **while .. do** and **repeat .. until**. The assignment operator is denoted by the symbol '←'. The **return** statement denotes the end of execution of the current pseudo-code and the returning of a specified value.

The level of nesting in complex statements is denoted by the **begin .. end** statement. Necessary comments are printed in `teletype` font and preceded by the symbol '▷'. Consecutive operations are usually followed by a semicolon (';'). The last statement of an algorithm is followed by a dot ('.').

The **exit** statement will denote an immediate termination of execution of a loop and passing the control to the first statement after the loop. The instruction of printing a message will be denoted by **write**. The **stop** statement will denote unconditional halting of the execution of a given algorithm.

Remark 2.13. The formulations of algorithms presented in this book slightly differ from the original formulations. The reason for that is the desire to unify the notation and the way of presentation. Therefore, some variable names are changed, added or deleted, and conditional loops are used instead of unconditional jump statements.

2.2.3 Polynomial-time algorithms

We say that an algorithm is efficient if its time complexity is polynomially bounded with respect to the length of input. This means that the number of elementary operations performed by this algorithm for any instance of a given problem will be not greater than a polynomial of the length of input for this instance. Such an algorithm is called a *polynomial-time (polynomial)* *algorithm* and is defined as follows.

Definition 2.14. (A polynomial-time algorithm)
An algorithm that solves a problem is said to be a polynomial-time (poly-nomial) algorithm *if there exists a polynomial q such that for any instance I of the problem the number of elementary operations performed by the algorithm is bounded from above by $q(|I|_b)$.*

In other words, the time complexity of a polynomial-time algorithm for an input of length $|I|_b$ is of order $O(q(|I|_b))$ for a certain polynomial q. Throughout this book, we will say that a problem is *polynomially solvable (computationally tractable)* if there exists a polynomial-time algorithm for this problem.

Remark 2.15. The size of the input of an algorithm depends on numbers in the input. If the numbers are distinct, they are represented as described above. In such a case, the size of the input is bounded from above by the sum of rounded binary logarithms of all the numbers. If the numbers are not distinct, we deal with *high multiplicity problems*. In this case, the size of the input is smaller than previously and bounded from above by the sum of rounded logarithms of distinct numbers and rounded logarithms of their multiplicities.

2.2.4 Strongly and weakly polynomial-time algorithms

If the input of a polynomial-time algorithm is composed of numbers encoded in binary encoding scheme, then it is possible to classify this algorithm as a *strongly polynomial* or *weakly polynomial* one.

Definition 2.16. (Strongly polynomial vs. weakly polynomial algorithms)
Let the input of a polynomial-time algorithm be composed of n integer numbers encoded in binary encoding scheme.
(a) The algorithm is said to be a strongly polynomial algorithm *if it uses only elementary arithmetic operations, the total number of arithmetic operations is bounded by a polynomial of n and the algorithm runs in polynomial space with respect to the size of the input of this algorithm.*
(b) The algorithm is said to be a weakly polynomial algorithm *if it runs in polynomial time but is not strongly polynomial.*

2.2.5 Exponential-time algorithms

An algorithm which is not a polynomial-time algorithm is called an *exponential-time* (*exponential*) *algorithm*. This means that the number of elementary operations performed by such an algorithm cannot be bounded from above by any polynomial of the length of input. The time complexity of an exponential algorithm either is an exponential function or grows at least as quickly as an exponential function. (Notice that the time complexity function of an exponential algorithm need not be an exponential function in the strict sense.) Since exponential functions grow faster than polynomials, the exponential-time algorithms are not efficient. The problems for which only exponential-time algorithms are known will be called *computationally intractable*.

2.2.6 Pseudo-polynomial algorithms

Algorithms which are polynomial with respect to both the length of input and the maximum value in the input are called *pseudo-polynomial algorithms*. Since polynomial and pseudo-polynomial algorithms are related in a certain way, we shall make a few remarks concerning the relation between them.

Remark 2.17. By Definition 2.14, any polynomial algorithm is a pseudo-polynomial algorithm as well.

Remark 2.18. Pseudo-polynomial algorithms are not polynomial, since the maximum value in input exponentially depends on the representation of this value both in the binary encoding scheme and in any other concise encoding scheme (cf. Remark 2.6).

Remark 2.19. Pseudo-polynomial algorithms would be polynomial if either the unary encoding scheme was used or the maximum value was bounded from above by a polynomial of the length of input.

Since the existence of a pseudo-polynomial algorithm for a computationally intractable problem has important consequences, we will come back to the concept of pseudo-polynomial algorithm in Sect. 3.1.

2.2.7 Offline algorithms vs. online algorithms

We have assumed so far that all input data of an algorithm are known at the moment of the start of execution of the algorithm. However, such a complete knowledge is not always possible. Therefore, we can divide all algorithms into *offline* algorithms and *online* algorithms, depending on whether the whole input data are available or not when an algorithm begins its execution. These two classes of algorithms are defined as follows.

Definition 2.20. (Offline algorithm and online algorithm)
(*a*) *An algorithm is called an* offline algorithm *if it processes its input as one unit and the whole input data are available at the moment of the start of execution of this algorithm.*
(*b*) *An algorithm is called an* online algorithm *if it processes its input piece by piece and only a part of the input is available at the moment of the start of execution of this algorithm.*

Remark 2.21. Sometimes, even the data concerning a piece of input of an online algorithm may be known only partially. In this case, the online algorithm is called a *semi-online algorithm*.

An online (a semi-online) algorithm can be evaluated by its competitive ratio. The ratio is a counterpart of the worst-case ratio for an approximation algorithm (cf. Definition 2.25) and it can be defined as follows.

Definition 2.22. (c-competitive algorithm and competitive ratio)
Let $A(I)$ and $OPT(I)$ denote the solutions generated by an online (semi-online) algorithm A and by an optimal offline algorithm, respectively, for a given instance I of a minimization problem.
(*a*) *Algorithm A is called c-competitive if there exist constant values c and k such that $A(I) \leqslant c \cdot OPT(I) + k$ for all I.*
(*b*) *The constant c defined as above is called the* competitive ratio *of algorithm A.*

Remark 2.23. Definition 2.22 (a) concerns an online minimization algorithm. An online maximization algorithm is c-competitive if there exist constant values c and k such that $A(I) \geqslant c \cdot OPT(I) + k$ for all I.

Remark 2.24. In general, the constant k from Definition 2.22 is a non-zero value. In the case of the problems considered in the book, we have $k = 0$.

2.3 Main types of exact algorithms

The solutions generated by an algorithm may or may not be exact. For example, if different schedules for a set of jobs exist, the schedule which meets all requirements and which has the smallest value of criterion function is the *exact (optimal)* solution, while the schedule which meets all requirements but has a cost greater than the cost of the optimal schedule is only a *feasible (suboptimal)* solution. Throughout this book, by an *exact algorithm* we will understand such an algorithm that finds the exact solution. Polynomial-time algorithms are examples of exact algorithms that find the exact solution in polynomial time. For the problems for which no polynomial-time algorithms are known, exact solutions can be found by enumeration, branch-and-bound or dynamic programming algorithms.

2.3.1 Enumeration algorithms

An *enumeration algorithm* directly enumerates all possible solutions. For example, if a problem consists in finding such a permutation of n elements of a set that for this permutation the value of a function is minimal, then a simple enumeration algorithm generates all possible permutations and selects the optimal one. However, since there are $n! > 2^n$ permutations of set I_n, this algorithm runs in exponential time.

2.3.2 Branch-and-bound algorithms

A *branch-and-bound algorithm* (cf. Chap. 13) finds an optimal solution by indirect enumeration of all possible solutions through examination of smaller and smaller subsets of the set of all solutions.

Branch-and-bound algorithms are more effective than enumeration algorithms when lower bounds allow us to cut off a large part of the tree of partial solutions. Then, such algorithms are capable of solving small and medium size instances of $\mathcal{NP}$-hard problems.

2.3.3 Dynamic programming algorithms

An exact algorithm may also be constructed by *dynamic programming*. In this case, the optimal solution is generated by a multi-stage decision process, which, starting from an *initial state*, constructs subsequent partial solutions from previously generated *states* in a step-by-step manner. The initial state is defined by some *initial conditions*, the subsequent states are defined by a *recursive formula*, and the final state is the *goal* we want to achieve. This final state corresponds to an optimal (exact) solution.

The theoretical foundation of dynamic programming is given by the so-called *Bellman's principle of optimality* (cf. [27, Chap. III, §3]). The principle says that whatever the initial state and initial decision are, the remaining decisions must constitute an optimal solution with respect to the state resulting from the first decision.

According to this principle, at each stage of the process, the decisions that lead to subsequent partial solutions are ranked in some way and the ones with the highest rank are taken into account in subsequent stages until the final (optimal) solution is achieved.

2.4 Main types of approximation algorithms

Not all problems encountered in practice have polynomial-time algorithms. Many problems have been proved to be *computationally intractable*, which means that for such problems polynomial algorithms probably do not exist.

However, even if it is known that a problem is computationally intractable, there still remains the question how to find a solution to the problem. If all known algorithms for the problem are inefficient, we may apply an *approximation algorithm* to solve this problem. In this section, we discuss the main types of such algorithms.

2.4.1 Approximation algorithms

An approximation algorithm is a polynomial algorithm which generates an *approximate* (*suboptimal*) solution that is close (in the sense defined below) to an optimal solution.

Since the solution generated by an approximation algorithm is only a sub-optimal solution, it is useful to know how close to the optimal one the solution is. A measure of the closeness is the *worst-case ratio* of the algorithm.

Definition 2.25. (An approximation algorithm and its worst-case ratio)
Let $A(I)$ and $OPT(I)$ denote a solution generated by an algorithm A and an optimal solution to a given instance I of a minimization (maximization) problem, respectively. Let $\epsilon > 0$ and $r = 1 + \epsilon$ ($r = 1 - \epsilon$). An algorithm A is said to be an r-approximation algorithm for a problem P if for any instance $I \in D_P$ the inequality $|A(I) - OPT(I)| \leqslant \epsilon \cdot OPT(I)$ holds. The value r is called the worst-case ratio *of the algorithm A.*

From Definition 2.25 it follows that if the algorithm A solves a minimization problem, then $A(I) \leqslant (1 + \epsilon) \cdot OPT(I) = r \cdot OPT(I)$ and the worst-case ratio $r \in \langle 1, +\infty)$. (If the algorithm A solves a maximization problem, we have $A(I) \geqslant (1 - \epsilon) \cdot OPT(I) = r \cdot OPT(I)$ and $r \in \langle 0, 1\rangle$.) In other words, an r-approximation algorithm generates a solution which is at most r times worse than the optimal one.

Remark 2.26. If I is an arbitrary instance of a minimization problem and A is an approximation algorithm for the problem, for this instance we can calculate the *absolute ratio* $R_A^a(I)$ of the value of the solution $A(I)$ generated by the algorithm to the value of the optimal solution $OPT(I)$, $R_A^a(I) := \frac{A(I)}{OPT(I)}$. (For a maximization problem we have $R_A^a(I) := \frac{OPT(I)}{A(I)}$.) The worst-case ratio r from Definition 2.25 is an infimum (a supremum for a maximization problem) over the ratios $R_A^a(I)$ over all possible instances I of a given problem.

Remark 2.27. Apart from the absolute ratio $R_A^a(I)$, for an instance I we can calculate the *relative ratio* $R_A^r(I) := \frac{A(I) - OPT(I)}{OPT(I)}$. (For a maximization problem we have $R_A^r(I) := \frac{OPT(I) - A(I)}{A(I)}$.) It is easy to notice that $R_A^r(I) = R_A^a(I) - 1$.

2.4.2 Approximation schemes

For some computationally intractable problems, it is possible to construct a family of approximation algorithms which generate solutions as close to the optimal one as it is desired. Such a family of algorithms is called an *approximation scheme*. There exist two types of approximation schemes: *polynomial-time* and *fully polynomial-time*.

Definition 2.28. (Approximation schemes)
(a) *A family of r-approximation algorithms is called a* polynomial-time approximation scheme *if for an arbitrary $\epsilon > 0$ any algorithm from this family has polynomial time complexity.*
(b) *If a polynomial-time approximation scheme is running in polynomial time with respect to $\frac{1}{\epsilon}$, it is called a* fully polynomial-time approximation scheme.

Throughout this book, a polynomial-time approximation scheme and a fully polynomial-time approximation scheme will be called a *PTAS* and an *FPTAS*, respectively. In Sect. 3.1 we will specify the conditions which have to be satisfied for a problem to have a PTAS (an FPTAS).

2.5 Main types of heuristic algorithms

Sometimes, it is difficult to establish the worst-case ratio (the competitive ratio) of an algorithm. If solutions generated by the algorithm are of a good quality, it may be used though its worst-case performance (competitiveness) is unknown. In this section, we discuss the main types of such algorithms.

2.5.1 Heuristic algorithms

An algorithm is called a *heuristic algorithm* (a *heuristic*, in short) if its worst-case (competitive) ratio is unknown. This means that one cannot predict the behaviour of this algorithm for all instances of the considered problem.

The efficiency of a heuristic algorithm can be evaluated with a *computational experiment*. In the experiment a set of *test instances* is generated and the solutions obtained by the heuristic under evaluation are compared with optimal solutions found by an exact algorithm.

2.5.2 Greedy algorithms

A huge spectrum of heuristics is known. An example of a simple heuristic is the so-called greedy algorithm.

A *greedy algorithm* repeatedly executes a procedure which tries to construct a solution by choosing a locally best partial solution at each step. In some cases, such a strategy leads to finding optimal solutions. In general, however, greedy algorithms produce only relatively good suboptimal solutions. Therefore, for more complex problems more sophisticated algorithms such as local search algorithms have been proposed.

2.5.3 Local search algorithms

Other examples of heuristics are *local search algorithms*. These algorithms start from an *initial solution* and iteratively generate a *neighbourhood* of the solution which is currently the best one. The neighbourhood is a set of all solutions which can be obtained from the current solution by feasible *moves*. The moves are performed by different *operators* whose definitions depend on a particular problem. The aim of an operator is to produce a new feasible solution from another feasible solution.

Given a neighbourhood and all feasible moves, a local search algorithm finds a new solution by using a *strategy* of local search of the neighbourhood.

The above described procedure of finding a new solution by a local search algorithm is performed until a *stop condition* is met. In this case, the algorithm stops, since the further decrease (increase) of the minimized (maximized) objective function is very unlikely.

2.5.4 Meta-heuristic algorithms

An important group of heuristic algorithms is composed of meta-heuristic algorithms (meta-heuristics). A *meta-heuristic algorithm (meta-heuristic)* is a template of a local search algorithm. The template includes a number of control parameters which have an impact on the quality of the solutions that are generated by the given meta-heuristic and the conditions that cause the termination of its execution.

Examples of meta-heuristics are *simulated annealing*, *tabu search* and *evolutionary algorithms*. In the meta-heuristics, different complicated strategies are applied to construct a new solution from a current solution.

With these remarks, we end the presentation of fundamental concepts concerning algorithms. In Chap. 3, closing the part, we will introduce the basic definitions and results concerning $\mathcal{NP}$-complete problems.

Bibliographic notes

Decision and optimization problems

The Turing machine was introduced by Turing [14]. Aho et al. [1], Hopcroft and Ullman [6] and Lewis and Papadimitriou [8] give a detailed description of the Turing machine and its variants.

The unary and the binary encoding schemes are discussed by Garey and Johnson [5, Chapter 2].

The undecidability of the halting problem was proved by Turing [14] in 1936. The undecidability of Hilbert's tenth problem was proved by Matiyasevich [9] in 1970. Davies [3, 4], Hopcroft and Ullman [6], Lewis and Papadimitriou [8], Papadimitriou [11] and Rogers [13] discuss the undecidability and its consequences for computer science.

Different aspects of combinatorial optimization problems are studied by Cook et al. [2], Lawler [7], Nemhauser and Wolsey [10], and Papadimitriou and Steiglitz [12].

Basic concepts related to algorithms

Euclid's algorithm is widely considered as one of the oldest algorithms. A detailed analysis of different variations of this algorithm is presented by Knuth [23, Vol. 1].

The basic concepts related to algorithms are presented by Aho et al. [15], Atallah [16], Cormen et al. [20], Knuth [23] and Skiena [25].

Strongly and weakly polynomial algorithms are discussed by Grötschel et al. [21, Sect. 1.3] and Schrijver [24, Sect. 4.12]. Strongly polynomial algorithms for high multiplicity scheduling problems are proposed, e.g., by Brauner et al. [17, 18], Clifford and Posner [19] and Hochbaum and Shamir [22].

Main types of exact algorithms

Exact exponential algorithms are considered by Fomin and Kratsch [28], Kohler and Steiglitz [30] and Woeginger [39].

Branch-and-bound algorithms are discussed by Lawler and Woods [31], Papadimitriou and Steiglitz [35] and Walukiewicz [38]. Mitten [33] and Rinnooy Kan [36] give axioms for branch-and-bound algorithms.

Jouglet and Carlier [29] review the applications of dominance rules in combinatorial optimization.

Dynamic programming and its applications are studied by Bellman [26], Bellman and Dreyfuss [27], Lew and Mauch [32], Nemhauser [34] and Snedovich [37].

Main types of approximation algorithms

Du et al. [42], Graham [44], Graham et al. [45], Hochbaum [46], Klein and Young [48], Shachnai and Tamir [55], Schuurman and Woeginger [56], Williamson and Shmoys [59] and Vazirani [57] discuss at length different aspects concerning approximation algorithms and approximation schemes.

Online algorithms are studied by Albers [40], Borodin and El-Yaniv [41], Fiat and Woeginger [43], Irani and Karlin [47], Phillips and Westbrook [52] and Pruhs et al. [53].

A greedy algorithm is optimal for a problem if the set of all solutions to the problem constitutes the so-called *matroid*. Matroids are studied in detail by Lawler [50], Oxley [51], Revyakin [54] and Welsh [58].

Applications of matroids in mathematics and computer science are presented by White [60]. Generalizations of matroids, *greedoids*, are studied by Korte et al. [49].

Main types of heuristic algorithms

Heuristic algorithms and their applications are discussed by Gigerenzer and Todd [68], Martí et al. [73], Pearl [78] and Salhi [80].

General properties and applications of meta-heuristics are discussed by Blum and Roli [63], Bozorg-Haddad et al. [64], Gendreau and Potvin [67], Glover and Kochenberger [69], Gonzalez [71], Michiels et al. [76], Osman and Kelly [77] and Voss et al. [81].

Simulated annealing is considered by Aarts and Korst [61] and Salamon et al. [79].

Tabu search is studied by Glover and Laguna [70].

Evolutionary algorithms are discussed by Bäck [62], Eiben et al. [65], Fogel [66], Hart et al. [72] Michalewicz [74] and Michalewicz and Fogel [75].

References

Decision and optimization problems

1. A. V. Aho, J. E. Hopcroft and J. D. Ullman, *The Design and Analysis of Computer Algorithms.* Reading: Addison-Wesley 1974.
2. W. J. Cook, W. H. Cunningham, W. R. Pulleyblank and A. Schrijver, *Combinatorial Optimization.* New York-Toronto: Wiley 1998.
3. M. Davies, *Computability and Undecidability.* New York: McGraw-Hill 1958.
4. M. Davies, *The Undecidable.* Nashville: Raven Press 1965.
5. M. R. Garey and D. S. Johnson, *Computers and Intractability: A Guide to the Theory of NP-Completeness.* San Francisco: Freeman 1979.
6. J. E. Hopcroft and J. D. Ullman, *Introduction to Automata Theory, Languages and Computation.* Reading: Addison-Wesley 1979.
7. E. L. Lawler, *Combinatorial Optimization: Networks and Matroids.* New York: Holt, Rinehart and Winston 1976.
8. H. R. Lewis and C. H. Papadimitriou, *Elements of the Theory of Computation*, 2nd ed. Upper Saddle River: Prentice-Hall 1998.
9. Yu. V. Matiyasevich, Solution to the tenth problem of Hilbert. *Matematik Lapok* **21** (1970), 83–87.
10. G. L. Nemhauser and L. A. Wolsey, *Integer and Combinatorial Optimization.* New York: Wiley 1988.
11. C. H. Papadimitriou, *Computational Complexity.* Reading: Addison-Wesley 1994.
12. C. H. Papadimitriou and K. Steiglitz, *Combinatorial Optimization: Algorithms and Complexity.* Englewood Cliffs: Prentice-Hall 1982.
13. H. Rogers, jr, *Theory of Recursive Functions and Effective Computability.* New York: McGraw-Hill 1967.
14. A. Turing, On computable numbers, with an application to the Entscheidungsproblem. *Proceedings of the London Mathematical Society* **42** (1936), no. 1, 230–265. (Erratum: *Proceedings of the London Mathematical Society* **43** (1937), no. 6, 544–546.)

Basic concepts related to algorithms

15. A. V. Aho, J. E. Hopcroft and J. D. Ullman, *The Design and Analysis of Computer Algorithms*. Reading: Addison-Wesley 1974.
16. M. J. Atallah (ed.), *Algorithms and Theory of Computation Handbook*. Boca Raton-Washington: CRC Press 1999.
17. N. Brauner, Y. Crama, A. Grigoriev and J. van de Klundert, A framework for the complexity of high-multiplicity scheduling problems. *Journal of Combinatorial Optimization* **9** (2005), no. 3, 313–323.
18. N. Brauner, Y. Crama, A. Grigoriev and J. van de Klundert, Multiplicity and complexity issues in contemporary production scheduling. *Statistica Neerlandica* **61** (2007), 75–91.
19. J. J. Clifford and M. E. Posner, Parallel machine scheduling with high multiplicity. *Mathematical Programming, Series A*, **89** (2001), no. 3, 359–383.
20. T. H. Cormen, C. E. Leiserson and R. L. Rivest, *Introduction to Algorithms*, 13th printing. Cambridge: MIT Press 1994.
21. M. Grötschel, L. Lovász and A. Schrijver, *Geometric Algorithms and Combinatorial Optimization*, 2nd corrected ed. Berlin-Heidelberg: Springer 1993.
22. D. S. Hochbaum and R. Shamir, Strongly polynomial algorithms for the high multiplicity scheduling problem. *Operations Research* **39** (1991), no. 4, 648–653.
23. D. E. Knuth, *The Art of Computer Programming*, vol. 1–3. Reading: Addison-Wesley 1967–1969.
24. A. Schrijver, *Combinatorial Optimization: Polyhedra and Efficiency*, vol. A-C. Berlin-Heidelberg: Springer 2004.
25. S. Skiena, *The Algorithm Design Manual*, 2nd ed. Berlin-Heidelberg: Springer 2012.

Main types of exact algorithms

26. R. Bellman, *Dynamic Programming*. Princeton: Princeton University Press 1957.
27. R. Bellman and S. E. Dreyfus, *Applied Dynamic Programming*. Princeton: Princeton University Press 1962.
28. F. V. Fomin and D. Kratsch, *Exact Exponential Algorithms*. Berlin-Heidelberg: Springer 2010.
29. A. Jouglet and J. Carlier, Dominance rules in combinatorial optimization problems. *European Journal of Operational Research* **212** (2011), no. 3, 433–444.
30. W. H. Kohler and K. Steiglitz, Enumerative and iterative computational approaches. In: E. G. Coffman, jr, (ed.), *Computer and Job-Shop Scheduling Theory*. New York: Wiley 1976.
31. E. L. Lawler and D. E. Woods, Branch-and-bound methods: a survey. *Operations Research* **14** (1966), no. 4, 699–719.
32. A. Lew and H. Mauch, *Dynamic Programming: A Computational Tool*. Berlin-Heidelberg: Springer 2007.
33. L. Mitten, Branch-and-bound methods: general formulation and properties. *Operations Research* **18** (1970), no. 1, 24–34. (Erratum: *Operations Research* **19** (1971), no. 2, 550.)
34. G. L. Nemhauser, *Introduction to Dynamic Programming*. New York: Wiley 1966.

35. C. H. Papadimitriou and K. Steiglitz, *Combinatorial Optimization: Algorithms and Complexity*. Englewood Cliffs: Prentice-Hall 1982.
36. A. H. G. Rinnooy Kan, On Mitten's axioms for branch-and-bound. *Operations Research* **24** (1976), no. 6, 1176–1178.
37. M. Snedovich, *Introduction to Dynamic Programming*. New York: Marcel Dekker 1992.
38. S. Walukiewicz, *Integer Programming*. Warszawa: Polish Scientific Publishers 1991.
39. G. J. Woeginger, Exact algorithms for NP-hard problems: a survey. *Lecture Notes in Computer Science* **2570** (2003), 187–205.

Main types of approximation algorithms

40. S. Albers, Online algorithms. In: D. Goldin, S. A. Smolka and P. Wegner (eds.), *Interactive Computation: The New Paradigm*. Berlin-Heidelberg: Springer 2006, pp. 143–164.
41. A. Borodin and R. El-Yaniv, *Online Computation and Competitive Analysis*. Cambridge: Cambridge University Press 1998.
42. D-Z. Du, K-I. Ko and X-D. Hu, *Design and Analysis of Approximation Algorithms*. Berlin-Heidelberg: Springer 2012.
43. A. Fiat and G. Woeginger (eds.), *Online Algorithms*. Berlin: Springer 1998.
44. R. L. Graham, Estimation of accuracy of scheduling algorithms. In: E. G. Coffman, jr, (ed.), *Computer and Job-Shop Scheduling Theory*. New York: Wiley 1976.
45. R. L. Graham, E. L. Lawler, J. K. Lenstra and A. H. G. Rinnooy Kan, Optimization and approximation in deterministic sequencing and scheduling: a survey. *Annals of Discrete Mathematics* **5** (1979), 287–326.
46. D. Hochbaum (ed.), *Approximation Algorithms for NP-hard Problems*. Boston: PWS Publishing 1998.
47. S. Irani and A. Karlin, Online computation. In: D. Hochbaum (ed.), *Approximation Algorithms for NP-hard Problems*. Boston: PWS Publishing 1998.
48. P. N. Klein and N. E. Young, Approximation algorithms. In: M. J. Attallah (ed.), *Algorithms and Theory of Computation Handbook*. Boca Raton-Washington: CRC Press 1999.
49. B. Korte, L. Lovasz and R. Schrader, *Greedoids*. Berlin-Heidelberg: Springer 1991.
50. E. L. Lawler, *Combinatorial Optimization: Networks and Matroids*. New York: Holt, Rinehart and Winston 1976.
51. J. G. Oxley, *Matroid Theory*. Oxford: Oxford University Press 1992.
52. S. Phillips and J. Westbrook, On-line algorithms: competitive analysis and beyond. In: M. J. Atallah (ed.), *Algorithms and Theory of Computation Handbook*. Boca Raton-Washington: CRC Press 1999.
53. K. Pruhs, E. Torng and J. Sgall, Online scheduling. In: J. Y.-T. Leung (ed.), *Handbook of Scheduling*. Boca Raton: Chapman and Hall/CRC 2004.
54. A. M. Revyakin, Matroids. *Journal of Mathematical Sciences* **108** (2002), no. 1, 71–130.
55. H. Shachnai and T. Tamir, Polynomial-time approximation schemes. In: T. F. Gonzalez (ed.), *Handbook of Approximation Algorithms and Metaheuristics*. Boca Raton: Chapmann & Hall/CRC 2007.

56. P. Schuurman and G. Woeginger, Approximation schemes – a tutorial. See http://wwwhome.cs.utwente.nl/~woegingergj/papers/ptas.pdf.
57. V. V. Vazirani, *Approximation Algorithms.* 2nd ed. Berlin-Heidelberg: Springer 2003.
58. D. J. A. Welsh, *Matroid theory.* London: Academic Press 1976.
59. D. P. Williamson and D. Shmoys, *The Design of Approximation Algorithms.* Cambridge: Cambridge University Press 2011.
60. N. White (ed.), *Matroid Applications.* Cambridge: Cambridge University Press 1992.

Main types of heuristic algorithms

61. E. Aarts and J. Korst, *Simulated Annealing and Boltzmann Machines: A Stochastic Approach to Combinatorial Optimization and Neural Computing.* New York: Wiley 1989.
62. T. Bäck, *Evolutionary Algorithms in Theory and Practice: Evolution Strategies, Evolutionary Programming, Genetic Algorithms.* Oxford: Oxford University Press 1996.
63. C. Blum and A. Roli, Metaheuristics in combinatorial optimization: overview and conceptual comparison. *ACM Computing Surveys* **35** (2003), no. 3, 268–308.
64. O. Bozorg-Haddad, M. Solgi and H. A. Loáiciga, *Meta-Heuristic and Evolutionary Algorithms for Engineering Optimization.* New York: Wiley 2017.
65. A. E. Eiben and J. E. Smith, *Introduction to Evolutionary Computing,* 2nd ed. Berlin: Springer 2007.
66. L. J. Fogel, *Intelligence Through Simulated Evolution: Forty Years of Evolutionary Programming.* New York: Wiley 1999.
67. M. Gendreau and J-Y. Potvin, Metaheuristics in combinatorial optimization. *Annals of Operations Research* **140** (2005), no. 1, 189–213.
68. G. Gigerenzer and P. M. Todd, *Simple Heuristics That Make Us Smart.* New York: Oxford University Press 2001.
69. F. W. Glover and G. A. Kochenberger (eds.), *Handbook of Metaheuristics.* International Series in Operations Research & Management Science **57**. Berlin: Springer 2003.
70. F. W. Glover and M. Laguna, *Tabu Search.* Dordrecht: Kluwer 1996.
71. T. F. Gonzalez (ed.), *Handbook of Approximation Algorithms and Metaheuristics.* Boca Raton: Chapmann & Hall/CRC 2007.
72. E. Hart, P. Ross and D. Corne, Evolutionary scheduling: a review. *Genetic Programming and Evolvable Machines* **6** (2005), no. 2, 191–220.
73. R. Martí, P. M. Pardalos and M. G. C. Resende (eds.), *Handbook of Heuristics.* Cham: Springer, 2018.
74. Z. Michalewicz, *Genetic Algorithms + Data Structures = Evolution Programs.* Berlin: Springer 1994.
75. Z. Michalewicz and D. B. Fogel, *How to Solve It: Modern Heuristics.* Berlin: Springer 2004.
76. W. Michiels, E. Aarts and J. Korst, *Theoretical Aspects of Local Search.* Berlin-Heidelberg: Springer 2007.
77. I. H. Osman and J. P. Kelly, *Meta-Heuristics: Theory and Applications.* Dordrecht: Kluwer 1996.

78. J. Pearl, *Heuristics: Intelligent Search Strategies for Computer Problem Solving.* Reading: Addison-Wesley 1984.
79. P. Salamon, P. Sibani and R. Frost, *Facts, Conjectures, and Improvements for Simulated Annealing.* SIAM Monographs on Mathematical Modeling and Computation **7**. Philadelphia: SIAM 2002.
80. S. Salhi, *Heuristic Search: The Emerging Science of Problem Solving.* Cham: Springer Nature, 2017.
81. S. Voss, S. Martello, I. H. Osman and C. Roucairol (eds.), *Meta-Heuristics: Advances and Trends in Local Search Paradigms for Optimization.* Dordrecht: Kluwer 1999.

3

$\mathcal{NP}$-complete problems

The theory of $\mathcal{NP}$-completeness has a great impact on the scheduling theory, since the knowledge about the complexity status of a problem allows us to facilitate further research on the problem. Therefore, in this chapter we recall the most fundamental concepts related to $\mathcal{NP}$-completeness.

Chapter 3 is composed of two sections. In Sect. 3.1, we recall the basic definitions and results concerning the theory of $\mathcal{NP}$-completeness. In Sect. 3.2, we formulate all $\mathcal{NP}$-complete problems which appear in $\mathcal{NP}$-completeness proofs presented in this book. The chapter is completed with bibliographic notes and a list of references.

3.1 Basic definitions and results

Let $\mathcal{P}$ ($\mathcal{NP}$) denote the class of all decision problems solved in polynomial time by a deterministic (non-deterministic) Turing machine.

If a decision problem $P \in \mathcal{P}$, it means that we can solve this problem by a polynomial-time algorithm. If a decision problem $P \in \mathcal{NP}$, it means that we can verify in polynomial time whether a given solution to P has the properties specified in the formulation of this problem. (Notice that if we know how to solve a decision problem in polynomial time, we can also verify in polynomial time any solution to the problem. Hence $\mathcal{P} \subseteq \mathcal{NP}$.)

3.1.1 The ordinary $\mathcal{NP}$-completeness

First, we introduce the concept of a polynomial-time reduction.

Definition 3.1. (A polynomial-time reduction)
A polynomial-time reduction of a decision problem P' into a decision problem P is a function $f : D_{P'} \to D_P$ satisfying the following conditions:
(a) the function can be computed in polynomial time;
(b) for all instances $I \in D_{P'}$, there exists a solution to I if and only if there exists a solution to $f(I) \in D_P$.

© Springer-Verlag GmbH Germany, part of Springer Nature 2020
S. Gawiejnowicz, *Models and Algorithms of Time-Dependent Scheduling*,
Monographs in Theoretical Computer Science. An EATCS Series,
https://doi.org/10.1007/978-3-662-59362-2_3

If there exists a polynomial-time reduction of the problem P' to the problem P (i.e., if P' is *polynomially reducible* to P), we will write $P' \propto P$.

Definition 3.2. (An $\mathcal{NP}$-complete problem)
A decision problem P is said to be $\mathcal{NP}$-complete if $P \in \mathcal{NP}$ and $P' \propto P$ for any $P' \in \mathcal{NP}$.

In other words, a problem P is $\mathcal{NP}$-complete if any solution to P can be verified in polynomial time and any other problem from the $\mathcal{NP}$ class is polynomially reducible to P.

Since the notion of an $\mathcal{NP}$-complete problem is one of the fundamental notions in the complexity theory, some remarks are necessary.

Remark 3.3. If for a problem P and any $P' \in \mathcal{NP}$ we have $P' \propto P$, the problem P is said to be $\mathcal{NP}$-*hard*.

Remark 3.4. The problems which are $\mathcal{NP}$-complete ($\mathcal{NP}$-hard) with respect to the binary encoding scheme become polynomial with respect to the unary encoding scheme. Therefore, such problems are also called $\mathcal{NP}$-*complete* ($\mathcal{NP}$-*hard*) *in the ordinary sense*, ordinary $\mathcal{NP}$-*complete* ($\mathcal{NP}$-*hard*) or *binary* $\mathcal{NP}$-*complete* ($\mathcal{NP}$-*hard*) problems.

Remark 3.5. All $\mathcal{NP}$-complete problems are related to each other in the sense that a polynomial-time algorithm which would solve at least one $\mathcal{NP}$-complete problem, would solve all $\mathcal{NP}$-complete problems. Therefore, $\mathcal{NP}$-complete problems most probably do not belong to the $\mathcal{P}$ class.

Remark 3.6. Since $\mathcal{P} \subseteq \mathcal{NP}$ and since no polynomial-time algorithm has been found so far for any problem from the $\mathcal{NP}$ class, the question whether $\mathcal{NP} \subseteq \mathcal{P}$ is still open. This fact implies conditional truth of $\mathcal{NP}$-completeness results: they hold, unless $\mathcal{P} = \mathcal{NP}$.

Remark 3.7. $\mathcal{NP}$-completeness of a problem is a very strong argument for the conjecture that this problem cannot be solved by a polynomial-time algorithm, unless $\mathcal{P} = \mathcal{NP}$. (An $\mathcal{NP}$-complete problem, however, may be solved by a pseudo-polynomial algorithm.)

The class of all $\mathcal{NP}$-complete problems will be denoted by $\mathcal{NPC}$.

Proving the $\mathcal{NP}$-completeness of a decision problem immediately from Definition 3.2 is usually a difficult task, since we have to show that any problem from the $\mathcal{NP}$ class is polynomially reducible to our problem. Hence, in order to prove that a decision problem is $\mathcal{NP}$-complete, the following result is commonly used.

Lemma 3.8. (Basic properties of the $\propto$ relation)
(a) *The relation $\propto$ is transitive, i.e., if $P_1 \propto P_2$ and $P_2 \propto P_3$, then $P_1 \propto P_3$.*
(b) *If $P_1 \in \mathcal{P}$ and $P_1 \propto P_2$, then $P_2 \in \mathcal{P}$.*
(c) *If P_1 and P_2 belong to $\mathcal{NP}$, P_1 is $\mathcal{NP}$-complete and $P_1 \propto P_2$, then P_2 is $\mathcal{NP}$-complete.*

Proof. (a) See Garey and Johnson [10, Chap. 2, Lemma 2.2].

(b) See Garey and Johnson [10, Chap. 2, Lemma 2.1].

(c) See Garey and Johnson [10, Chap. 2, Lemma 2.3].   ◇

Now, by Lemma 3.8 (c), in order to prove that a decision problem P is $\mathcal{NP}$-complete, it is sufficient to show that $P \in \mathcal{NP}$ and that there exists an $\mathcal{NP}$-complete problem P' that is polynomially reducible to P.

3.1.2 The strong $\mathcal{NP}$-completeness

In Sect. 2.2, the so-called pseudo-polynomial algorithms were defined. Pseudo-polynomial algorithms exist only for some $\mathcal{NP}$-complete problems. Hence, there is a need to characterize in more detail those problems from the $\mathcal{NPC}$ class for which such algorithms exist.

Given a decision problem P, let $Length : D_P \to \mathbb{Z}_+$ denote the function returning the number of symbols used to describe any instance $I \in D_P$ and let $Max : D_P \to \mathbb{Z}_+$ denote the function returning the magnitude of the largest number in any instance $I \in D_P$.

For any problem P and any polynomial q over $\mathbb{Z}$, let P_q denote the sub-problem of P obtained by restricting P to instances I satisfying the inequality $Max(I) \leqslant q(Length(I))$.

Definition 3.9. (A strongly $\mathcal{NP}$-complete problem)
A decision problem P is said to be strongly $\mathcal{NP}$-complete ($\mathcal{NP}$-complete in the strong sense) *if $P \in \mathcal{NP}$ and if there exists a polynomial q over $\mathbb{Z}$ for which the problem P_q is $\mathcal{NP}$-complete.*

In other words, a decision problem is $\mathcal{NP}$-complete in the strong sense if the problem is $\mathcal{NP}$-complete in the ordinary sense, even if we restrict it to these instances in which the maximum value in input is polynomially bounded with respect to the length of the input.

Since the notion of an $\mathcal{NP}$-complete problem in the strong sense is as important as the notion of an $\mathcal{NP}$-complete problem, a few remarks are necessary.

Remark 3.10. If for a problem P there exists a polynomial q over $\mathbb{Z}$ for which the problem P_q is $\mathcal{NP}$-hard, the problem P is said to be *strongly $\mathcal{NP}$-hard* (*$\mathcal{NP}$-hard in the strong sense*).

Remark 3.11. Problems which are $\mathcal{NP}$-complete ($\mathcal{NP}$-hard) in the strong sense with respect to the binary encoding scheme, will remain $\mathcal{NP}$-complete ($\mathcal{NP}$-hard) also with respect to the unary encoding scheme. Hence, the problems are sometimes called *unary $\mathcal{NP}$-complete* (*$\mathcal{NP}$-hard*).

Remark 3.12. The notion of strong $\mathcal{NP}$-completeness allows us to divide all problems from the $\mathcal{NPC}$ class into the problems which can be solved by a pseudo-polynomial algorithm and the problems which cannot be solved by such an algorithm.

Remark 3.13. A decision problem which is $\mathcal{NP}$-complete in the strong sense cannot be solved by a pseudo-polynomial algorithm unless $\mathcal{P} = \mathcal{NP}$ and unless this problem is a *number problem*.

Definition 3.14. (A number problem)
A problem P is called a number problem *if there exists no polynomial p such that $Max(I) \leqslant p(Length(I))$ for any instance $I \in D_P$.*

The $\mathcal{NP}$-complete problems given in Section 3.2 are examples of number problems.

Remark 3.15. If a problem is $\mathcal{NP}$-complete in the ordinary sense, then for the problem there is no difference between a pseudo-polynomial and a polynomial algorithm. Such a distinction, however, exists for a problem that is $\mathcal{NP}$-complete in the strong sense.

The class of all strongly $\mathcal{NP}$-complete problems will be denoted by $\mathcal{SNPC}$.

Since in order to prove that an optimization problem is ordinary (strongly) $\mathcal{NP}$-hard it is sufficient to show that its decision counterpart is ordinary (strongly) $\mathcal{NP}$-complete, from now on we will mainly consider the problems which are ordinary (strongly) $\mathcal{NP}$-complete.

As in the case of $\mathcal{NP}$-complete problems, it is not easy to prove that a decision problem is $\mathcal{NP}$-complete in the strong sense, using Definition 3.9. The notion of a pseudo-polynomial reduction is commonly used instead.

Definition 3.16. (A pseudo-polynomial reduction)
A pseudo-polynomial reduction from a decision problem P to a decision problem P' is a function $f : D_P \to D_{P'}$ such that
(a) for all instances $I \in D_P$ there exists a solution to I if and only if there exists a solution to $f(I) \in D_{P'}$;
(b) f can be computed in time which is polynomial with respect to $Max(I)$ and $Length(I)$;
(c) there exists a polynomial q_1 such that for all instances $I \in D_P$ the inequality $q_1(Length'(f(I))) \geqslant Length(I)$ holds;
(d) there exists a two-variable polynomial q_2 such that for all instances $I \in D_P$ the inequality $Max'(f(I)) \leqslant q_2(Max(I), Length(I))$ holds, where functions $Length()$ and $Max()$ correspond to the problem P and functions $Length()'$ and $Max()'$ correspond to the problem P'.

The application of the pseudo-polynomial reduction simplifies proofs of the strong $\mathcal{NP}$-completeness, since the following result holds.

Lemma 3.17. (Basic property of the pseudo-polynomial reduction)
If $P' \in \mathcal{NP}$, if P is $\mathcal{NP}$-complete in the strong sense and if there exists a pseudo-polynomial reduction from P to P', then P' is $\mathcal{NP}$-complete in the strong sense.

Proof. See Garey and Johnson [10, Chap. 4, Lemma 4.1].   ◇

3.1.3 Coping with $\mathcal{NP}$-completeness

The existence of polynomial algorithms for intractable ($\mathcal{NP}$-complete or $\mathcal{NP}$-hard) problems is very unlikely unless $\mathcal{P} = \mathcal{NP}$ (cf. Remark 3.6). Since exact algorithms for such problems run in exponential time, the only reasonable approach to their solution is to apply approximation algorithms or approximation schemes. Therefore, we complete the section with results concerning the application of approximation algorithms and approximation schemes for $\mathcal{NP}$-complete problems.

Theorem 3.18. *If there exists a polynomial q of two variables such that for any instance $I \in D_P$ the inequality*

$$OPT(I) < q(Length(I), Max(I))$$

holds, then from the existence of an FPTAS for a problem P there follows the existence of a pseudo-polynomial approximation algorithm for this problem.

Proof. See Garey and Johnson [10, Chap. 6].   ◇

Lemma 3.19. *Let P be an optimization problem with integer solutions and let the assumptions of Theorem 3.18 be satisfied. If P is $\mathcal{NP}$-hard in the strong sense, then it cannot be solved by an FPTAS unless $\mathcal{P} = \mathcal{NP}$.*

Proof. The result is a corollary from Theorem 3.18.   □

3.2 $\mathcal{NP}$-complete problems

The following additive and multiplicative $\mathcal{NP}$-complete problems will be used in proofs of $\mathcal{NP}$-completeness of scheduling problems presented in this book.

3.2.1 Additive $\mathcal{NP}$-complete problems

Additive problems are expressed in terms of additive operations such as addition or subtraction, and are used in $\mathcal{NP}$-completeness proofs of scheduling problems with fixed job processing times. This is caused by the fact that the formulae describing job completion times and the value of criteria functions for such problems are sums of certain expressions.

In $\mathcal{NP}$-completeness proofs discussed in this book, the following additive $\mathcal{NP}$-complete problems are used.

PARTITION PROBLEM (PP): Given a set $X = \{x_1, x_2, \ldots, x_k\}$ of positive integers such that $\sum_{i=1}^{k} x_i$ is an even number, does there exist a subset $X' \subset X$ such that $\sum_{x_i \in X'} x_i = \sum_{x_i \in X \setminus X'} x_i$?

The PP problem is $\mathcal{NP}$-complete in the ordinary sense (see Garey and Johnson [24, Chap. 3, Theorem 3.5]).

EVEN-ODD PARTITION (EOP): A variant of the PP problem in which we require that $|X| = 2k$, elements of set X are ordered as $x_1, x_2, \ldots, x_{2n}$ and subset X' contains exactly one of x_{2i-1}, x_{2i} for $1 \leqslant i \leqslant k$.

The EOP problem is $\mathcal{NP}$-complete in the ordinary sense (see Garey and Johnson [24, Sect. A3.2]).

MODIFIED EVEN-ODD PARTITION (MEOP): A variant of the EOP problem in which to the input is added a rational number $q > 1$, and we require that elements of set X satisfy inequalities $x_{2i+1} > B_i$ and $x_{2i+2} > x_{2i+1}$ for $i = 0, 1, \ldots, k - 1$, where $B_0 = 0$, $B_i = \frac{1+q}{2} q^{k-i} x_{2i}$ for $i = 0, 1, \ldots, k - 1$.

The MEOP problem is $\mathcal{NP}$-complete in the ordinary sense (see Jaehn and Sedding [25, Theorem 1]).

SUBSET SUM (SS): Given $C \in \mathbb{Z}_+$, a set $R = \{1, 2, \ldots, r\}$ and a value $u_i \in \mathbb{Z}_+$ for each $i \in R$, does there exist a subset $R' \subseteq R$ such that $\sum_{i \in R'} u_i = C$?

The SS problem is $\mathcal{NP}$-complete in the ordinary sense (see Karp [26]).

3-PARTITION (3-P): Given $K \in \mathbb{Z}_+$ and a set $C = \{c_1, c_2, \ldots, c_{3h}\}$ of $3h$ integers such that $\frac{K}{4} < c_i < \frac{K}{2}$ for $1 \leqslant i \leqslant 3h$ and $\sum_{i=1}^{3h} c_i = hK$, can C be partitioned into disjoint sets $C_1, C_2, \ldots, C_h$ such that $\sum_{c_i \in C_j} c_i = K$ for each $1 \leqslant j \leqslant h$?

The 3-P problem is $\mathcal{NP}$-complete in the strong sense (see Garey and Johnson [24, Sect. 4.2, Theorem 4.4]).

NON-NUMBER 3-PARTITION (N3P): Given $Z \in \mathbb{Z}_+$ and $3w$ positive integers $z_1, z_2, \ldots, z_{3w}$ such that $\sum_{i=1}^{3w} z_i = wZ$, where $\frac{Z}{4} < z_i < \frac{Z}{2}$ for $1 \leqslant i \leqslant 3w$ and z_i is bounded by a polynomial of w, does there exist a partition of set $\{1, 2, \ldots, 3w\}$ into w disjoint subsets $Z_1, Z_2, \ldots, Z_w$ such that $\sum_{i \in Z_j} z_i = Z$ for $1 \leqslant j \leqslant w$?

The N3P problem is $\mathcal{NP}$-complete in the strong sense (see Garey and Johnson [24]).

KNAPSACK (KP): Given $U, W \in \mathbb{Z}_+$, a set $K = \{1, 2, \ldots, k\}$ and values $u_i \in \mathbb{Z}_+$ and $w_i \in \mathbb{Z}_+$ for each $k \in K$, does there exist $K' \subseteq K$ such that $\sum_{k \in K'} u_k \leqslant U$ and $\sum_{k \in K'} w_k \geqslant W$?

The KP problem is $\mathcal{NP}$-complete in the ordinary sense (see Karp [26]).

BIN PACKING (BP): Given $T, V \in \mathbb{Z}_+$, a set $L = \{1, 2, \ldots, l\}$ and values $u_i \in \mathbb{Z}_+$ for each $l \in L$, does there exist a partition of L into disjoint sets $L_1, L_2, \ldots, L_V$ such that $\sum_{l \in L_k} u_l \leqslant T$ for $1 \leqslant k \leqslant V$?

The BP problem is $\mathcal{NP}$-complete in the strong sense (see Garey and Johnson [24]).

3.2.2 Multiplicative $\mathcal{NP}$-complete problems

Multiplicative problems are expressed in terms of multiplicative operations such as multiplication or division, and are used in $\mathcal{NP}$-completeness proofs of scheduling problems with proportional, proportional-linear or linear time-dependent job processing times. This is caused by the fact that the formulae describing job completion times and the value of criteria functions for such problems are products of certain expressions.

In $\mathcal{NP}$-completeness proofs discussed in this book, the following multiplicative $\mathcal{NP}$-complete problems are used.

SUBSET PRODUCT (SP): Given $B \in \mathbb{Z}_+$, a set $P = \{1, 2, \ldots, p\}$ and a value $y_i \in \mathbb{Z}_+$ for each $i \in P$, does there exist a subset $P' \subseteq P$ such that $\prod_{i \in P'} y_i = B$?

The SP problem is $\mathcal{NP}$-complete in the ordinary sense (see Johnson [29]).

EQUAL PRODUCTS PROBLEM (EPP): Given a set $Q = \{1, 2, \ldots, q\}$ and a value $z_i \in \mathbb{Z}_+$ for each $i \in Q$ such that $\prod_{i \in Q} z_i$ is the square of an integer, does there exist a subset $Q' \subset Q$ such that $\prod_{i \in Q'} z_i = \prod_{i \in Q \setminus Q'} z_i$?

In order to illustrate the main steps of a typical $\mathcal{NP}$-completeness proof, we will show that the EPP problem is computationally intractable (an alternative proof one can find in Agnetis et al. [27]).

Lemma 3.20. *The EPP problem is $\mathcal{NP}$-complete in the ordinary sense.*

Proof. We will show that the SP problem is polynomially reducible to the EPP problem and therefore, by Lemma 3.8 (c), the latter problem is $\mathcal{NP}$-complete in the ordinary sense.

Let us consider the following reduction from the SP problem: $q = p + 2$, $z_i = y_i$ for $1 \leqslant i \leqslant p$, $z_{p+1} = 2\frac{Y}{B}$, $z_{p+2} = 2B$, the threshold value is $G = 2Y$, where $Y = \prod_{i \in P} y_i$.

First, let us notice that the above reduction is polynomial. Second, since for a given $Q' \subset Q$ we can check in polynomial time whether $\prod_{i \in Q'} z_i = \prod_{i \in Q \setminus Q'} z_i$, we have $EPP \in \mathcal{NP}$.

Hence, to end the proof it is sufficient to show that the SP problem has a solution if and only if the EPP problem has a solution.

If the SP problem has a solution, define $Q' := P' \cup \{z_{p+1}\}$. Then it is easy to check that $\prod_{i \in Q'} z_i = 2Y = G$. Hence the EPP problem has a solution.

If the EPP problem has a solution, then there exists a set $Q' \subset Q$ such that $\prod_{i \in Q'} z_i \leqslant G = 2Y$. Since $\prod_{i \in Q} z_i = 4Y^2$ and $Q' \cap (Q \setminus Q') = \emptyset$, the inequality $\prod_{i \in Q'} z_i < 2Y$ does not hold. Hence it must be $\prod_{i \in Q'} z_i = 2Y$.

Since $z_{p+1} \times z_{p+2} = 4Y$, the elements z_{p+1} and z_{p+2} cannot both belong to set Q'. Let us assume first that $z_{p+1} \in Q'$. Then $\prod_{i \in Q' \cup \{p+1\}} z_i = \prod_{i \in Q'} z_i \times z_{p+1} = \prod_{i \in Q'} \frac{2Y}{B} \times z_{p+1} = 2Y$. Hence $\prod_{i \in Q'} z_i = \prod_{i \in Q'} y_i = B$ and the SP problem has a solution. If $z_{m+1} \in Q \setminus Q'$, then by similar reasoning as above we have $\prod_{i \in Q \setminus Q' \setminus \{p+1\}} y_i = B$ and the SP problem has a solution as well. ∎

The EPP problem is $\mathcal{NP}$-complete in the strong sense (see Ng et al. [31]).

4-PRODUCT (4-P): Given $D \in \mathbb{Q}_+$, a set $N = \{1, 2, \ldots, 4p\}$ and a value $D^{\frac{1}{5}} < u_i < D^{\frac{1}{3}} \in \mathbb{Q}_+$ for each $i \in N$, $\prod_{i \in N} u_i = D^p$, do there exist disjoint subsets $N_1, N_2, \ldots, N_p$ such that $\bigcup_{i=1}^{p} N_i = N$ and $\prod_{i \in N_j} u_i = D$ for $1 \leqslant j \leqslant p$?

The 4-P problem, which is a multiplicative version of the 4-PARTITION problem (see Garey and Johnson [28, Sect. 4.2, Theorem 4.3]), is $\mathcal{NP}$-complete in the strong sense (see Kononov [30]).

With this remark, we end the first part of the book, devoted to presentation of the basic definitions and fundamental results concerning $\mathcal{NP}$-complete problems. In the next part, we will introduce the basic definitions, notions and models of the scheduling theory.

Bibliographic notes

The definitions and notions of the theory of $\mathcal{NP}$-completeness presented in this chapter are expressed in terms of decision problems and reductions between these problems. Alternatively, these definitions and notions can be expressed in terms of *languages* and reductions between languages.

Ausiello et al. [1], Bovet and Crescenzi [2], Bürgisser et al. [4], Calude [5], Davis et al. [7], Downey and Fellows [8], Du and Ko [9], Garey and Johnson [10], Homer and Selman [11], Hopcroft and Ullman [12], Papadimitriou [17], Pudlák [18], Rudich and Widgerson [19], Savage [20], Sipser [21], Wagner and Wechsung [22], Wegener [23] present the theory of $\mathcal{NP}$-completeness from different perspectives.

From Definition 3.2, it does not follow that $\mathcal{NP}$-complete problems exist at all. The fact that $\mathcal{NPC} \neq \emptyset$ was proved independently by Cook [6] in 1971 and Levin [16] in 1973.

The classes $\mathcal{P}$, $\mathcal{NP}$ and $\mathcal{NPC}$ have a fundamental meaning in the complexity theory. Johnson [13] presents a detailed review of other complexity classes.

The list of $\mathcal{NP}$-complete problems, initiated by Cook [6], Karp [15] and Levin [16], contains a great number of problems and is still growing. Ausiello et al. [1], Brucker and Knust [3], Garey and Johnson [10] and Johnson [14] present extensive excerpts from this list, including problems from different domains of computer science and discrete mathematics.

The functions $Length()$ and $Max()$ are discussed in detail by Garey and Johnson [10, Sect. 4.2].

References

Basic definitions and results

1. G. Ausiello, P. Crescenzi, G. Gambosi, V. Kann, A. Marchetti-Spaccamela and M. Protasi, *Complexity and Approximation: Combinatorial Optimization Problems and Their Approximability Properties*. Berlin-Heidelberg: Springer 1999.
2. D. P. Bovet and P. Crescenzi, *Introduction to Theory of Complexity*. Harlow: Prentice-Hall 1994.
3. P. Brucker and S. Knust, Complexity results for scheduling problems. See http://www.mathematik.uni-osnabrueck.de/research/OR/index.shtml.
4. P. Bürgisser, M. Clausen and M. A. Shokrollahi, *Algebraic Complexity Theory*. Berlin-Heidelberg: Springer 1997.
5. C. Calude, *Theories of Computational Complexity*. Amsterdam: North-Holland 1988.
6. S. A. Cook, The complexity of theorem-proving procedures. *Proceedings of the 3rd ACM Symposium on the Theory of Computing*, 1971, 151–158.
7. M. D. Davis, R. Sigal and E. J. Weyuker, *Computability, Complexity, and Languages: Fundamentals of Theoretical Computer Science*, 2nd ed. San Diego: Academic Press 1994.
8. R. G. Downey and M. R. Fellows, *Fundamentals of Parametrized Complexity*. Berlin-Heidelberg: Springer 2013.
9. D-Z. Du and K-I. Ko, *Theory of Computational Complexity*. New York: Wiley 2000.
10. M. R. Garey and D. S. Johnson, *Computers and Intractability: A Guide to the Theory of NP-Completeness*. San Francisco: Freeman 1979.
11. S. Homer and A. L. Selman, *Computability and Complexity Theory*, 2nd ed. New York: Springer Science+Business Media 2011.
12. J. E. Hopcroft and J. D. Ullman, *Introduction to Automata Theory, Languages and Computation*. Reading: Addison-Wesley 1979.
13. D. S. Johnson, A catalog of complexity classes. In: J. van Leeuwen (ed.), *Handbook of Theoretical Computer Science: Algorithms and Complexity*. Amsterdam-Cambridge: Elsevier/MIT 1990, pp. 67–161.
14. D. S. Johnson, The NP-completeness column: an ongoing guide. *Journal of Algorithms* **2** (1982), no. 4, 393–405.
15. R. M. Karp, Reducibility among combinatorial problems. In: R. E. Miller and J. W. Thatcher (eds.), *Complexity of Computer Computations*. New York: Plenum Press 1972, pp. 85–103.
16. L. A. Levin, Universal sorting problems. *Problemy Peredachi Informatsii* **9** (1973), no. 3, 115–116 (in Russian).
    English translation: *Problems of Information Transmission* **9** (1973), no. 3, 265–266.
17. C. H. Papadimitriou, *Computational Complexity*. Reading, MA: Addison-Wesley 1994.
18. P. Pudlák, *Logical Foundations of Mathematics and Computational Complexity: A Gentle Introduction*. Heidelberg: Springer 2013.
19. S. Rudich and A. Widgerson (eds.), *Computational Complexity Theory*. Rhode Island: American Mathematical Society 2004.
20. J. E. Savage, *The Complexity of Computing*. New York: Wiley 1976.

21. M. Sipser, *Introduction to the Theory of Computation*. Boston: PWS Publishing 1997.
22. K. Wagner and G. Wechsung, *Computational Complexity*. Berlin: Deutscher Verlag der Wissenschaften 1986.
23. I. Wegener, *Complexity Theory: Exploring the Limits of Efficient Algorithms*. Berlin-Heidelberg: Springer 2005.

Additive $\mathcal{NP}$-complete problems

24. M. R. Garey and D. S. Johnson, *Computers and Intractability: A Guide to the Theory of NP-Completeness*. San Francisco: Freeman 1979.
25. F. Jaehn and H. A. Sedding, Scheduling with time-dependent discrepancy times. *Journal of Scheduling* **19** (2016), no. 6, 737–757.
26. R. M. Karp, Reducibility among combinatorial problems. In: R. E. Miller and J. W. Thatcher (eds.), *Complexity of Computer Computations*. New York: Plenum Press 1972, pp. 85–103.

Multiplicative $\mathcal{NP}$-complete problems

27. A. Agnetis, P. Detti, M. Pranzo and. M. S. Sodhi, Sequencing unreliable jobs on parallel machines. *Journal of Scheduling* **12** (2009), no. 1, 45–54.
28. M. R. Garey and D. S. Johnson, *Computers and Intractability: A Guide to the Theory of NP-Completeness*. San Francisco: Freeman 1979.
29. D. S. Johnson, The NP-completeness column: an ongoing guide. *Journal of Algorithms* **2** (1982), no. 4, 393–405.
30. A. Kononov, Combinatorial complexity of scheduling jobs with simple linear deterioration. *Discrete Analysis and Operations Research* **3** (1996), no. 2, 15–32 (in Russian).
31. C. T. Ng, M. S. Barketau, T-C. E. Cheng and M. Y. Kovalyov, "Product-Partition" and related problems of scheduling and systems reliability: computational complexity and approximation. *European Journal of Operational Research* **207** (2010), no. 2, 601–604.

SCHEDULING MODELS

SCHEDULING MODELS

4

The classical scheduling theory

Time-dependent scheduling is one of research domains in the scheduling theory, rich in various scheduling models. Though this domain possesses its own notions and methods, it is also based on those which are common for the models studied in other domains of the scheduling theory. Therefore, in the second part of the book we recall the basic facts concerning the main models of the scheduling theory.

This part of the book is composed of three chapters. In Chap. 4, we recall the basic definitions and notions of the classical scheduling theory model. The main models in the modern scheduling theory are discussed in Chap. 5. The principal concepts and definitions of the time-dependent scheduling model are presented in Chap. 6.

Chapter 4 is composed of six sections. In Sect. 4.1, we introduce the notions of scheduling model and scheduling problem. In Sect. 4.2, we formulate the basic assumptions of the classical scheduling theory. In Sect. 4.3, we define the parameters which describe any problem which can be considered in the framework of the scheduling theory. In Sect. 4.4, we introduce the notion of schedule, the main notion of the scheduling theory. In Sect. 4.5, we define the criteria of optimality of a schedule. In Sect. 4.6, we introduce the $\alpha|\beta|\gamma$ notation, used for symbolic description of scheduling problems. The chapter is completed with bibliographic notes and a list of references.

4.1 Models and problems of the scheduling theory

We begin this chapter with some general remarks on the basic notions of the scheduling theory and their mutual relations.

4.1.1 Scheduling models

The scheduling theory arose to solve resource allocation problems encountered in industrial practice (Conway et al. [12], Muth and Thompson [35]). During

© Springer-Verlag GmbH Germany, part of Springer Nature 2020
S. Gawiejnowicz, *Models and Algorithms of Time-Dependent Scheduling*,
Monographs in Theoretical Computer Science. An EATCS Series,
https://doi.org/10.1007/978-3-662-59362-2_4

more than six decades of development, in this theory appeared numerous research domains, focusing on different classes of scheduling problems. Though each such class has its own set of basic notions, two notions are common to the whole scheduling theory: *scheduling model* and *scheduling problem*.

Definition 4.1. (Scheduling model)
A scheduling model *is a partial abstract description of a problem related to scheduling, where mathematical notions and data structures are used to represent selected entities or relations appearing in the problem.*

A scheduling model concerns some of the parts of a scheduling problem description such as definitions of job processing times or machine speeds. Different models of job processing times are the most common examples of scheduling models in the scheduling theory.

Example 4.2. (a) An example of a scheduling model is the classical scheduling model of fixed processing times (discussed in the chapter), where job processing times are integer or rational numbers.
(b) Examples of scheduling models are also the modern scheduling models of variable job processing times, where job processing times are functions of resources allocated to the jobs or positions of the jobs in the schedule (see Chap. 5) or their starting times (see Chap. 6). ◆

Each scheduling model allows us to gain practical knowledge about a scheduling problem, since based on the properties of the model we construct algorithms used to solve the issues appearing in the problem. Therefore, the creation of a model is usually the first step in the analysis of the scheduling problem under consideration.

However, knowing a particular scheduling model, we possess only a partial image of the studied scheduling problem, since in such a model the notion of solution optimality is not defined. This means that scheduling models do not specify details concerning the optimality, leaving this issue open.

4.1.2 Scheduling problems

A notion with a narrower meaning than scheduling model is the notion of scheduling problem.

Definition 4.3. (Scheduling problem)
A scheduling problem *is a full description of a given problem related to scheduling, expressed in terms of a particular scheduling model and specifying the way in which solutions to the problem will be evaluated.*

Scheduling models and scheduling problems are related, because we always consider a given scheduling problem in the context of a particular scheduling model. Hence, similar scheduling problems defined in different scheduling models may be solved using different algorithms or have a different complexity.

Example 4.4. (*a*) A single machine scheduling problem with the classical scheduling model of fixed job processing times and the maximum completion time criterion is solvable in $O(n)$ time by scheduling jobs in an arbitrary order (Baker [2]). A similar scheduling problem with the modern scheduling model of linear time-dependent processing times is solvable in $O(n \log n)$ time by scheduling jobs in non-increasing order of some ratios (Gawiejnowicz and Pankowska [21], Gupta and Gupta [23], Tanaev et al. [46], Wajs [49]).

(*b*) A parallel machine scheduling problem with the classical scheduling model of fixed job processing times and the total completion time criterion is solvable in $O(n \log n)$ time by scheduling jobs using the SPT (Shortest Processing Time first) algorithm (Conway et al. [12]). A similar scheduling problem with the modern scheduling model of proportional time-dependent job processing times is $\mathcal{NP}$-hard already for two machines (Chen [9], Kononov [30]).   ♦

In subsequent chapters, we will consider the scheduling models of the classical and modern scheduling theory. We begin with the scheduling model of the classical scheduling theory.

4.2 Basic assumptions of the classical scheduling theory

The classical scheduling theory is based on a set of assumptions concerning the set of machines and the set of jobs. The following are the most important assumptions of this theory, where assumptions (M1)–(M3) concern the set of machines, while assumptions (J1)–(J4) concern the set of jobs:

(M1) each machine is continuously available;
(M2) each machine can handle at most one job at a time;
(M3) machine speeds are fixed and known in advance;
(J1) jobs are strictly-ordered sequences of operations;
(J2) each job may be performed only by one machine;
(J3) the processing times of successive operations of a job do not overlap;
(J4) job parameters are numbers known in advance.

Throughout this book, the scheduling theory with assumptions (M1)–(J4) will be called the *classical scheduling theory*, as opposed to the *modern scheduling theory*, where at least one of these assumptions has been changed.

4.3 Formulation of classical scheduling problems

Regardless of its nature, every scheduling problem $\mathcal{S}$ can be formulated as a quadruple, $\mathcal{S} = (\mathcal{J}, \mathcal{M}, \mathcal{R}, \varphi)$, where $\mathcal{J}$ is a set of pieces of work to be executed, $\mathcal{M}$ is a set of entities that will perform the pieces of work, $\mathcal{R}$ is a set of additional entities needed for performing these pieces of work and φ is a function that is used as a measure of quality of solutions to the problem under consideration. We start this section with a brief description of the parameters of the quadruple.

4.3.1 Parameters of the set of jobs

The elements of set $\mathcal{J}$ are called *jobs*. Unless otherwise specified, we will assume that $|\mathcal{J}| = n$, i.e., there are n jobs. We will denote jobs by $J_1, J_2, \ldots, J_n$. The set of indices of jobs from the set $\mathcal{J}$ will be denoted by $N_{\mathcal{J}}$.

Job J_j, $1 \leqslant j \leqslant n$, consists of n_j *operations*, $O_{1j}, O_{2j}, \ldots, O_{n_j,j}$. For each operation, we define the *processing time* of the operation, i.e., the time needed for processing this operation.

Remark 4.5. If a job consists of one operation only, we will identify the job with the operation. In this case, the processing time of the operation is the processing time of the job.

The processing time of job J_j (operation O_{ij}) will be denoted by p_j (p_{ij}), where $1 \leqslant j \leqslant n$ $(1 \leqslant i \leqslant n_j$ and $1 \leqslant j \leqslant n)$.

For job J_j, $1 \leqslant j \leqslant n$, there may be defined a *ready time*, r_j, a *deadline*, d_j, and a *weight*, w_j. The first operation of job J_j cannot be started before the ready time r_j and the last operation of the job cannot be completed after the deadline d_j. We will say that there are no ready times (deadlines) if $r_j = 0$ $(d_j = +\infty)$ for all j. The weight w_j indicates the importance of job J_j, compared to other jobs. We will assume that $w_j = 1$ for all j, unless otherwise stated.

Throughout the book, unless otherwise stated, we will assume that job (operation) parameters are positive integer numbers, i.e., p_j (p_{ij}), $r_j, d_j, w_j \in \mathbb{Z}^+$ for $1 \leqslant j \leqslant n$ $(1 \leqslant i \leqslant n_j$ and $1 \leqslant j \leqslant n)$.

Example 4.6. Let the set $\mathcal{J}$ be composed of 4 jobs, $\mathcal{J} = \{J_1, J_2, J_3, J_4\}$, such that $p_1 = 1, p_2 = 2, p_3 = 3$ and $p_4 = 4$, with no ready times and deadlines, and with unit job weights. Then $r_j = 0$, $d_j = +\infty$ and $w_j = 1$ for $1 \leqslant j \leqslant 4$. ♦

There may be also defined *precedence constraints* among jobs, which reflect the fact that some jobs have to be executed before others. The precedence constraints correspond to a partial order $\prec \subseteq \mathcal{J} \times \mathcal{J}$. We will assume that precedence constraints between jobs can be given in the form of a set of chains, a tree, a series-parallel digraph or an arbitrary acyclic digraph.

If precedence constraints are defined on the set of jobs, we will call the jobs *dependent*; otherwise we will say that they are *independent*.

Example 4.7. The jobs from Example 4.6 are independent, $\prec = \emptyset$. If we assume that job precedence constraints in the set are as in Fig. 1.1b or Fig. 1.2d, then the jobs are dependent. ♦

Jobs can be *preemptable* or *non-preemptable*. If a job is preemptable, then the execution of this job can be interrupted at any time without any cost, and resumed at a later time on the machine on which it was executed before the preemption, or on another one. Otherwise, the job is non-preemptable.

In this book, we will mainly consider scheduling problems with non-preemptable independent jobs. We will also assume that there are no ready times, no deadlines and all job weights are equal, unless otherwise specified.

4.3.2 Parameters of the set of machines

Jobs are performed by elements of set $\mathcal{M}$, called *machines*. We will assume that $|\mathcal{M}| = m$, i.e., we are given m machines. The machines will be denoted by $M_1, M_2, \ldots, M_m$.

Remark 4.8. Sometimes elements of set $\mathcal{M}$ have other names than 'machines'. For example, in scheduling problems which arise in computer systems, the elements of set $\mathcal{M}$ are called *processors*. Throughout the book, we will use the term 'machine' to denote a single element of set $\mathcal{M}$.

In the simplest case, when $m = 1$, we deal with a *single machine*. Despite its simplicity, this case is worth considering, since it appears in more complex machine environments described below.

If all $m \geqslant 2$ machines are of the same kind and have the same processing speed, we deal with *parallel identical machines*. In this case, job J_j with the processing time p_j, $1 \leqslant j \leqslant n$, can be performed by any machine and its execution on the machine will take p_j units of time.

If among the available machines there is a slowest machine, M_1, and any other machine, $M_k \neq M_1$, has a speed s_k that is a multiple of the speed s_1 of machine M_1, we deal with *parallel uniform machines*. In this case, the execution of job J_j on machine M_k will take $\frac{p_j}{s_k}$ units of time.

Finally, if the machines differ in speed but the speeds depend on the performed job, we deal with *parallel unrelated machines*. In this case, the symbol p_{ij} is used for denoting the processing time of job J_j on machine M_i, where $1 \leqslant i \leqslant m$ and $1 \leqslant j \leqslant n$.

So far, we assumed that all machines in a machine environment perform the same functions, i.e., any job that consists of only one operation can be performed on any machine. If the available machines have different functions, i.e., some of them cannot perform some jobs, we deal with *dedicated machines*. In this case, job J_j consists of $n_j \geqslant 1$ different operations, which are performed by different machines, $1 \leqslant j \leqslant n$. We will consider three main types of such a machine environment: flow shop, open shop and job shop.

A *flow shop* consists of $m \geqslant 2$ machines, $M_1, M_2, \ldots, M_m$. Each job consists of m operations, $n_j = m$ for $1 \leqslant j \leqslant n$. The ith operation of any job has to be executed by machine $M_i, 1 \leqslant i \leqslant m$. Moreover, this operation can start only if the previous operation of this particular job has been completed. All jobs follow the same route from the first machine to the last one. (In other words, precedence constraints between operations of any job in a flow shop are in the form of a chain whose length is equal to the number of machines in the flow shop.)

An *open shop* consists of $m \geqslant 2$ machines. Each job consists of m operations, $n_j = m$ for $1 \leqslant j \leqslant n$, but the order of processing of operations can be different for different jobs. This means that each job has to go through all machines but the route can be arbitrary. (In other words, the operations of any job in an open shop are independent and the number of the operations is equal to the number of machines in the open shop.)

A *job shop* consists of $m \geqslant 2$ machines. Each job can consist of n_j operations, where not necessarily $n_j = m$ for $1 \leqslant j \leqslant n$.. Moreover, each job has its own route of performing its operations, and it can visit a certain machine more than once or may not visit some machines at all. (In other words, precedence constraints between operations of any job in a jobs shop are in the form of a chain and the number of operations may be arbitrary.)

Remark 4.9. In some dedicated-machine environments additional constraints, which restrict the job flow through the machines, may be imposed on available machines. For example, a flow shop may be of the 'no-wait' type. The *no-wait* constraint means that buffers between machines are of zero capacity and a job after completion of its processing on one machine must immediately start on the next (consecutive) machine.

In all the above cases of dedicated machine environments, the symbol p_{ij} will be used for denoting the processing time of operation O_{ij} of job J_j, where $1 \leqslant i \leqslant n_j \leqslant m$ and $1 \leqslant j \leqslant n$ for flow shop and open shop problems, and $1 \leqslant i \leqslant n_j$ and $1 \leqslant j \leqslant n$ for a job shop problem.

Remark 4.10. Throughout the book, unless otherwise stated, we will assume that jobs are processed on machines which are *continuously available*. In some applications, however, it is required to consider *machine non-availability periods* in which the machines are not available for processing due to maintenance operations, rest periods or machine breakdowns. We will come back to scheduling problems with machine non-availability periods in Sect. 21.1.

We will not define other types of machine environments, since, in this book, we will consider scheduling problems on a single machine, on parallel machines and on dedicated machines only.

4.3.3 Parameters of the set of resources

In some problems, the execution of jobs requires additional entities other than machines. The entities, elements of set $\mathcal{R}$, are called *resources*. The resources may be continuous or discrete. A resource is *continuous* if it can be allocated to a job in an arbitrary amount. A resource is *discrete* if it can be allocated to a job only in a non-negative integer number of units.

Example 4.11. Energy, gas and power are continuous resources. Tools, robots and automated guided vehicles are discrete resources. ◆

In real-world applications, the available resources are usually constrained. A resource is *constrained* if it can be allocated only in an amount which is between the minimum and the maximum number of units of the resource; otherwise, it is *unconstrained*. Example 4.11 concerns constrained resources. There also exist certain applications in which constrained resources are available in a huge number of units.

Example 4.12. Virtual memory in computer systems, manpower in problems of scheduling very-large-scale projects and money in some finance management problems are unconstrained resources. ♦

Since, in this book, we consider the scheduling problems in which jobs do not need additional resources for execution, $\mathcal{R} = \emptyset$, we omit a more detailed description of the parameters of set $\mathcal{R}$.

4.4 The notion of schedule

The value of a criterion function may be calculated once a solution to the instance of a particular scheduling problem is known. Before we define possible forms of the criterion function, we describe the solution.

Given a description of sets $\mathcal{J}$ and $\mathcal{M}$ for a scheduling problem, we can start looking for a solution to the problem. Roughly speaking, a solution to a scheduling problem is an assignment of machines to jobs in time that satisfies some (defined below) requirements. The solution will be called a *schedule*. For the purpose of this book, we assume the following definition of the notion.

Definition 4.13. (A schedule)
A schedule is an assignment of machines (and possibly resources) to jobs in time such that the following conditions are satisfied:
(a) at every moment of time, each machine is assigned to at most one job and each job is processed by at most one machine;
(b) job J_j, $1 \leqslant j \leqslant n$, is processed in time interval $\langle r_j, +\infty \rangle$;
(c) all jobs are completed;
(d) if there exist precedence constraints for some jobs, then the jobs are executed in the order consistent with these constraints;
(e) if there exist resource contraints, then they are satisfied;
(f) if jobs are non-preemptable, then no job is preempted; otherwise the number of preemptions of each job is finite.

Since the notion of a schedule plays a fundamental role in the scheduling theory, we will now add a few remarks to Definition 4.13.

Remark 4.14. An arbitrary schedule specifies two sets of time intervals. The first set consists of the time intervals in which available machines perform some jobs. In every interval of this kind, a job is executed by a machine. If no job was preempted, then only one time interval corresponds to each job; otherwise, a number of intervals correspond to each job. The first set is always non-empty and the intervals are not necessarily disjoint. The second set, which may be empty, consists of the time intervals in which the available machines do not work. These time intervals will be called *idle times* of the machines.

Remark 4.15. In some dedicated machine environments, the available machines may have some limitations which concern idle times. For example, a flow shop may be of the 'no-idle' type. The *no-idle* constraint means that each machine, once it has commenced its work, must process all operations assigned to it without idle times. (Another constraint concerning the flow shop environment, 'no-wait', is described in Remark 4.9.)

Remark 4.16. An arbitrary schedule is composed of a number of partial schedules that correspond to particular machines. The partial schedules will be called *subschedules*. The number of subschedules of a schedule is equal to the number of machines in the schedule. Note that a schedule for a single machine is identical with its subschedule.

Remark 4.17. In some cases (e.g., no preemptions, no idle times, ready times of all jobs are equal) a schedule may be fully described by the permutations of indices of jobs that are assigned to particular machines in that schedule. The permutation corresponding to such a schedule (subschedule) will be called a *job sequence (subsequence)*.

4.4.1 The presentation of schedules

Schedules are usually presented by Gantt charts. A *Gantt chart* is a two-dimensional diagram composed of a number of labelled rectangles and a number of horizontal axes. When the rectangles represent jobs (operations) and the axes correspond to machines, we say that the Gantt chart is *machine-oriented*. When the rectangles represent machines and the axes correspond to jobs (operations), the Gantt chart is *job-oriented*.

Throughout this book, we will use machine-oriented Gantt charts.

Example 4.18. Consider the set of jobs $\mathcal{J}$ defined as in Example 4.6. Since the jobs are independent and there are no ready times and deadlines, any sequence of the jobs corresponds to a schedule. An example schedule for this set of jobs, corresponding to sequence (J_4, J_3, J_1, J_2), is given in Fig. 4.1.

◆

Remark 4.19. The schedule presented in Fig. 4.1 is an example of a schedule in which no machine is kept idle when it could start the processing of some job (operation). Such schedules are called *non-delay* schedules.

Fig. 4.1: A schedule for the set of jobs from Example 4.18

4.4.2 Parameters of job in a schedule

If we know a schedule σ for an instance of a scheduling problem, then for any job J_j, $1 \leqslant j \leqslant n$, we may calculate the values of parameters characterizing this job in schedule σ. Examples of parameters of the job J_j in schedule σ are:

- *the starting time* $S_j(\sigma)$,
- *the completion time* $C_j(\sigma) = S_j(\sigma) + p_j$,
- *the waiting time* $W_j(\sigma) = C_j(\sigma) - r_j - p_j$,
- *the lateness* $L_j(\sigma) = C_j(\sigma) - d_j$ and
- *the tardiness* $T_j(\sigma) = \max\{0, L_j(\sigma)\}$.

If it is clear which schedule we will consider, we will omit the symbol σ and write S_j, C_j, W_j, L_j and T_j, respectively.

Example 4.20. Consider the Gantt chart given in Fig. 4.1. From the chart we can read, e.g., that $S_4 = 0$ and $C_4 = 4$, while $S_1 = 7$ and $C_1 = 8$.   ◆

4.4.3 Types of schedules

As a rule, there can be found different schedules for a given scheduling problem S. The set of all schedules for a given S will be denoted by $\mathcal{Z}(S)$. The schedules which compose the set $\mathcal{Z}(S)$ can be of different types. Throughout this book, we will distinguish the following types of schedules.

Definition 4.21. (A feasible schedule)
(a) *A schedule is said to be* feasible *if it satisfies all conditions of Definition 4.13, and if other conditions specific for a given problem are satisfied.*
(b) *A feasible schedule is said to be* non-preemptive *if no job has been preempted; otherwise, it is* preemptive.

Example 4.22. The schedule given in Fig. 4.1 is a feasible schedule for the set of jobs from Example 4.18. Moreover, any other schedule obtained from the schedule by a rearrangement of jobs is also a feasible schedule.   ◆

The set of all feasible schedules for a given scheduling problem S will be denoted by $\mathcal{Z}_{feas}(S)$.

Definition 4.23. (A semi-active schedule)
A schedule is said to be semi-active *if it is obtained from any feasible schedule by shifting all jobs (operations) to start as early as possible but without changing any job sequence.*

In other words, a schedule is semi-active if jobs (operations) in the schedule cannot be shifted to start earlier without changing the job sequence, violating precedence constraints or ready times.

Example 4.24. The schedule given in Figure 4.1 is a semi-active schedule for the set of jobs from Example 4.18, since no job can be shifted to start earlier without changing the job sequence. ♦

Example 4.25. The schedule presented in Figure 4.2 is another feasible schedule for the set of jobs from Example 4.18. The schedule, however, is not a semi-active schedule, since we can shift job J_2 one unit of time to the left. ♦

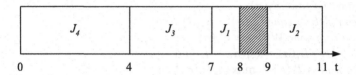

Fig. 4.2: A feasible schedule for Example 4.18

Remark 4.26. Since each job (operation) starts its execution as early as possible in any semi-active schedule, each semi-active schedule is completely characterized by the job sequence (subsequences) corresponding to the schedule (subschedules).

Remark 4.27. A schedule which is completely specified by the job sequence(s) is called a *permutation schedule.*

Remark 4.28. For any job sequence, there exists only one semi-active schedule.

The set of all semi-active schedules for a given scheduling problem S will be denoted by $\mathcal{Z}_{s-act}(S)$.

Definition 4.29. (An active schedule)
A schedule is said to be active *if it is obtained from any semi-active schedule by shifting all jobs (operations) to start as early as possible even if the shifting causes a change in some job sequence(s).*

In other words, a schedule is active if jobs (operations) in the schedule cannot be shifted to start earlier without violating precedence constraints or ready times.

Example 4.30. The semi-active schedule from Fig. 4.1 is also an active schedule for the set of jobs from Example 4.18. ♦

The set of all active schedules for a given scheduling problem $\mathcal{S}$ will be denoted by $\mathcal{Z}_{act}(\mathcal{S})$.

Definition 4.31. (An optimal schedule)
A schedule is said to be optimal *if the value of optimality criterion for the schedule is optimal.*

The set of all optimal schedules for a given scheduling problem $\mathcal{S}$ will be denoted by $\mathcal{Z}_{opt}(\mathcal{S})$.

Remark 4.32. The criteria of optimality of a schedule will be defined in Definition 4.36.

Remark 4.33. The optimal schedule need not be unique, i.e., $|\mathcal{Z}_{opt}(\mathcal{S})| \geqslant 1$.

Definition 4.34. (A dominant set)
A set of schedules is said to be dominant *if it contains at least one optimal schedule.*

Remark 4.35. Notice that there hold the inclusions $\mathcal{Z}_{act}(\mathcal{S}) \subseteq \mathcal{Z}_{s-act}(\mathcal{S}) \subset \mathcal{Z}_{feas}(\mathcal{S}) \subset \mathcal{Z}(\mathcal{S})$ and $\mathcal{Z}_{opt}(\mathcal{S}) \subset \mathcal{Z}_{s-act}(\mathcal{S})$.

In this book, we will consider mainly non-preemptive semi-active schedules.

4.5 The criteria of schedule optimality

In general, the criterion of optimality in a scheduling problem can be an arbitrary function, which has real values and has been defined on the set of all feasible schedules for the problem. In this book, we will consider mainly the following optimality criteria.

Definition 4.36. (Criteria of optimality of a schedule)
Let $C_1, C_2, \ldots, C_n$ be the completion times of jobs in a schedule. The criteria of the maximum completion time ($C_{\max}$), the maximum lateness ($L_{\max}$), the maximum tardiness ($T_{\max}$), the maximum cost ($f_{\max}$), the total completion time ($\sum C_j$), the total general completion time ($\sum C_j^k$), the total weighted completion time ($\sum w_j C_j$), the total machine load ($\sum C_{\max}^{(k)}$), the number of tardy jobs ($\sum U_j$) *and the* total cost ($\sum f_j$) *are defined as follows:*

(a) $C_{\max} := \max\limits_{1 \leqslant j \leqslant n} \{C_j\} \equiv \max\limits_{1 \leqslant k \leqslant m} \{C_{\max}^{(k)}\}$,

(b) $L_{\max} := \max\limits_{1 \leqslant j \leqslant n} \{L_j\} := \max\limits_{1 \leqslant j \leqslant n} \{C_j - d_j\}$,

(c) $T_{\max} := \max\limits_{1 \leqslant j \leqslant n} \{T_j\} := \max\limits_{1 \leqslant j \leqslant n} \{\max\{0, L_j\}\}$,

(d) $f_{\max} := \max\limits_{1 \leqslant j \leqslant n} \{f_j(C_j)\}$, *where* $f_1, f_2, \ldots, f_n$ *are given cost functions,*

(e1) $\sum C_j := \sum_{j=1}^{n} C_j$, (e2) $\sum C_j^k := \sum_{j=1}^{n} C_j^k$, *where $k > 0$ is a constant,*

(f) $\sum w_j C_j := \sum_{j=1}^{n} w_j C_j$,

(g) $\sum C_{\max}^{(k)} := \sum_{j=1}^{m} C_{\max}^{(k)}$, where $C_{\max}^{(k)}$ denotes the maximum completion time over all jobs assigned to machine $M_k, 1 \leqslant k \leqslant m$;

(h) $\sum U_j := \sum_{j=1}^{n} U_j$, where $U_j := 0$ if $L_j \leqslant 0$ and $U_j := 1$ if $L_j > 0$,

(i) $\sum f_j := \sum_{j=1}^{n} f_j(C_j)$, where $f_1, f_2, \ldots, f_n$ are given cost functions.

Example 4.37. The schedule given in Fig. 4.2 is an optimal schedule for the set of jobs from Example 4.18 with respect to the $C_{\max}$ criterion. This schedule, however, is not optimal with respect to the $\sum C_j$ criterion.   ◆

Since the criteria of optimality have a fundamental meaning to the scheduling theory, we shall make a few remarks concerning the above definition.

Remark 4.38. The criteria defined in Def. 4.36 can be divided into the *maximal criteria* ($C_{\max}$, $L_{\max}$, $T_{\max}$ and $f_{\max}$) and *total criteria* ($\sum C_j$, $\sum w_j C_j$, $\sum C_{\max}^{(k)}$, $\sum U_j$ and $\sum f_j$).

Remark 4.39. Some of the above criteria are special cases of the l_p norm (cf. Definition 1.19). For example, if $C = [C_1, C_2, \ldots, C_n]$ is the vector of job completion times in a schedule, then $\|C\|_1 \equiv \sum_{j=1}^{n} C_j$ and $\|C\|_\infty \equiv C_{\max}$. We will come back to this topic in Chap. 19.

Remark 4.40. Some criteria may also be defined in terms of other criteria. For example, $C_{\max} := \max_{1 \leqslant j \leqslant n} \{C_j\} \equiv \max_{1 \leqslant k \leqslant m} \{C_{\max}^{(k)}\}$.

Remark 4.41. The completion time of a job is the basic parameter characterizing the job in a schedule, since the optimality criteria from Definition 4.36 are functions of the job completion times. Given any feasible schedule, the starting time of a job in the schedule is the startpoint of the first time interval corresponding to the job. Similarly, the completion time of a job in the schedule is the endpoint of the last time interval corresponding to the job.

Remark 4.42. The criteria of optimality of a schedule given in Definition 4.36 are single-valued functions. In Chap. 20, we will consider *bi-criteria* scheduling problems. In the chapter, we will extend the definition to include schedule optimality criteria composed of two single-valued functions.

As a rule, the applied criterion of optimality of a schedule is minimized. Therefore, the criteria that have properties which allow to find the optimum are of practical interest. Examples of such criteria are *regular criteria*, introduced by Conway et al. [12].

Definition 4.43. (A regular criterion)
Let $C = [C_1, C_2, \ldots, C_n]$ be the vector of job completion times and let $\varphi : C \to \mathbb{R}$ be a criterion function. Criterion φ is said to be a regular criterion if for any other vector of job completion times, $C' = [C_1', C_2', \ldots, C_n']$, the inequality $\varphi(C') \geqslant \varphi(C)$ holds if and only if there exists an index k, $1 \leqslant k \leqslant n$, such that $C_k' \geqslant C_k$.

In other words, φ is a regular criterion only if it is a non-decreasing function with respect to job completion times.

Example 4.44. The $C_{\max}$ criterion (also called the *schedule length* or *makespan*) and other criteria from Definition 4.36 are regular criteria.   ♦

Example 4.45. An example of a *non-regular criterion* is *the total absolute deviation of job completion times*, $\sum |C_j - d_j| \equiv \sum_{j=1}^{n} |C_j - d_j| = \sum_{j=1}^{n} (E_j + T_j)$, where $E_j = \max\{0, d_j - C_j\}$, for $1 \leqslant j \leqslant n$, is the earliness of job J_j.   ♦

Throughout this book, the value of the criterion function φ for a schedule σ will be denoted by $\varphi(\sigma)$. The optimal schedule and the optimal value of the criterion φ will be denoted by $\sigma^\star$ and $\varphi^\star := \varphi(\sigma^\star)$, respectively.

The following result shows the importance of regular criteria.

Lemma 4.46. (A dominant set for regular criteria)
The set $\mathcal{Z}_{s-act}(\mathcal{S})$ is dominant for regular criteria of optimality of schedule.

Proof. See Baker [2, Theorem 2.1], Conway et al. [12, Section 6.5].   ◇

From now on, we will assume that a *scheduling problem* is defined if the sets $\mathcal{J}, \mathcal{M}, \mathcal{R}$ have been described using the parameters given in Sect. 4.3 and if the form of the criterion φ is known. If the parameters have been assigned specific values, we deal with an *instance* of a given scheduling problem.

An algorithm which solves a scheduling problem will be called a *scheduling algorithm*. Scheduling algorithms can be divided into offline and online algorithms. In the case of an *offline scheduling algorithm*, the input data concerning all jobs to be scheduled are given in advance, and the schedule constructed by the algorithm exploits the knowledge about the whole set of jobs. An *online scheduling algorithm* generates a schedule in a job-by-job manner using only partial data concerning a single job that is currently processed.

In this book, we will consider offline and online scheduling algorithms.

4.6 Notation of scheduling problems

For simplicity of presentation, we will denote the scheduling problems using the $\alpha|\beta|\gamma$ notation, introduced by Graham et al. [22]. We will restrict our attention only to the symbols used in this book.

The description of any scheduling problem in the $\alpha|\beta|\gamma$ notation is a complex symbol composed of three fields, separated by the character '|'.

The first field, α, refers to the machine environment and is composed of two symbols, $\alpha = \alpha_1 \alpha_2$. Symbol α_1 characterizes the type of machine. If α_1 is an empty symbol, we deal with a single machine case. Otherwise, symbols P, Q, R, F, O and J denote parallel identical machine, parallel uniform machine, parallel unrelated machine, flow shop, open shop and job shop environment, respectively. Symbol α_2 denotes the number of machines. If α_1 is not an empty symbol and $\alpha_2 = m$, we deal with $m \geqslant 2$ machines. If α_1 is an empty symbol and $\alpha_2 = 1$, we deal with a single machine.

Field β describes the parameters of the set of jobs. In this field, we will use the following symbols (in parentheses we give the meaning of a particular symbol): *pmtn* (job preemption is allowed), *chains, tree, ser-par, prec* (precedence constraints among jobs are in the form of a set of chains, a tree, a series-parallel digraph or an arbitrary acyclic digraph), r_j (a ready time is defined for each job), d_j (a deadline is defined for each job).

If no symbol appears in β, *default values* are assumed: no preemption, arbitrary (but fixed) job processing times, no additional resources, no precedence constraints, no ready times and no deadlines.

Field γ contains the form of the criterion function, expressed in terms of the symbols from Definition 4.36. The dash symbol '$-$' in this field means that testing for the existence of a feasible schedule is considered.

Example 4.47. (a) Symbol $1|prec|\sum w_j C_j$ denotes a single machine scheduling problem with arbitrary job processing times, arbitrary precedence constraints, arbitrary job weights, no ready times, no deadlines and the total weighted completion time criterion.

(b) Symbol $Pm|r_j = r, p_j = 1, tree|C_{\max}$ denotes an m-identical-machine scheduling problem with unit processing time jobs, a common ready time for all jobs, no deadlines, precedence constraints among jobs in the form of a tree and the maximum completion time criterion.

(c) Symbol $F2|no\text{-}wait|\sum C_j$ denotes a two-machine 'no-wait' flow shop problem (see Remark 4.9 for the description of the 'no-wait' constraint), with arbitrary job processing times, no precedence constraints, no ready times, arbitrary deadlines and the total completion time criterion.

(d) Symbol $O3||L_{\max}$ denotes a three-machine open shop problem with arbitrary job processing times, no precedence constraints, no ready times, arbitrary deadlines and the maximum lateness criterion. (Notice that since deadlines d_j are used in definition of the $L_{\max}$ criterion, d_j does not appear in the field β.)

(e) Symbol $Jm||C_{\max}$ denotes an m-machine job shop problem with arbitrary job processing times, no precedence constraints, no ready times, no deadlines and the maximum completion time criterion.   ◆

Remark 4.48. In Sect. 6.4, we will extend the $\alpha|\beta|\gamma$ notation to include the symbols describing time-dependent job (operation) processing times. In Chap. 7, we extend this notation to denote time-dependent batch scheduling problems. In Chaps. 20 and 21, we will extend the notation further to include the symbols describing bi-criteria time-dependent scheduling, time-dependent scheduling on machines with limited availability, time-dependent two-agent scheduling, time-dependent scheduling with job rejection and time-dependent scheduling under mixed deterioration.

Bibliographic notes

Błażewicz et al. [6, Chap. 3], Brucker [7, Chap. 1] and Leung [33, Chap. 1] present a general description of scheduling problems. Brucker [7, Chap. 1] presents a detailed description of machine environments. Błażewicz et al. [5] give a comprehensive description of the set $\mathcal{R}$ of resources and problems of scheduling under resource constraints.

Definitions of criteria other than these from Definition 4.36 may be found in Błażewicz et al. [6, Chap. 3] and Brucker [7, Chapter 1].

Numerous books have been published on the scheduling theory, including Agnetis et al. [1], Baker [2], Baker and Trietsch [3], Baruah et al. [4], Błażewicz et al. [5, 6], Brucker [7], Chen and Guo [8], Chrétienne et al. [10], Coffman [11], Conway et al. [12], Dempster et al. [13], Drozdowski [15], Elmaghraby [16], Emmons and Vairaktarakis [17], Framinan et al. [18], French [19], Gawiejnowicz [20], Hall [24], Hartmann [25], Józefowska [29], Leung [33], Morton and Pentico [34], Muth and Thompson [35], Parker [37], Pinedo [38, 39], Rinnooy Kan [40], Robert and Vivien [41], Słowiński and Hapke [42], Strusevich and Rustogi [43], Sule [44], Suwa and Sandoh [45], Tanaev et al. [46, 47]. These books cover a huge spectrum of different aspects of scheduling with single-valued criteria and may serve as excellent references on the theory.

Problems of scheduling with multiple criteria are discussed by Dileepan and Sen [14], Hoogeveen [26, 27], Lee and Vairaktarakis [32], Nagar et al. [36] and T'kindt and Billaut [48].

Full explanation of the $\alpha|\beta|\gamma$ notation may be found, e.g., in Błażewicz et al. [6, Chap. 3], Graham et al. [22] and Lawler et al. [31]. Extensions of the $\alpha|\beta|\gamma$ notation are presented, e.g., in Agnetis et al. [1], Drozdowski [15], Gawiejnowicz [20] and Strusevich and Rustogi [43].

References

1. A. Agnetis, J-C. Billaut, S. Gawiejnowicz, D. Pacciarelli and A. Soukhal, *Multi-agent Scheduling: Models and Algorithms*. Berlin-Heidelberg: Springer 2014.
2. K. R. Baker, *Introduction to Sequencing and Scheduling*. New York: Wiley 1974.
3. K. R. Baker and D. Trietsch, *Principles of Sequencing and Scheduling*. New York: Wiley 2009.
4. S. Baruah, M. Bertogna and G. Butazzo, *Multiprocessor Scheduling for Real-Time Systems*. Cham: Springer 2015.
5. J. Błażewicz, W. Cellary, R. Słowiński and J. Węglarz, *Scheduling Under Resource Constraints: Deterministic Models*. Basel: Baltzer 1986.
6. J. Błażewicz, K. H. Ecker, E. Pesch, G. Schmidt and J. Węglarz, *Handbook on Scheduling: From Theory to Applications*. Berlin-Heidelberg: Springer 2007.
7. P. Brucker, *Scheduling Algorithms*, 4th ed. Berlin-Heidelberg: Springer 2004.
8. Q. Chen and M. Guo, *Task Scheduling for Multi-core and Parallel Architectures: Challenges, Solutions and Perspectives*. Singapore: Springer Nature 2017.

9. Z-L. Chen, Parallel machine scheduling with time dependent processing times. *Discrete Applied Mathematics* **70** (1996), no. 1, 81–93. (Erratum: *Discrete Applied Mathematics* **75** (1997), no. 1, 103.)

10. P. Chrétienne, E. G. Coffman, jr, J. K. Lenstra and Z. Liu (eds.), *Scheduling Theory and its Applications*. New York: Wiley 1995.

11. E. G. Coffman, jr, (ed.), *Computer and Job-shop Scheduling Theory*. New York: Wiley 1976.

12. R. W. Conway, W. L. Maxwell and L. W. Miller, *Theory of Scheduling*. Reading: Addison-Wesley 1967.

13. M. A. H. Dempster, J. K. Lenstra and A. H. G. Rinnooy Kan, *Deterministic and Stochastic Scheduling*. Dordrecht: Reidel 1982.

14. P. Dileepan and T. Sen, Bicriterion static scheduling research for a single machine. *Omega* **16** (1988), no. 1, 53–59.

15. M. Drozdowski, *Scheduling for Parallel Processing*. London: Springer 2009.

16. S. E. Elmaghraby (ed.), *Symposium on the Theory of Scheduling and its Applications*. Lecture Notes in Economics and Mathematical Systems **86**. Berlin: Springer 1973.

17. H. Emmons and G. Vairaktarakis, *Flow Shop Scheduling: Theoretical Results, Algorithms, and Applications*. New York: Springer 2013.

18. J. Framinan, R. Leisten and R. Ruiz García, *Manufacturing Scheduling Systems: An Integrated View on Models, Methods and Tools*. London: Springer 2014.

19. S. French, *Sequencing and Scheduling: An Introduction to the Mathematics of the Job-Shop*. Chichester: Horwood 1982.

20. S. Gawiejnowicz, *Time-Dependent Scheduling*. Berlin-Heidelberg: Springer 2008.

21. S. Gawiejnowicz and L. Pankowska, Scheduling jobs with varying processing times. *Information Processing Letters* **54** (1995), no. 3, 175–178.

22. R. L. Graham, E. L. Lawler, J. K. Lenstra and A. H. G. Rinnooy Kan, Optimization and approximation in deterministic sequencing and scheduling: a survey. *Annals of Discrete Mathematics* **5** (1979), 287–326.

23. J. N. D. Gupta and S. K. Gupta, Single facility scheduling with nonlinear processing times. *Computers and Industrial Engineering* **14** (1988), no. 4, 387–393.

24. R. Hall (ed.), *Handbook of Healthcare System Scheduling*. New York: Springer 2012.

25. S. Hartmann, *Project Scheduling under Limited Resources*. Lecture Notes in Economical and Mathematical Systems **478**. Berlin-Heidelberg: Springer 1999.

26. H. Hoogeveen, Multicriteria scheduling. *European Journal of Operational Research* **167** (2005), no. 3, 592–623.

27. J. A. Hoogeveen, *Single machine bicriteria scheduling*. Amsterdam: CWI 1992.

28. S. M. Johnson, Optimal two and three stage production schedules with setup times included. *Naval Research Logistics Quarterly* **1** (1954), no. 1, 61–68.

29. J. Józefowska, *Just-in-time Scheduling: Models and Algorithms for Computer and Manufacturing Systems*. New York: Springer 2007.

30. A. Kononov, Scheduling problems with linear increasing processing times. In: U. Zimmermann et al. (eds.), *Operations Research 1996*. Berlin-Heidelberg: Springer 1997, pp. 208–212.

31. E. L. Lawler, J. K. Lenstra, A. H. G. Rinnooy Kan and D. B. Shmoys, Sequencing and scheduling: algorithms and complexity. In: S. C. Graves, A. H. G. Rinnooy Kan and P. H. Zipkin (eds.), *Logistics of Production and Inventory*. Handbooks in Operations Research and Management Science **4**. Amsterdam: North–Holland 1993, pp. 445–522.

32. C-Y. Lee and G. Vairaktarakis, Complexity of single machine hierarchical scheduling: a survey. In: P. M. Pardalos (ed.), *Complexity in Numerical Optimization*. Singapore: World Scientific 1993, pp. 269–298.

33. J. Y-T. Leung (ed.), *Handbook of Scheduling*. Chapman and Hall/CRC, Boca Raton, 2004.

34. T. E. Morton and D. W. Pentico, *Heuristic Scheduling Systems with Applications to Production Systems and Project Management*. New York: Wiley 1993.

35. J. F. Muth and G. L. Thompson, *Industrial Scheduling*. Englewood Cliffs: Prentice-Hall 1963.

36. A. Nagar, J. Haddock and S. Heragu, Multiple and bicriteria scheduling: a literature survey. *European Journal of the Operational Research* **81** (1995), no. 1, 88–104.

37. R. G. Parker, *Deterministic Scheduling Theory*. London: Chapman and Hall 1995.

38. M. Pinedo, *Scheduling: Theory, Algorithms, and Systems*. Berlin-Heidelberg: Springer 2016.

39. M. Pinedo, *Planning and Scheduling in Manufacturing and Services*, 2nd ed. Berlin-Heidelberg: Springer 2009.

40. A. H. G. Rinnooy Kan, *Machine Scheduling Problems: Classification, Complexity and Computations*. The Hague: Nijhoff 1976.

41. Y. Robert and F. Vivien, *Introduction to Scheduling*. Boca Raton: CRC Press 2010.

42. R. Słowiński and M. Hapke (eds.), *Scheduling Under Fuzziness*. Heidelberg: Physica-Verlag 2000.

43. V. A. Strusevich and K. Rustogi, *Scheduling with Times-Changing Effects and Rate-Modifying Activities*. Berlin-Heidelberg: Springer 2017.

44. D. R. Sule, *Industrial Scheduling*. Boston-Toronto: PWS Publishing Company 1997.

45. H. Suwa and H. Sandoh, *Online Scheduling in Manufacturing: A Cumulative Delay Approach*. London: Springer 2013.

46. V. S. Tanaev, V. S. Gordon and Y. M. Shafransky, *Scheduling Theory: Single-Stage Systems*. Dordrecht: Kluwer 1994.

47. V. S. Tanaev, Y. N. Sotskov and V. A. Strusevich, *Scheduling Theory: Multi-Stage Systems*. Dordrecht: Kluwer 1994.

48. V. T'kindt and J-C. Billaut, *Multicriteria Scheduling*. Berlin-Heidelberg: Springer 2002.

49. W. Wajs, Polynomial algorithm for dynamic sequencing problem. *Archiwum Automatyki i Telemechaniki* **31** (1986), no. 3, 209–213 (in Polish).

5

The modern scheduling theory

In the previous chapter, we recalled the basic notions and definitions of the classical scheduling theory, where one assumes that *fixed* job processing times are numbers. Though the scope of applicability of the classical scheduling theory is impressive, the above assumption does not allow us to consider problems in which jobs have *variable* processing times. In this chapter, we present the main models of the *modern scheduling theory*, where scheduling problems with variable job processing times are considered.

The models discussed in the chapter are *time-independent*, i.e., job processing times do not depend on the job starting times. Time-dependent models of job processing times, where the processing times of jobs are functions of the job starting times, are discussed in Chap. 6.

Chapter 5 is composed of two sections. In Sect. 5.1, we introduce the reader to scheduling with variable job processing times. In Sect. 5.2, we describe the models of position-dependent and controllable job processing times. The chapter is completed with bibliographic notes and a list of references.

5.1 Main directions in the modern scheduling theory

In the period of over six decades that elapsed since the classical scheduling theory was formulated, numerous practical problems have appeared which could not be solved in the framework of this theory. The main reason of that was a certain restrictiveness of assumptions (M1)–(M3) and (J1)–(J4) listed in Sect. 4.2. For example, a machine may have a variable processing speed due to the changing state of this machine or job processing times may increase due to job deterioration. In order to overcome these difficulties and to adapt the theory to cover new problems, assumptions (M1)–(M3) and (J1)–(J4) were repeatedly modified. This, in turn, led to new research directions in the scheduling theory, such as scheduling multiprocessor tasks, scheduling on machines with variable processing speed and scheduling jobs with variable

© Springer-Verlag GmbH Germany, part of Springer Nature 2020
S. Gawiejnowicz, *Models and Algorithms of Time-Dependent Scheduling*,
Monographs in Theoretical Computer Science. An EATCS Series,
https://doi.org/10.1007/978-3-662-59362-2_5

processing times. For the completeness of further presentation, we will now shortly describe each of these directions.

5.1.1 Scheduling multiprocessor tasks

In this case, assumptions (M2) and (J2) have been modified: the same operations (called *tasks*) may be performed in the same time by two or more different machines (processors).

The applications of scheduling multiprocessor tasks concern reliable computing in fault-tolerant systems, which are able to detect errors and recover the status of the systems from before the error. Examples of fault-tolerant systems are aircraft control systems, in which the same tasks are executed by two or more machines simultaneously in order to increase the safety of the systems. Other examples of applications of scheduling multiprocessor tasks are modeling the work of parallel computers, problems of dynamic bandwidth allocation in communication systems, berth allocation and loom scheduling in the textile industry.

5.1.2 Scheduling on machines with variable processing speeds

In this case, assumption (M3) has been modified: the machines have variable processing speeds, i.e., the speeds change in time.

There are three main approaches to the phenomenon of variable processing speeds. In the first approach, the speed is described by a differential equation and depends on a continuous resource. Alternatively, the speed is described by a continuous (the second approach) or a discrete (the third approach) function. In both cases the speed depends on the amount of allotted resource which is continuous or discrete.

Scheduling with continuous resources has applications in such production environments in which jobs are executed on machines driven by a common power source, for example common mixing machines or refueling terminals. Scheduling with discrete resources is applied in modern manufacturing systems, in which jobs to be executed need machines as well as other resources such as robots or automated guided vehicles.

5.1.3 Scheduling jobs with variable processing times

In this case, assumption (J4) has been modified: the processing times of jobs are variable and can change in time.

The variability of job processing times can be modeled in different ways. For example, one can assume that the processing time of a job is a fuzzy number, a function of a continuous resource, a function of the job waiting time, a function of the position of the job in a schedule, or is varying in some interval between a certain minimum and maximum value.

Scheduling with variable job processing times has numerous applications, e.g., in the modeling of the forging process in steel plants, manufacturing of preheated parts in plastic molding or in silverware production, financial management, and in scheduling maintenance or learning activities.

The time-dependent scheduling problems which we will consider in this book are scheduling problems with variable job processing times.

5.2 Main models of variable job processing times

The variability of job processing times may be modeled in various ways. In this section, we consider the following models of variable job processing times described by functions or intervals:

- models of *position-dependent* job processing times – the processing time of a job is a function of the position of the job in a schedule;
- models of *controllable* job processing times – the processing time of a job is varying in some interval between a certain minimum and maximum value.

In subsequent sections, we briefly describe the above two time-independent models of variable job processing times. Time-dependent models of job processing times will be discussed in Chap. 6.

5.2.1 Models of position-dependent job processing times

Models of position-dependent job processing times compose the first, most popular group of models of variable job processing times. In this case, job processing times are functions of the job positions in a schedule. Scheduling problems of this type are considered in the research domain in the scheduling theory called *position-dependent scheduling*.

Position-dependent job processing times occur in manufacturing systems, in which the assembly time of a product is a function of skills of the worker who is involved in the process of making the product. Because the skills have an impact on the process, the assembly time is a function of the worker's experience. Moreover, since the latter can be expressed by the number of products made by the worker earlier, this time is a function of the product position in a schedule.

There are two main groups of models of position-dependent job processing times. The most popular is the first of these groups, where the processing time of a job is a non-increasing (decreasing) function of the job position in a schedule. Since models from this group are related to the so-called *learning effect*, scheduling problems with these models of job processing times are called *scheduling problems with learning effect*.

The simplest (and the most popular) model of learning effect is *log-linear learning effect* in which the processing time p_{jr} of job J_j scheduled in position r is in the form of

$$p_{jr} = p_j r^\alpha, \tag{5.1}$$

where $1 \leqslant j, r \leqslant n$. Here p_j is the *basic processing time* of job J_j, $\alpha < 0$ is the *learning index*, and r denotes the position of J_j in the schedule.

Example 5.1. (Agnetis et al. [31]) Let us consider position-dependent log-linear job processing times with learning effect defined by (5.1).

In this case, as shown in Table 5.1, we have the following job processing times p_{jr}, where $1 \leqslant j, r \leqslant n$, in which the job processing time in position j, r is equal to the basic job processing time p_j divided by $r^{|\alpha|}$:

Table 5.1: Possible values of job processing times in the form of $p_{jr} = p_j r^\alpha$

r	p_{1r}	p_{2r}	p_{3r}	$\cdots$	p_{nr}
1	p_1	p_2	p_3	$\cdots$	p_n
2	$\frac{p_1}{2^{|\alpha|}}$	$\frac{p_2}{2^{|\alpha|}}$	$\frac{p_3}{2^{|\alpha|}}$	$\cdots$	$\frac{p_n}{2^{|\alpha|}}$
3	$\frac{p_1}{3^{|\alpha|}}$	$\frac{p_2}{3^{|\alpha|}}$	$\frac{p_3}{3^{|\alpha|}}$	$\cdots$	$\frac{p_n}{3^{|\alpha|}}$
$\cdots$	$\cdots$	$\cdots$	$\cdots$	$\cdots$	$\cdots$
n	$\frac{p_1}{n^{|\alpha|}}$	$\frac{p_2}{n^{|\alpha|}}$	$\frac{p_3}{n^{|\alpha|}}$	$\cdots$	$\frac{p_n}{n^{|\alpha|}}$

◆

Remark 5.2. Gawiejnowicz [38] proposed position-dependent job processing times in the form of

$$p_{jk} = \frac{p_j}{v(k)}, \tag{5.2}$$

where the actual processing time p_{jk} of job J_j depends on an arbitrary function $v(k)$ of the number k of jobs executed before the job J_j started, and proved that problem $1|p_{jk} \equiv (5.2)|C_{\max}$ can be solved in $O(n \log n)$ time for an arbitrary function $v(k)$. Let us notice that if we assume that job J_j is scheduled in position r, then $k = r - 1$ and model (5.1) is a special case of model (5.2) for $v(k) = \frac{1}{(k+1)^a}$, where $1 \leqslant k, r \leqslant n$ and $a < 0$.

Another model of learning effect is the *linear learning effect*. In this case,

$$p_{jr} = p_j - \beta_j r, \tag{5.3}$$

where p_j is the basic processing time of job J_j, β_j is the *learning factor* and $0 < \beta_j < \frac{p_j}{n}$ for $1 \leqslant j \leqslant n$.

Throughout the chapter, we say that position-dependent jobs have *log-linear* or *linear processing times with the learning effect* if the processing times are in the form of (5.1) or (5.3), respectively.

A separate subgroup of models of learning effect takes into account the impact of previously scheduled jobs. The models from this subgroup are called models with the *past-sequence-dependent learning effect* and may be in a few different forms. For example, job processing times with the past-sequence-dependent learning effect may be in the form of

$$p_{jr} = p_j \max\left\{ \left(1 + \sum_{k=1}^{r-1} p_{[k]}\right)^{\alpha}, \beta \right\},$$

(5.4)

$$p_{jr} = p_j \left(1 + \sum_{k=1}^{r-1} p_{[k]}\right)^{\alpha},$$

(5.5)

or

$$p_{jr} = p_j \left(\frac{1 + \sum_{k=1}^{r-1} p_{[k]}}{1 + \sum_{k=1}^{r-1} p_k}\right)^{\alpha},$$

(5.6)

where p_j, $p_{[k]}$, $\alpha < 0$ and $0 < \beta < 1$ denote the basic processing time of job J_j, the processing time of the job in position k, the *learning index* and the *truncation parameter*, respectively.

Throughout the chapter, we say that position-dependent jobs have *past-sequence-dependent processing times with the learning effect* if the processing times are in the form of (5.4), (5.5) or (5.6).

The second group of models of position-dependent variable job processing times comprises the models where the processing time of a job is a non-decreasing (or an increasing) function of the job position in a schedule. Since models from this group are related to the so-called *ageing effect*, scheduling problems with these models of job processing times are called *scheduling problems with ageing effect*.

The simplest (and the most popular) model of the ageing effect is the *log-linear ageing effect* that has the same form as the learning effect (5.1) but with a positive exponent. In this case,

$$p_{jr} = p_j r^{\beta},$$

(5.7)

where p_j is the basic processing time of job J_j and the *ageing index* $\beta > 0$.

A similar change of the sign of a parameter that has an impact on the job processing time may also concern other models of learning effect. For example, job processing times with the ageing effect may be in the form of

$$p_{jr} = p_j + \beta_j r,$$

(5.8)

$$p_{jr} = p_j \left(1 + \sum_{k=1}^{r-1} p_{[k]}\right)^{\beta},$$

(5.9)

or

$$p_{jr} = p_j \left(\frac{1 + \sum_{k=1}^{r-1} p_{[k]}}{1 + \sum_{k=1}^{r-1} p_k}\right)^{\beta},$$

(5.10)

where p_j is the basic processing time of job J_j and $\beta > 0$.

Example 5.3. (Agnetis et al. [31]) Let us consider position-dependent log-linear job processing times with the ageing effect (5.7). In this case, as shown in Table 5.2, the job processing time in position j, r where $1 \leqslant j, r \leqslant n$, is equal to the basic processing time p_j multiplied by r^β:

Table 5.2: Possible values of job processing times in the form of $p_{jr} = p_j r^\beta$

r	p_{1r}	p_{2r}	p_{3r}	$\cdots$	p_{nr}
1	p_1	p_2	p_3	$\cdots$	p_n
2	$p_1 2^\beta$	$p_2 2^\beta$	$p_3 2^\beta$	$\cdots$	$p_n 2^\beta$
3	$p_1 3^\beta$	$p_2 3^\beta$	$p_3 3^\beta$	$\cdots$	$p_n 3^\beta$
$\cdots$	$\cdots$	$\cdots$	$\cdots$	$\cdots$	$\cdots$
n	$p_1 n^\beta$	$p_2 n^\beta$	$p_3 n^\beta$	$\cdots$	$p_n n^\beta$

◆

Notice that ageing effects of the forms (5.7), (5.8), (5.9) and (5.10) are counterparts of learning effects of the forms (5.1), (5.3), (5.5) and (5.6), respectively.

Throughout the chapter, we say that position-dependent jobs have *log-linear* or *linear processing times with the ageing effect* if the processing times are in the form of (5.7) or (5.8), respectively. Similarly, we say that position-dependent jobs have *past-sequence-dependent processing times with the ageing effect* if the processing times are in the form of (5.9) or (5.10).

Notation of position-dependent scheduling problems

We will denote scheduling problems with position-dependent job processing times using the modified $\alpha|\beta|\gamma$ notation (Agnetis et al. [31]), where in the β field the form of job processing times will be specified.

Example 5.4. Symbol $1|p_{jr} = p_j \left(1 + \sum_{k=1}^{r-1} p_{[k]}\right)| \sum C_j$ denotes a single machine scheduling problem with the past-sequence-dependent ageing effect with $\beta = 1$ and the objective of minimizing the $\sum C_j$. ◆

For clarity, we use different symbols for different forms of position-dependent job processing times. For example, in order to distinguish between different forms of learning and ageing effects, we denote negative and positive exponents appearing in these forms as α and β, respectively. Thus, in the second field of the three-field notation, we do not further specify assumptions on the parameters which are given in the description of the considered problem.

Example 5.5. (a) The symbol $1|p_{jr} = p_j r^\beta| \sum C_j$ denotes a single machine scheduling problem with the log-linear ageing effect and the objective of minimizing $\sum C_j$. Since the positive exponent β is specific for the ageing effect, we do not specify in the second field that $\beta > 0$.

(b) The symbol $1|p_{jr} = p_j r^\alpha|C_{\max}$ denotes a single machine scheduling problem with the log-linear learning effect and the objective of minimizing $C_{\max}$. Similarly to the previous case, we omit the assumption $\alpha < 0$, since the negative exponent α is specific for this form of learning effect.

(c) The symbol $1|p_{jr} = p_j - \beta r|f_{\max}$ denotes a single machine scheduling problem with the linear learning effect and the objective of minimizing $f_{\max}$. Since the coefficient β in this form of learning effect is positive, we do not specify in the second field that $0 < \beta < \frac{p_j}{n}$ for $1 \leqslant j \leqslant n$.

(d) The symbol $1|p_{jr} = p_j \left(1 + \sum_{k=1}^{r-1} p_{[k]}\right)^\beta| \sum U_j$ denotes a single machine scheduling problem with the past-sequence-dependent ageing effect and the objective of minimizing $\sum U_j$. As in Examples 5.5 (a) (c), we do not specify the assumption $\beta > 0$ in the second field. ◆

Example results on position-dependent job scheduling

The first group of results in this domain concerns position-dependent log-linear job processing times (5.1).

Theorem 5.6. (Gawiejnowicz [17], Alidaee and Ahmadian [12], Biskup [36], Mosheiov [41], Bachman and Janiak [13], Zhao et al. [45])
(a) *Problem $1|p_{jr} = p_j r^\alpha|C_{\max}$ is solvable in $O(n \log n)$ time by scheduling jobs in non-decreasing order of the basic processing times p_j and*

$$C_{\max}(\sigma) = t_0 + \sum_{j=1}^n \left(p_{[j]} j^\alpha\right). \tag{5.11}$$

(b) *If basic job processing times and due dates are* agreeable, *i.e., $p_i \leqslant p_j$ implies $d_i \geqslant d_j$ for $1 \leqslant i \neq j \leqslant n$, then problem $1|p_{jr} = p_j r^\alpha|L_{\max}$ is solvable in $O(n \log n)$ time by scheduling jobs in non-decreasing order of due dates d_j.*
(c) *Problem $1|p_{jr} = p_j r^\alpha| \sum C_j$ is solvable in $O(n \log n)$ time by scheduling jobs in non-decreasing order of the basic processing times p_j and*

$$\sum_{j=1}^n C_j(\sigma) = nt_0 + \sum_{j=1}^n (n - j + 1)p_{[j]} j^\alpha. \tag{5.12}$$

(d1) *If all basic job processing times are equal, i.e., $p_j = p$ for $1 \leqslant j \leqslant n$, then problem $1|p_{jr} = p_j r^\alpha| \sum w_j C_j$ is solvable in $O(n \log n)$ time by scheduling jobs in non-decreasing order of the job weights w_j.*
(d2) *If job weights satisfy the equality $w_j = wp_j$ for $1 \leqslant j \leqslant n$, then problem*

$1|p_{jr} = p_j r^\alpha| \sum w_j C_j$ is solvable in $O(n \log n)$ time by scheduling jobs in non-decreasing order of the basic processing times p_j.

(d3) If basic job processing times and job weights are agreeable, i.e., $p_i \leqslant p_j$ implies $w_i \geqslant w_j$ for $1 \leqslant i \neq j \leqslant n$, then problem $1|p_{jr} = p_j r^\alpha| \sum w_j C_j$ is solvable in $O(n \log n)$ time by scheduling jobs in non-decreasing order of the ratios $\frac{p_j}{w_j}$.

(e) Problem $1|p_j = p_j r^\alpha| \sum T_j$ is $\mathcal{NP}$-hard.

Proof. (a) By the adjacent job interchange technique (Biskup [36], Mosheiov [41]) or by applying Lemma 1.2 (b) (Gawiejnowicz [17]). Formula (5.11) can be proved by mathematical induction with respect to the number of jobs.

(b) By the adjacent job interchange technique (Wu et al. [44]).

(c) By the adjacent job interchange technique (Biskup [36]) or by Lemma 1.2(b) (Alidaee and Ahmadian [12]). Formula (5.12) can be proved by mathematical induction with respect to the number of jobs.

(d1), (d2) and (d3) By the adjacent job interchange technique (Bachman and Janiak [13]).

(e) The result follows from the $\mathcal{NP}$-hardness of problem $1||T_{\max}$ (Du and Leung [37]), which is a special case of problem $1|p_j = p_j r^\alpha| \sum T_j$ when $\alpha = 0$. □

Remark 5.7. The time complexity of problem $1|p_{jr} = p_j r^\alpha| \sum T_j$ for $\alpha \neq 0$ is unknown. Similarly, the time complexity of problems $1|p_{jr} = p_j r^\alpha|L_{\max}$, $1|p_{jr} = p_j r^\alpha|f_{\max}$, $1|p_{jr} = p_j r^\alpha| \sum w_j C_j$ and $1|p_{jr} = p_j r^\alpha| \sum U_j$ without additional assumptions on α is unknown.

Similar results to those of Theorem 5.6 hold for the log-linear ageing effect (5.7), since it differs from the log-linear learning effect only by the replacement of the negative exponent α by the positive exponent β.

The next group of results on position-dependent job processing times concerns linear processing times (5.3) with $\beta_j = \beta$ for $1 \leqslant j \leqslant n$.

Theorem 5.8. (a) Problem $1|p_{jr} = p_j - \beta r|C_{\max}$ is solvable in $O(n)$ time and

$$C_{\max}(\sigma) = t_0 + \sum_{j=1}^{n} p_{[j]} - \frac{n(n+1)}{2}\beta \tag{5.13}$$

does not depend on schedule σ.

(b) Problem $1|p_{jr} = p_j - \beta r| \sum C_j$ is solvable in $O(n \log n)$ time by scheduling jobs in non-decreasing order of the basic processing times p_j, and

$$\sum_{j=1}^{n} C_j(\sigma) = nt_0 + \sum_{j=1}^{n}(n-j+1)p_{[j]} - \frac{n(n+1)(n+2)}{6}\beta. \tag{5.14}$$

Proof. (a) and (b) By direct computation. Formulae (5.13) and (5.14) can be proved by mathematical induction with respect to the number of jobs. □

Remark 5.9. Counterparts of Theorem 5.8 (a), (b) for job processing times in the form of $p_{jr} = p_j - \beta_j r$, where $1 \leqslant j \leqslant n$, are given by Bachman and Janiak [34].

Similar results to those of Theorem 5.8 hold for the linear ageing effect (5.8), since it differs from the linear learning effect only by the sign before the ageing factor β.

The last group of results on position-dependent job processing times concerns past-sequence-dependent processing times (5.5).

Theorem 5.10. (Kuo and Yang [39, 40])
(a) *Problem* $1|p_{jr} = p_j \left(1 + \sum_{k=1}^{r-1} p_{[k]}\right)^{\alpha} |C_{\max}$ *is solvable in* $O(n \log n)$ *time by scheduling jobs in non-decreasing order of the basic processing times* p_j.
(b) *Problem* $1|p_{jr} = p_j \left(1 + \sum_{k=1}^{r-1} p_{[k]}\right)^{\alpha} |\sum C_j$ *is solvable in* $O(n \log n)$ *time by scheduling jobs in non-decreasing order of the basic processing times* p_j.
(c) *If basic job processing times and job weights are agreeable as in Theorem 5.6 (d3), then problem* $1|p_{jr} = p_j \left(1 + \sum_{k=1}^{r-1} p_{[k]}\right)^{\alpha} |\sum w_j C_j$ *is solvable in* $O(n \log n)$ *time by scheduling jobs in non-decreasing order of the ratios* $\frac{p_j}{w_j}$.

Proof. (a) By the adjacent job interchange technique (Kuo and Yang [39]).
       (b) By the adjacent job interchange technique (Kuo and Yang [40]).   □

The time complexity of problem $1|p_{jr} = p_j \left(1 + \sum_{k=1}^{r-1} p_{[k]}\right)^{\alpha} |\sum w_j C_j$ without additional assumptions is unknown. The same concerns the problems $1|p_{jr} = p_j \left(1 + \sum_{k=1}^{r-1} p_{[k]}\right)^{\alpha} |L_{\max}$ and $1|p_{jr} = p_j \left(1 + \sum_{k=1}^{r-1} p_{[k]}\right)^{\alpha} |\sum U_j$.

Similar results to those of Theorem 5.10 hold also for position-dependent job processing times in the form of (5.9).

5.2.2 Models of controllable job processing times

Models of controllable job processing times compose the second group of models of variable job processing times. In this case, job processing times are described by *intervals*, i.e., the processing time of a job varies between a minimum value and a maximal value that are specific for each job. Moreover, the processing time of a job is expressed by a non-increasing function of the amount of a *resource* allocated to a given job. This resource may be continuous or discrete and, in most cases, it is non-renewable and its availability is limited by an upper bound. Scheduling problems of this type are considered in the so-called *resource-dependent scheduling*.

Controllable job processing times appear in some industrial applications such as the problem of organizing the production at a blacksmith's division in a steel mill. In this case, the job processing times vary within certain limits and they require resources such as gas or power. Allocated resources may change job processing times but since the former are scarce, a *cost* has to be paid for each unit of the resource employed.

There are two main models of controllable job processing times. The first model, called *convex*, assumes that the processing time p_j of job J_j is a convex function of the resource amount u_j allocated to J_j.

The second model is called *linear*. In this case, the processing time p_j of job J_j is a linear function of the amount u_j of a resource allocated to the job,

$$p_j = \overline{p}_j - \nu_j u_j, \tag{5.15}$$

where $0 \leqslant u_j \leqslant \overline{u}_j \leqslant \frac{\overline{p}_j}{\nu_j}$. The values $\overline{p}_j$, $\overline{u}_j$ and $\nu_j > 0$ are called the *non-compressed (maximal) processing time*, the upper bound on the amount of an allocated resource and the *compression rate* of job J_j, $1 \leqslant j \leqslant n$, respectively.

There is one more possible model of linear controllable job processing times. In this model, actual processing time $p_j \in [\underline{p}_j, \overline{p}_j]$, where $\underline{p}_j \leqslant \overline{p}_j$ for $1 \leqslant j \leqslant n$. The maximal processing time can be *compressed* (decreased) at the cost $c_j x_j$, where

$$x_j = \overline{p}_j - p_j \tag{5.16}$$

is the *amount of compression of job* J_j and c_j is the *compression cost per unit time*. This implies that

$$p_j = \overline{p}_j - x_j, \tag{5.17}$$

where $\overline{p}_j$ and x_j are defined as in (5.15) and (5.16), respectively. The cost of compression is measured by the total compression *cost function* $\sum_j c_j x_j$.

Throughout the chapter, we say that controllable jobs have *linear processing times* if the processing times are in the form of (5.15) or (5.17).

Example 5.11. (Agnetis et al. [31]) Let us consider $n = 3$ jobs with processing times (5.17): $\overline{p}_1 = 2$, $\overline{p}_2 = 1$, $\overline{p}_3 = 3$, $c_1 = 1$, $c_2 = 3$, $c_3 = 2$ and $t_0 = 0$.

Table 5.3: Schedules for an instance of problem $1|p_j = \overline{p}_j - x_j| \sum C_j + \sum c_j x_j$

Sequence σ	$p_{[1]}$	$C_{[1]}$	$p_{[2]}$	$C_{[2]}$	$p_{[3]}$	$C_{[3]}$	$\sum C_j(\sigma) + \sum c_j x_j(\sigma)$
(J_1, J_2, J_3)	1	1	$\frac{1}{2}$	$\frac{3}{2}$	$\frac{3}{2}$	3	11
(J_1, J_3, J_2)	1	1	$\frac{3}{2}$	$\frac{5}{2}$	$\frac{1}{2}$	3	12
(J_2, J_1, J_3)	$\frac{1}{2}$	$\frac{1}{2}$	1	$\frac{3}{2}$	$\frac{3}{2}$	3	$\frac{21}{2}$
(J_2, J_3, J_1)	$\frac{1}{2}$	$\frac{1}{2}$	$\frac{3}{2}$	2	1	3	11
(J_3, J_1, J_2)	$\frac{3}{2}$	$\frac{3}{2}$	1	$\frac{5}{2}$	$\frac{1}{2}$	3	$\frac{25}{2}$
(J_3, J_2, J_1)	$\frac{3}{2}$	$\frac{3}{2}$	$\frac{1}{2}$	2	1	3	12

Let us assume that all jobs have been crashed by 50%, i.e., when $x_1 = 1$, $x_2 = \frac{1}{2}$ and $x_3 = \frac{3}{2}$. Then, job processing times, job completion times and the values of the $\sum C_j + \sum c_j x_j$ criterion are as in Table 5.3. ◆

The level at which jobs are crashed affects the processing times of the jobs.

Example 5.12. (Agnetis et al. [31]) Let us consider the instance from Example 5.11. Notice that if jobs are not compressed at all, i.e., if $x_1 = x_2 = x_3 = 0$, sequence (J_2, J_1, J_3) gives the minimum total completion time (equal to 10).

If all jobs are crashed by 50%, i.e., if $x_1 = 1$, $x_2 = \frac{1}{2}$ and $x_3 = \frac{3}{2}$, the minimum value of the objective $\sum C_j(\sigma) + \sum c_j x_j(\sigma)$ is obtained for two sequences, (J_1, J_2, J_3) and (J_2, J_3, J_1).

If all jobs have been crashed by 75%, i.e., if $x_1 = \frac{3}{2}$, $x_2 = \frac{3}{4}$ and $x_3 = \frac{9}{4}$, the values of job processing times, job completion times and the $\sum C_j + \sum c_j x_j$ criterion for all possible sequences for the instance are as follows:

Table 5.4: Schedules for an instance of problem $1|p_j = \overline{p}_j - x_j|\sum C_j + \sum c_j x_j$

Sequence σ	$p_{[1]}$	$C_{[1]}$	$p_{[2]}$	$C_{[2]}$	$p_{[3]}$	$C_{[3]}$	$\sum C_j(\sigma) + \sum c_j x_j$
(J_1, J_2, J_3)	$\frac{1}{2}$	$\frac{1}{2}$	$\frac{1}{4}$	$\frac{3}{4}$	$\frac{3}{4}$	$\frac{3}{2}$	11
(J_1, J_3, J_2)	$\frac{1}{2}$	$\frac{1}{2}$	$\frac{3}{4}$	$\frac{5}{4}$	$\frac{1}{4}$	$\frac{3}{2}$	$\frac{23}{2}$
(J_2, J_1, J_3)	$\frac{1}{4}$	$\frac{1}{4}$	$\frac{1}{2}$	$\frac{3}{4}$	$\frac{3}{4}$	$\frac{3}{2}$	$\frac{43}{4}$
(J_2, J_3, J_1)	$\frac{1}{4}$	$\frac{1}{4}$	$\frac{3}{4}$	1	$\frac{1}{2}$	$\frac{3}{2}$	11
(J_3, J_1, J_2)	$\frac{3}{4}$	$\frac{3}{4}$	$\frac{1}{2}$	$\frac{5}{4}$	$\frac{1}{4}$	$\frac{3}{2}$	$\frac{47}{4}$
(J_3, J_2, J_1)	$\frac{3}{4}$	$\frac{3}{4}$	$\frac{1}{4}$	1	$\frac{1}{2}$	$\frac{3}{2}$	$\frac{23}{2}$

The minimum value of the $\sum C_j(\sigma) + \sum c_j x_j$ criterion, obtained for sequence (J_2, J_1, J_3), is equal to $\frac{43}{4}$. ◆

Notice that any solution to a scheduling problem with controllable job processing times has two components. Namely, if job processing times are in the form of (5.15), a solution to the problem is specified by (*i*) a vector of resource amounts allocated to each job and (*ii*) a schedule. Similarly, if job processing times are in the form of (5.17), the solution is specified by (*i*) a vector of job processing time compressions and (*ii*) a schedule.

Moreover, in scheduling problems with controllable job processing times, two criteria are used to measure the quality of a schedule, namely a scheduling criterion and a cost function measuring the cost of job compression in the evaluated schedule.

Notation of controllable scheduling problems

We will denote scheduling problems with controllable job processing times using similar rules as mentioned earlier.

Example 5.13. (*a*) Symbol $1|p_j = \overline{p}_j - x_j|f_{\max}$ denotes a single machine scheduling problem with linear controllable job processing times and the objective of minimizing $f_{\max}$.
(*b*) Symbol $1|p_j = \overline{p}_j - x_j|\sum C_j + \sum c_j x_j$ denotes a single machine scheduling problem with linear controllable job processing times and the objective of minimizing the sum of the total completion time and the total compression cost, $\sum_j C_j + \sum_j c_j x_j$. ◆

Example results on controllable job scheduling

In this section, we recall basic results concerning the single-agent single-machine controllable job scheduling problem $1|p_j = \overline{p}_j - x_j|\sum w_j C_j + \sum c_j x_j$.
  The above problem was formulated and studied for the first time by Vickson [53, 54]. Recalling that x_j denotes the shortening of job J_j, we have the following two properties.

Property 5.14. (Vickson [53]) For problem $1|p_j = \overline{p}_j - x_j|\sum w_j C_j + \sum c_j x_j$, there exists an optimal schedule such that for all $1 \leqslant j \leqslant n$ either $x_j = 0$ or $x_j = \overline{p}_j$.

Proof. By direct computation; see [53, Sect. 1].   ◇

Property 5.15. (Vickson [53]) In an optimal schedule for problem $1|p_j = \overline{p}_j - x_j|\sum w_j C_j + \sum c_j x_j$, jobs are arranged in non-decreasing order of the ratios $\frac{p_j}{w_j}$.

Proof. The result follows from the result for problem $1||\sum w_j C_j$ by Smith [52]. ☐

  The complexity of the problem has been established by Wan et al. [55] and Hoogeveen and Woeginger [48].

Theorem 5.16. (Wan et al. [55], Hoogeveen and Woeginger [48]) *Problem* $1|p_j = \overline{p}_j - x_j|\sum w_j C_j + \sum c_j x_j$ *is $\mathcal{NP}$-hard.*

Proof. The proof is based on a complex reduction from the EOP problem (cf. Sect. 3.2); see Wan et al. [55, Sect. 3] for details.   ◇

Theorem 5.17. (Vickson [54]) *Problem* $1|p_j = \overline{p}_j - x_j|T_{\max} + \sum c_j x_j$ *is solvable in $O(n^2)$ time.*

Proof. Problem $1|p_j = \overline{p}_j - x_j|T_{\max} + \sum c_j x_j$ can be reformulated as a production inventory problem with no backlogging, production capacity constraints and an unlimited entering inventory allowed. Since the latter problem can be solved in $O(n^2)$ time, the result follows.   □

A discrete variation of problem $1|p_j = \overline{p}_j - x_j|T_{\max} + \sum c_j x_j$, i.e., the case when the used resource is discrete and each job may have only a finite number of possible processing times, is intractable.

Theorem 5.18. (Vickson [54]) *A discrete variation of problem* $1|p_j = \overline{p}_j - x_j|T_{\max} + \sum c_j x_j$ *is ordinarily $\mathcal{NP}$-hard.*

Proof. The proof is based on a reduction from the KP problem; see Vickson [54, Appendix] for details.   ◇

Bibliographic notes

The literature on modern scheduling problems is quite rich. Below we list the main references on scheduling multiprocessor tasks, resource-dependent scheduling, scheduling on machines with variable speed, and different forms of variable, position-dependent and controllable job processing times.

Scheduling multiprocessor tasks

The problems of scheduling multiprocessor tasks are reviewed in detail by Drozdowski [1] and Lee et al. [2].

Resource-dependent scheduling

The problems of scheduling with continuous resources are discussed in detail by Błażewicz et al. [3, Chapter 12] and Gawiejnowicz [4].

Scheduling on machines with variable speed

The problems of scheduling on machines with variable speed are considered, e.g., by Dror et al. [5], Gawiejnowicz [6, 7], Megow and Verschae [8], Meilijson and Tamir [9], Timmermans and Vredeveld [10] and Trick [11].

Variable job processing times

The variability of the processing time of a job can be modeled in many different ways. The job processing time can be, e.g., a function of the job waiting time (see, e.g., Barketau et al. [14], Finke and Jiang [15], Finke et al. [16], Leung et al. [21], Lin and Cheng [22], Sriskandarajah and Goyal [27]), a function of a continuous resource (see, e.g., Janiak [18, 19]) or a fuzzy number (see, e.g., Słowiński and Hapke [26]).

The processing time of a job may also depend on the number of executed jobs (Alidaee and Ahmadian [12], Gawiejnowicz [17]), the position of the job in a schedule (Bachman and Janiak [13], Strusevich and Rustogi [28]), the length of a machine non-availability period (Lahlou and Dauzère-Pérès [20]) or it may vary in some interval between a certain minimum and maximum value (see, e.g., Nowicki and Zdrzałka [23, 24], Shakhlevich and Strusevich [25], and Vickson [29, 30]).

Position-dependent job scheduling problems

The first review on position-dependent scheduling problems, Bachman and Janiak [34], covers the literature up to 2001 and presents complexity results on scheduling position-dependent log-linear and linear jobs.

The next review on position-dependent scheduling, Biskup [35], discusses papers published up to 2007 and covers single- and multiple-machine scheduling problems, in identical parallel and dedicated parallel environments, with many forms of position-dependent job processing times.

The review by Anzanello and Fogliatto [33] includes a detailed analysis of curve shapes describing different forms of the learning effect. A critical discussion of existing literature on scheduling problems with position-dependent job processing times and a proposal of a unifying view on some of the problems can be found in the review by Rustogi and Strusevich [42].

Some position-dependent scheduling problems are also discussed in Chap. 6 of the monograph by Agnetis et al. [31] and in the monograph by Strusevich and Rustogi [43].

The most recent review of the literature on position-dependent scheduling, Azzouz et al. [32], covers papers published up to 2018.

Controllable job scheduling problems

The first survey on scheduling problems with controllable job processing times, Nowicki and Zdrzałka [49], concerns mainly single-machine problems, though it also addresses flow shop and parallel-machine problems. This survey covers papers published up to 1988.

The most recent review on the subject, Shabtay and Steiner [50], follows the classification scheme introduced in [49] and presents a detailed discussion of single-, parallel- and dedicated-machine scheduling problems. This review covers literature of the subject up to 2006.

Some controllable job scheduling problems are also discussed in reviews by Chen et al. [46] and Hoogeveen [47].

Solution approaches to scheduling problems with controllable job processing times are discussed by Shioura et al. [51].

References

Scheduling multiprocessor tasks

1. M. Drozdowski, *Scheduling for Parallel Processing*. Berlin: Springer 2009.
2. C-Y. Lee, L. Lei and M. Pinedo, Current trends in deterministic scheduling. *Annals of Operations Research* **70** (1997), no. 1, 1–41.

Resource-dependent scheduling

3. J. Błażewicz, K. H. Ecker, E. Pesch, G. Schmidt and J. Węglarz, *Handbook on Scheduling: From Theory to Applications*. Berlin-Heidelberg: Springer 2007.
4. S. Gawiejnowicz, Brief survey of continuous models of scheduling. *Foundations of Computing and Decision Sciences* **21** (1996), no. 2, 81–100.

Scheduling on machines with variable speed

5. M. Dror, H. I. Stern and J. K. Lenstra, Parallel machine scheduling: processing rates dependent on number of jobs in operation. *Management Science* **33** (1987), no. 8, 1001–1009.
6. S. Gawiejnowicz, A note on scheduling on a single processor with speed dependent on a number of executed jobs. *Information Processing Letters* **57** (1996), no. 6, 297–300.
7. S. Gawiejnowicz, Minimizing the flow time and the lateness on a processor with a varying speed. *Ricerca Operativa* **21** (1997), no. 83, 53–58.
8. N. Megow and J. Verschae, Dual techniques for scheduling on a machine with varying speed. *Lecture Notes in Computer Science* **7965** (2013), 745–756.
9. I. Meilijson and A. Tamir, Minimizing flow time on parallel identical processors with variable unit processing time. *Operations Research* **32** (1984), no. 2, 440–448.
10. V. Timmermans and T. Vredeveld, Scheduling with state-dependent machine speed. *Lecture Notes in Computer Science* **9499** (2015), 196–208.
11. M. A. Trick, Scheduling multiple variable-speed machines. *Operations Research* **42** (1994), no. 2, 234–248.

Scheduling jobs with variable processing times

12. B. Alidaee and A. Ahmadian, Scheduling on a single processor with variable speed. *Information Processing Letters* **60** (1996), no. 4, 189–193.
13. A. Bachman and A. Janiak, Scheduling jobs with position-dependent processing times. *Journal of the Operational Research Society* **55** (2004), no. 3, 257–264.
14. M. S. Barketau, T-C. E. Cheng, C-T. Ng, V. Kotov and M. Y. Kovalyov, Batch scheduling of step deteriorating jobs. *Journal of Scheduling* **11** (2008), no. 1, 17–28.
15. G. Finke and H. Jiang, A variant of the permutation flow shop model with variable processing times. *Discrete Applied Mathematics* **76** (1997), no. 1–3, 123–140.

16. G. Finke, M. L. Espinouse and H. Jiang, General flowshop models: job dependent capacities, job overlapping and deterioration. *International Transactions in Operational Research* **9** (2002), no. 4, 399–414.

17. S. Gawiejnowicz, A note on scheduling on a single processor with speed dependent on a number of executed jobs. *Information Processing Letters* **57** (1996), no. 6, 297–300.

18. A. Janiak, Time-optimal control in a single machine problem with resource constraints. *Automatica* **22** (1986), no. 6, 745–747.

19. A. Janiak, General flow-shop scheduling with resource constraints. *International Journal of Production Research* **26** (1988), no. 6, 1089–1103.

20. C. Lahlou and S. Dauzère-Pérès, Single-machine scheduling with time window-dependent processing times. *Journal of the Operational Research Society* **57** (2006), no. 2, 133-139.

21. J. Y-T. Leung, C-T. Ng and T-C. E. Cheng, Minimizing sum of completion times for batch scheduling of jobs with deteriorating processing times. *European Journal of Operational Research* **187** (2008), no. 3, 1090–1099.

22. B. M-T. Lin and T-C. E. Cheng, Two-machine flowshop scheduling with conditional deteriorating second operations. *International Transactions in Operations Research* **13** (2006), no. 2, 91–98.

23. E. Nowicki and S. Zdrzałka, A survey of results for sequencing problems with controllable processing times. *Discrete Applied Mathematics* **26** (1990), no. 2–3, 271–287.

24. E. Nowicki and S. Zdrzałka, Optimal control of a complex of independent operations. *International Journal of Systems Sciences* **12** (1981), no. 1, 77–93.

25. N. V. Shakhlevich and V. A. Strusevich, Pre-emptive scheduling problems with controllable processing times. *Journal of Scheduling* **8** (2005), no. 3, 233–253.

26. R. Słowiński and M. Hapke (eds.), *Scheduling Under Fuzziness*. Heidelberg: Physica-Verlag 2000.

27. C. Sriskandarajah and S.K. Goyal, Scheduling of a two-machine flowshop with processing time linearly dependent on job waiting-time. *Journal of the Operational Research Society* **40** (1989), no. 10, 907–921.

28. V. A. Strusevich and K. Rustogi, *Scheduling with Times-Changing Effects and Rate-Modifying Activities*. Berlin-Heidelberg: Springer 2017.

29. R. Vickson, Choosing the job sequence and processing times to minimize total processing plus flow cos on a single machine. *Operations Research* **28** (1980), no. 5, 1155–1167.

30. R. Vickson, Two single-machine sequencing problems involving controllable job processing times. *AIIE Transactions* **12** (1980), no. 3, 258–262.

Position-dependent job scheduling problems

31. A. Agnetis, J-C. Billaut, S. Gawiejnowicz, D. Pacciarelli and A. Soukhal, *Multi-agent Scheduling: Models and Algorithms*. Berlin-Heidelberg: Springer 2014.

32. A. Azzouz, M. Ennigrou and L. Ben Said, Scheduling problems under learning effects: classification and cartography. *International Journal of Production Research* **56** (2018), no. 4, 1642–1661.

33. M. J. Anzanello and F. S. Fogliatto, Learning curve models and applications: Literature review and research directions. *International Journal of Industrial Ergonomics* **41** (2011), no. 5, 573–583.

34. A. Bachman and A. Janiak, Scheduling jobs with position-dependent processing times. *Journal of the Operational Research Society* **55** (2004), no. 3, 257–264.
35. D. Biskup, A state-of-the-art-review on scheduling with learning effects. *European Journal of Operational Research* **188** (2008), no. 2, 315–329.
36. D. Biskup, Single-machine scheduling with learning considerations. *European Journal of Operational Research* **115** (1999), no. 1, 173–178.
37. J. Du and J.Y-T. Leung, Minimizing total tardiness on one machine is NP-hard. *Mathematics of Operations Research* **15** (1990), no. 3, 483–495.
38. S. Gawiejnowicz, A note on scheduling on a single processor with speed dependent on a number of executed jobs. *Information Processing Letters* **57** (1996), no. 6, 297–300.
39. W-H. Kuo and D-L. Yang, Minimizing the makespan in a single machine scheduling problem with a time-based learning effect. *Information Processing Letters* **97** (2006), no. 2, 64–67.
40. W-H. Kuo and D-L. Yang, Minimizing the total completion time in a single-machine scheduling problem with a time-dependent learning effect. *European Journal of Operational Research* **174** (2006), no. 2, 1184–1190.
41. G. Mosheiov, Scheduling problems with a learning effect. *European Journal of Operational Research* **132** (2001), no. 3, 687–693.
42. K. Rustogi and V. A. Strusevich, Simple matching vs linear assignment in scheduling models with positional effects: a critical review. *European Journal of Operational Research* **222** (2012), no. 3, 393–407.
43. V. A. Strusevich and K. Rustogi, *Scheduling with Times-Changing Effects and Rate-Modifying Activities*. Berlin-Heidelberg: Springer 2017.
44. C-C. Wu, W-C. Lee and T. Chen, Heuristic algorithms for solving the maximum lateness scheduling problem with learning considerations. *Computers and Industrial Engineering* **52** (2007), no. 1, 124–132.
45. C-L. Zhao, Q-L. Zhang and H-Y. Tang, Machine scheduling problems with a learning effect. *Dynamics of Continuous, Discrete and Impulsive Systems, Series A: Mathematical Analysis* **11** (2004), 741–750.

Controllable job scheduling problems

46. B. Chen, C. N. Potts and G. J. Woeginger, A review of machine scheduling: complexity and approximability. In: D. Z. Du and P. M. Pardalos (eds.), *Handbook of Combinatorial Optimization*, Dordrecht: Kluwer 1998, pp. 21–169.
47. H. Hoogeveen, Multicriteria scheduling. *European Journal of Operational Research* **167** (2005), no. 3, 592–623.
48. H. Hoogeveen and G. J. Woeginger, Some comments on sequencing with controllable processing times. *Computing* **68** (2002), no. 2, 181–192.
49. E. Nowicki and S. Zdrzałka, A survey of results for sequencing problems with controllable processing times. *Discrete Applied Mathematics* **26** (1990), no. 2–3, 271–287.
50. D. Shabtay and G. Steiner, A survey of scheduling with controllable processing times. *Discrete Applied Mathematics* **155** (2007), no. 13, 1643–1666.
51. A. Shioura, N. V. Shakhlevich and V. A. Strusevich, Preemptive models of scheduling with controllable processing times and of scheduling with imprecise computation: a review of solution approaches. *European Journal of Operational Research* **266** (2018), no. 3, 795–818.

52. W. E. Smith, Various optimizers for single-stage production. *Naval Research Logistics Quarterly* **3** (1956), no. 1–2, 59–66.

53. R. Vickson, Choosing the job sequence and processing times to minimize total processing plus flow cost on a single machine. *Operations Research* **28** (1980), no. 5, 1155–1167.

54. R. Vickson, Two single-machine sequencing problems involving controllable job processing times. *AIIE Transactions* **12** (1980), no. 3, 258–262.

55. G. Wan, B. P. C. Yen and C-L. Li, Single machine scheduling to minimize total compression plus weighted flow cost is NP-hard. *Information Processing Letters* **79** (2001), no. 6, 273–280.

6

The time-dependent scheduling

This chapter completes the second part of the book, where the main scheduling theory models are considered. In this chapter, we introduce the model of time-dependent job processing times, where the processing time of a job is a function of the job starting time. We also describe three basic variations of the model, introduce the terminology and notation used in the book, and review the applications of time-dependent scheduling.

Chapter 6 is composed of six sections. In Sect. 6.1, we introduce the terminology used in the book. In Sects. 6.2 and 6.3, we describe the pure and mixed models of job processing times in time-dependent scheduling problems, respectively. In Sect. 6.4, we clarify the notation used for symbolic description of time-dependent scheduling problems. In Sect. 6.5, we briefly review the mathematical background of time-dependent scheduling. In Sect. 6.6, we discuss applications of time-dependent scheduling. The chapter is completed with bibliographic notes and a list of references.

6.1 Terminology of time-dependent scheduling

The models of time-dependent job processing times comprise the third group of models of variable job processing times. In this case, the processing times of jobs are functions of the job starting times. Scheduling problems with job processing times of this type are considered in a dynamically developing research domain of the modern scheduling theory, called *time-dependent scheduling*.

There are three main groups of models of job processing times in time-dependent scheduling, each having its own specificity and possible application areas. Although under some assumptions the models from the first two groups are mutually related, at present all the three groups are developing separately.

The first group, more often encountered in the literature and easier to study, concerns scheduling problems in which job processing times are *increasing* (or *non-decreasing*) functions of the job starting times. This means that job processing times *deteriorate* in time, i.e., a job that is started later

© Springer-Verlag GmbH Germany, part of Springer Nature 2020
S. Gawiejnowicz, *Models and Algorithms of Time-Dependent Scheduling*,
Monographs in Theoretical Computer Science. An EATCS Series,
https://doi.org/10.1007/978-3-662-59362-2_6

has longer (not lower) processing time than the same job started earlier. Jobs with this model of processing times are called *deteriorating* jobs. Most of the time-dependent scheduling literature concerns scheduling deteriorating jobs.

The second group may cause some problems already at the stage of problem formulation, since we have to make some additional assumptions in order to avoid the negativity of job processing times. In this case job processing times are *decreasing* (or *non-increasing*) functions of the job starting times. This means that job processing times *shorten* in time, i.e., the processing time of a job becomes shorter if the job is started later. Jobs with time-dependent processing times of this type are called *shortening* jobs. Scheduling problems with shortening jobs are less explored than those with deteriorating jobs.

The third group of models concerns scheduling problems with the *alteration* of job processing times. In this case, the functions describing job processing times are not monotonic and the processing time of each job varies in time in a non-monotonic way. Jobs with time-dependent processing times of this type are called *alterable* jobs. Scheduling problems with alterable jobs are yet less explored than those with shortening jobs.

Remark 6.1. Regardless of the type of functions which we have chosen to describe job processing times in our problem, we still deal with *deterministic* scheduling, since all parameters of the problem are assumed to be known in advance. This clarification is important, since *stochastic* scheduling problems with deteriorating jobs are also known (see, e.g., Glazebrook [91]).

6.2 Pure models of time-dependent processing times

In this section, we describe the first group of models of time-dependent job processing times. The models will be called *pure models*, since job processing times considered in these models are functions of the job starting times only. First, we focus on general models of time-dependent job processing times.

6.2.1 General models of time-dependent processing times

The general model of time-dependent processing times depends on the machine environment. In time-dependent parallel machine scheduling problems the processing time p_j of job J_j is a function of the starting time $S_j \geqslant S_0$ of the job,

$$p_j(S_j) = g_j(S_j), \tag{6.1}$$

where $S_0 \geqslant 0$ is the time at which the first scheduled job starts and g_j is an arbitrary non-negative function of S_j for $1 \leqslant j \leqslant n$.

In time-dependent dedicated machine scheduling problems all jobs are composed of operations, and the processing time of the ith operation of the jth job is in the form of

$$p_{ij}(S_{ij}) = g_{ij}(S_{ij}), \tag{6.2}$$

where g_{ij} is an arbitrary non-negative function of the starting time $S_{ij} \geqslant S_0$ of the operation, $1 \leqslant i \leqslant m$ and $1 \leqslant j \leqslant n$.

Less general variants of models (6.1) and (6.2),

$$p_j(S_j) = b_j f_j(S_j) \tag{6.3}$$

and

$$p_{ij}(S_{ij}) = b_{ij} f_{ij}(S_{ij}), \tag{6.4}$$

define time-dependent job (operation) processing times as arbitrary non-negative functions $f_j(S_j)$ $(f_{ij}(S_{ij}))$ of the job (operation) starting times $S_j \geqslant S_0$ $(S_{ij} \geqslant S_0)$, multiplied by rates $b_j > 0$ $(b_{ij} > 0)$, where $1 \leqslant j \leqslant n$ $(1 \leqslant i \leqslant m, 1 \leqslant j \leqslant n)$.

The second general model of time-dependent job processing times,

$$p_j(S_j) = a_j + f_j(S_j), \tag{6.5}$$

where $a_j \geqslant 0$ and function f_j is an arbitrary non-negative function of $S_j \geqslant 0$ for $1 \leqslant j \leqslant n$, is more often encountered because it indicates both fixed and variable parts of the processing times.

Similarly, the following model of the processing time of an operation,

$$p_{ij}(S_{ij}) = a_{ij} + f_{ij}(S_{ij}), \tag{6.6}$$

where $a_{ij} \geqslant 0$ and f_{ij} are arbitrary non-negative functions of $S_{ij} \geqslant 0$ for $1 \leqslant i \leqslant m$ and $1 \leqslant j \leqslant n$, is more common than the model (6.2). The constant part of a job (operation) processing time, a_j (a_{ij}), is called the *basic processing time* of the job (operation).

Remark 6.2. The assumption that functions $g_j(S_j)$ and $f_j(S_j)$ $(g_{ij}(S_{ij})$ and $f_{ij}(S_{ij}))$ are non-negative for non-negative arguments is essential and from now on, unless otherwise stated, it will be always satisfied.

Remark 6.3. Since the models (6.5) and (6.6) of job processing times give us more information, in further considerations we will mainly use the functions $f_j(S_j)$ and $f_{ij}(S_{ij})$.

If we assume that functions f_j in model (6.5) are strictly increasing, then the following result holds.

Lemma 6.4. (Gawiejnowicz and Lin [17]) *Let $f_j(S_j)$ be a strictly increasing function of S_j, $f_j(S_j) > 0$ for $S_j > 0$ and let γ be a regular criterion. Then an optimal schedule for a single machine problem with job processing times in the form of $p_j = a_j + f_j(S_j)$ and criterion γ is a non-delay schedule.*

Proof. Assume that in an optimal schedule for the problem there exist a number of idle times. Then, moving the jobs following each of the idle times to the left as much as possible, we obtain a schedule with a smaller value of criterion γ than the initial one. We achieve a contradiction with the assumption.   □

Taking into account Lemma 6.4, unless otherwise stated, we will identify a single machine schedule and an appropriate sequence of job indices.

Remark 6.5. For brevity of notation, and to indicate that the starting time S_j is the variable on which the processing time p_j depends, in subsequent sections we will write $p_j(t)$ and $f_j(t)$ instead of $p_j(S_j)$ and $f_j(S_j)$, respectively. Similarly, we will write $p_{ij}(t)$ and $f_{ij}(t)$ instead of $p_{ij}(S_{ij})$ and $f_{ij}(S_{ij})$, respectively.

Other parameters which describe a time-dependent scheduling problem, such as job ready times, deadlines, weights, precedence constraints or the optimality criterion, are as those in the classical scheduling (cf. Chap. 4).

6.2.2 Specific models of deteriorating processing times

There are known several specific models of time-dependent job processing times, differing in the form of functions describing the processing times. First let us describe the specific models of job deterioration.

The simplest model of job deterioration is *proportional deterioration*, introduced by Mosheiov [79]. In this case, when we deal with a single or parallel machine time-dependent scheduling problem, job processing times $p_j(t)$ are in the form of

$$p_j(t) = b_j t, \tag{6.7}$$

where $b_j > 0$ for $1 \leqslant j \leqslant n$ and t denotes the starting time of job J_j. Coefficient b_j is called the *deterioration rate* of job J_j, $1 \leqslant j \leqslant n$. Moreover, in order to avoid the trivial case when all processing times of jobs executed on a machine are equal to zero, we assume that the first scheduled job starts at time $t_0 > 0$. (If no value is specified explicitly, we assume that $t_0 = 1$.)

Example 6.6. Let $n = 2$, $b_1 = 3$, $b_2 = 1$ and $t_0 = 1$. Then $p_1(t) = 3t$ and $p_2(t) = t$. Since the processing times $p_1(t)$ and $p_2(t)$ are strictly increasing functions of the job starting times, they deteriorate in time. In view of monotonicity of the processing times, an optimal schedule for the instance is a non-delay schedule. Hence, we can identify a sequence of job indices and the schedule corresponding to this sequence. In our case, there exist two non-delay schedules starting at $t_0 = 1$, $\sigma_1 = (1, 2)$ and $\sigma_2 = (2, 1)$, such that $p_1(1) = 3$ and $p_2(4) = 4$ in schedule σ_1, while $p_2(1) = 1$ and $p_1(2) = 6$ in schedule σ_2. Let us notice that $C_{\max}(\sigma_1) = C_{\max}(\sigma_2) = 8$. ♦

Remark 6.7. In time-dependent scheduling problems the processing time of the same job may be different in different schedules. For example, consider schedules σ_1 and σ_2 from Example 6.6. The processing time of job J_1 in schedule σ_1 is equal to 3, while in schedule σ_2 it is equal to 6.

In case of dedicated machine problems, we deal with operations instead of jobs and processing times (6.7) take the form of

$$p_{ij}(t) = b_{ij}t, \tag{6.8}$$

where $b_{ij} > 0$ for $1 \leqslant i \leqslant m$, $1 \leqslant j \leqslant n$, and t denotes the starting time of an operation. Coefficients b_{ij} are called the *deterioration rates* of operations of job J_j, $1 \leqslant j \leqslant n$.

A more general model of job deterioration is *proportional-linear deterioration*, introduced by Kononov [94]. In this case, for a single or parallel machine problem, job processing times are in the form of

$$p_j(t) = b_j(a + bt), \tag{6.9}$$

where $t_0 = 0$, $b_j > 0$ for $1 \leqslant j \leqslant n$, $a \geqslant 0$ and $b \geqslant 0$.

Example 6.8. Let $n = 2$, $b_1 = 3$, $b_2 = 2$, $a = 2$, $b = 3$ and $t_0 = 0$. Then $p_1(t) = 3(2 + 3t)$ and $p_2(t) = 2(2 + 3t)$. Hence, $p_1(0) = 6$ and $p_2(6) = 40$ in schedule σ_1, while $p_2(0) = 4$ and $p_1(4) = 42$ in schedule σ_2, where schedules σ_1 and σ_2 are defined as in Example 6.6. Let us notice that $C_{\max}(\sigma_1) = C_{\max}(\sigma_2) = 46$. ◆

In case of linear-proportional deterioration and dedicated machine problems, operation processing times (6.9) take the form of

$$p_{ij}(t) = b_{ij}(a + bt), \tag{6.10}$$

where $t_0 = 0$, $b_{ij} > 0$ for $1 \leqslant i \leqslant m$ and $1 \leqslant j \leqslant n$, $a \geqslant 0$ and $b \geqslant 0$.

The next model of job deterioration is *linear deterioration*, introduced by Tanaev et al. [98]. In case of a single or parallel machine problem, deteriorating job processing times in this model are linear functions of the job starting times,

$$p_j(t) = a_j + b_j t, \tag{6.11}$$

where $t_0 = 0$, $a_j > 0$ and $b_j > 0$ for $1 \leqslant j \leqslant n$. Coefficients a_j and b_j are called, respectively, the *basic processing time* and the *deterioration rate* of job J_j, $1 \leqslant j \leqslant n$.

Example 6.9. Let $n = 2$, $a_1 = 3$, $b_1 = 1$, $a_2 = 2$, $b_2 = 3$ and $t_0 = 0$. Then $p_1(t) = 3 + t$ and $p_2(t) = 2 + 3t$. Hence, $p_1(0) = 3$ and $p_2(3) = 11$ in schedule σ_1, while $p_2(0) = 2$ and $p_1(2) = 5$ in schedule σ_2, where schedules σ_1 and σ_2 are defined as in Examples 6.6 and 6.8. Let us notice that $C_{\max}(\sigma_1) = 14 \neq C_{\max}(\sigma_2) = 7$. ◆

In case of linear deterioration and dedicated machine problems, operation processing times are in the form of

$$p_{ij}(t) = a_{ij} + b_{ij}t, \tag{6.12}$$

where $t_0 = 0$, $a_j > 0$, $a_{ij} > 0$ and $b_{ij} > 0$ for $1 \leqslant i \leqslant m$, $1 \leqslant j \leqslant n$. Coefficients a_{ij} and b_{ij} are called, respectively, the *basic processing times* and *deterioration rates* of operations of job J_j, $1 \leqslant i \leqslant m$, $1 \leqslant j \leqslant n$.

Throughout the book, we will write that deteriorating jobs (operations) have *proportional, proportional-linear* or *linear processing times* if the processing times are in the form of (6.7)–(6.8), (6.9)–(6.10) or (6.11)–(6.12), respectively. Alternatively, we will write that we deal with *proportional* (*proportional-linear, linear*) or *proportionally* (*proportional-linearly, linearly*) *deteriorating jobs* (*operations*).

A more general model of job deterioration than the linear one is the *simple general non-linear deterioration*, in which job processing times are in the form of

$$p_j(t) = a_j + f(t), \tag{6.13}$$

where $a_j > 0$ for $1 \leqslant j \leqslant n$ and $f(t)$ is an arbitrary function such that

$$f(t) \geqslant 0 \quad \text{for} \quad t \geqslant 0. \tag{6.14}$$

This model was introduced by Melnikov and Shafransky [96] and is a special case of model (6.5), with $f_j(t) = f(t)$ for $1 \leqslant j \leqslant n$.

Remark 6.10. Strusevich and Rustogi [97] considered a combination of model (6.13) with a monotone position-dependent function $g(r)$ such that $g(1) = 1$,

$$p_j(t, r) = (a_j + f(t)) g(r), \tag{6.15}$$

where $f(t)$ is a non-negative differentiable function for $t \geqslant t_0$ such that $f(0) = 0$ and r is the position of the jth job in schedule, $1 \leqslant r \leqslant n$; see [97, Chap. 8] for details.

A similar model of simple general job deterioration, where function f depends on time t_0 and is multiplied by a positive coefficient,

$$p_j(t) = a_j + b_j f(t, t_0), \tag{6.16}$$

where $f(t, t_0) = 0$ for $t \leqslant t_0$, $f(t, t_0) \nearrow$ for $t > t_0$ and $b_j > 0$ for $1 \leqslant j \leqslant n$, was introduced by Cai et al. [71].

Another model of simple general non-linear deterioration is the one in which job processing times proportionally deteriorate according to a certain function,

$$p_j(t) = b_j f(t), \tag{6.17}$$

where $f(t)$ is a convex (concave) function for $t \geqslant t_0$. This model of job deterioration is a special case of model (6.3) and was introduced by Kononov [94].

Remark 6.11. Yin and Xu [100] introduced a combination of the model (6.17) with a monotone position-dependent function $g(r)$,

$$p_j(t, r) = b_j f(t) g(r), \tag{6.18}$$

where r, $f(t)$ and $g(r)$ are defined as in model (6.15) with the difference that $f(0) = 1$; see [97, Chap. 9] for details.

Throughout the book, we will write that the jobs (operations) have *simple general non-linear deteriorating processing times* if the processing times are in the form (6.13) or (6.17). Alternatively, we will call them in short *general non-linearly deteriorating jobs (operations)*.

There exist several specific models of general non-linear job deterioration.

Gupta and Gupta [92] introduced *polynomial* job deterioration. In this case, the processing times of jobs are in the form of

$$p_j(t) = a_j + b_j t + c_j t^2 + \cdots + m_j t^m, \tag{6.19}$$

where $a_j, b_j \ldots, m_j$ are positive constants for $1 \leqslant j \leqslant n$ and integer $m \geqslant 1$. The authors also introduced [92] *quadratic* job deterioration which is a special case of polynomial deterioration (6.19) for $m = 2$,

$$p_j(t) = a_j + b_j t + c_j t^2, \tag{6.20}$$

where a_j, b_j, c_j are positive constants for $1 \leqslant j \leqslant n$.

Strusevich and Rustogi [97] generalized model (6.9) with $a = 1$ and introduced non-linear time-dependent job processing times in the form of

$$p_j(t) = b_j(1 + bt)^A, \tag{6.21}$$

where $b_j > 0$ for $1 \leqslant j \leqslant n$, $b > 0$, $-\frac{1}{b_{\max} b} \leqslant A < 0$ or $A > 0$ and $b_{\max} := \max_{1 \leqslant j \leqslant n}\{b_j\}$. This model is a counterpart of model (6.40).

Another special case of general non-linear job deterioration is the *exponential* deterioration. This model of job deterioration can have a few distinct forms. One of them, introduced by Alidaee [82], is the following one:

$$p_j(t) = e^{b_j t}, \tag{6.22}$$

where $0 < b_j < 1$ for $1 \leqslant j \leqslant n$.

Janiak and Kovalyov [93] introduced general exponential deterioration of job processing times in the form of

$$p_j(t) = a_j 2^{b_j(t - r_j)}, \tag{6.23}$$

where $a_j, b_j, r_j \geqslant 0$ for $1 \leqslant j \leqslant n$.

Kunnathur and Gupta [25] introduced *unbounded step-linear* processing times

$$p_j(t) = \begin{cases} a_j, & \text{if } t \leqslant d_j, \\ a_j + b_j(t - d_j), & \text{if } t > d_j, \end{cases} \tag{6.24}$$

where $1 \leqslant j \leqslant n$. In this case, job processing times can grow up to infinity after some prespecified time has elapsed.

Remark 6.12. The equivalent form of job processing times (6.24) is $p_j = a_j + \max\{0, b_j(t - d_j)\}$ for all j. We will use both these forms interchangeably.

Sundararaghavan and Kunnathur [21] introduced the model of *step* deterioration, where job processing times

$$p_j(t) = \begin{cases} a, & \text{if } t \leqslant D, \\ a + b_j, & \text{if } t > D, \end{cases} \tag{6.25}$$

$a > 0$, $b_j > 0$ for $1 \leqslant j \leqslant n$ and $D > 0$ is the *common critical start time* for all jobs in a schedule.

Mosheiov [19] introduced another model of *step* deterioration, where job processing times are in the form of

$$p_j(t) = \begin{cases} a_j, & \text{if } t \leqslant d_j, \\ b_j, & \text{if } t > d_j, \end{cases} \tag{6.26}$$

and

$$b_j \geqslant a_j \text{ for } 1 \leqslant j \leqslant n. \tag{6.27}$$

Remark 6.13. Without loss of generality, we can also assume that in this case $d_1 \leqslant d_2 \leqslant \cdots \leqslant d_n$.

Remark 6.14. Non-linear job processing times (6.26) will be denoted in short as $p_j \in \{a_j, b_j : a_j \leqslant b_j\}$.

Mosheiov [19] extended the step deterioration (6.26), introducing *multi-step* deterioration. In this case, we have processing times in the form of

$$p_j^k(t) = \begin{cases} a_j^k, & \text{if } d_j^{k-1} < S_j \leqslant d_j^k, \\ a_j^{m+1}, & \text{if } S_j > d_j^m, \end{cases} \tag{6.28}$$

where $d_j^0 = 0$,

$$a_j^1 < a_j^2 < \cdots < a_j^{m+1} \tag{6.29}$$

and

$$d_j^1 < d_j^2 < \cdots < d_j^{m+1} \tag{6.30}$$

for $1 \leqslant j \leqslant n$.

Kubiak and van de Velde [95] introduced the *bounded step-linear* deterioration, in which the processing time of each job can grow only up to some limit value. Formally,

$$p_j(t) = \begin{cases} a_j, & \text{if } t \leqslant d, \\ a_j + b_j(t - d), & \text{if } d < t < D, \\ a_j + b_j(D - d), & \text{if } t \geqslant D, \end{cases} \tag{6.31}$$

where d and D, $d < D$, are *common critical date* and *common maximum deterioration date*, respectively. We will also assume that $\sum_{j=1}^{n} a_j > d$, since otherwise the problem is trivial (all jobs can start by time d).

Remark 6.15. In the case of job processing times given by (6.31), the deterioration will be called *bounded* if $D < +\infty$, and it will be called *unbounded* if $D = +\infty$.

6.2.3 Specific models of shortening processing times

The simplest form of job processing time shortening is *proportional-linear shortening*, introduced by Wang and Xia [99]. In the case of this model of processing time shortening and a single or parallel machine problem, job processing times are in the form of

$$p_j(t) = b_j(a - bt), \tag{6.32}$$

where *shortening rates* b_j are rational, they satisfy the condition

$$0 < b_j b < 1, \tag{6.33}$$

the condition

$$b\left(\sum_{i=1}^{n} b_i - b_j\right) < a \tag{6.34}$$

holds for $1 \leqslant j \leqslant n$, $a > 0$ and $0 < b < 1$. Conditions (6.33) and (6.34) ensure that job processing times (6.32) are positive in any non-idle schedule.

Example 6.16. Let $n = 2$, $b_1 = \frac{2}{3}$, $b_2 = \frac{3}{4}$, $a = 2$, $b = \frac{1}{5}$ and $t_0 = 0$. Then $p_1(t) = \frac{2}{3}(2 - \frac{1}{5}t)$ and $p_2(t) = \frac{3}{4}(2 - \frac{1}{5}t)$. Hence, $p_1(0) = \frac{4}{3}$ and $p_2(\frac{4}{3}) = \frac{13}{10}$ in schedule σ_1, while $p_2(0) = \frac{3}{2}$ and $p_1(\frac{3}{2}) = \frac{17}{15}$ in schedule σ_2, where schedules σ_1 and σ_2 are defined as in Examples 6.6, 6.8 and 6.9. Let us notice that $C_{\max}(\sigma_1) = C_{\max}(\sigma_2) = \frac{79}{30}$. $\blacklozenge$

In case of dedicated machine problems, shortening processing times (6.32) take the form of

$$p_{ij}(t) = b_{ij}(a - bt), \tag{6.35}$$

where *shortening rates* b_{ij} are rational, they satisfy the condition

$$0 < b_{ij}b < 1 \tag{6.36}$$

and the condition

$$b\left(\sum_{k=1}^{n} b_{ik} - b_{ij}\right) < a \tag{6.37}$$

holds for $1 \leqslant i \leqslant m$ and $1 \leqslant j \leqslant n$. As previously, conditions (6.36) and (6.37) ensure that processing times (6.35) are positive in any non-idle schedule.

A special case of proportional-linear shortening (6.32) is the case when $a = 1$, i.e., when job processing times are in the form of

$$p_j(t) = b_j(1 - bt), \tag{6.38}$$

where $b_j > 0$ for $1 \leqslant j \leqslant n$ and $0 < b < 1$. In this case, condition (6.34) takes the form of

$$b\left(\sum_{j=1}^{n} b_j - b_{\min}\right) < 1, \tag{6.39}$$

where $b_{\min} := \min_{1 \leqslant j \leqslant n}\{b_j\}$.

Example 6.17. Let $n = 2$, $b_1 = \frac{1}{3}$, $b_2 = \frac{1}{4}$, $a = 1$, $b = \frac{1}{2}$ and $t_0 = 0$. Then $p_1(t) = \frac{1}{3}(1 - \frac{1}{2}t)$ and $p_2(t) = \frac{1}{4}(1 - \frac{1}{2}t)$. Hence, $p_1(0) = \frac{1}{3}$ and $p_2(\frac{1}{3}) = \frac{5}{24}$ in schedule σ_1, while $p_2(0) = \frac{1}{4}$ and $p_1(\frac{1}{4}) = \frac{7}{24}$ in schedule σ_2, where schedules σ_1 and σ_2 are defined as in Examples 6.6, 6.8, 6.9 and 6.16. Let us notice that $C_{\max}(\sigma_1) = C_{\max}(\sigma_2) = \frac{13}{24}$. ◆

Strusevich and Rustogi [97] introduced non-linear time-dependent job processing times in the form of

$$p_j(t) = b_j(1 - bt)^A, \tag{6.40}$$

where $b_j > 0$ for $1 \leqslant j \leqslant n$, $bt < 1$ for all t, $A < 0$ or $1 \leqslant A \leqslant \frac{1}{b_{\max}b}$ and $b_{\max}$ is defined as in Sect. 6.2.2. This model generalizes model (6.38) and is a counterpart of model (6.21).

Counterparts of job processing times (6.38) for dedicated machine problems are operation processing times in the form of

$$p_{ij}(t) = b_{ij}(1 - bt), \tag{6.41}$$

where $b_{ij} > 0$ for $1 \leqslant i \leqslant m$, $1 \leqslant j \leqslant n$. In this case, condition (6.34) is in the form of

$$b \left(\sum_{j=1}^{n} b_{ij} - b_{i,\min} \right) < 1, \tag{6.42}$$

where $1 \leqslant i \leqslant m$, $1 \leqslant j \leqslant n$ and $b_{i,\min} := \min_{1 \leqslant j \leqslant n}\{b_{ij}\}$.

Ho et al. [24] introduced the next type of job shortening called *linear shortening*, in which job processing times are decreasing linear functions of the job starting times. In this case, job processing times are in the form of

$$p_j(t) = a_j - b_j t, \tag{6.43}$$

where *shortening rates* b_j are rational, and conditions

$$0 < b_j < 1 \tag{6.44}$$

and

$$b_j \left(\sum_{i=1}^{n} a_i - a_j \right) < a_j \tag{6.45}$$

hold for $1 \leqslant j \leqslant n$.

The conditions (6.44) and (6.45) are counterparts of conditions (6.33) and (6.34), respectively.

Example 6.18. Let $n = 2$, $a_1 = 2$, $b_1 = \frac{1}{7}$, $a_2 = 3$, $b_2 = \frac{1}{5}$ and $t_0 = 0$. Then $p_1(t) = 2 - \frac{1}{7}t$ and $p_2(t) = 3 - \frac{1}{5}t$. Hence, $p_1(0) = 2$ and $p_2(2) = \frac{13}{5}$ in schedule σ_1, while $p_2(0) = 3$ and $p_1(3) = \frac{11}{7}$ in schedule σ_2, where σ_1 and σ_2 are defined as in Examples 6.6, 6.8, 6.9, 6.16 and 6.17. Let us notice that $C_{\max}(\sigma_1) = \frac{23}{5} \neq C_{\max}(\sigma_2) = \frac{32}{7}$. ◆

Counterparts of shortening job processing times (6.43) for dedicated machine problems are operation processing times in the form of

$$p_{ij}(t) = a_{ij} - b_{ij}t, \tag{6.46}$$

where *shortening rates* b_{ij} are rational, and conditions

$$0 < b_{ij} < 1 \tag{6.47}$$

and

$$b_{ij}\left(\sum_{k=1}^{n} a_{kj} - a_{ij}\right) < a_{ij} \tag{6.48}$$

hold for $1 \leqslant i \leqslant m$ and $1 \leqslant j \leqslant n$.

Throughout the book, we will write that shortening jobs (operations) have *proportional, proportional-linear* or *linear processing times* if the processing times are in the form of (6.32) or (6.35), (6.38) or (6.41) or (6.43) or (6.46), respectively. Alternatively, we will write that we deal with *proportional, (proportional-linear, linear)* or *proportionally (proportional-linearly, linearly) shortening jobs (operations)*.

Similarly to non-linearly deteriorating job processing times, there exist a few specific models of non-linear job shortening.

Cheng et al. [85] introduced the *step-linear* shortening job processing times. In this case, the processing time p_{ij} of the job J_i scheduled on machine M_j, $1 \leqslant i \leqslant n$ and $1 \leqslant j \leqslant m$, is as follows:

$$p_{ij}(t) = \begin{cases} a_{ij}, & \text{if } t \leqslant y, \\ a_{ij} - b_{ij}(t - y), & \text{if } y < t < Y, \\ a_{ij} - b_{ij}(Y - y), & \text{if } t \geqslant Y, \end{cases} \tag{6.49}$$

where $a_{ij} > 0, b_{ij} > 0, y \geqslant 0$ and $Y \geqslant y$ are the *basic* processing time, the *shortening rate*, the *common initial shortening date* and the *common final shortening date*, respectively. It is also assumed that

$$0 < b_{ij} < 1 \tag{6.50}$$

and

$$a_{ij} > b_{ij}\left(\min\left\{\sum_{k=1}^{n} a_{ik} - a_{ij}, Y\right\} - y\right) \tag{6.51}$$

for $1 \leqslant j \leqslant n$ and $1 \leqslant i \leqslant m$.

Let us notice that the decreasing step-linear processing times (6.49) are counterparts of the increasing step-linear processing times (6.31). Conditions (6.50) and (6.51) are generalizations of conditions (6.44) and (6.45) for the case of parallel machines.

Cheng et al. [87] introduced *step* shortening job processing times

$$p_{ij}(t) = \begin{cases} a_j, & \text{if } t < D, \\ a_j - b_j, & \text{if } t \geqslant D, \end{cases} \tag{6.52}$$

where $\sum_{j=1}^{n} a_j \geqslant D$ and $0 \leqslant b_j \leqslant a_j$ for $1 \leqslant j \leqslant n$.

We will write that the jobs (operations) have *non-linear shortening processing times* if the processing times are in the form of (6.49) or (6.52), and we will call them, in short, *non-linearly shortening jobs (operations)*.

6.2.4 Specific models of alterable processing times

Jaehn and Sedding [27] proposed alterable, *non-monotonic piecewise-linear* time-dependent job processing times in the form of

$$p_j(t) = a_j + b\,|f_j(t) - T_j|\,, \tag{6.53}$$

where $a_j \geqslant 0$, $0 < b < 2$, $f_j(t) = \frac{t+C_j}{2}$, $T_j > 0$ and

$$C_j = \begin{cases} t + \dfrac{a_j - b(t-T_j)}{1+\frac{b}{2}}, & \text{if } t \leqslant T_j - \frac{a_j}{2}, \\[2mm] t + \dfrac{a_j + b(t-T_j)}{1-\frac{b}{2}}, & \text{if } t > T_j - \frac{a_j}{2} \end{cases}$$

for $1 \leqslant j \leqslant n$.

Example 6.19. Let $n = 2$, $a_1 = 1$, $a_2 = 3$, $T_1 = 1$, $T_2 = 3$, $b = \frac{3}{2}$ and $t_0 = 0$. Then $p_1(t) = 1 + \frac{3}{2}|f_1(t) - 1|$, $p_2(t) = 3 + \frac{3}{2}|f_2(t) - 3|$, where $f_1(t) = \frac{t+C_1}{2}$, $f_2(t) = \frac{t+C_2}{2}$,

$$C_1 = \begin{cases} \frac{10}{7} + \frac{1}{7}t, & \text{if } t \leqslant \frac{1}{4}, \\ -2 + 7t, & \text{if } t > \frac{1}{4}, \end{cases}$$

and

$$C_2 = \begin{cases} \frac{30}{7} + \frac{1}{7}t, & \text{if } t \leqslant \frac{3}{2}, \\ -6 + 7t, & \text{if } t > \frac{3}{2}. \end{cases}$$

Therefore, $p_1(0) = \frac{10}{7}$ and $p_2(\frac{10}{3}) = \frac{150}{49}$ in schedule σ_1, while $p_2(0) = \frac{30}{7}$ and $p_1(\frac{30}{7}) = \frac{142}{7}$ in schedule σ_2, where σ_1 and σ_2 are defined as in Examples 6.6, 6.8, 6.9, 6.16, 6.17 and 6.18. Let us notice that $C_{\max}(\sigma_1) = \frac{220}{49} \neq C_{\max}(\sigma_2) = \frac{172}{7}$. ◆

Sedding [28] proposed alterable time-dependent job processing times in the form of

$$p_j = a_j + \max\{-a\,(t - T_0),\, b\,(t - T_0)\}, \tag{6.54}$$

where $a \in [0, 1]$ and $b \in [0, \infty)$ denote rational slopes of each job, while $T_0 \geqslant 0$ and $a_j \in \mathbb{Q}_{\geqslant 0}$, $1 \leqslant j \leqslant n$, denote the *ideal starting time* and basic processing time of job J_j, respectively.

We will write that jobs (operations) have *alterable processing times*, if the processing times are in the form of (6.53) or (6.54).

6.3 Mixed models of time-dependent job processing times

In *pure models* of job processing times considered in Sect. 6.2, the processing
time of each job is a single variable function. Besides the models, in the modern scheduling theory there are also studied *mixed models* of job processing
times, where the processing time of a job is a function of both the starting
time of the job and some other variable, e.g. the position of the job in the
schedule (see Sect. 21.4 for examples). Because scheduling problems with job
processing times depending on the job positions in schedule are called scheduling problems with a *learning effect* (cf. Sect. 5.2.1), we will refer to them as
problems of *time-dependent scheduling with a learning effect*. This domain of
time-dependent scheduling was initiated by Wang [47].

The literature on time-dependent scheduling with learning effects is relatively rich and it concerns, among others,

- due-date assignment problems (Kuo and Yang [34], Wang and Guo [51],
  Wang and Wang [54], Yang et al. [66]);
- scheduling problems with resource-dependent processing times of jobs (Lu
  et al. [36]);
- batch scheduling problems (Pei et al. [37], Yang [65], Zhan and Yan [69]);
- single machine scheduling (Bai et al. [30], Gordon et al. [31], Huang
  et al. [32], Jiang et al. [33], Lee and Lai [35], Shen et al. [38], Strusevich
  and Rustogi [39], Sun [40], Toksarı [41], Toksarı et al. [46], Wang [48, 49],
  Wang and Cheng [50, 59], Wang et al. [52, 56, 57, 58], Wu et al. [61], Wu
  et al. [60], Yang and Kuo [64, 62], Yang and Yang [63], Yin and Xu [67],
  Yin et al. [68], Zhang et al. [70]);
- parallel machine scheduling (Toksarı and Güner [42, 43, 44, 45], Strusevich
  and Rustogi [39]) and
- dedicated machine scheduling problems (Wang and Liu [53], Wang and
  Wang [55]).

Remark 6.20. In this book, we focus only on pure models of time-dependent
job processing times (cf. Sect. 6.2). Therefore, other models of time-dependent
job processing times, e.g. those with a learning effect, are beyond the scope
of this book and will be not considered in detail.

6.4 Notation of time-dependent scheduling problems

Generally, the time-scheduling problems considered in this book will be de-
noted using the $\alpha|\beta|\gamma$ notation (see Sect. 4.6 for details). Each problem will
be denoted by $\alpha_1\alpha_2|p_j(t) = a_j + f_j(t)|\varphi$ or $\alpha_1\alpha_2|p_{ij}(t) = a_{ij} + f_{ij}(t)|\varphi$, where
$\alpha_1\alpha_2$, $f_j(t)$ and φ denote the machine environment, the form of the variable
part of job processing time and the criterion function, respectively.

Remark 6.21. We will use the $\alpha|\beta|\gamma$ notation, if it will yield a simple notation for the considered scheduling problem. In some cases, however, we will resign from the notation in favour of the description by words, if the descriptive approach will be more readable.

The short form of symbol $\alpha_1\alpha_2|p_j(t) = a_j + f_j(t)|\varphi$ is symbol $\alpha_1\alpha_2|p_j = a_j + f_j(t)|\varphi$. The short form of symbol $\alpha_1\alpha_2|p_{ij}(t) = a_{ij} + f_{ij}(t)|\varphi$ is symbol $\alpha_1\alpha_2|p_{ij} = a_{ij} + f_{ij}(t)|\varphi$. Throughout this book we will use the short form of the symbols denoting time-dependent scheduling problems.

If the same function $f(t)$ is used for all jobs, $f_j(t) = f(t)$ for $1 \leqslant j \leqslant n$ or $f_{ij}(t) = f(t)$ for $1 \leqslant i \leqslant n$ and $1 \leqslant j \leqslant k_i$, we will speak about *simple deterioration (shortening)* of job processing times. In the opposite case, we will speak about *general deterioration (shortening)* of job processing times.

We illustrate the application of the $\alpha|\beta|\gamma$ notation by several examples. We begin with single machine time-dependent scheduling problems.

Example 6.22. (a) Symbol $1|p_j = b_j t|C_{\max}$ will denote a single machine scheduling problem with proportional job processing times and the $C_{\max}$ criterion.
(b) Symbol $1|p_j = a_j + b_j t|L_{\max}$ will denote a single machine scheduling problem with linear job processing times and the $L_{\max}$ criterion.
(c) Symbol $1|p_j = b_j(a + bt)|\sum w_j C_j$ will denote a single machine scheduling problem with proportional-linear job processing times and the $\sum w_j C_j$ criterion. ◆

In a similar way will be denoted parallel machine time-dependent scheduling problems.

Example 6.23. (a) Symbol $P2|p_j = b_j t|\sum C_j$ will denote a two identical parallel machine scheduling problem with proportionally deteriorating jobs and the $\sum C_j$ criterion.
(b) Symbol $Pm|p_j = a_j + f(t)|C_{\max}$ will denote a multiple identical machine scheduling problem with simple general deterioration of jobs and the $C_{\max}$ criterion.
(c) Symbol $Pm|p_j = a_j + bt|\sum w_j C_j$ will denote a multiple identical machine scheduling problem with linearly deteriorating jobs with the same deterioration rate for all jobs and the $\sum w_j C_j$ criterion. ◆

Unlike the single and parallel machine time-dependent scheduling problems, where a single index is necessary for denoting job processing times, in the case of dedicated machine time-dependent scheduling problems we use two indices for operation processing times.

Example 6.24. (a) Symbol $F2|p_{ij} = a_{ij} + b_{ij}t|L_{\max}$ will denote a two-machine flow shop problem with linear operation processing times and the $L_{\max}$ criterion.
(b) Symbol $O3|p_{ij} = b_{ij}t, b_{3i} = b|C_{\max}$ will denote a three-machine open shop

problem with proportional operation processing times such that deterioration rates of all operations on machine M_3 are equal to each other, and with the $C_{\max}$ criterion.

(c) Symbol $J2|p_{ij} = b_{ij}t|C_{\max}$ will denote a two-machine job shop problem with proportional operation processing times and the $C_{\max}$ criterion. ♦

6.5 Mathematical background of time-dependent scheduling

Time-dependent scheduling is a domain with a rich mathematical background of twofold nature. On the one hand, the vast of time-dependent scheduling problems can be solved using the classical scheduling theory methods such as

- the pairwise job interchange argument,
- mathematical induction,
- proof by contradiction,
- priority-generating functions,
- minimizing functions over a set of all permutations.

Numerous classical scheduling problems can also be solved using the methods of graph theory and computational complexity theory. We refer the reader to the literature (see, e.g., Agnetis et al. [1, Chaps. 1–2], Błażewicz et al. [2, Chap. 2], Garey and Johnson [4, Chaps. 2–4], Gawiejnowicz [5, Chaps. 1 and 3], Leung [10, Chap. 2], Parker [12, Chap. 2], Pinedo [13], Strusevich and Rustogi [14, Chaps. 1–3], Tanaev et al. [15, Chap. 3]) for details.

On the other hand, time-dependent scheduling problems are solved using the methods specific for modern scheduling problems, such as

- signatures,
- the matrix approach,
- solution methods for multiplicative problems,
- properties of mutually related scheduling problems,
- new methods of proving $\mathcal{NP}$-completeness,
- composition operator properties.

We refer the reader to the literature (see, e.g., Cheng et al. [3], Gawiejnowicz [5, Chaps. 10 and 12], Gawiejnowicz and Kononov [6], Gawiejnowicz et al. [7, 8], Kawase et al. [9], Ng et al. [11]) for details.

6.6 Applications of time-dependent scheduling

The motivation for research into time-dependent scheduling follows from the existence of many real-life problems which can be formulated in terms of scheduling jobs with time-dependent processing times. Such problems appear in all cases in which any delay in processing causes an increase (a decrease) of the processing times of executed jobs. If job processing times increase, we deal with deteriorating job processing times; if they decrease, we deal with shortening job processing times; if they vary in a non-monotonic way, we deal with alterable job processing times. In this section, we give several examples of problems which can be modeled in time-dependent scheduling.

6.6.1 Scheduling problems with deteriorating jobs

Gupta et al. [18] consider the problem of the *repayment of multiple loans*. We have to repay n loans, $L_1, L_2, \ldots, L_n$. A loan may represent an amount of borrowed cash or a payment to be made for a credit purchase. Loan L_k qualifies for a discount u_k if it is paid on or before a specified time b_k. A penalty at the rate v_k per day is imposed if the loan is not paid by due date d_k, $1 \leqslant k \leqslant n$. The debtor earmarks a constant amount of q dollars per day, $q < v_k$, for repayment of the loans. Cash flows are continuously discounted with discount factor $(1 + r)^{-1}$. The aim is to find an optimal repayment schedule which minimizes the present value PV of all cash outflows, $PV := \sum_{k=1}^{n} \frac{A_k}{(1+r)^{T_k}}$, where A_k and T_k denote, respectively, the actual amount paid for loan L_k and the time at which the loan L_k is repaid, $1 \leqslant k \leqslant n$. This problem can be modeled as a single-machine scheduling problem with time-dependent job processing times and the PV criterion.

Mosheiov [19] considers the following problem of *scheduling maintenance procedures*. A set of n maintenance procedures P_k, $1 \leqslant k \leqslant n$, has to be executed by $m \geqslant 1$ machines. A maintenance procedure P_k has to take place before a specified deadline d_k. The procedure consists of a series of actions which last altogether p_k^1 time units. If the procedure does not complete by the deadline, several additional actions are required. The new processing time of procedure P_k is $p_k^2 > p_k^1$ time units. The aim is to find an order of execution of maintenance procedures $P_1, P_2, \ldots, P_n$ which minimizes the maximum completion time of the last executed procedure. This problem can be modeled as a single- or multiple-machine scheduling problem with two-step deteriorating job processing times.

Gawiejnowicz et al. [16] consider the following problem of *scheduling derusting operations*. We are given n items (e.g. parts of devices), which are subject to maintenance (e.g. they should be cleared from rust). This maintenance is performed by a single worker, who can himself determine the sequence of maintenance procedures. All procedures are non-preemptable, i.e., no maintenance procedure can be interrupted once it has started. At the moment $t = 0$ all items need the same amount of time for maintenance, e.g. one unit of time. As time elapses, each item corrodes at a rate which depends on the kind of the material from which the particular item is made. The rate of corrosion for the jth item is equal to b_j, $1 \leqslant j \leqslant n$, and the time needed for the maintenance of each item grows proportionally to the time which elapsed from the moment $t = 0$. The problem is to choose such a sequence of the maintenance procedures that minimizes the total completion time of maintenance of all items. This problem can be modeled as the single-machine time-dependent scheduling problem $1|p_j = 1 + b_j t| \sum C_j$.

Rachaniotis and Pappis [20] consider the problem of *scheduling a single fire-fighting resource* in the case when there are several fires to be controlled. The aim is to find such an order of suppressing n existing fires that the total damage caused by the fires is minimized. The problem can be modeled as a single machine scheduling problem with time-dependent processing times and the total cost minimization criterion.

Gawiejnowicz and Lin [17] address the problem of *modeling the work of a team of cleaning workers*, in which are engaged fully fledged workers, average workers and beginners. This team has to clean a set of rooms, devices or similar objects. Each cleaning task requires a certain working time and depends on when the task started, i.e., if the task starts later, it will take a longer time to complete. The most experienced workers do the cleaning tasks in a constant time; the average staff need a proportional time, while for the beginners a linear time is necessary, with respect to the time when the task starts. The aim is to find an optimal schedule for the cleaning tasks with respect to a criterion function. This problem can be formulated as a single machine time-dependent scheduling problem with mixed deterioration.

Wu et al. [22] consider the problem of *timely medical treatment* of a set of patients, using a medical device. The time needed to perform each patient treatment is known in advance, and for each of them there is defined a deadline by which the treatment must complete. The medical device needs a cyclic maintenance procedure whose length linearly depends on the starting time of the procedure. The aim is to find a schedule of the patients, minimizing the number of tardy treatments. The problem can be modeled as a single machine time-dependent scheduling problem with linear jobs and the $\sum U_j$ criterion.

Zhang et al. [23] consider a few variations of the problem of *handling surgical patients*. For each of the patients waiting for a surgery treatment, there is defined a weight and a deadline. The time of treatment of the jth patient is equal to $p_j(t) = a_j + \alpha \max\{0, t - \beta\}$, where $a_j > 0$, $\alpha \geqslant 0$ and $\beta > 0$ denote the basic processing time, a common deterioration rate and a deferred deterioration rate of the patient, respectively. The aim is to find a schedule of the patients, minimizing a given optimality criterion. The problems are modeled as single machine time-dependent scheduling problems with non-linear job processing times and the $C_{\max}$, $L_{\max}$ and $\sum w_j C_j$ criteria.

6.6.2 Scheduling problems with shortening jobs

Ho et al. [24] consider the problem of *recognizing aerial threats*. A radar station recognizes some aerial threats approaching the station. The time required to recognize the threats decreases as they get closer. The aim is to find an optimal order of recognizing the threats which minimizes the maximum completion time. This problem can be modeled as a single machine scheduling problem with shortening job processing times and the $C_{\max}$ criterion.

Kunnathur and Gupta [25] and Ng et al. [26] consider the problem of *producing ingots in a steel mill*. A set of ingots has to be produced in a steel mill. After being heated in a blast furnace, hot liquid metal is poured into steel ladles and next into ingot moulds, where it solidifies. Next, after the ingot stripper process, the ingots are segregated into batches and transported to the soaking pits, where they are preheated up to a certain temperature. Finally, the ingots are hot-rolled on the blooming mill. If the temperature of an ingot, while waiting in a buffer between the furnace and the rolling machine, has dropped below a certain value, then the ingot needs to be reheated to the temperature required for rolling. The reheating time depends on the time spent by the ingot in the buffer. The problem is to find a sequence of preheating the ingots which minimizes the maximum completion time of the last ingot produced. This problem can be modeled as a single machine scheduling problem with shortening job processing times and the $C_{\max}$ criterion.

6.6.3 Scheduling problems with alterable jobs

Jaehn and Sedding [78] analyzed the following problem in *production planning for car assembly lines*. Given a production facility consisting of a conveyor, moving steadily along a straight line, and shelves, placed statically along the conveyor, the aim is to minimize the total distance covered by a single worker, working on one car. The authors modeled the problem as a single machine time-dependent scheduling problem with alterable non-linear job processing times and the $C_{\max}$ criterion. Ph. D. dissertation by Sedding [29] includes a detailed discussion of this problem.

6.6.4 Scheduling problems with time-dependent parameters

The time dependence may concern not only job processing times but also other parameters of a scheduling problem. For example, Cai et al. [71] consider the following *crackdown scheduling problem*. There are n illicit drug markets, all of which need to be brought down to a negligible level of activity. Each market is eliminated by a procedure consisting in a crackdown phase and a maintenance phase. The crackdown phase utilizes all the available resources until the market is brought down to the desired level. The maintenance phase, which follows after the crackdown phase and uses a significantly smaller amount of resources, maintains the market at this level. The aim is to find an order of elimination of the drug markets which minimizes the total time spent in eliminating all drug markets. The problem can be modeled as a single-machine scheduling problem of minimizing the total cost $\sum f_j$, where f_j are monotonically increasing time-dependent cost functions.

Other examples of scheduling problems in which some parameters are time-dependent include multiprocessor tasks scheduling (Bampis and Kononov [72]), scheduling in a contaminated area (Janiak and Kovalyov [74, 73]), multicriteria project sequencing (Klamroth and Wiecek [75]) and scheduling jobs with deteriorating job values (Voutsinas and Pappis [76]).

6.6.5 Other problems with time-dependent parameters

Time-dependent parameters also appear in problems not related to scheduling. Shakeri and Logendran [80] consider the following problem of *maximizing satisfaction level* in a multitasking environment. Several plates are spinning on vertical poles. An operator has to ensure all plates spin as smoothly as possible. A value, called the *satisfaction level*, can be assigned to each plate's spinning state. The satisfaction level of a plate is ranging from 0% (i.e., the plate is not spinning) up to 100% (the plate is spinning perfectly). The objective is to maximize the average satisfaction level of all plates over time.

The above problem is applicable to multitasking environments in which we cannot easily determine the completion time of any job. Examples of such environments are the environments of the control of a plane's flight parameters, monitoring air traffic or the work of nuclear power plants. A special case of the problem, when a 100% satisfaction level is equivalent to the completion of a job, is a single-machine time-dependent scheduling problem.

Other examples of practical problems which can be modeled in terms of time-dependent scheduling include the control of queues in communication systems in which jobs deteriorate as they wait for processing (Browne and Yechiali [77]), search for an object in worsening weather or growing darkness, performance of medical procedures under deterioration of the patient conditions and repair of machines or vehicles under deteriorating mechanical conditions (Mosheiov [79]).

This section ends the second part of the book, devoted to the presentation of the main models of the scheduling theory. In the next part, we discuss polynomially solvable time-dependent scheduling problems.

Bibliographic notes

The literature on time-dependent scheduling contains to date about 350 papers and books. The major part of the literature, about 60% of the whole, concerns single machine time-dependent scheduling problems. The literature on parallel machine and dedicated machine time-dependent scheduling encompasses about 25% and 15% of the whole time-dependent scheduling literature.

Single machine time-dependent scheduling problems

Single machine time-dependent scheduling problems are reviewed in four papers and four books. The first review, Gawiejnowicz [88], discusses papers on single machine time-dependent scheduling problems published up to 1995. This review presents time-dependent scheduling in terms of *discrete-continuous scheduling* where, apart from machines which are *discrete resources*, jobs also need *continuous resources* such as fuel or power. The second review, Alidaee and Womer [83], focuses on single machine time-dependent scheduling literature up to 1998. The third review, by Cheng et al. [86], discusses papers on single machine time-dependent scheduling, published up to 2003. The last review, by Gawiejnowicz [90], presents the main results, open problems and new directions in single machine time-dependent scheduling, based on the literature published up to 2018.

There are also four books in which single machine time-dependent scheduling problems are discussed. The main reference on the subject is the monograph by Gawiejnowicz [89], where Chap. 6 is devoted to single machine problems. The monograph by Agnetis et al. [81] focuses on multi-agent scheduling problems; however, Chap. 6 of this book includes a wider discussion of scheduling problems with time-dependent job processing times, illustrated by examples. Single machine time-dependent scheduling problems are also discussed in Chaps. 8 and 9 of the monograph by Strusevich and Rustogi [97], devoted to scheduling problems with time-changing effects and *rate-modifying activities* (see Sect. 21.1 for more details), and in Chap. 12 of the monograph by Błażewicz et al. [84], devoted to the theory and applications of the classical and modern scheduling.

Parallel machine time-dependent scheduling problems

Parallel machine time-dependent scheduling problems are briefly reviewed in the paper by Cheng et al. [102]. A more detailed presentation is given in Chap. 7 of the monograph by Gawiejnowicz [103]. The most recent review of parallel machine time-dependent scheduling is given by Gawiejnowicz [104]. Some parallel-machine time-dependent scheduling problems are also considered in Chap. 11 of the monograph by Strusevich and Rustogi [105], and in Chap. 12 of the monograph by Błażewicz et al. [101].

Dedicated machine time-dependent scheduling problems

A brief review of dedicated machine time-dependent scheduling literature is given in the paper by Cheng et al. [107]. Some time-dependent flow shop and open shop problems are mentioned in the book by Tanaev et al. [110] and in Chap. 12 of the monograph by Błażewicz et al. [106]. A detailed presentation of dedicated machine time-dependent scheduling is given in Chap. 8 of the monograph by Gawiejnowicz [108]. An updated review of this domain is given in the paper by Gawiejnowicz [109].

References

Mathematical background of time-dependent scheduling

1. A. Agnetis, J.-C. Billaut, S. Gawiejnowicz, D. Pacciarelli and A. Soukhal, *Multi-agent Scheduling: Models and Algorithms*. Berlin-Heidelberg: Springer 2014.

2. J. Błażewicz, K. Ecker, E. Pesch, G. Schmidt and J. Węglarz, *Handbook on Scheduling*. Berlin-Heidelberg: Springer 2007.
3. T-C. E. Cheng, Y. Shafransky and C-T.Ng, An alternative approach for proving the NP-hardness of optimization problems. *European Journal of Operational Research* **248** (2016), 52–58.
4. M. R. Garey and D. S. Johnson, *Computers and Intractability: A Guide to the Theory of NP-Completeness*. San Francisco: Freeman 1979.
5. S. Gawiejnowicz, *Time-Dependent Scheduling*. Berlin-Heidelberg: Springer 2008.
6. S. Gawiejnowicz and A. Kononov, Isomorphic scheduling problems. *Annals of Operations Research* **213** (2014), 131–145.
7. S. Gawiejnowicz, W. Kurc and L. Pankowska, Equivalent time-dependent scheduling problems. *European Journal of Operational Research* **196** (2009), 919–929.
8. S. Gawiejnowicz, W. Kurc and L. Pankowska, Conjugate problems in time-dependent scheduling. *Journal of Scheduling* **12** (2009), 543–553.
9. Y. Kawase, K. Makino and K. Seimi, Optimal composition ordering problems for piecewise linear functions. *Algorithmica* **80** (2018), 2134–2159.
10. J. Y-T. Leung (ed.), *Handbook of Scheduling: Models, Algorithms and Performance Analysis*. New York: Chappman & Hall/CRC 2004.
11. C-T. Ng, M. S. Barketau, T-C. E. Cheng and M. Y. Kovalyov, "Product Partition" and related problems of scheduling and systems reliability: computational complexity and approximation. *European Journal of Operational Research* **207** (2010), 601–604.
12. R. G. Parker, *Deterministic Scheduling Theory*. London: Chapman & Hall 1995.
13. M. L. Pinedo, *Scheduling: Theory, Algorithms, and Systems*, 5th ed. Berlin-Heidelberg: Springer 2016.
14. V. A. Strusevich and K. Rustogi, *Scheduling with Time-Changing Effects and Rate-Modifying Activities*. Berlin-Heidelberg: Springer 2017.
15. V. S. Tanaev, V. S. Gordon and Y. M. Shafransky, *Scheduling Theory: Single-Stage Systems*. Dordrecht: Kluwer 1994.

Scheduling problems with deteriorating jobs

16. S. Gawiejnowicz, W. Kurc and L. Pankowska, Pareto and scalar bicriterion scheduling of deteriorating jobs. *Computers and Operations Research* **33** (2006), no. 3, 746–767.
17. S. Gawiejnowicz and B. M-T. Lin, Scheduling time-dependent jobs under mixed deterioration. *Applied Mathematics and Computation* **216** (2010), no. 2, 438–447.
18. S. K. Gupta, A. S. Kunnathur and K. Dandapani, Optimal repayment policies for multiple loans. *Omega* **15** (1987), no. 4, 323–330.
19. G. Mosheiov, Scheduling jobs with step-deterioration: minimizing makespan on a single- and multi-machine. *Computers and Industrial Engineering* **28** (1995), no. 4, 869–879.
20. N. P. Rachaniotis and C. P. Pappis, Scheduling fire-fighting tasks using the concept of 'deteriorating jobs'. *Canadian Journal of Forest Research* **36** (2006), no. 3, 652–658.
21. P. S. Sundararaghavan and A. S. Kunnathur, Single machine scheduling with start time dependent processing times: some solvable cases. *European Journal of Operational Research* **78** (1994), no. 3, 394–403.

22. Y. Wu, M. Dong and Z. Zheng, Patient scheduling with periodic deteriorating maintenance on single medical device. *Computers and Operations Research* **49** (2014), 107–116.
23. X. Zhang, H. Wang and X. Wang, Patients scheduling problems with deferred deteriorated functions. *Journal of Combinatorial Optimization* **30** (2015), no. 4, 1027–1041.

Scheduling problems with shortening jobs

24. K. I-J. Ho, J. Y-T. Leung and W-D. Wei, Complexity of scheduling tasks with time-dependent execution times. *Information Processing Letters* **48** (1993), no. 6, 315–320.
25. A. S. Kunnathur and S. K. Gupta, Minimizing the makespan with late start penalties added to processing times in a single facility scheduling problem. *European Journal of Operational Research* **47** (1990), no. 1, 56–64.
26. C-T. Ng, T-C. E. Cheng, A. Bachman and A. Janiak, Three scheduling problems with deteriorating jobs to minimize the total completion time. *Information Processing Letters* **81** (2002), no. 6, 327–333.

Scheduling problems with alterable jobs

27. F. Jaehn and H. A. Sedding, Scheduling with time-dependent discrepancy times. *Journal of Scheduling* **19** (2016), no. 6, 737–757.
28. H. A. Sedding, Scheduling non-monotonous convex piecewise-linear time-dependent processing times of a uniform shape, *The 2nd International Workshop on Dynamic Scheduling Problems*, 2018, 79–84.
29. H. A. Sedding, *Time-Dependent Path Scheduling: Algorithmic Minimization of Walking Time at the Moving Assembly Line.* Wiesbaden: Springer Vieweg 2020.

Time-dependent scheduling problems with a learning effect

30. J. Bai, Z-R. Li and X. Huang, Single-machine group scheduling with general deterioration and learning effects. *Applied Mathematical Modelling* **36** (2012), no. 3, 1267–1274.
31. V. S. Gordon, C. N. Potts and V. A. Strusevich, Single machine scheduling models with deterioration and learning: handling precedence constraints via priority generation. *Journal of Scheduling* **11** (2008), no. 5, 357–370.
32. X. Huang, J-B. Wang, L-Y. Wang, W-J. Gao and X-R. Wang, Single machine scheduling with time-dependent deterioration and exponential learning effect. *Computers and Industrial Engineering* **58** (2010), no. 1, 58–63.
33. Z. Jiang, F. Chen and H. Kang, Single-machine scheduling problems with actual time-dependent and job-dependent learning effect. *European Journal of Operational Research* **227** (2013), no. 1, 76–80.
34. W-H. Kuo and D-L. Yang, A note on due-date assignment and single-machine scheduling with deteriorating jobs and learning effects. *Journal of the Operational Research Society* **62** (2011), no. 1, 206–210.
35. W-C. Lee and P-J. Lai, Scheduling problems with general effects of deterioration and learning. *Information Sciences* **181** (2011), no. 6, 1164–1170.

36. Y-Y. Lu, J. Jin, P. Ji and J-B. Wang, Resource-dependent scheduling with deteriorating jobs and learning effects on unrelated parallel machine. *Neutral Computing and Applications* **27** (2016), no. 7, 1993–2000.

37. J. Pei, X. Liu, P. M. Pardalos, A. Migdalas and S. Yang, Serial-batching scheduling with time-dependent setup time and effects of deterioration and learning on a single-machine. *Journal of Global Optimization* **67** (2017), no. 1–2, 251–262.

38. P. Shen, C-M. Wei and Y-B. Wu, A note on deteriorating jobs and learning effects on a single-machine scheduling with past-sequence-dependent setup times. *International Journal of Advanced Manufacturing Technology* **58** (2012), no. 5–8, 723–725.

39. V. A. Strusevich and K. Rustogi, *Scheduling with Time-Changing Effects and Rate-Modifying Activities*. Berlin-Heidelberg: Springer 2017.

40. L. Sun, Single-machine scheduling problems with deteriorating jobs and learning effects. *Computers and Industrial Engineering* **57** (2009), no. 3, 843–846.

41. M. D. Toksarı, A branch and bound algorithm for minimizing makespan on a single machine with unequal release times under learning effect and deteriorating jobs. *Computers and Operations Research* **38** (2011), no. 9, 1361–1365.

42. M. D. Toksarı and E. Güner, Minimizing the earliness/tardiness costs on parallel machine with learning effects and deteriorating jobs: a mixed nonlinear integer programming approach. *The International Journal of Advanced Manufacturing Technology* **38** (2008), no. 7–8, 801–808.

43. M. D. Toksarı and E. Güner, Parallel machine earliness/tardiness scheduling problem under the effects of position based learning and linear/nonlinear deterioration. *Computers and Operations Research* **36** (2009), no. 8, 2394–2417.

44. M. D. Toksarı and E. Güner, Parallel machine scheduling problem to minimize the earliness/tardiness costs with learning effect and deteriorating jobs. *The International Journal of Advanced Manufacturing Technology* **21** (2010), no. 6, 843–851.

45. M. D. Toksarı and E. Güner, The common due-date early/tardy scheduling problem on a parallel machine under the effects of time-dependent learning and linear and nonlinear deterioration. *Expert Systems with Applications* **37** (2010), no. 7, 92–112.

46. M. D. Toksarı, D. Oron and E. Güner, Single machine scheduling problems under the effects of nonlinear deterioration and time-dependent learning. *Mathematical and Computer Modelling* **50** (2009), no. 3–4, 401–406.

47. J-B. Wang, A note on scheduling problems with learning effect and deteriorating jobs. *International Journal of Systems Science* **37** (2006), no. 12, 827–833.

48. J-B. Wang, Single-machine scheduling problems with the effects of learning and deterioration. *Omega* **35** (2007), no. 4, 397–402.

49. J-B. Wang, Single machine scheduling with a time-dependent learning effect and deteriorating jobs. *Journal of the Operational Research Society* **60** (2009), no. 4, 583–586.

50. J-B. Wang and T-C. E. Cheng, Scheduling problems with the effects of deterioration and learning, *Asia-Pacific Journal of Operational Research* **24** (2007), no. 2, 245–261.

51. J-B. Wang and Q. Guo, A due-date assignment problem with learning effect and deteriorating jobs. *Applied Mathematical Modelling* **34** (2010), no. 2, 309–313.

52. J-B. Wang, C-J. Hsu and D-L. Yang, Single-machine scheduling with effects of exponential learning and general deterioration. *Applied Mathematical Modelling* **37** (2013), no. 4, 2293–2299.

53. J-B. Wang and L-L. Liu, Two-machine flow shop problem with effects of deterioration and learning. *Computers and Industrial Engineering* **57** (2009), no. 3, 1114–1121.

54. J-B. Wang and C. Wang, Single-machine due-window assignment problem with learning effect and deteriorating jobs. *Applied Mathematical Modelling* **35** (2011), no. 8, 4017–4022.

55. J-B. Wang and J-J. Wang, Flowshop scheduling with a general exponential learning effect. *Computers and Operations Research* **43** (2014), 292–308.

56. J-B. Wang, M-Z. Wang and P. Ji, Single machine total completion time minimization scheduling with a time-dependent learning effect and deteriorating jobs. *International Journal of Systems Science* **43** (2012), no. 5, 861–868.

57. L-Y. Wang and E-M. Feng, A note on single-machine scheduling problems with the effects of deterioration and learning. *International Journal of Advanced Manufacturing Technology* **59** (2012), no. 5–8, 539–545.

58. L-Y. Wang, J-B. Wang, W-J. Gao, X. Huang and E-M. Feng, Two single-machine scheduling problems with the effects of deterioration and learning. *The International Journal of Advanced Manufacturing Technology* **46** (2010), no. 5–8, 715–720.

59. X. Wang and T-C. E. Cheng, Single-machine scheduling with deteriorating jobs and learning effects to minimize the makespan. *European Journal of Operational Research* **178** (2007), no. 1, 57–70.

60. Y-B. Wu, X. Huang, L. Li and P. Ji, A note on 'Scheduling problems with the effects of deterioration and learning'. *Asia-Pacific Journal of Operational Research* **31** (2014), no. 3, 1450011.

61. Y-B. Wu, M-Z. Wang and J-B. Wang, Some single-machine scheduling with both learning and deterioration effects. *Applied Mathematical Modelling* **35** (2011), no. 8, 3731–3736.

62. D-L. Yang and W-H. Kuo, Some scheduling problems with deteriorating jobs and learning effects. *Computers and Industrial Engineering* **58** (2010), no. 1, 25–28.

63. S-J. Yang and D-L. Yang, Single-machine group scheduling problems under the effects of deterioration and learning. *Computers and Industrial Engineering* **58** (2010), no. 4, 754–758.

64. D-L. Yang and W-H. Kuo, Single-machine scheduling with both deterioration and learning effects. *Annals of Operations Research* **172** (2009), no. 1, 315–327.

65. S-J. Yang, Group scheduling problems with simultaneous considerations of learning and deterioration effects on a single-machine. *Applied Mathematical Modelling* **35** (2011), no. 8, 4008–4016.

66. S-W. Yang, L. Wan and Y. Na, Research on single machine SLK/DIF due window assignment problem with learning effect. *Applied Mathematical Modelling* **39** (2015), 4593–4598.

67. Y. Yin and D. Xu, Some single-machine scheduling problems with general effects of learning and deterioration. *Computers and Mathematics with Applications* **61** (2011), no. 1, 100–108.

68. Y. Yin, D. Xu and X. Huang, Notes on 'Scheduling problems with general effects of deterioration and learning'. *Information Sciences* **195** (2012), 296–297.

69. X. Zhan and G. Yan, Single-machine group scheduling problems with deteriorated and learning effect. *Applied Mathematics and Computation* **216** (2010), no. 4, 1259–1266.

70. X. Zhang, G. Yan, W. Huang and G. Tang, Single-machine scheduling problems with time and position dependent processing times. *Annals of Operations Research* **186** (2011), no. 1, 345–356.

Scheduling problems with time-dependent parameters

71. P. Cai, J-Y. Cai and A.V. Naik, Efficient algorithms for a scheduling problem and its applications to illicit drug market crackdowns. *Journal of Combinatorial Optimization* **1** (1998), no. 4, 367–376.
72. E. Bampis and A. Kononov, On the approximability of scheduling multiprocessor tasks with time-dependent processor and time requirements. In: *Proceedings of the 15th International Symposium on Parallel and Distributed Processing*, IEEE Press, 2001, pp. 2144–2151.
73. A. Janiak and M. Y. Kovalyov, Scheduling jobs in a contaminated area: a model and heuristic algorithms. *Journal of the Operational Research Society* **57** (2008), no. 7, 977–987.
74. A. Janiak and M. Y. Kovalyov, Scheduling in a contaminated area: a model and polynomial algorithms. *European Journal of Operational Research* **173** (2006), no. 1, 125–132.
75. K. Klamroth and M. Wiecek, A time-dependent multiple criteria single-machine scheduling problem. *European Journal of Operational Research* **135** (2001), no. 1, 17–26.
76. T. G. Voutsinas and C. P. Pappis, Scheduling jobs with values exponentially deteriorating over time. *International Journal of Production Economics* **79** (2002), no. 3, 163–169.

Other problems with time-dependent parameters

77. S. Browne and U. Yechiali. Scheduling deteriorating jobs on a single processor. *Operations Research* **38** (1990), no. 3, 495–498.
78. F. Jaehn and H. A. Sedding, Scheduling with time-dependent discrepancy times. *Journal of Scheduling* **19** (2016), 737–757.
79. G. Mosheiov, Scheduling jobs under simple linear deterioration. *Computers and Operations Research* **21** (1994), no. 6, 653–659.
80. S. Shakeri and R. Logendran, A mathematical programming-based scheduling framework for multitasking environments. *European Journal of the Operational Research* **176** (2007), 193–209.

Single machine time-dependent scheduling problems

81. A. Agnetis, J-C. Billaut, S. Gawiejnowicz, D. Pacciarelli and A. Soukhal, *Multiagent Scheduling: Models and Algorithms*. Berlin-Heidelberg: Springer 2014.
82. B. Alidaee, A heuristic solution procedure to minimize makespan on a single machine with non-linear cost functions. *Journal of the Operational Research Society* **41** (1990), no. 11, 1065–1068.
83. B. Alidaee and N. K. Womer, Scheduling with time dependent processing times: Review and extensions. *Journal of the Operational Research Society* **50** (1999), no. 6, 711–720.

84. J. Błażewicz, K. Ecker, E. Pesch, G. Schmidt, M. Sterna and J. Węglarz, *Handbook of Scheduling: From Applications to Theory.* Berlin-Heidelberg: Springer 2019.

85. T.-C. E. Cheng, Q. Ding, M. Y. Kovalyov, A. Bachman and A. Janiak, Scheduling jobs with piecewise linear decreasing processing times. *Naval Research Logistics* **50** (2003), no. 6, 531–554.

86. T.-C. E. Cheng, Q. Ding and B. M. T. Lin, A concise survey of scheduling with time-dependent processing times. *European Journal of Operational Research* **152** (2004), no. 1, 1–13.

87. T.-C. E. Cheng, Y. He, H. Hoogeveen, M. Ji and G. Woeginger, Scheduling with step-improving processing times. *Operations Research Letters* **34** (2006), no. 1, 37–40.

88. S. Gawiejnowicz, Brief survey of continuous models of scheduling. *Foundations of Computing and Decision Sciences* **21** (1996), no. 2, 81–100.

89. S. Gawiejnowicz, *Time-Dependent Scheduling.* Berlin-Heidelberg: Springer 2008.

90. S. Gawiejnowicz, A review of four decades of time-dependent scheduling: main results, new directions, and open problems. *Journal of Scheduling,* 2020, https://doi.org/10.1007/s10951-019-00630-w.

91. K. D. Glazebrook, Single-machine scheduling of stochastic jobs subject to deterioration or delay. *Naval Research Logistics* **39** (1992), no. 5, 613–633.

92. J. N. D. Gupta and S. K. Gupta, Single facility scheduling with nonlinear processing times. *Computers and Industrial Engineering* **14** (1988), no. 4, 387–393.

93. A. Janiak and M. Y. Kovalyov, Job sequencing with exponential functions of processing times. *Informatica* **17** (2006), no. 1, 13–24.

94. A. Kononov, Single machine scheduling problems with processing times proportional to an arbitrary function. *Discrete Analysis and Operations Research* **5** (1998), no. 3, 17–37 (in Russian).

95. W. Kubiak and S. L. van de Velde, Scheduling deteriorating jobs to minimize makespan. *Naval Research Logistics* **45** (1998), no. 5, 511–523.

96. O. I. Melnikov and Y. M. Shafransky, Parametric problem of scheduling theory. *Kibernetika* **6** (1979), 53–57 (in Russian). English translation: *Cybernetics and System Analysis* **15** (1980), 352–357.

97. V. A. Strusevich and K. Rustogi, *Scheduling with Times-Changing Effects and Rate-Modifying Activities.* Berlin-Heidelberg: Springer 2017.

98. V. S. Tanaev, V. S. Gordon and Y. M. Shafransky, *Scheduling Theory: Single-Stage Systems.* Dordrecht: Kluwer 1994.

99. J-B. Wang and Z-Q. Xia, Scheduling jobs under decreasing linear deterioration. *Information Processing Letters* **94** (2005), no. 2, 63–69.

100. Y.-Q. Yin, D.-H. Xu, Some single-machine scheduling problems with general effects of learning and deterioration, Comput. Math. Appl., 61, 2011, 100–108.

Parallel machine time-dependent scheduling problems

101. J. Błażewicz, K. Ecker, E. Pesch, G. Schmidt, M. Sterna and J. Węglarz, *Handbook of Scheduling: From Applications to Theory.* Berlin-Heidelberg: Springer 2019.

102. T.-C. E. Cheng, Q. Ding and B. M. T. Lin, A concise survey of scheduling with time-dependent processing times. *European Journal of Operational Research* **152** (2004), no. 1, 1–13.

103. S. Gawiejnowicz, *Time-Dependent Scheduling*. Berlin-Heidelberg: Springer 2008.

104. S. Gawiejnowicz, A review of four decades of time-dependent scheduling: main results, new directions, and open problems. *Journal of Scheduling*, 2020, https://doi.org/10.1007/s10951-019-00630-w.

105. V. A. Strusevich and K. Rustogi, *Scheduling with Times-Changing Effects and Rate-Modifying Activities*. Berlin-Heidelberg: Springer 2017.

Dedicated machine time-dependent scheduling problems

106. J. Błażewicz, K. Ecker, E. Pesch, G. Schmidt, M. Sterna and J. Węglarz, *Handbook of Scheduling: From Applications to Theory*. Berlin-Heidelberg: Springer 2019.

107. T-C. E. Cheng, Q. Ding and B. M. T. Lin, A concise survey of scheduling with time-dependent processing times. *European Journal of Operational Research* **152** (2004), no 1, 1–13.

108. S. Gawiejnowicz, *Time-Dependent Scheduling*. Berlin-Heidelberg: Springer 2008.

109. S. Gawiejnowicz, A review of four decades of time-dependent scheduling: main results, new directions, and open problems. *Journal of Scheduling*, 2020, https://doi.org/10.1007/s10951-019-00630-w.

110. V. S. Tanaev, Y. N. Sotskov and V. A. Strusevich, *Scheduling Theory: Multistage Systems*. Dordrecht: Kluwer 1994.

POLYNOMIAL PROBLEMS

Polynomial single machine problems

This chapter opens the third part of the book, devoted to polynomially solvable time-dependent scheduling problems. The results presented in this part constitute the backbone of time-dependent scheduling and are often used as tools for solving various time-dependent scheduling problems.

This part of the book is composed of three chapters. In Chap. 7, we focus on single machine time-dependent scheduling problems. In Chaps. 8 and 9, we discuss parallel machine and dedicated machine time-dependent scheduling problems, respectively.

Chapter 7 is composed of four sections. In Sect. 7.1, we present the results concerning minimization of the $C_{\max}$ criterion. In Sect. 7.2, we address the problems of minimization of the $\sum C_j$ criterion. In Sect. 7.3, we study the results concerning minimization of the $L_{\max}$ criterion. In Sect. 7.4, we focus on the results concerning minimization of other criteria than $C_{\max}$, $\sum C_j$ and $L_{\max}$. The chapter is completed with a list of references.

7.1 Minimizing the maximum completion time

In this section, we present the polynomially solvable results concerning minimization of the $C_{\max}$ criterion.

7.1.1 Proportional deterioration

First we consider the single machine time-dependent scheduling problems with proportional job processing times (6.7), in which neither non-zero ready times nor finite deadlines have been defined, i.e., $r_j = 0$ and $d_j = +\infty$ for $1 \leqslant j \leqslant n$.

Theorem 7.1. (Mosheiov [25]) *Problem* $1|p_j = b_j t|C_{\max}$ *is solvable in* $O(n)$ *time, and the maximum completion time does not depend on the schedule of jobs.*

© Springer-Verlag GmbH Germany, part of Springer Nature 2020
S. Gawiejnowicz, *Models and Algorithms of Time-Dependent Scheduling*,
Monographs in Theoretical Computer Science. An EATCS Series,
https://doi.org/10.1007/978-3-662-59362-2_7

Proof. Let σ be an arbitrary schedule and let $[i]$ be the index of the ith job in σ. Since

$$C_{[j]}(\sigma) = S_1 \prod_{i=1}^{j}(1 + b_{[i]}) = t_0 \prod_{i=1}^{j}(1 + b_{[i]}), \qquad (7.1)$$

we have

$$C_{\max} = C_{[n]}(\sigma) = t_0 \prod_{i=1}^{n}(1 + b_{[i]}). \qquad (7.2)$$

Since the product $\prod_{i=1}^{n}(1 + b_{[i]})$ can be calculated in $O(n)$ time and since it is independent of the schedule, the result follows.

An alternative proof uses the pairwise job interchange argument. We consider schedule σ' in which job $J_{[i]}$ is immediately followed by job $J_{[j]}$, and schedule σ'' in which the jobs are in the reverse order. Since we have $C_{[j]}(\sigma') - C_{[i]}(\sigma'') = 0$, the result follows. ∎

Remark 7.2. For simplicity of presentation, unless otherwise stated, in order to identify the position of a job in a schedule we will write j instead of $[j]$. Hence, e.g., symbols $b_j \equiv b_{[j]}$ and $C_j \equiv C_{[j]}$ will denote, respectively, deterioration rate and the completion time of a job at the jth position in a schedule.

Problem $1|p_j = b_j t|C_{\max}$ can be a basis for more general problems. Below we consider a few of them.

Cheng and Sun [5] reformulated problem $1|p_j = b_j t|C_{\max}$ to obtain the following *time-dependent batch scheduling problem*. We are given n jobs $J_1, J_2, \ldots, J_n$ which are available starting from time $t_0 = 0$. The jobs are classified into m groups $G_1, G_2, \ldots, G_m$. Group G_i is composed of k_i jobs, where $1 \leqslant i \leqslant m$ and $\sum_{i=1}^{m} k_i = n$. Jobs in the same group G_i, $1 \leqslant i \leqslant m$, are processed consecutively and without idle times. The setup time θ_i precedes the processing of the group G_i, $1 \leqslant i \leqslant m$. The processing time of the jth job in group G_i is in the form of $p_{ij} = b_{ij} t$, where $b_{ij} > 0$ for $1 \leqslant i \leqslant m$ and $1 \leqslant j \leqslant k_i$. The aim is to find a sequence of groups and a sequence of jobs in each group which together minimize the $C_{\max}$ criterion.

Remark 7.3. The assumption that jobs in the same group are processed consecutively and without idle times is called *group technology*; see Potts and Kovalyov [26], Tanaev et al. [29] for more details.

Remark 7.4. Batch scheduling problems with time-dependent job processing times are also considered by Barketau et al. [2] and Leung et al. [23], where the processing time of a job depends on the *waiting time* of the job.

Remark 7.5. If we put symbols GT and θ_i in the second field of a symbol in the $\alpha|\beta|\gamma$ notation (cf. Sect. 4.6), the whole symbol will denote a batch scheduling problem with group technology and setup times. For example, symbol $1|p_{ij} = b_{ij}t, \theta_i, GT|C_{\max}$ will denote the above described time-dependent batch scheduling problem.

For problem $1|p_{ij} = b_{ij}t, \theta_i, GT|C_{\max}$, Cheng and Sun [5] proposed the following algorithm.

Algorithm 7.1. for problem $1|p_{ij} = b_{ij}t, \theta_i, GT|C_{\max}$

---

1 **Input** : sequences $(\theta_1, \theta_2, \ldots, \theta_m)$, (b_{ij}) for $1 \leqslant i \leqslant m$ and
$\qquad 1 \leqslant j \leqslant k_i$

2 **Output:** an optimal schedule
$\quad \triangleright$ Step 1

3 **for** $i \leftarrow 1$ **to** m **do**

4 $\qquad B_i \leftarrow \prod\limits_{j=1}^{k_i} (1 + b_{ij})$;

   end
$\quad \triangleright$ Step 2

5 Schedule groups of jobs in non-decreasing order of the ratios $\frac{\theta_i B_i}{B_i - 1}$;
$\quad \triangleright$ Step 3

6 **for** $i \leftarrow 1$ **to** m **do**

7 $\qquad$ Schedule jobs in group G_i in an arbitrary order;

   end

8 **return.**

Theorem 7.6. (Cheng and Sun [5]) *Problem $1|p_{ij} = b_{ij}t, \theta_i, GT|C_{\max}$ is solvable in $O(n \log n)$ time by Algorithm 7.1.*

Proof. Let G_i, $1 \leqslant i \leqslant m$, denote the ith group of jobs. Notice that the completion time C_{i,k_i} of the last job in group G_i is given by the equation

$$C_{i,k_i} = (C_{i-1,k_{i-1}} + \theta_i) \prod_{j=1}^{k_i} (1 + b_{ij}),$$

where $1 \leqslant i \leqslant m$ and $C_{0,0} := 0$.

Let $\sigma^1 = (G_1, G_2, \ldots, G_{i-1}, G_i, G_{i+1}, \ldots, G_m)$ be a schedule such that

$$\frac{\theta_i \prod\limits_{j=1}^{k_i} (1 + b_{ij})}{\prod\limits_{j=1}^{k_i} (1 + b_{ij}) - 1} \geqslant \frac{\theta_{i+1} \prod\limits_{j=1}^{k_{i+1}} (1 + b_{i+1,j})}{\prod\limits_{j=1}^{k_{i+1}} (1 + b_{i+1,j}) - 1} \tag{7.3}$$

for some $1 \leqslant i \leqslant m - 1$. Let $\sigma^2 = (G_1, G_2, \ldots, G_{i-1}, G_{i+1}, G_i, \ldots, G_m)$ be the schedule obtained from σ^1 by mutual exchange of groups G_i and G_{i+1}. Since

$$C_{i+1,k_{i+1}}(\sigma^1) - C_{i,k_i}(\sigma^2) = \left(\prod_{j=1}^{k_{i+1}}(1 + b_{i+1,j}) - 1\right)\left(\prod_{j=1}^{k_i}(1 + b_{ij}) - 1\right) \times$$

$$\times \left(\frac{\theta_i \prod_{j=1}^{k_i}(1+b_{ij})}{\prod_{j=1}^{k_i}(1+b_{ij})-1} - \frac{\theta_{i+1} \prod_{j=1}^{k_{i+1}}(1+b_{ij})}{\prod_{j=1}^{k_{i+1}}(1+b_{ij})-1}\right),$$

by (7.3) the difference is non-positive. Hence σ^2 is better than σ^1.

Repeating, if necessary, the above described exchange, we obtain a schedule in which all groups of jobs are in non-decreasing order of the ratios $\dfrac{\theta_i \prod_{j=1}^{k_i} b_{ij}}{\prod_{j=1}^{k_i} b_{ij}-1}$.

To complete the proof, it is sufficient to note that by Theorem 7.1 the sequence of jobs in each group is immaterial. The overall time complexity of Algorithm 7.1 is $O(n \log n)$, since Step 1 needs $O(m \log m)$ time, Step 2 needs $O(n)$ time and $m = O(n)$. ∎

Theorem 7.6 was generalized by Wu et al. [36], who considered time-dependent setup times, $\theta_i = \delta_i t$, where $\delta_i > 0$ for $1 \leqslant i \leqslant m$.

Theorem 7.7. (Wu et al. [36]) *Problem* $1|GT, p_{ij} = b_{ij}t, \theta_i = \delta_i|C_{\max}$ *is solvable in* $O(n)$ *time, and the maximum completion time does not depend either on the schedule of jobs in a group or on the order of groups.*

Proof. Consider an arbitrary schedule σ for problem $1|GT, p_{ij} = b_{ij}t, \theta_i = \delta_i|C_{\max}$. Without loss of generality we can assume that σ is in the form of $(\sigma_{1,[1]}, \sigma_{1,[2]}, \ldots, \sigma_{1,[n_1]}, \sigma_{2,[1]}, \sigma_{2,[2]}, \ldots, \sigma_{2,[n_2]}, \ldots, \sigma_{m,[1]}, \sigma_{m,[2]}, \ldots, \sigma_{m,[n_m]})$, where $\sum_{i=1}^m n_i = n$. Since, by Theorem 7.1,

$$C_{\max}(\sigma) = t_0 \prod_{i=1}^m (1 + \delta_i) \prod_{i=1}^m \prod_{j=1}^{n_i} (1 + b_{i,[j]}) \tag{7.4}$$

and since the value of the right side of (7.4) does not depend on the order of jobs, the result follows. ∎

Remark 7.8. Scheduling deteriorating jobs with setup times and batch scheduling of deteriorating jobs are rarely studied in time-dependent scheduling. In the classical scheduling, both these topics have been studied since early 1960s and have an extensive literature (see, e.g., the reviews by Allahverdi et al. [1], Potts and Kovalyov [26] and Webster and Baker [35] and the book by Tanaev et al. [29] for details). ◇

7.1.2 Proportional-linear deterioration

Theorem 7.1 can be generalized for the case of proportional-linear processing times (6.9).

Theorem 7.9. (Kononov [18]) *If inequalities*

$$a + bt_0 > 0 \tag{7.5}$$

and

$$1 + b_j b > 0 \quad \text{for} \quad 1 \leqslant j \leqslant n \tag{7.6}$$

hold, then problem $1|p_j = b_j(a + bt)|C_{\max}$ *is solvable in* $O(n)$ *time, the maximum completion time*

$$C_{\max} = \begin{cases} t_0 + a \sum\limits_{j=1}^{n} b_j, & \text{if } b = 0, \\ (t_0 + \frac{a}{b}) \prod\limits_{j=1}^{n} (1 + b_j b) - \frac{a}{b}, & \text{if } b \neq 0, \end{cases} \tag{7.7}$$

and it does not depend on the schedule of jobs.

Proof. By mathematical induction with respect to the number of jobs. □

Remark 7.10. A slightly different form of Theorem 7.9, without conditions (7.5) and (7.6) but with assumptions $a > 0, b > 0$, $b_j > 0$ for $1 \leqslant j \leqslant n$, was given by Zhao et al. [37, Theorem 1].

Theorem 7.9 was generalized for case when $p_j = b_j(1 + B_j t)$.

Theorem 7.11. (Strusevich and Rustogi [27]) *Problem* $1|p_j = b_j(1+B_j t)|C_{\max}$ *is solvable in* $O(n \log n)$ *time by scheduling jobs in non-increasing order of the deterioration rates* B_j.

Proof. The result can be proved by showing that function $\omega(j) = B_j$ is a 1-priority-generating function for the considered problem; see [27, Sect. 9.1] for details. ◇

Guo and Wang [14] generalized Theorem 7.6 to the case of proportional-linear job processing times. For problem $1|p_{ij} = b_{ij}(a + bt), \theta_i, GT|C_{\max}$, the authors proposed the following algorithm.

Algorithm 7.2. for problem $1|p_{ij} = b_{ij}(a + bt), \theta_i, GT|C_{\max}$

---

1 **Input** : sequences $(\theta_1, \theta_2, \ldots, \theta_m)$, (b_{ij}) for $1 \leqslant i \leqslant m$ and
          $1 \leqslant j \leqslant k_i$, numbers a, b

2 **Output:** an optimal schedule

▷ Step 1

3 Schedule groups of jobs in non-decreasing order of the ratios $\dfrac{\theta_i \prod_{j=1}^{k_i} b_{ij}}{\prod_{j=1}^{k_i} b_{ij} - 1}$;

▷ Step 2

4 **for** $i \leftarrow 1$ **to** m **do**

5 | Schedule jobs in group G_i in an arbitrary order;
  end

6 **return.**

Theorem 7.12. (Guo and Wang [14]) *Problem* $1|p_{ij} = b_{ij}(a+bt), \theta_i, GT|C_{\max}$ *is solvable in* $O(n \log n)$ *time by Algorithm 7.2.*

Proof. Similar to the proof of Theorem 7.6.   □

Theorems 7.7 and 7.12 were generalized by Wang et al. [32]. The authors assumed that job processing times are in the form of (6.9) and setup times are proportional-linear, i.e., for $1 \leqslant i \leqslant m$ we have

$$\theta_i = \delta_i(a + bt). \tag{7.8}$$

Theorem 7.13. (Wang et al. [32]) *Problem* $1|p_{ij} = b_{ij}(a + bt), \theta_i = \delta_i(a + bt),$ $GT|C_{\max}$ *is solvable in* $O(n)$ *time, and the maximum completion time does not depend either on the schedule of jobs in a group or on the order of groups.*

Proof. Similar to the proof of Theorem 7.7; see [32, Theorem 1].   ◇

7.1.3 Linear deterioration

Time-dependent scheduling problems with linear job processing times (6.11) constitute a significant part of the time-dependent scheduling literature. We start this section with the case when all ready times and deadlines are equal.

The following result has been obtained independently by several authors. Since the authors used different proof techniques, we will shortly describe the approaches which have been applied in order to prove the result.

Theorem 7.14. (Gawiejnowicz and Pankowska [13]; Gupta and Gupta [15]; Tanaev et al. [28]; Wajs [30]) *Problem* $1|p_j = a_j + b_j t|C_{\max}$ *is solvable in* $O(n \log n)$ *time by scheduling jobs in non-increasing order of the ratios* $\frac{b_j}{a_j}$.

Proof. Let us notice that for a given schedule σ the equality

$$C_j(\sigma) = \sum_{i=1}^{j} a_{\sigma_i} \prod_{k=i+1}^{j} (1 + b_{\sigma_k}), \tag{7.9}$$

holds, where $1 \leqslant j \leqslant n$. (Eq. (7.9) can be proved by mathematical induction with respect to j.) Therefore, $C_{\max}(\sigma) = C_n(\sigma) = \sum_{i=1}^{n} a_{\sigma_i} \prod_{k=i+1}^{n}(1 + b_{\sigma_k})$.

The first and simplest proof of Theorem 7.14 uses the pairwise job interchange argument: we assume that in schedule σ' job J_i precedes job J_j and in schedule σ'' job J_i follows job J_j. Next, we calculate the difference between $C_{\max}(\sigma') \equiv C_n(\sigma')$ and $C_{\max}(\sigma'') \equiv C_n(\sigma'')$. Finally, we show that the difference does not depend on time. (This approach has been used by Gupta and Gupta [15] and Wajs [30].)

In the second proof, Theorem 7.14 follows from (7.9) and Lemma 1.2 (a) as a corollary.

The third proof of Theorem 7.14 uses a priority-generating function (see Definition 1.20). Let $C_{\max}(\sigma)$ denote the length of a schedule for a given job sequence σ. It has been proved (see [28, Chap. 3]) that function

$$\omega(\sigma) = \frac{\Psi(\sigma)}{C_{\max}(\sigma)},$$

where

$$\Psi(\sigma) = \sum_{j=1}^{n} b_{\pi_j}(1 + x_{\sigma_j}), x_{\sigma_1} = 0, x_{\sigma_j} = \sum_{i=1}^{j-1} b_{\sigma_i}(1 + x_{\sigma_i}),$$

is a priority function and the $C_{\max}$ criterion is a priority-generating function for problem $1|p_j = a_j + b_j t|C_{\max}$. Thus, by Remark 1.21 and Theorem 1.25, the optimal schedule for the problem can be obtained in $O(n \log n)$ time by scheduling jobs in non-increasing order of their priorities. (This approach was used by Tanaev et al. [28].)

The fourth way of proving Theorem 7.14 uses the following idea. Let us denote the set of linear functions which describe job processing times by $F := \{f_1, f_2, \ldots, f_n\}$, where $f_j = a_j + b_j t$ for $j \in J$. Let $\sigma = (\sigma_1, \sigma_2, \ldots, \sigma_n)$ and $\pi = (\pi_1, \pi_2, \ldots, \pi_n)$, $\sigma \neq \pi$, be permutations of elements of set I_n, and let $\preceq$ be an ordering relation on set J such that $i \preceq j \Leftrightarrow a_i b_j - a_j b_i \leqslant 0$. Let sequence $p_{\pi_1}, p_{\pi_2}, \ldots, p_{\pi_n}$ be defined in the following way: $p_{\pi_1} = f_{\pi_1}(0)$, $p_{\pi_2} = f_{\pi_2}(p_{\pi_1}), \ldots, p_{\pi_n} = f_{\pi_n}(\sum_{j=1}^{n-1} p_{\pi_j})$, where $f_{\pi_i} \in F$. Then

$$\sum_{j=1}^{n} p_{\sigma_j} = \min_{\pi \in \mathfrak{S}_n} \sum_{j=1}^{n} p_{\pi_j} \quad \Leftrightarrow \quad \sigma_1 \preceq \sigma_2 \preceq \cdots \preceq \sigma_n. \tag{7.10}$$

In other words, the optimal schedule for problem $1|p_j = a_j + b_j t|C_{\max}$ (equivalently, the permutation of job indices) is generated by non-decreasing sorting of job indices according to the $\preceq$ relation. The rest of this proof is technical: the main idea is to consider an expanded form of the formulae which describe the length of a schedule. (This approach, exploiting an ordering relation in the set of functions which describe job processing times, was applied by Gawiejnowicz and Pankowska [13].) ∎

Remark 7.15. Let us notice that by formula (7.9) we can easily prove that sequence $(C_j)_{j=0}^{n}$ is non-decreasing, since $C_{\sigma_{j-1}} - C_{\sigma_{j-1}} = a_{\sigma_j} + (1 + b_{\sigma_j})C_{\sigma_{j-1}} - C_{\sigma_{j-1}} = a_{\sigma_j} + b_{\sigma_j}C_{\sigma_{j-1}} > 0$ for $j = 1, 2, \ldots, n$.

Remark 7.16. Formula (7.9) is an extension of formula (7.1) to the case where $a_{\sigma_j} \neq 0$ for $1 \leqslant j \leqslant n$.

Remark 7.17. Formula (7.9), in turn, is a special case of the formula

$$C_j(\sigma) = \sum_{i=1}^{j} a_{\sigma_i} \prod_{k=i+1}^{j} (1 + b_{\sigma_k}) + t_0 \prod_{i=1}^{j}(1 + b_{\sigma_i}), \tag{7.11}$$

which describes $C_j(\sigma)$ in the case when $t_0 > 0$. Some authors give special cases of (7.11); see, e.g., Zhao et al. [38, Lemma 2].

Remark 7.18. Lemma 1.2 (a) was used by Browne and Yechiali [3] for deriving the expected value and the variance of a single machine schedule length for linearly deteriorating jobs. Namely, scheduling jobs in non-decreasing order of $\frac{E(a_j)}{b_j}$ minimizes the expected maximum completion time, and scheduling jobs in non-decreasing order of $\frac{Var(a_j)}{(1+b_j)^2-1}$ minimizes the variance of the maximum completion time, where $E(a_j)$ and $Var(a_j)$ are the expected maximum completion time and the variance of the maximum completion time for $1 \leqslant j \leqslant n$, respectively; see [3, Sect. 1].   ◇

By Theorem 7.14, we can construct the following scheduling algorithm.

Algorithm 7.3. for problem $1|p_j = a_j + b_j t|C_{\max}$

1 **Input** : sequence $((a_1, b_1), (a_2, b_2), \ldots, (a_n, b_n))$
2 **Output:** an optimal schedule
3 ▷ Step 1
4 Schedule jobs in non-increasing order of the ratios $\frac{b_j}{a_j}$;
5 **return** .

Algorithm 7.3 works only if $a_j \neq 0$ for $1 \leqslant j \leqslant n$. There arises a question whether we can consider a more general case of problem $1|p_j = a_j + b_j t|C_{\max}$, when some of the coefficients a_j and b_j are equal to 0 (but not both simultaneously). The next algorithm, proposed by Gawiejnowicz and Pankowska [13], which additionally covers the case when in (6.11) for $1 \leqslant j \leqslant n$ we have either $a_j = 0$ or $b_j = 0$, gives an affirmative answer to this question.

Algorithm 7.4. for problem $1|p_j = a_j + b_j t|C_{\max}$

1 **Input** : sequence $((a_1, b_1), (a_2, b_2), \ldots, (a_n, b_n))$
2 **Output:** an optimal schedule
  ▷ Step 1
3 Schedule jobs with $a_j = 0$ in an arbitrary order;
  ▷ Step 2
4 Schedule jobs with $a_j, b_j > 0$ in non-increasing order of the ratios $\frac{b_j}{a_j}$;
  ▷ Step 3
5 Schedule jobs with $b_j = 0$ in an arbitrary order;
6 **return** .

Example 7.19. Let us assume that jobs $J_1, J_2, \ldots, J_6$ have the processing times $p_1 = 2$, $p_2 = 3$, $p_3 = 3 + 4t$, $p_4 = 1 + 2t$, $p_5 = 3t$ and $p_6 = 2t$.

Since jobs with $a_j = 0$ and jobs with $b_j = 0$ can be scheduled in an arbitrary order, and since jobs with $a_j, b_j > 0$ must be scheduled in non-increasing order of $\frac{b_j}{a_j}$ ratios, Algorithm 7.4 generates four schedules: $(1, 2, 4, 3, 5, 6)$, $(2, 1, 4, 3, 5, 6)$, $(1, 2, 4, 3, 6, 5)$ and $(2, 1, 4, 3, 6, 5)$.   ◆

There exists yet another algorithm for problem $1|p_j = a_j + b_j t|C_{\max}$, also proposed by Gawiejnowicz and Pankowska [12] (see Algorithm 7.5).

Algorithm 7.5. for problem $1|p_j = a_j + b_j t|C_{\max}$

---

1   **Input**   : sequences $(a_1, a_2, \ldots, a_n), (b_1, b_2, \ldots, b_n)$
2   **Output:** an optimal schedule $\sigma^\star$
    ▷ Step 1
3   **for** $i \leftarrow 1$ **to** n **do**
4   **for** $j \leftarrow 1$ **to** n **do**
5       $A_{ij} \leftarrow a_j * b_i$;
      end
    end
    ▷ Step 2
6   **for** $i \leftarrow 1$ **to** n **do**
7   **for** $j \leftarrow 1$ **to** n **do**
8   **if** $(i \neq j)$ **then**
9         $B_{ij} \leftarrow 0$;
      else
10         $B_{ij} \leftarrow 1$;
      end
      end
    end
    ▷ Step 3
11   **for** $i \leftarrow 1$ **to** n **do**
12   **for** $j \leftarrow 1$ **to** n **do**
13   **if** $(A_{ij} \neq A_{ji})$ **then**
14   **if** $(A_{ij} < A_{ji})$ **then**
          $B_{ji} \leftarrow 0$;
        else
15           $B_{ij} \leftarrow 1$;
        end
      end
      end
    end
    ▷ Step 4
16   **for** $i \leftarrow 1$ **to** n **do**
17     $\sigma_i^\star \leftarrow 0$;
    for $j \leftarrow 1$ **to** n **do**
18       $\sigma_i^\star \leftarrow \sigma_i^\star + B_{ij}$;
    end
    end
19   **return** $\sigma^\star$.

Algorithm 7.5 uses two matrices, A and B. Matrix A contains all products in the form of $a_i * b_j$, $1 \leqslant i, j \leqslant n$. Matrix B is a $\{0,1\}$-matrix in which

$$B_{ij} = \begin{cases} 0 & \text{if } A_{ij} = a_i * b_j < a_j * b_i = A_{ji}, \\ 1 & \text{otherwise.} \end{cases}$$

The algorithm is based on equivalence (7.10) and hence it does not use the division operation. This is important in the case when some a_j are very small numbers, since then the calculation of the quotients $\frac{b_j}{a_j}$ may lead to numerical errors. Though the algorithm runs in $O(n^2)$ time, it needs only $O(k \log k)$ time in the case of adding k new jobs to set $\mathcal{J}$, while Algorithms 7.3 and 7.4 need $O((n+k)\log(n+k))$ time. (This is caused by the fact that we do not need to fill the whole matrices A and B again but only their new parts.)

Example 7.20. Let jobs J_1, J_2, J_3 have the processing times in the form of $p_1 = 1 + 2t$, $p_2 = 2 + t$, $p_3 = 5$. Then

$$A = \begin{bmatrix} 2 & 1 & 0 \\ 4 & 2 & 0 \\ 10 & 5 & 0 \end{bmatrix} \quad \text{and} \quad B = \begin{bmatrix} 1 & 0 & 0 \\ 1 & 1 & 0 \\ 1 & 1 & 1 \end{bmatrix}.$$

Hence, $\sigma^* = (1, 2, 3)$ is an optimal schedule, with $C_{\max}(\sigma^*) = 9$. The schedule is presented in Fig. 7.1. ◆

Fig. 7.1: Gantt chart for schedule σ^* in Example 7.20

Example 7.21. Let jobs J_1, J_2, J_3 now have the processing times $p_1 = 1 + 2t$, $p_2 = 2 + t$, $p_3 = 4 + 2t$. In this case

$$A = \begin{bmatrix} 2 & 1 & 2 \\ 4 & 2 & 4 \\ 8 & 4 & 8 \end{bmatrix} \quad \text{and} \quad B = \begin{bmatrix} 1 & 0 & 0 \\ 1 & 1 & 0 \\ 1 & 0 & 1 \end{bmatrix}.$$

The sums of elements of the second and the third rows of B are the same and equal to 2. This means that in the optimal schedule the order of jobs J_2 and J_3 is immaterial. Therefore, there are two optimal schedules, $\sigma^{*'} = (1, 2, 3)$ and $\sigma^{*''} = (1, 3, 2)$, which both are of length equal to 16. The schedules are presented in Fig. 7.2a and Fig. 7.2b, respectively. ◆

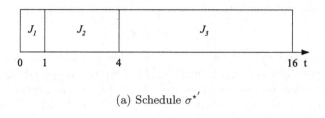

(a) Schedule $\sigma^{*'}$

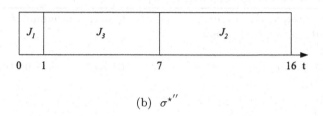

(b) $\sigma^{*''}$

Fig. 7.2: Optimal schedules in Example 7.21

Theorem 7.14 can be generalized to the case when $t_0 > 0$.

Theorem 7.22. (Gawiejnowicz and Lin [11]) *If $t_0 > 0$, then the optimal schedule for problem $1|p_j = a_j + b_j t|C_{\max}$ can be obtained by scheduling jobs in non-increasing order of the ratios $\frac{b_j}{a_j}$.*

Proof. Let σ be a schedule of linear jobs starting from time $t_0 > 0$. Then, by mathematical induction with respect to the value of i, we can show that the weighted completion time of the ith job in σ, $1 \leqslant i \leqslant n$, is equal to

$$w_{\sigma_i} C_i(\sigma) = w_{\sigma_i} \left(\sum_{j=1}^{i} A_{\sigma_j} \prod_{k=j+1}^{i} (1 + B_{\sigma_k}) + t_0 \prod_{j=1}^{i} (1 + B_{\sigma_j}) \right). \quad (7.12)$$

Assuming in (7.12) that $w_{\sigma_i} = 1$ for $1 \leqslant i \leqslant n$, the maximum completion time for schedule σ of linear jobs is equal to

$$C_{\max}(\sigma) = C_n(\sigma) = \sum_{j=1}^{n} a_{\sigma_j} \prod_{k=j+1}^{n} (1 + b_{\sigma_k}) + t_0 \prod_{j=1}^{n} (1 + B_{\sigma_j}). \quad (7.13)$$

Since the term $t_0 \prod_{j=1}^{n} (1 + b_{\sigma_j})$ has the same value for all possible job sequences, the value of $C_{\max}$ is minimized when the sum $\sum_{j=1}^{n} a_{\sigma_j} \prod_{k=j+1}^{n} (1 + b_{\sigma_k})$ is minimized. By Lemma 1.2 (a), the minimum is obtained when the elements of the sum are in non-increasing order of the ratios $\frac{b_j}{a_j}$. ∎

Simplifying $\mathcal{NP}$-hard problem $1|p_j = a_j + b_j t, d_j|C_{\max}$ (cf. Theorem 10.4), we can obtain polynomially solvable cases. Let $b_j = b$ for $1 \leqslant j \leqslant n$, i.e.,

$$p_j = a_j + bt, \tag{7.14}$$

where $b > 0$ and $a_j > 0$ for $1 \leqslant j \leqslant n$. This problem was considered for the first time by Cheng and Ding [6]. The authors proposed the following algorithm for job processing times given by (7.14) and for distinct deadlines. For a given $\sigma^i \in \tilde{\mathfrak{S}}_n$, let $J(\sigma^i)$ denote a set of jobs with indices from σ^i.

Algorithm 7.6. for problem $1|p_j = a_j + bt, d_j|C_{\max}$

1 **Input** : sequences $(a_1, a_2, \ldots, a_n)$, $(d_1, d_2, \ldots, d_n)$, number b
2 **Output:** an optimal schedule
  ▷ Step 1
3 Arrange jobs in non-decreasing order of the basic processing times a_j;
4 $\sigma^1 \leftarrow ([1], [2], \ldots, [n])$;
5 $\sigma^2 \leftarrow (\phi)$;
6 $C_{[0]} \leftarrow 0$;
7 $C \leftarrow 0$;
  ▷ Step 2
8 **for** $i \leftarrow 1$ **to** n **do**
9 $\quad \mid \quad C_{[i]} \leftarrow (1+b)C_{[i-1]} + a_{[i]}$;
  end
  ▷ Step 3
10 **while** $(C \neq C_{[n]})$ **do**
11 $\quad \mid \quad s \leftarrow C_{[n]}$;
12 $\quad \mid \quad$ **for** $i \leftarrow n$ **downto** 1 **do**
13 $\quad \mid \quad \quad \mid \quad t \leftarrow \max\{C_{[i]}, s\}$;
14 $\quad \mid \quad \quad \mid \quad$ Find job $J_{(i)} \in J(\sigma^1)$ with maximal $a_{(i)}$ and $d_{(i)} \geqslant t$;
15 $\quad \mid \quad \quad \mid \quad$ **if** (there exists no such $J_{(i)}$) **then**
16 $\quad \mid \quad \quad \mid \quad \quad \mid \quad$ write 'There exists no feasible schedule';
17 $\quad \mid \quad \quad \mid \quad \quad \mid \quad$ stop;
  $\quad \mid \quad \quad \mid \quad$ **else**
18 $\quad \mid \quad \quad \mid \quad \quad \mid \quad s \leftarrow \frac{t - a_{(i)}}{1+b}$;
19 $\quad \mid \quad \quad \mid \quad \quad \mid \quad \sigma^1 \leftarrow \sigma^1 \setminus \{(i)\}$;
20 $\quad \mid \quad \quad \mid \quad \quad \mid \quad \sigma^2 \leftarrow \sigma^2 \cup \{(i)\}$;
  $\quad \mid \quad \quad \mid \quad$ **end**
21 $\quad \mid \quad \quad \mid \quad C \leftarrow C_{[n]}$;
22 $\quad \mid \quad \quad \mid \quad \sigma^1 \leftarrow \sigma^2$;
23 $\quad \mid \quad \quad \mid \quad$ **for** $i \leftarrow 1$ **to** n **do**
24 $\quad \mid \quad \quad \mid \quad \quad \mid \quad C_{[i]} \leftarrow (1+b)C_{[i-1]} + a_{[i]}$;
  $\quad \mid \quad \quad \mid \quad$ **end**
  $\quad \mid \quad$ **end**
  end
25 **return** .

Theorem 7.23. (Cheng and Ding [6]) *Problem $1|p_j = a_j + bt, d_j|C_{\max}$ is solvable in $O(n^5)$ time by Algorithm 7.6.*

Proof. The optimality of the schedule generated by Algorithm 7.6 follows from some results concerning the so-called *canonical* schedules for the problem (see Definition 7.50 and [8, Sect. 3] for details).

The time complexity of Algorithm 7.6 can be established as follows. Step 1 and Step 2 can be completed in $O(n \log n)$ and $O(n)$ time, respectively. The 'while' loop in Step 3 needs $O(n^2)$ time, while the loop 'for' in this step needs $O(n^2)$ time. Finding job $J_{(i)} \in \sigma^1$ with maximal $a_{(i)}$ and $d_{(i)} \geqslant t$ needs $O(n)$ time. Therefore, the overall time complexity of Algorithm 7.6 is $O(n^5)$. ∎

If we simplify the latter problem further, it can be solved more efficiently.

Theorem 7.24. (Cheng and Ding [8]) *Problem $1|p_j = a + b_j t, d_j, b_j \in \{B_1, B_2\}|C_{\max}$ is solvable in $O(n \log n)$ time by an appropriately modified Algorithm 7.8.*

Proof. Similar to the proof of Theorem 7.53. □

7.1.4 Simple non-linear deterioration

In this section, we consider non-linear forms of job deterioration. The first result concerns simple general non-linear deterioration.

From the point of view of applications, the most interesting case is when function $f(t)$ is a non-decreasing function, i.e., when

$$\text{if } t_1 \leqslant t_2, \text{ then } f(t_1) \leqslant f(t_2), \tag{7.15}$$

In this case, the following result holds.

Theorem 7.25. (Melnikov and Shafransky [24]) *If $f(t)$ is an arbitrary function satisfying conditions (6.14)–(7.15), then problem $1|p_j = a_j + f(t)|C_{\max}$ is solvable in $O(n \log n)$ time by scheduling jobs in non-decreasing order of the basic processing times a_j.*

Proof. We apply the pairwise job interchange argument. Let schedule σ',

$$\sigma' = (\sigma_1, \ldots, \sigma_{i-1}, \sigma_i, \sigma_{i+1}, \sigma_{i+2}, \ldots, \sigma_n),$$

start from time $t_0 > 0$ and let two jobs, J_{σ_i} and $J_{\sigma_{i+1}}$, be such that $a_{\sigma_i} \geqslant a_{\sigma_{i+1}}$. Consider now schedule σ'',

$$\sigma'' = (\sigma_1, \ldots, \sigma_{i-1}, \sigma_{i+1}, \sigma_i, \sigma_{i+2}, \ldots, \sigma_n),$$

differing from σ' only in the order of jobs J_{σ_i} and $J_{\sigma_{i+1}}$. We will show that $C_{\max}(\sigma'') \leqslant C_{\max}(\sigma')$.

Note that since the first $i-1$ jobs in schedule σ' are the same as the first $i-1$ jobs in schedule σ'', job J_{σ_i} in schedule σ' and job $J_{\sigma_{i+1}}$ in schedule σ'' start at the same time t_i. Let us calculate the completion times $C_{\sigma_i}(\sigma')$ and $C_{\sigma_{i+1}}(\sigma'')$. We then have $C_{\sigma_i}(\sigma') = t_i + p_{\sigma_i}(t_i) = t_i + a_{\sigma_i} + f(t_i)$ and $C_{\sigma_{i+1}}(\sigma') = C_{\sigma_i}(\sigma') + p_{\sigma_{i+1}}(C_{\sigma_i}(\sigma')) = t_i + a_{\sigma_i} + f(t_i) + a_{\sigma_{i+1}} + f(t_i + a_{\sigma_i} + f(t_i))$.

Next, we have $C_{\sigma_{i+1}}(\sigma'') = t_i + p_{\sigma_{i+1}}(t_i) = t_i + a_{\sigma_{i+1}} + f(t_i)$ and $C_{\sigma_i}(\sigma'') = C_{\sigma_{i+1}}(\sigma'') + p_{\sigma_i}(C_{\sigma_{i+1}}(\sigma'')) = t_i + a_{\sigma_{i+1}} + f(t_i) + a_{\sigma_i} + f(t_i + a_{\sigma_{i+1}} + f(t_i))$. Therefore,

$$C_{\sigma_i}(\sigma'') - C_{\sigma_{i+1}}(\sigma') = f(t_i + a_{\sigma_{i+1}} + f(t_i)) - f(t_i + a_{\sigma_i} + f(t_i)) \leqslant 0,$$

since $f(t)$ is an increasing function and $a_{\sigma_i} \geqslant a_{\sigma_{i+1}}$ by assumption. From that it follows that $C_{\max}(\sigma'') - C_{\max}(\sigma') \leqslant 0$, since the $C_{\max}$ criterion is regular. Repeating, if necessary, the above mutual exchange for other pairs of jobs, we will obtain an optimal schedule in which all jobs are scheduled in non-decreasing order of the basic processing times a_j.   ∎

Remark 7.26. A special case of Theorem 7.25, for $f(t) := \sum_{i=1}^{m} \lambda_i t^{r_i}$ and $r_i \in [0, +\infty)$ for $1 \leqslant i \leqslant n$ is considered by Kuo and Yang [21, Props. 1–2]. ◇

If we assume that $f(t)$ is an arbitrary non-increasing function, i.e.,

$$\text{if } t_1 \leqslant t_2, \text{ then } f(t_1) \geqslant f(t_2), \tag{7.16}$$

$f(t)$ is differentiable and its first derivative is bounded,

$$\left| \frac{df}{dt} \right| \leqslant 1, \tag{7.17}$$

then the following result holds.

Theorem 7.27. (Melnikov and Shafransky [24]) *If $f(t)$ is an arbitrary differentiable function satisfying conditions (6.14), (7.16) and (7.17), then problem $1|p_j = a_j + f(t)|C_{\max}$ is solvable in $O(n \log n)$ time by scheduling jobs in non-increasing order of the basic processing times a_j.*

Proof. Since condition (7.17) holds, for any $t \geqslant 0$ the inequality

$$|f(t + \Delta t) - f(t)| \leqslant \Delta t$$

holds. Repeating the reasoning from the proof of Theorem 7.25, we obtain the result.   □

For simple general non-linear deterioration given by (6.17) the following result is known.

Theorem 7.28. (Kononov [18]) *If $f(t)$ is a convex (concave) function for $t \geqslant 0$ and conditions*

$$f(t_0) > 0 \qquad (7.18)$$

and

$$t_1 + b_j f(t_1) \leqslant t_2 + b_j f(t_2) \text{ for all } t_2 > t_1 \geqslant t_0 \text{ and all } J_j \in \mathcal{J} \qquad (7.19)$$

hold, then problem $1|p_j = b_j f(t)|C_{\max}$ is solvable in $O(n \log n)$ time by scheduling jobs in non-decreasing (non-increasing) order of the deterioration rates b_j.

Proof. The main idea is to prove that function $\omega_i = -b_i$ ($\omega_i = b_i$) is a 1-priority-generating function for the considered problem with convex (concave) function $f(t)$. Then, by Theorem 1.25, the result follows. □

Remark 7.29. The 1-priority-generating functions and priority functions were introduced in Definition 1.20.

Remark 7.30. Special cases of Theorem 7.28, for $f(t) := 1 + \sum_{i=1}^{m} \lambda_i t^{r_i}$ and $r_i \in [1, +\infty)$ or $r_i \in (0, 1)$ for $1 \leqslant i \leqslant n$ are considered by Kuo and Yang [21, Props. 3–5]. ◇

Remark 7.31. Polynomial algorithm for $f(t) = (a + bt)^k$, where $k \geqslant 0$, was proposed by Wang and Wang [33]. ◇

Remark 7.32. Polynomial algorithm for less general case than in [33], $f(t) = (1 + t)^k$, where $k > 0$, was proposed by Kuo and Yang [22]. ◇

7.1.5 General non-linear deterioration

In this section, we consider these forms of non-linear deterioration in which different job processing times are described by different functions.

Kunnathur and Gupta [20] proved a few properties of the single-machine time-dependent scheduling problem with job processing times given by (6.24).

Property 7.33. (Kunnathur and Gupta [20]) Let T be the earliest possible starting time for jobs in the set of unscheduled jobs, U, for problem $1|p_j = a_j + \max\{0, b_j(t - d_j)\}|C_{\max}$. If $d_j < T$ for all $J_k \in U$, then scheduling jobs from U in non-decreasing order of the ratios $\frac{a_j - b_j d_j}{b_j}$ minimizes $C_{\max}(U)$.

Proof. By the pairwise job interchange argument. □

Theorem 7.34. (Kunnathur and Gupta [20]) *Problem $1|p_j = a_j + \max\{0, b(t - d)\}|C_{\max}$ is solvable in $O(n \log n)$ time by scheduling jobs in non-decreasing order of the basic processing times a_j.*

Proof. The result is a corollary from Property 7.33 for $b_j = b$ and $d_j = d$ for $1 \leqslant j \leqslant n$. □

Property 7.35. (Kunnathur and Gupta [20]) Let S be a set of scheduled jobs, $U := \mathcal{J} \setminus S$ and $c_1 := C_{\max}(S)$. Then the maximum completion time $C_{\max}(S|U)$ for problem $1|p_j = a_j + \max\{0, b_j(t - d_j)\}|C_{\max}$ is not less than $c_1 + \sum_{J_j \in U}(a_j + \max\{0, b_j(c_1 - d_j)\})$.

Proof. The result follows from the fact that for any U the schedule $(S|U)$ starts with a subschedule S, and from the definition of job processing times (6.24). □

Property 7.36. (Kunnathur and Gupta [20]) Let S'' be the set of jobs scheduled after jobs from the set $S' := \mathcal{J} \setminus S''$. Then the maximum completion time for problem $1|p_j = a_j + \max\{0, b_j(t - d_j)\}|C_{\max}$ for sequences in the form of $(S'|S'')$ is not less than $\sum_{J_j \in S'} a_j + \sum_{J_j \in S''}(a_j + \max\{0, b_j(t - d_j)\})$.

Proof. Similar to the proof of Property 7.35. □

Property 7.37. (Kunnathur and Gupta [20]) Let S be the set of jobs for problem $1|p_j = a_j + \max\{0, b_j(t - d_j)\}|C_{\max}$, scheduled in non-decreasing order of the sums $a_j + d_j$. If $S_j > d_j$ for any job $J_j \in S$, then there does not exist a schedule σ such that $C_{\max}(\sigma) = \sum_{J_j \in \mathcal{J}} a_j$.

Proof. Since $C_{\max}(\sigma) = \sum_{J_j \in \mathcal{J}} a_j$ holds only for the schedules in which no job is tardy, the result follows. ■

Property 7.38. (Kunnathur and Gupta [20]) Let S be the set of jobs for problem $1|p_j = a_j + \max\{0, b_j(t - d_j)\}|C_{\max}$, scheduled in non-decreasing order of the sums $a_j + d_j$, let S_j be computed assuming that $b_j = 0$ for all $J_j \in S$ and let S' be the set S with job J_j such that $b_j > 0$. Then $C_{\max}(S')$ is not less than $a_j + \max_{J_j \in S}\{0, (S_j - d_j)\} \min_{J_j \in \mathcal{J}}\{b_j\}$.

Proof. If $b_j = 0$ for all $J_j \in S$, then scheduling jobs in non-decreasing order of the deadlines d_j minimizes the maximum tardiness. Since $\max_{J_j \in S}\{0, (S_j - d_j)\}$ equals the minimum value of the maximum tardiness and, simultaneously, equals the minimum value of the maximum delay in the starting time of any job in the case when $b_j > 0$ for a job $J_j \in \mathcal{J}$, the result follows. ■

Remark 7.39. Non-decreasing order of the deadlines d_j is alternatively called the *EDD order*.

Cai et al. [4] considered deteriorating job processing times given by (6.16) in the case when $f(t, t_0) := \max\{t - t_0, 0\}$.

Lemma 7.40. (Cai et al. [4]) *For a given schedule* $\sigma = (\sigma_1, \sigma_2, \ldots, \sigma_n)$ *for problem* $1|p_j = a_j + b_j \max\{t - t_0, 0\}|C_{\max}$, *the maximum completion time is equal to*

$$C_{\max}(\sigma) = t_0 + \left(\sum_{j=1}^{k} a_{\sigma_j} - t_0\right) \prod_{r=k+1}^{n}(1 + b_{\sigma_r}) + \sum_{j=k+1}^{n} a_{\sigma_j} \prod_{r=j+1}^{n}(1 + b_{\sigma_r}).$$

Proof. By mathematical induction with respect to n.   □

Theorem 7.41. (Cai et al. [4]) *Given a set of jobs that start at $t \leqslant t_0$, the maximum completion time for problem $1|p_j = a_j + b_j \max\{t - t_0, 0\}|C_{\max}$ is minimized by scheduling jobs that start after time t_0 in non-decreasing order of the ratios $\frac{a_j}{b_j}$.*

Proof. The result follows from Lemma 7.40 and Lemma 1.2 (a).   □

Theorem 7.42. (Cai et al. [4])
(a) *Problem $1|p_j = a + b_j \max\{t - t_0, 0\}|C_{\max}$ is solvable in $O(n \log n)$ time by scheduling jobs in non-increasing order of the deterioration rates b_j.*
(b) *Problem $1|p_j = a_j + b \max\{t - t_0, 0\}|C_{\max}$ is solvable in $O(n \log n)$ time by scheduling jobs in non-decreasing order of the basic processing times a_j.*
(c) *If $a_j = k b_j$ for $1 \leqslant j \leqslant n$ and a constant $k > 0$, then problem $1|p_j = a_j + b_j \max\{t - t_0, 0\}|C_{\max}$ is solvable in $O(n \log n)$ time by scheduling jobs in non-decreasing order of the deterioration rates b_j.*

Proof. (a),(b) The results are corollaries from Theorem 7.41.   □
(c) See [4, Theorem 4].   ◇

For deteriorating job processing times (6.26), in the case when $d_j = D$ for $1 \leqslant j \leqslant n$, Cheng and Ding [7] proposed dividing the set of jobs $\mathcal{J}$ into a number of chains, using the approach applied for problems $1|p_j \in \{a_j, b_j : a_j \leqslant b_j\}, d_j = D| \sum C_j$ (cf. Sect. 7.2.5) and $1|p_j \in \{a_j, b_j : a_j \leqslant b_j\}, d_j = D| \sum w_j C_j$ (cf. Sect. 7.4.5).

Theorem 7.43. (Cheng and Ding [7]) *If an instance of problem $1|p_j \in \{a_j, b_j : a_j \leqslant b_j\}, d_j = D|C_{\max}$ has a fixed number of chains, then the instance can be solved in polynomial time.*

Proof. Similar to the proof of Theorem 7.77.   □

Remark 7.44. Let us notice that since in problem $1|p_j \in \{a_j, b_j : a_j \leqslant b_j\}$, $d_j = D|C_{\max}$ jobs are independent, the chains from Theorem 7.43 are not related to job precedence constraints in the sense defined in Sect. 4.3.1.

7.1.6 Proportional-linear shortening

Wang and Xia [34] considered job processing times given by (6.32) with $a = 1$. Let $b_{\min}$ be defined as in inequality (6.39).

Theorem 7.45. (Wang and Xia [34]) *Problem $1|p_j = b_j(1 - bt)$, $b > 0$,*

$$b\left(\sum_{j=1}^{n} b_j - b_{\min}\right) < 1|C_{\max} \text{ is solvable in } O(n) \text{ time and the maximum com-}$$

pletion time $C_{\max}(\sigma) = \frac{1}{b}\left(1 - \prod_{j=1}^{n}(1 - b_{\sigma_j} b)\right)$ does not depend on the schedule of jobs.

Proof. By mathematical induction with respect to n.   □

Theorem 7.45 was generalized for case when $p_j = b_j(1 - B_j t)$.

Theorem 7.46. [97] *Problem $1|p_j = b_j(1-B_j t)|C_{\max}$ is solvable in $O(n \log n)$ time by scheduling jobs in non-decreasing order of deterioration rates B_j.*

Proof. The result can be proved by showing that $w(j) := -B_j$ is a 1-priority function for the considered problem (see [97, Sect. 9.1] for details).   ◇

7.1.7 Linear shortening

For problem $1|p_{ij} = a_{ij} - b_{ij}t, \theta_i, GT|C_{\max}$, Wang et al. [31] proposed the following algorithm.

Algorithm 7.7. for problem $1|p_{ij} = a_{ij} - b_{ij}t, \theta_i, GT|C_{\max}$

> **Input** : sequences (a_{ij}), (b_{ij}), (θ_i) for $1 \leqslant i \leqslant m$ and $1 \leqslant j \leqslant k_i$
> **Output:** an optimal schedule
> ▷ Step 1
1 **for** $i \leftarrow 1$ **to** m **do**
2 $\quad$ Arrange jobs in group G_i in non-increasing order of the ratios $\frac{b_{ij}}{a_{ij}}$;
3 $\quad$ Call the sequence $\sigma^{(i)}$;
> end
> ▷ Step 2
4 **for** $i \leftarrow 1$ **to** m **do**
5 $\quad$ Calculate $\rho(G_i) := \dfrac{\sum\limits_{j=1}^{k_i} a_{ij} \prod\limits_{k=j+1}^{k_i}(1-b_{ik}) + \theta_i \prod\limits_{j=1}^{k_i}(1-b_{ij})}{1 - \prod\limits_{j=1}^{k_i}(1-b_{ij})}$;
> end
> ▷ Step 3
6 **for** $i \leftarrow 1$ **to** m **do**
7 $\quad$| Schedule groups in non-increasing order of the ratios $\rho(G_i)$;
> end
> ▷ Step 4
8 $\sigma^\star \leftarrow (\sigma^{([1])}|\sigma^{([2])}|\ldots|\sigma^{([m])})$;
9 **return** $\sigma^\star$.

Theorem 7.47. (Wang et al. [31]) *Problem $1|p_{ij} = a_{ij} - b_{ij}t, \theta_i, GT|C_{\max}$ is solvable in $O(n \log n)$ time by Algorithm 7.7.*

Proof. By the pairwise job interchange; see [31, Thms. 1 and 2] for details. ◇

Remark 7.48. Cheng and Sun [5, Theorem 9] considered a special case of Theorem 7.47, with $a_{ij} = 1$ for $1 \leqslant i \leqslant m$ and $1 \leqslant j \leqslant k_i$, where $\sum_{i=1}^m k_i = n$.

Theorem 7.49. (Ho et al. [16]) *If conditions* (6.44), (6.45) *and* $b_j d_j \leqslant a_j \leqslant d_j$ *hold for* $1 \leqslant j \leqslant n$, *then problem* $1|p_j = a_j - b_j t, d_j = D|-$ *is solvable in* $O(n \log n)$ *time by scheduling jobs in non-increasing order of the ratios* $\frac{a_j}{b_j}$.

Proof. Since $d_j = D$ for $1 \leqslant j \leqslant n$, it is sufficient to start scheduling jobs from time $t = 0$, to apply Theorem 7.14 to construct a schedule and to check if the maximum completion time for the schedule does not exceed time $t = D$. $\square$

For the case when in job processing times given by (6.43) all shortening rates are equal,

$$p_j = a_j - bt \tag{7.20}$$

for $1 \leqslant j \leqslant n$, several results are known.

If we simplify problem $1|p_j = a_j - b_j t, d_j|C_{\max}$, assuming that $a_j = a$ and $b_j \in \{B_1, B_2\}$ for $1 \leqslant j \leqslant n$, then the new problem can be solved in polynomial time. Before we state the result, we introduce a new notion proposed by Cheng and Ding [8].

Definition 7.50. (A canonical schedule)
Given an instance of problem $1|p_j = a_j - b_j t, d_j, b_j \in \{B_1, B_2\}|C_{\max}$ *with* m *distinct deadlines* $D_1 < D_2 < \cdots < D_m$, *a schedule* σ *is called canonical if the jobs with the same* b_j *are scheduled in* σ *in the EDD order and the jobs in sets* $R_j := \{J_j \in \mathcal{J} : D_{j-1} < C_j \leqslant D_j\}$, *where* $1 \leqslant j \leqslant m$ *and* $D_0 := 0$, *are scheduled in non-decreasing order of the deterioration rates* b_j.

Lemma 7.51. (Cheng and Ding [8]) *If there exists a feasible (optimal) schedule for an instance of problem* $1|p_j = a_j - b_j t, d_j, b_j \in \{B_1, B_2\}|C_{\max}$, *then there exists a canonical feasible (optimal) schedule for the instance.*

Proof. See [8, Lemma 1]. ◇

Now we briefly describe the construction of a schedule for problem $1|p_j = a_j - b_j t, d_j, b_j \in \{B_1, B_2\}|C_{\max}$, [8]. The construction is composed of the following three steps.

In the first step, we schedule jobs with the same b_j value in the EDD order, obtaining two chains of jobs, $\mathcal{C}_1$ and $\mathcal{C}_2$.

In the second step, we construct a *middle* schedule $M(\mathcal{C}_1)$. To this end, we insert jobs from $\mathcal{C}_2$ into time interval $\langle 0, D_m \rangle$ backwards as follows. We assign the jobs with the largest deadline D_m to a subschedule σ^{1m} of $M(\mathcal{C}_1)$ in such a way that the jobs are started at some time T_m^1 and completed at D_m. Next, we assign the jobs with deadline D_{m-1} to a subschedule $\sigma^{1,m-1}$ in such a way that the jobs are started at some time T_{m-1}^1 and completed at $\min\{D_{m-1}, T_m^1\}$. We continue until all jobs have been assigned to subschedules σ^{1j} for $1 \leqslant j \leqslant k \leqslant m$. (Note that it must be $T_j^1 \geqslant D_{j-1}$, where $1 \leqslant j \leqslant k$ and $D_0 := 0$; hence, if some jobs cannot be put in $\sigma^{1,j}$, we move them to the set of late jobs $L(\mathcal{C}_1)$.) After the completion of this step we obtain the schedule $M(\mathcal{C}_1) := (\sigma^{11}|\sigma^{12}|\ldots|\sigma^{1k})$ in the form of $(\mathcal{I}_1|\mathcal{W}_1|\mathcal{I}_2|\mathcal{W}_2|\ldots|\mathcal{I}_k|\mathcal{W}_k)$,

$k \leqslant m$, where $\mathcal{I}_j$ and $\mathcal{W}_j$ denote the jth idle time and the jth period of continuous machine working time, $1 \leqslant j \leqslant k$, respectively.

In the third step, we construct the *final* schedule $F(\mathcal{C}_1, \mathcal{C}_2)$. To this end we insert jobs from $\mathcal{C}_1$ into the idle times of $M(\mathcal{C}_1)$ as follows. Starting from the beginning of $\mathcal{I}_1$, we assign jobs to $\mathcal{I}_1$ until an assigned job cannot meet its deadline. This job is moved to the set of late jobs $L(\mathcal{C}_2)$. We continue the procedure with the remaining jobs until the end of $\mathcal{I}_1$. Then we shift σ^{11} forward to connect it with σ^{21}, where σ^{21} denotes the schedule of jobs assigned to $\mathcal{I}_1$. We continue until all jobs have been assigned to idle times $\mathcal{I}_j$, $2 \leqslant j \leqslant k$. After the completion of this step we obtain the schedule $F(\mathcal{C}_1, \mathcal{C}_2) := (\sigma^{21}|\sigma^{11}|\sigma^{22}|\sigma^{12}|\cdots|\sigma^{2k}|\sigma^{1k})$ and the set of late jobs $L(\mathcal{C}_1, \mathcal{C}_2) := L(\mathcal{C}_1) \cup L(\mathcal{C}_2)$.

Lemma 7.52. Cheng and Ding [8]) *Let I be an instance of problem $1|p_j = a_j - b_j t, d_j, b_j \in \{B_1, B_2\}|C_{\max}$. Then*
(a) *if $L(I) = \emptyset$, then the final schedule is optimal;*
(b) *if $L(I) \neq \emptyset$, then for I there does not exist a feasible schedule.*

Proof. (a) See [8, Lemma 5].
   (b) See [8, Lemma 6].   ◇

On the basis of Lemma 7.52, Cheng and Ding proposed the following optimal algorithm for problem $1|p_j = a - b_j t, d_j, b_j \in \{B_1, B_2\}|C_{\max}$.

Algorithm 7.8. for problem $1|p_j = a - b_j t, d_j, b_j \in \{B_1, B_2\}|C_{\max}$

---

1 **Input** : numbers a, B_1, B_2, sequence $(d_1, d_2, \ldots, d_n)$
2 **Output:** an optimal schedule $\sigma^\star$
   ▷ Step 1
3 $\mathcal{C}_1 \leftarrow \{J_j \in \mathcal{J} : b_j = B_1\}$;
4 $\mathcal{C}_2 \leftarrow \{J_j \in \mathcal{J} : b_j = B_2\}$;
   ▷ Step 2
5 Construct schedules $M(\mathcal{C}_1)$, $F(\mathcal{C}_1, \mathcal{C}_2)$ and set $L(\mathcal{C}_1, \mathcal{C}_2)$;
6 $\sigma^\star \leftarrow (F|L)$;
   ▷ Step 3
7 **if** $(L \neq \emptyset)$ **then**
8 | write 'there exists no feasible schedule';
9 | return .
   else
10 | return $\sigma^\star$.
   end

Theorem 7.53. (Cheng and Ding [8]) *Problem $1|p_j = a - b_j t, d_j, b_j \in \{B_1, B_2\}|C_{\max}$ is solvable by Algorithm 7.8 in $O(n \log n)$ time.*

Proof. The correctness of Algorithm 7.8 follows from Lemma 7.52. Since no step needs more than $O(n \log n)$ time, the overall time complexity of the algorithm is $O(n \log n)$.   □

The problem when $d_j = +\infty$, $b_j = b$ and $r_j \neq 0$ for $1 \leqslant j \leqslant n$ has been considered by Cheng and Ding [9]. The authors proposed the following algorithm for this problem.

Algorithm 7.9. for problem $1|p_j = a_j - bt, r_j|C_{\max}$

---

1 **Input** : sequences $(a_1, a_2, \ldots, a_n)$, $(r_1, r_2, \ldots, r_n)$, number $b := \frac{v}{u}$
2 **Output:** an optimal schedule
3 $B_1 \leftarrow \max\limits_{1 \leqslant j \leqslant n} \{r_j\}$;

4 $B_2 \leftarrow u^{n-1} \left(\sum\limits_{j=1}^{n} a_j + B_1 \right)$;

▷ Step 2
5 **while** $(B_2 - B_1 > 1)$ **do**
6     $G' \leftarrow \lceil \frac{B_1 + B_2}{2} \rceil$;
7     $G \leftarrow \frac{G'}{u^{n-1}}$;
8   Use Lemma 19.43 to construct instance I of problem
       $1|p_j = a_j + bt, d_j|C_{\max}$;
9   Apply Algorithm 7.6 to instance I to find an optimal schedule;
10   **if** (*there exists an optimal schedule for I*) **then**
11 | $B_1 \leftarrow G'$;
    else
12 | $G \leftarrow \frac{B_2}{u^{n-1}}$;
    end
end
return.

Theorem 7.54. (Cheng and Ding [9]) *Problem* $1|p_j = a_j - bt, r_j|C_{\max}$ *is solvable in* $O(n^6 \log n)$ *time by Algorithm 7.9.*

Proof. Let us consider any instance I of problem $1|p_j = a_j - bt, r_j|C_{\max}$. By Lemma 19.43, for this instance there exists an instance I' of problem $1|p_j = a_j + bt, d_j|C_{\max}$. Changing iteratively the threshold value G and applying Algorithm 7.6 to I', we can check whether there exists an optimal schedule σ' for I'. Due to the symmetry between problems $1|p_j = a_j - bt, r_j|C_{\max}$ and $1|p_j = a_j + bt, d_j|C_{\max}$, from σ' we can construct an optimal schedule σ for I.

The value of G can be determined in at most $\log(B_2 - B_1) \equiv O(n \log n)$ time. In each iteration of the loop '**while**' at most $O(n^5)$ time is needed for execution of Algorithm 7.6. Therefore, the overall time complexity of Algorithm 7.9 is $O(n^6 \log n)$. ∎

7.1.8 Non-linear shortening

In this section, we consider single-machine time-dependent scheduling problems with non-linear shortening job processing times and the $C_{\max}$ criterion.

Cheng et al. [10] proved a few properties of optimal schedules of problem $1|p_j = a_j - b_j(t-y), y > 0, 0 < b_j < 1, Y < +\infty|C_{\max}$, where $\sum_{j=1}^{n} a_j > y$. The terminology used in formulations of the properties is equivalent to the terminology introduced by Kubiak and van de Velde [19] for bounded step-linear processing times (6.31), provided that $y \equiv d$ and $Y \equiv D$.

Property 7.55. (Cheng et al. [10]) The order of the early jobs and the order of the suspended jobs are immaterial.

Proof. Since both the early and suspended jobs have fixed processing times and the completion time of the last early (suspended) job does not depend on the order of the early (suspended) jobs, the result follows. ∎

Property 7.56. (Cheng et al. [10]) The tardy jobs are sequenced in non-increasing order of the ratios $\frac{a_j}{b_j}$.

Proof. The result is a corollary from Theorem 7.49 (c). □

Property 7.57. (Cheng et al. [10]) The pivotal job has a processing time not larger than that of any of the early jobs.

Proof. By the pairwise job interchange argument. □

Property 7.58. (Cheng et al. [10]) If $a_i \geqslant a_j$ and $b_i \leqslant b_j$, then job J_i precedes job J_j.

Proof. By the pairwise job interchange argument. □

Theorem 7.59. (Cheng et al. [10])
Problem $1|p_j = a_j - b_j(t-y), y > 0, 0 < b_j < 1, Y < +\infty|C_{\max}$ is solvable in $O(n \log n)$ time by scheduling jobs in (a) *non-increasing order of the basic processing times a_j if $b_j = b$ for $1 \leqslant j \leqslant n$;* (b) *non-decreasing order of the deterioration rates b_j if $a_j = a$ for $1 \leqslant j \leqslant n$.*

Proof. (a), (b) The results are corollaries from Property 7.58. □

7.1.9 Simple alteration

Jaehn and Sedding [17] proved that a special case of problem $1|p_j = a_j + b|m_j - M||C_{\max}$ is polynomially solvable.

Theorem 7.60. Jaehn and Sedding [17] *Problem $1|p_j = a_j + b|m_j - M||C_{\max}$ with non-fixed starting time of the first job is solvable in $O(n \log n)$ time.*

Proof. See [17, Sect. 6]. ◇

7.2 Minimizing the total completion time

In this section, we consider polynomially solvable single machine time-dependent scheduling problems with the $\sum C_j$ criterion.

7.2.1 Proportional deterioration

The single machine time-dependent scheduling problem with proportional processing times given by (6.7), equal ready times and equal deadlines is polynomially solvable.

Theorem 7.61. (Mosheiov [44]) *Problem* $1|p_j = b_j t| \sum C_j$ *is solvable in* $O(n \log n)$ *time by scheduling jobs in non-decreasing order of the deterioration rates* b_j.

Proof. First, note that by adding $C_j \equiv C_{[j]}(\sigma)$ from (7.1) for $1 \leqslant j \leqslant n$ we obtain the formula for the total completion time:

$$\sum_{j=1}^{n} C_j = S_1 \sum_{j=1}^{n} \prod_{k=1}^{j} (1 + b_{[k]}) = t_0 \sum_{j=1}^{n} \prod_{k=1}^{j} (1 + b_{[k]}). \qquad (7.21)$$

Now we can prove that the right side of formula (7.21) is minimized by sequencing b_j in non-decreasing order in two different ways.

The first way is to apply the pairwise job interchange argument. To this end we calculate the total completion time for schedule σ' in which job J_i is followed by job J_j and $a_i > a_j$.

Next, we show that schedule σ'', which is obtained from σ' by changing the order of jobs J_i and J_j, has a lower total completion time than σ'. Repeating the above reasoning for all such pairs of jobs, we obtain the result.

The second way of proving Theorem 7.61 is to apply Lemma 1.2 (a) to formula (7.21). ☐

Remark 7.62. The algorithm which schedules proportionally deteriorating jobs in non-decreasing order of the deterioration rates b_j is called *SDR* (*Smallest Deterioration Rate first*) algorithm.

Example 7.63. Let us consider an instance of problem $1|p_j = b_j t| \sum C_j$, where $n = 3$, $t_0 = 1$, $b_1 = 3$, $b_2 = 1$ and $b_1 = 2$. Then, the optimal schedule generated by algorithm SDR for this instance is $\sigma^* = (2, 3, 1)$. The optimal total completion time is $\sum C_j(\sigma^*) = 32$. ◆

Wu et al. [51] considered problem $1|p_j = b_j t| \sum C_j$ in a batch environment (cf. Theorem 7.7).

Theorem 7.64. (Wu et al. [51]) *In the optimal schedule for problem* $1|GT$, $p_{ij} = b_{ij} t, \theta_i = \delta_i | \sum C_j$ *in each group jobs are scheduled in non-decreasing order of the deterioration rates* b_{ij} *and groups are scheduled in non-decreasing order of the ratios* $\frac{(1+\delta_i) A_{n_i} - 1}{(1+\delta_i) B_{n_i}}$, *where* $A_{n_i} := \prod_{j=1}^{n_i} (1 + b_{ij})$ *and* $B_{n_i} := \sum_{k=1}^{n_i} \prod_{j=1}^{k} (1 + b_{ij})$ *for* $1 \leqslant i \leqslant m$.

Proof. The first part of the theorem is a corollary from Theorem 7.1. The second part can be proved by contradiction; see [51, Theorem 2].   ◇

On the basis of Theorem 7.64, Wu et al. proposed the following algorithm.

Algorithm 7.10. for problem $1|GT, p_{ij} = b_{ij}t, \theta_i = \delta_i| \sum C_j$

1 **Input** : sequences $(\delta_1, \delta_2, \ldots, \delta_m)$, (b_{ij}) for $1 \leqslant i \leqslant m$ and $1 \leqslant j \leqslant k_i$
2 **Output:** an optimal schedule σ^*
  ▷ Step 1
3 **for** $i \leftarrow 1$ **to** n **do**
4 | Arrange jobs in group G_i in non-decreasing order of the
  | deterioration rates b_{ij};
end
  ▷ Step 2
5 Arrange groups $G_1, G_2, \ldots, G_m$ in non-decreasing order of the ratios
  $\frac{(1+\delta_i)A_{n_i}-1}{(1+\delta_i)B_{n_i}}$;
  ▷ Step 3
6 $\sigma^* \leftarrow$ the schedule obtained in Step 2;
7 **return** σ^*.

Theorem 7.65. (Wu et al. [51]) *Problem $1|GT, p_{ij} = b_{ij}t, \theta_i = \delta_i| \sum C_j$ is solvable in $O(n \log n)$ time by Algorithm 7.10.*

Proof. The result is a consequence of Theorem 7.64 and the fact that Step 1 and Step 2 need $O(n \log n)$ and $O(m \log m)$ time, respectively.   □

If all jobs have distinct deadlines, the problem of minimization of the total completion time is polynomially solvable by Algorithm 7.11, provided that there exists a schedule in which all jobs are completed before their deadlines.

Algorithm 7.11. for problem $1|p_j = b_jt, d_j| \sum C_j$

1 **Input** : sequences $(b_1, b_2, \ldots, b_n)$, $(d_1, d_2, \ldots, d_n)$
2 **Output:** an optimal schedule σ^*
  ▷ Step 1
3 $B \leftarrow \prod_{j=1}^{n}(b_j + 1)$; $N_{\mathcal{J}} \leftarrow \{1, 2, \ldots, n\}$; $k \leftarrow n$; $\sigma^* \leftarrow (\phi)$;
  ▷ Step 2
4 **while** $(N_{\mathcal{J}} \neq \emptyset)$ **do**
5 | $\mathcal{J}_B \leftarrow \{J_i \in \mathcal{J} : d_i \geqslant B\}$;
6 | Choose job $J_j \in \mathcal{J}_B$ with maximal b_j;
7 | Schedule J_j in σ^* in position k;
8 | $k \leftarrow k - 1$; $N_{\mathcal{J}} \leftarrow N_{\mathcal{J}} \setminus \{j\}$; $B \leftarrow \frac{B}{b_j+1}$;
end
  ▷ Step 3
return σ^*.

Theorem 7.66. (Cheng et al. [39]) *Problem* $1|p_j = b_j t, d_j| \sum C_j$ *is solvable in* $O(n \log n)$ *time by Algorithm 7.11.*

Proof. The result follows from the fact that Algorithm 7.11 is an adaptation of Smith's backward scheduling rule for problem $1|d_j| \sum C_j$ (cf. Smith [47]) to problem $1|p_j = b_j t, d_j| \sum C_j$. □

Remark 7.67. Cheng et al. [39] give the result without a proof. The formulation of Algorithm 7.11 comes from the present author.

7.2.2 Proportional-linear deterioration

The problem of minimizing the $\sum C_j$ criterion with job processing times given by (6.8) is polynomially solvable.

Theorem 7.68. (Strusevich and Rustogi [48]) *Problem* $1|p_j = b_j(a+bt)| \sum C_j$ *is solvable in* $O(n \log n)$ *time by scheduling jobs in non-decreasing order of the deterioration rates* b_j.

Proof. The result can be proved by showing that function $w(j) = b_j$ is a 1-priority-generating function for the considered problem; see [48, Sect. 9.1] for details. ◇

The total completion time for a given schedule σ for the problem equals

$$\sum C_j(\sigma) = \left(t_0 + \frac{a}{b}\right) \sum_{j=1}^{n} \prod_{k=1}^{j} \left(1 + b_{[k]} b\right) - \frac{na}{b}. \tag{7.22}$$

Formula (7.22) can be proved by mathematical induction with respect to the number of jobs.

7.2.3 Linear deterioration

Though problem $1|p_j = a_j + b_j t| \sum C_j$ is much more difficult than its counterpart with the $C_{\max}$ criterion and its time complexity is still unknown (see Sect. A.1 for details), the following result can be shown.

Lemma 7.69. *For a schedule* σ *for problem* $1|p_j = a_j + b_j t| \sum C_j$ *the formula*

$$\sum C_j(\sigma) = \sum_{i=1}^{n} \sum_{j=1}^{i} a_{\sigma_j} \prod_{k=j+1}^{i} \left(1 + b_{\sigma_k}\right) + t_0 \sum_{i=1}^{n} \prod_{j=1}^{i} (1 + b_{\sigma_j}) \tag{7.23}$$

holds.

Proof. Adding to each other formulae (7.11) for $1 \leqslant j \leqslant n$, the formula (7.23) follows. □

Ocetkiewicz [46] considered a polynomial case of problem $1|p_j = 1 + b_jt| \sum C_j$. The author showed that if for any i and j such that $1 \leqslant i \neq j \leqslant n$ the implication

$$b_i > b_j \implies b_i \geqslant \frac{b_{\min} + 1}{b_{\min}} b_j + \frac{1}{b_{\min}} \qquad (7.24)$$

holds, where $b_{\min}$ is defined as in inequality (6.39), the optimal V-shaped job sequence for the problem (cf. Definition 1.3) can be constructed by a polynomial algorithm. The pseudo-code of the algorithm (see Algorithm 7.12) can be formulated as follows.

Algorithm 7.12. for problem $1|p_j = 1 + b_jt| \sum C_j$

---

1 **Input** : sequence $(b_1, b_2, \ldots, b_n)$
2 **Output:** an optimal schedule $\sigma^\star$
  ▷ Step 1
3 Arrange jobs in non-decreasing order of the deterioration rates b_j;
4 $L \leftarrow (1 + b_{[n-1]}); \sigma^{(L)} \leftarrow (b_{[n-1]}); R \leftarrow 0; \sigma^{(R)} \leftarrow (\phi)$;
  ▷ Step 2
5 **for** $i \leftarrow n - 2$ **downto** 2 **do**
6 | **if** $(L > R)$ **then**
7 | | $\sigma^{(R)} \leftarrow (b_{[i]}|\sigma^{(R)}); R \leftarrow (R + 1)(1 + b_{[i]})$;
  | **else**
8 | | $\sigma^{(L)} \leftarrow (\sigma^{(L)}|b_{[i]}); L \leftarrow (L + 1)(1 + b_{[i]})$;
  | **end**
  end
  ▷ Step 3
9 $\sigma^\star \leftarrow (b_{[n]}|\sigma^{(L)}|b_{[1]}|\sigma^{(R)})$;
  return $\sigma^\star$.

Algorithm 7.12 exploits the fact that implication (7.24) allows us to indicate to which branch of the constructed V-shaped sequence a given ratio b_i should be added. Therefore, the following result holds.

Theorem 7.70. (Ocetkiewicz [46]) *If all jobs have distinct deterioration rates and if for any $1 \leqslant i \neq j \leqslant n$ the implication (7.24) holds, then problem $1|p_j = 1 + b_jt| \sum C_j$ is solvable by Algorithm 7.12 in $O(n \log n)$ time.*

Proof. By direct computation; see [46, Sect. 2].   ◇

If deadlines are arbitrary and $b_j = b$ for $1 \leqslant j \leqslant n$, then the following result for problem $1|p_j = a_j + bt, d_j|C_{\max}$ holds.

Lemma 7.71. (Cheng and Ding [40]) *Problem $1p_j = a_j + bt, d_j|C_{\max}$ is equivalent to problem $1|p_j = a_j + bt, d_j| \sum C_j$.*

Proof. Let σ be an arbitrary schedule for problem $1p_j = a_j + bt, d_j|C_{\max}$. Then

$$C_i(\sigma) = a_i + Ba_{i-1} + \cdots + B^{i-1}a_1 = \sum_{k=1}^{i} B^{i-k}a_k,$$

where $B = 1 + b$.

Since $C_i(\sigma) = \sum_{j=1}^{i} p_j$ and $p_j = a_j + bS_j$, we have

$$C_i(\sigma) = \sum_{j=1}^{i}(a_j + bS_j) = \sum_{j=1}^{i} a_j + b\sum_{j=1}^{i} S_j = \sum_{j=1}^{i} a_j + b\sum_{j=1}^{i} \tfrac{1}{B}(C_j(\sigma) - a_j) = $$
$$\tfrac{1}{B}\sum_{j=1}^{i} a_j + \tfrac{b}{B}\sum_{j=1}^{i} C_j.$$

Hence, in this case, the problem of minimizing the maximum completion time is equivalent to the problem of minimizing the total completion time. ∎

Remark 7.72. Lemma 7.71 gives an example of a pair of time-dependent scheduling problems that are *equivalent* in some sense. In Sect. 19.4, we will consider other pairs of equivalent time-dependent scheduling problems.

Theorem 7.73. (Cheng and Ding [42]) *Problem* $1|p_j = a_j + bt, d_j|\sum C_j$ *is solvable in* $O(n^5)$ *time by Algorithm 7.6.*

Proof. By Lemma 7.71 and Theorem 7.23, the result follows. □

7.2.4 Simple non-linear deterioration

A single-machine time-dependent scheduling problem with simple non-linear job processing times given by (6.13), equal ready times and equal deadlines is polynomially solvable.

Theorem 7.74. (Gawiejnowicz [43]) *Problem* $1|p_j = a_j + f(t)|\sum C_j$, *where* $f(t)$ *is an arbitrary increasing function, is optimally solved in* $O(n \log n)$ *time by scheduling jobs in non-decreasing order of the basic processing times* a_j.

Proof. Since $\sum_{j=1}^{n} C_j = \sum_{j=1}^{n}(n - j + 1)(a_j + f(S_j))$, and since sequence $(n - j + 1)$ is decreasing, the result follows from Lemma 1.2 (b). □

Theorem 7.75. (Wang and Wang [50]) *Problem* $1|p_j = b_j(a + bt)^k|\sum C_j$, *where* $k \geqslant 1$, *is solvable in* $O(n \log n)$ *time by scheduling jobs in non-decreasing order of the deterioration rates* b_j.

Proof. By the pairwise job interchange argument. □

7.2.5 General non-linear deterioration

Cheng and Ding [41] identified some polynomially solvable cases of problem $1|p_j \in \{a_j, b_j : a_j \leqslant b_j\}, d_j = D|\sum C_j$.

Assume that conjunction $a_i \neq a_j \wedge b_i \neq b_j$ for any $1 \leqslant i, j \leqslant n$ holds. Let $E := \{J_k \in \mathcal{J} : C_k \leqslant d\}$ and $L := \mathcal{J} \setminus E$.

Lemma 7.76. (Cheng and Ding [41]) *If $a_i \leqslant a_j$ and $a_i + b_i \leqslant a_j + b_j$, then job J_i precedes job J_j in any optimal schedule for problem $1|p_j \in \{a_j, b_j : a_j \leqslant b_j\}| \sum C_j$.*

Proof. By the pairwise job interchange argument.   □

Based on Lemma 7.76, Cheng and Ding [41] proposed an exact algorithm for $d_j = D$ for $1 \leqslant j \leqslant n$ (see Chap. 13, Algorithm 13.3).

If k is a fixed number, then the following result holds.

Theorem 7.77. (Cheng and Ding [41]) *If in an instance of problem $1|p_j \in \{a_j, b_j : a_j \leqslant b_j\}, d_j = D| \sum C_j$ the number of chains k is fixed, then Algorithm 13.3 is a polynomial-time algorithm for this instance.*

Proof. For a fixed k, the total running time of Algorithm 13.3 becomes polynomial with respect to n.   ■

7.2.6 Proportional-linear shortening

Time-dependent scheduling problems with proportional-linearly shortening job processing times given by (6.32) are of similar difficulty as those with proportional-linearly deteriorating job processing times given by (6.9).

Lemma 7.78. *For a schedule σ for problem $1|p_j = b_j(1 - bt)| \sum C_j$, the formula*

$$\sum C_j(\sigma) = \left(t_0 - \frac{1}{b}\right) \sum_{j=1}^{n} \prod_{k=1}^{n} (1 - b_{[k]}b) + \frac{n}{b} \qquad (7.25)$$

holds.

Proof. Formula (7.25) is a variant of formula (7.22) in which a and b are replaced with 1 and $-b$, respectively.   □

In view of Lemma 7.78, the following result holds.

Theorem 7.79. (Ng et al. [45]) *Problem $1|p_j = b_j(1 - bt)| \sum C_j$ is solvable in $O(n \log n)$ time by scheduling jobs in non-decreasing order of the deterioration rates b_j.*

Proof. By the pairwise job interchange argument.   □

7.2.7 Linear shortening

We complete this section with a few results concerning polynomially solvable single-machine time-dependent scheduling problems with shortening job processing times, equal ready times and deadlines.

We consider shortening job processing times given by (6.43), where for $1 \leqslant j \leqslant n$ conditions (6.44) and (6.45) hold. The assumptions eliminate some trivial cases and ensure that the constructed instances make sense from the practical point of view.

Property 7.80. (Ng et al. [45]) Problem $1|p_j = a_j - bt, 0 < b < 1|\sum C_j$ is solvable in $O(n \log n)$ time by scheduling jobs in non-decreasing order of the basic processing times a_j.

Proof. By the pairwise job interchange argument.   □

Wang et al. [49] considered the problem of single-machine batch scheduling with job processing times (6.46), where condition (6.47) and

$$b_{ij}\left(\sum_{i=1}^{m}\sum_{j=1}^{k_i}(\theta_i + a_{ij}) - a_{ij}\right) < a_{ij} \tag{7.26}$$

are satisfied for $1 \leqslant i \leqslant m$ and $1 \leqslant j \leqslant k_i$, $\sum_{i=1}^{m} k_i = n$.

Remark 7.81. Conditions (6.47) and (7.26) for time-dependent scheduling problems with group technology are counterparts of conditions (6.44) and (6.45), respectively, for time-dependent scheduling problems without batching.

For problem $1|p_{ij} = a_{ij} - b_{ij}t, \theta_i, GT|\sum C_j$, with $b_{ij} = b_i$ for all j, Wang et al. [49] proposed the following algorithm.

Algorithm 7.13. for problem $1|p_{ij} = a_{ij} - b_it, \theta_i, GT|\sum C_j$

1 **Input** : sequences (a_{ij}), (b_i), (θ_i) for $1 \leqslant i \leqslant m$ and $1 \leqslant j \leqslant k_i$
2 **Output:** an optimal schedule $\sigma^\star$
  ▷ Step 1
3 **for** $i \leftarrow 1$ **to** m **do**
4 | Arrange jobs in group G_i in non-increasing order of the basic
   | processing times a_{ij};
5 | Call the sequence $\sigma^{(i)}$;
  end
  ▷ Step 2
6 **for** $i \leftarrow 1$ **to** m **do**
7 | Calculate $\rho(G_i) := \dfrac{\theta_i(1-b_i)^{k_i} + \sum_{j=1}^{k_i} a_{ij}(1-b_i)^{k_i-j}}{\sum_{j=1}^{k_i} a_{ij}(1-b_i)^j}$;
  end
  ▷ Step 3
8 Schedule group in non-decreasing order of the ratios $\rho(G_i)$;
  ▷ Step 4
9 $\sigma^\star \leftarrow (\sigma^{([1])}|\sigma^{([2])}|\ldots|\sigma^{([m])})$;
  return $\sigma^\star$.

Theorem 7.82. (Wang et al. [49]) *Problem $1|p_{ij} = a_{ij} - b_it, \theta_i, GT|\sum C_j$ is solvable by Algorithm 7.13 in $O(n \log n)$ time.*

Proof. By the pairwise job interchange; see [49, Theorems 4–5] for details. ◇

Remark 7.83. Wang et al. [49] also considered the case when $b_{ij} = a_{ij}b$ for all i, j. This case is solved by an algorithm that is similar to Algorithm 7.13; see [49, Algorithm 3, Theorems 7–8] for details. ◇

7.3 Minimizing the maximum lateness

In this section, we consider polynomially solvable single machine time-dependent scheduling problems with the $L_{\max}$ criterion.

7.3.1 Proportional deterioration

The single machine time-dependent scheduling problem with proportional job processing times given by (6.7) is easy to solve.

Theorem 7.84. (Mosheiov [55]) *Problem* $1|p_j = b_j t|L_{\max}$ *is solvable in* $O(n \log n)$ *time by scheduling jobs in non-decreasing order of the deadlines* d_j.

Proof. The first proof is by the pairwise job interchange argument. Let us consider schedule σ' in which job J_i is followed by job J_j and $d_i > d_j$. Then $L_i(\sigma') = (1 + b_i)S_i - d_i$ and $L_j(\sigma') = (1 + b_j)(1 + b_i)S_i - d_j$.

Now let us consider schedule σ'', obtained from σ' by interchanging jobs J_i and J_j. Then $L_j(\sigma'') = (1+b_j)S_i - d_j$ and $L_i(\sigma'') = (1+b_i)(1+b_j)S_i - d_i$. Since $L_j(\sigma') > L_j(\sigma'')$ and $L_j(\sigma') > L_i(\sigma'')$, schedule σ'' is better than schedule σ'.

Repeating this reasoning for all other pairs of jobs which are scheduled in non-increasing order of d_j, we obtain an optimal schedule in which all jobs are scheduled in non-decreasing order of the deadlines d_j.   ∎

Example 7.85. (Agnetis et al. [52]) Let us consider an instance of problem $1|p_j = b_j t|L_{\max}$, where jobs J_1, J_2 and J_3 have the following processing times and deadlines: $p_1 = t$, $p_2 = 3t$, $p_3 = 2t$, $d_1 = 12$, $d_2 = 4$ and $d_3 = 24$.

The optimal schedule is $\sigma = (2, 1, 3)$ with $L_{\max}(\sigma) = 0$.   ◆

7.3.2 Proportional-linear deterioration

Theorem 7.84 was generalized by Kononov for proportional-linear job processing times (6.8).

Theorem 7.86. (Kononov [54]) *If inequalities* (7.5) *and* (7.6) *hold, then problem* $1|p_j = b_j(a + bt)|L_{\max}$ *is solvable in* $O(n \log n)$ *time by scheduling jobs in non-increasing order of the deadlines* d_j.

Proof. By the pairwise job interchange argument.   □

Remark 7.87. A version of Theorem 7.86, without conditions (7.5) and (7.6) but with assumptions $a > 0, b > 0$, $b_j > 0$ for $1 \leqslant j \leqslant n$, was given by Zhao et al. [56, Theorem 4].   ◇

7.3.3 Linear deterioration

Single machine scheduling of linearly deteriorating jobs with the $L_{\max}$ criterion is intractable (see Sect. 10.1.2 for details). However, if we assume that all jobs deteriorate at the same rate, then problem $1|p_j = a_j + b_j t|L_{\max}$ becomes polynomially solvable.

Algorithm 7.14. for problem $1|p_j = a_j + bt|L_{\max}$

---

1 **Input** : sequence $(a_1, a_2, \ldots, a_n)$, $(d_1, d_2, \ldots, d_n)$, number $b := \frac{v}{u}$
2 **Output:** the minimal value of $L_{\max}$
  ▷ Step 1

3 $B_1 \leftarrow u^{n-1}\left(\min_{1 \leqslant j \leqslant n}\{d_j - a_j\}\right)$;

4 $B_2 \leftarrow (u + v)^{n-1}\sum_{j=1}^{n} a_j$;

  ▷ Step 2
5 **while** $(B_2 - B_1 > 1)$ **do**
6 $\quad$ $L' \leftarrow \lceil\frac{B_1 + B_2}{2}\rceil$;
7 $\quad$ $L \leftarrow \frac{L'}{u^{n-1}}$;
8 $\quad$ **for** $i \leftarrow 1$ **to** n **do**
9 $\quad\quad$ $d_i' \leftarrow d_i + L$;
$\quad$ **end**
10 $\quad$ Apply Algorithm 7.6 to the instance I modified in lines 8–9;
11 $\quad$ **if** (*there exists an optimal schedule for* I) **then**
12 $\quad\quad$ $B_1 \leftarrow L'$;
$\quad$ **else**
13 $\quad\quad$ $B_2 \leftarrow L'$;
$\quad$ **end**
end
  ▷ Step 3
14 $L \leftarrow \frac{B_1}{u^{n-1}}$;
15 **return** L.

Theorem 7.88. (Cheng and Ding [53]) *Problem $1|p_j = a_j + bt|L_{\max}$ is solvable in $O(n^6 \log n)$ time by Algorithm 7.14.*

Proof. The optimality of the schedule generated by Algorithm 7.14 follows from the relation between problem $1|p_j = a_j + bt|L_{\max}$ and problem $1|p_j = a_j + bt, d_j|C_{\max}$; see [53, Section 5].

The time complexity of Algorithm 7.14 follows from the fact that the 'while' loop in Step 2 is executed at most $O(n \log n)$ times and each iteration of the loop needs $O(n^5)$ time due to the execution of Algorithm 7.6. ∎

7.3.4 Simple non-linear deterioration

Kononov [54] proved that single machine time-dependent scheduling with simple non-linearly deteriorating jobs is polynomially solvable for convex (concave) functions.

Theorem 7.89. (Kononov [54]) *If $f(t)$ is a convex (concave) function for $t \geqslant 0$ and if conditions (7.18) and (7.6) hold, then problem $1|p_j = b_j f(t)|L_{max}$ is solvable in $O(n \log n)$ time by scheduling jobs in non-decreasing (non-increasing) order of the sums $b_j + d_j$.*

Proof. The main idea is to prove that function $\omega_i = -b_i - d_i$ ($\omega_i = b_i + d_i$) is a 1-priority-generating function for the considered problem with convex (concave) function $f(t)$. Then, by Theorem 1.25, the result follows.   □

7.4 Other criteria

In this section, we consider polynomially solvable single machine time-dependent scheduling problems with criteria other than C_{max}, $\sum C_j$ or L_{max}.

7.4.1 Proportional deterioration

Single machine time-dependent scheduling problems with proportional job processing times are, usually, polynomially solvable.

Minimizing the total weighted completion time

For proportional job processing times given by (6.7), the following result is known.

Theorem 7.90. (Mosheiov [75]) *Problem $1|p_j = b_j t| \sum w_j C_j$ is solvable in $O(n \log n)$ time by scheduling jobs in non-decreasing order of the ratios $\frac{b_j}{(1+b_j)w_j}$.*

Proof. By the pairwise job interchange argument.   □

Remark 7.91. Note that if $w_j = 1$ for $1 \leqslant j \leqslant n$, the scheduling rule from Theorem 7.90 is reduced to the rule given in Theorem 7.21.

Example 7.92. Let us consider an instance of problem $1|p_j = b_j t| \sum w_j C_j$, where jobs J_1, J_2 and J_3 have the following processing times and weights: $p_1 = 2t$, $p_2 = t$, $p_3 = 3t$, $w_1 = 2$, $w_2 = 1$ and $w_3 = 3$. Let assume that $t_0 = 1$. Then, by Theorem 7.90, the optimal schedule is $\sigma = (3, 1, 2)$ with $\sum w_j C_j(\sigma) = 60$.   ◆

Minimizing the maximal cost

The problem of minimizing the maximum cost for proportionally deteriorating jobs is polynomially solvable as well.

Theorem 7.93. *Problem* $1|p_j = b_j t|f_{\max}$ *is solvable in* $O(n^2)$ *time by Algorithm 7.17.*

Proof. The result is a corollary from Theorem 7.111 for $A = 0$ and $B = 1$. ∎

Example 7.94. Let us consider an instance of problem $1|p_j = b_j t|f_{\max}$, where jobs J_1, J_2 and J_3 have the following processing times and cost functions: $p_1 = 3t$, $p_2 = t$, $p_3 = 2t$, $f_1 = 2C_1 + 3$, $f_2 = \frac{1}{2}C_2^2 + 1$ and $f_3 = \sqrt{C_3 + 12}$. Let assume that $t_0 = 1$. Then, by Theorem 7.90, the optimal schedule is $\sigma = (2, 1, 3)$ with $f_{\max}(\sigma) = 19$.   ◆

Minimizing the total lateness and the maximum tardiness

For proportional job processing times given by (6.7), the following results are known.

Theorem 7.95. (Mosheiov [75]) *Problem* $1|p_j = b_j t|\varphi$ *is solvable in* $O(n \log n)$ *time by scheduling jobs*
(a) *in non-decreasing order of the deterioration rates* b_j, *if* φ *is the total lateness criterion* $(\varphi \equiv \sum L_j)$;
(b) *in non-decreasing order of the deadlines* d_j, *if* φ *is the maximum tardiness criterion* $(\varphi \equiv T_{\max})$.

Proof. (a) Since total lateness $\sum_{j=1}^{n} L_j = \sum_{j=1}^{n}(C_j - d_j) = \sum_{j=1}^{n} C_j - \sum_{j=1}^{n} d_j$, and sum $\sum_{j=1}^{n} d_j$ is a constant value, the problem of minimizing sum $\sum_{j=1}^{n} L_j$ is equivalent to the problem of minimizing sum $\sum_{j=1}^{n} C_j$. The latter problem, by Theorem 7.21, is solvable in $O(n \log n)$ time by scheduling jobs in non-decreasing order of the deterioration rates b_j.
   (b) Since for any schedule σ we have $T_i(\sigma) = \max\{0, L_i(\sigma)\}$, $1 \leq i \leq n$, the result follows by the reasoning from the proof of Theorem 7.84.   □

Example 7.96. Let us consider an instance of problem $1|p_j = b_j t|\varphi$, where $\varphi \in \{\sum L_j, T_{\max}\}$. Let jobs J_1, J_2 and J_3 have the following processing times and deadlines: $p_1 = t$, $p_2 = 3t$, $p_3 = 2t$, $d_1 = 12$, $d_2 = 4$ and $d_3 = 24$. Let assume that $t_0 = 1$.
   Then, by Theorem 7.95 (a), the optimal schedule for problem $1|p_j = b_j t|\sum L_j$ is $\sigma = (1, 3, 2)$ with $\sum L_j(\sigma) = -8$.
   The optimal schedule for problem $1|p_j = b_j t|T_{\max}$, by Theorem 7.95 (b), is $\sigma = (2, 1, 3)$ with $T_{\max}(\sigma) = 0$.   ◆

Minimizing the total number of tardy jobs

The problem of minimizing the number of tardy jobs, which proportionally deteriorate, is optimally solved by the following algorithm, which is an adaptation of Moore–Hodgson's algorithm for problem $1||\sum U_j$, [73].

Algorithm 7.15. for problem $1|p_j = b_j t|\sum U_j$

1 **Input** : sequences $(b_1, b_2, \ldots, b_n)$, $(d_1, d_2, \ldots, d_n)$
2 **Output:** an optimal schedule $\sigma^\star$
   ▷ Step 1
3 Arrange jobs in non-decreasing order of the deadlines d_j;
4 Call the sequence $\sigma^\star$;
   ▷ Step 2
5 **while** $(TRUE)$ **do**
6   **if** (*no jobs in sequence $\sigma^\star$ are late*) **then**
7   exit;
        else
8   Find in $\sigma^\star$ the first late job;
9   Call the job $J_{[m]}$;
        end
10   Find job $J_{[k]}$ such that $b_{[k]} = \max\limits_{1 \leqslant i \leqslant m} \{b_{[i]}\}$;
11   Move job $J_{[k]}$ to the end of $\sigma^\star$;
   end
   ▷ Step 3
12 **return** $\sigma^\star$.

Remark 7.97. By constant $TRUE$ we denote the logical truth. Similarly, by $FALSE$ we denote the logical false.

Theorem 7.98. (Mosheiov [75]) *Problem* $1|p_j = b_j t|\sum U_j$ *is solvable in* $O(n \log n)$ *time by Algorithm 7.15.*

Proof. The proof of optimality of Algorithm 7.15 is similar to the original proof of optimality of Moore–Hodgson's algorithm, [73], and consists of two steps.

In the first step, we prove that if there exists a schedule with no late jobs, then there are no late jobs in the schedule obtained by arranging jobs in the EDD order.

In the second step, by mathematical induction, we prove that if Algorithm 7.15 generates a schedule σ that has k late jobs, then there does not exist another schedule, σ', with only $k - 1$ late jobs.   □

Minimizing the total deviation of completion times

Oron [76] considered the total deviation of job completion times criterion, $\sum(C_i - C_j) := \sum_{i=1}^{n} \sum_{k=i+1}^{n} (C_{[k]} - C_{[i]})$.

Notice that $\sum(C_i - C_j) \equiv \sum_{i=1}^{n}\sum_{k=i+1}^{n} C_{[k]} - \sum_{i=1}^{n}\sum_{k=i+1}^{n} C_{[i]} = \sum_{i=1}^{n}(i-1)C_{[i]} - \sum_{i=1}^{n}(n-i)C_{[i]} = \sum_{i=1}^{n}(2i-n-1)C_{[i]}$. Hence, by (7.1), $\sum(C_i - C_j) \equiv S_1 \sum_{i=1}^{n}(2i-n-1)\prod_{j=1}^{i}(1+b_{[j]})$.

Oron [76] proved a few properties of an optimal schedule for problem $1|p_j = b_j t| \sum(C_i - C_j)$.

Property 7.99. (Oron [76]) If $n \geqslant 2$, then there exists an optimal schedule for problem $1|p_j = b_j t| \sum(C_i - C_j)$ in which the job with the smallest deterioration rate is not scheduled as the first one.

Proof. Let $b_1 := \min_{1 \leqslant j \leqslant n}\{b_j\}$, $\sigma^1 := (1, [2], \ldots, [n])$ and $\sigma^2 := ([2], 1, \ldots, [n])$. Since $\sum(C_i - C_j)(\sigma^1) - \sum(C_i - C_j)(\sigma^2) = (n-1)(b_{[2]} - b_1) \geqslant 0$, the result follows. ∎

Property 7.100. (Oron [76]) If $n \geqslant 3$, then there exists an optimal schedule for problem $1|p_j = b_j t| \sum(C_i - C_j)$ in which the job with the smallest deterioration rate is not scheduled as the last one.

Proof. Similar to the proof of Property 7.99; see [76, Proposition 2].   ◇

Property 7.101. (Oron [76]) Let $J_{[i-1]}$, $J_{[i]}$ and $J_{[i+1]}$ be three consecutive jobs in a schedule for problem $1|p_j = b_j t| \sum(C_i - C_j)$. If $b_{[i]} > b_{[i-1]}$ and $b_{[i]} > b_{[i+1]}$, then this schedule cannot be optimal.

Proof. Similar to the proof of Property A.7; see [76, Proposition 3].   ◇

Property 7.102. (Oron [76]) Let $b_{[k]} := \min\{b_j : 1 \leqslant j \leqslant n\}$ and $b_{[l]} := \min\{b_j : 1 \leqslant j \neq k \leqslant n\}$ be the two smallest job deterioration rates.
(a) If n is even, then in optimal schedule for problem $1|p_j = b_j t| \sum(C_i - C_j)$ job $J_{[k]}$ is scheduled in position $\frac{n}{2} + 1$.
(b) If n is odd, then in optimal schedule for problem $1|p_j = b_j t| \sum(C_i - C_j)$ jobs $J_{[k]}$ and $J_{[l]}$ are scheduled in positions $\frac{n+1}{2}$ and $\frac{n+3}{2}$, respectively.

Proof. By direct computation; see [76, Propositions 4-5].   ◇

Property 7.103. (Oron [76]) The optimal value of the total deviation of job completion times for problem $1|p_j = b_j t| \sum(C_i - C_j)$ is not less than

$$\sum_{i=1}^{\frac{n}{2}}(2i-n-1)\prod_{j=1}^{i}(1+b_{[n+2-2j]}) + \sum_{i=\frac{n}{2}+1}^{n}(2i-n-1)\prod_{j=1}^{i}(1+b_{[j]}).$$

Proof. By direct computation; see [76, Proposition 8].   ◇

The author proved also that for problem $1|p_j = b_j t| \sum(C_i - C_j)$ the following counterpart of Theorem A.8 holds.

Theorem 7.104. (Oron [76]) *The optimal schedule for single machine problem $1|p_j = b_j t| \sum(C_i - C_j)$ is V-shaped with respect to the deterioration rates b_j.*

Proof. The result is a consequence of Properties 7.99–7.101.   □

7.4.2 Proportional-linear deterioration

In this section, we consider single machine time-dependent scheduling problems with proportional-linear job processing times. These problems are of similar nature as those with proportional job processing times.

Minimizing the total weighted completion time

Theorem 7.90 was generalized by Strusevich and Rustogi for proportionally-linear job processing times given by (6.8) with $A = 1$.

Theorem 7.105. (Strusevich and Rustogi [78]) *Problem* $1|p_j = b_j(1 + bt)|$ $\sum w_j C_j$ *is solvable in* $O(n \log n)$ *time by scheduling jobs in non-increasing order of the ratios* $\frac{w_j(1+b_j b)}{b_j b}$.

Proof. The result can be proved by showing that function $\omega(j) = \frac{w_j(1+b_j b)}{b_j b}$ is a 1-priority-generating function for the considered problem; see [78, Sect. 9.1] for details. □

Wang et al. [81] proposed for batch scheduling problem $1|p_{ij} = b_{ij}(a + bt),$ $\theta_i = \delta_i(a + bt), GT| \sum w_j C_j$ the following algorithm.

Algorithm 7.16. for problem $1|p_{ij} = b_{ij}(a + bt), \theta_i = \delta_i(a + bt),$ $GT| \sum w_j C_j$

---

1 **Input** : sequences $(\delta_1, \delta_2, \ldots, \delta_m)$, (b_{ij}), (w_{ij}) for $1 \leqslant i \leqslant m$ and
$\qquad 1 \leqslant j \leqslant k_i$,

2 **Output:** an optimal schedule σ^*
$\qquad \triangleright$ Step 1

3 **for** $i \leftarrow 1$ **to** m **do**

4 $\qquad$ Arrange jobs in group G_i in non-decreasing order of the ratios
$\qquad \frac{b_{ij}}{w_{ij}(1+b_{ij})}$;

5 $\qquad$ Call the sequence $\sigma^{(i)}$;

end
$\qquad \triangleright$ Step 2

6 **for** $i \leftarrow 1$ **to** m **do**

7 $\qquad$ Calculate $\rho(G_i) := \dfrac{(1+B\delta_i)\prod_{j=1}^{k_i}(1+b_{ij}b)-1}{(1+b\delta_i)\sum_{j=1}^{k_j} w_{ij}\prod_{l=1}^{j}(1+b_{il}b)}$;

end
$\qquad \triangleright$ Step 3

8 Schedule groups in non-decreasing order of the ratios $\rho(G_i)$;
$\qquad \triangleright$ Step 4

9 $\sigma^* \leftarrow (\sigma^{([1])}|\sigma^{([2])}|\ldots|\sigma^{([m])})$;

10 **return** σ^*.

Theorem 7.106. (Wang et al. [81]) *Problem* $1|GT, p_{ij} = b_{ij}(a + bt), \theta_i = \delta_i(a + bt)| \sum w_j C_j$ *is solvable by Algorithm 7.16 in* $O(n \log n)$ *time.*

Proof. By the pairwise job interchange; see [81, Theorem 2] for details.   ◇

Remark 7.107. A special case of Theorem 7.106, with $a = 0$, $b = 1$ and $\delta_i = 0$, was given by Cheng and Sun [62, Theorem 5].   ◇

Remark 7.108. A special case of Theorem 7.106, with $\theta_i = const$, was given by Xu et al. [83, Theorems 1–2].   ◇

Minimizing the total number of tardy jobs

Theorem 7.98 was generalized by Kononov [69] for proportionally-linear job processing times given by (6.8).

Theorem 7.109. (Kononov [69]) *If inequalities* (7.5) *and* (7.6) *hold, then problem* $1|p_j = b_j(a + bt)| \sum U_j$ *is solvable in* $O(n \log n)$ *time by Algorithm 7.15.*

Proof. Similar to the proof of Theorem 7.98.   □

Minimizing the maximum cost

The problem of minimizing the maximum cost for jobs with proportional-linear processing times is solved by the following algorithm.

Algorithm 7.17. for problem $1|p_j = b_j(a + bt)|f_{\max}$

---

1 **Input** : sequences $(b_1, b_2, \ldots, b_n)$, $(f_1, f_2, \ldots, f_n)$, numbers a, b
2 **Output:** an optimal schedule $\sigma^\star$
  ▷ Step 1
3 $\sigma^\star \leftarrow (\phi)$;
4 $N_{\mathcal{J}} \leftarrow \{1, 2, \ldots, n\}$;
5 $T \leftarrow (t_0 + \frac{a}{b}) \prod\limits_{j=1}^{n} (1 + b_j b) - \frac{a}{b}$;
  ▷ Step 2
6 **while** $(N_{\mathcal{J}} \neq \emptyset)$ **do**
7 | Find job J_k such that $f_k(T) = \min\{f_j(T) : j \in N_{\mathcal{J}}\}$;
8 | $\sigma^\star \leftarrow (k|\sigma^\star)$;
9 | $T \leftarrow \frac{T - ab_k}{1 + b_k b}$;
10 | $N_{\mathcal{J}} \leftarrow N_{\mathcal{J}} \setminus \{k\}$;
  end
  ▷ Step 3
11 **return** $\sigma^\star$.

Remark 7.110. Algorithm 7.17 is an adaptation of Lawler's algorithm for problem $1|prec|f_{\max}$, [72]. Since so far we assumed that there are no precedence constraints between jobs, Algorithm 7.17 is a simplified version of Lawler's algorithm. The full version of Algorithm 7.17, for dependent jobs with proportional-linear processing times, we present in Chap. 18.

Theorem 7.111. (Kononov [69]) *If inequalities* (7.5) *and* (7.6) *hold, problem* $1|p_j = b_j(a + bt)|f_{\max}$ *is solvable in* $O(n^2)$ *time by Algorithm 7.17.*

Proof. Similar to the proof of Theorem 18.17. $\qquad\qquad\qquad\qquad$ □

Remark 7.112. Kononov [69] proved Theorem 7.111 in a more general form, admitting arbitrary job precedence constraints in the problem. For simplicity of presentation (cf. Remark 7.110), we assumed no precedence constraints.

Remark 7.113. A version of Theorem 7.111, without job precedence constraints, without conditions (7.5) and (7.6) but with assumptions $a > 0, b > 0$, $b_j > 0$ for $1 \leqslant j \leqslant n$, was given by Zhao et al. [84, Theorem 3]. $\qquad$ ◇

7.4.3 Linear deterioration

In this section, we consider polynomially solvable single machine time-dependent scheduling problem with linear job processing times.

Minimizing the total weighted completion times

The problem of minimizing the total weighted completion time, $\sum w_j C_j$, for a single machine and linear deterioration given by (6.11) was considered for the first time by Mosheiov [74].

Remark 7.114. Browne and Yechiali [60] studied the problem earlier, but they considered the *expected* total weighted completion time. Namely, if $\frac{E(a_1)}{b_1} < \frac{E(a_1)}{b_1} < \cdots < \frac{E(a_n)}{b_n}$ and $\frac{b_1}{w_1(1+a_1)} < \frac{b_2}{w_2(1+a_2)} < \cdots < \frac{b_n}{w_n(1+a_n)}$, then schedule $(1, 2, \ldots, n)$ minimizes the expected value of $\sum w_j C_j$; see [60, Proposition 2].

Mosheiov [74] considered the weights of jobs which are proportional to the basic job processing times, i.e., $w_j = \delta a_j$ for a given constant $\delta > 0$ and $b_j = b$ for $1 \leqslant j \leqslant n$. The criterion function is in the form of

$$\sum_{j=1}^{n} w_j C_j \equiv \sum_{j=1}^{n} w_j C_j = \delta \sum_{j=1}^{n} a_j \sum_{k=1}^{j} a_k (1+b)^{j-k}.$$

The following properties of problem $1|p_j = a_j + bt|\sum w_j C_j$, with $w_j = \delta a_j$, are known. First, the symmetry property similar to Property A.3 holds.

Property 7.115. For any job sequence σ, let $\bar{\sigma}$ denote the sequence reverse to σ. Then the equality $\sum w_j C_j(\sigma) = \sum w_j C_j(\bar{\sigma})$ holds.

Proof. By direct computation; see [74, Proposition 1]. ◇

Property 7.116. (Mosheiov [74]) If $w_j = \delta a_j$ for $1 \leqslant j \leqslant n$, then the optimal schedule for problem $1|p_j = a_j + bt| \sum w_j C_j$, where $w_j = \delta a_j$, is Λ-shaped.

Proof. Let J_{i-1}, J_i, J_{i+1} be three consecutive jobs such that $a_i < a_{i-1}$ and $a_i < a_{i+1}$. Let us assume that job sequence $\sigma^* = (1, 2, \ldots, i-2, i-1, i, i+1, i+2, \ldots, n)$ is optimal. Let us consider job sequences $\sigma' = (1, 2, \ldots, i-2, i, i-1, i+1, i+2, \ldots, n)$ and $\sigma'' = (1, 2, \ldots, i-2, i-1, i+1, i, i+2, \ldots, n)$. Let us calculate differences $v_1 = \sum w_j C_j(\sigma^*) - \sum w_j C_j(\sigma')$ and $v_2 = \sum w_j C_j(\sigma^*) - \sum w_j C_j(\sigma'')$.

Since v_1 and v_2 cannot both be negative (see [74, Proposition 2]), either σ' or σ'' is a better sequence than σ^*. A contradiction. ■

By Property 7.116, the dominant set for problem $1|p_j = a_j + bt| \sum w_j C_j$, where $w_j = \delta a_j$, is composed of Λ-shaped schedules. Since there exist $O(2^n)$ Λ-shaped sequences for a given sequence $a = (a_1, a_2, \ldots, a_n)$, the problem seems to be intractable. However, the following result holds.

Property 7.117. (Mosheiov [74]) If $w_j = \delta a_j$ for $1 \leqslant j \leqslant n$ and if in an instance of problem $1|p_j = a_j + bt| \sum w_j C_j$ jobs are numbered in non-decreasing order of the basic processing times a_j, then the optimal permutation for this instance is in the form of $\sigma^1 = (1, 3, \ldots, n-2, n, n-1, n-3, \ldots, 4, 2)$ if n is odd, and it is in the form of $\sigma^2 = (1, 3, \ldots, n-1, n, n-2, \ldots, 4, 2)$ if n is even.

Proof. By direct computation; see [74, Proposition 3]. ◇

Theorem 7.118. *Problem* $1|p_j = a_j + b_j t| \sum w_j C_j$ *with equal deterioration rates and weights proportional to basic job processing times is solvable in* $O(n)$ *time.*

Proof. The result is a corollary from Property 7.117. □

Minimizing the maximal processing time

Alidaee and Landram [57] considered linear job processing times and the criterion of minimizing the *maximum processing time*, $P_{\max}$. The problem is to find such a schedule $\sigma \in \mathfrak{S}_n$ that minimizes $\max_{1 \leqslant j \leqslant n} \{p_{[j]}(\sigma)\}$. The following example shows that the problem of minimizing $P_{\max}$ is not equivalent to the problem of minimizing $C_{\max}$.

Example 7.119. (Alidaee and Landram [57]) Let $p_1 = 100 + \frac{1}{5}t$, $p_2 = 2 + \frac{2}{9}t$ and $p_3 = 70 + \frac{3}{10}$. Then schedule $(1, 3, 2)$ is optimal for the $P_{\max}$ criterion, while schedule $(2, 3, 1)$ is optimal for the $C_{\max}$ criterion. ◆

For the $P_{\max}$ criterion the following results are known.

Property 7.120. (Alidaee and Landram [57]) If $a_j > 0$ and $b_j \geqslant 1$ for $1 \leqslant j \leqslant n$, then for any sequence of jobs with processing times in the form of $p_j = a_j + b_j t$ the maximum processing time occurs for the last job.

Proof. The result follows from the fact that all processing times are described by increasing functions and the jobs start their execution at increasingly ordered starting times.   □

Remark 7.121. By Property 7.120 the problem of minimizing the maximum processing time is equivalent to the problem of minimizing the processing time of the last job in a schedule.

Alidaee and Landram [57] proposed for problem $1|p_j = g_j(t)|P_{\max}$ an algorithm based on Lawler's algorithm for problem $1|prec|f_{\max}$ [72].

Theorem 7.122. (Alidaee and Landram [57]) *If $a_j > 0$ and $b_j \geqslant 1$ for $1 \leqslant j \leqslant n$, then the problem of minimizing the processing time of the last job in a single machine schedule of a set of linearly deteriorating jobs is solvable in $O(n \log n)$ time.*

Proof. See [57, Proposition 2].   ◇

Another polynomially solvable case is when $p_j = a_j + bt$.

Theorem 7.123. (Alidaee and Landram [57]) *If $a_j > 0$ and $b_j = b$ for $1 \leqslant j \leqslant n$, then the problem of minimizing the maximum processing time is optimally solved by scheduling jobs in non-decreasing order of the basic processing times a_j.*

Proof. By the pairwise job interchange argument; see [57, Proposition 3].   ◇

Minimizing the total earliness and tardiness

Cheng et al. [63] considered the problem of minimizing the sum of earliness, tardiness and due-date penalties, and common due-date assignment.

Job processing times are special cases of linearly deteriorating processing times (6.11), i.e.,

$$p_j = a_j + bt,$$

where $b > 0$ and $a_j > 0$ for $1 \leqslant j \leqslant n$.

The authors proved the following properties of the problem.

Property 7.124. (Cheng et al. [63]) For any schedule σ for problem $1|p_j = a_j + bt| \sum(\alpha E_j + \beta T_j + \gamma d)$ there exists an optimal due-date $d^* = C_{[k]}$ such that $k = \lceil \frac{n\beta - n\gamma}{\alpha + \beta} \rceil$ and exactly k jobs are non-tardy.

Proof. It is an adaptation of the proof of [77, Lemma 1].   ◇

Property 7.125. (Cheng et al. [63]) If k is defined as in Property 7.124, then in any optimal schedule $\sigma^\star$ for problem $1|p_j = a_j + bt| \sum(\alpha E_j + \beta T_j + \gamma d)$ inequalities $a_{[k+1]} \leqslant a_{[k+2]} \leqslant \cdots \leqslant a_{[n]}$ hold.

Proof. By the pairwise job interchange argument.   □

Before we formulate the next property, we introduce new notation. Let

$$m_i := \begin{cases} b \sum_{j=1}^{k} (\alpha(j-1) + n\gamma)(1+b)^{j-1} + \\ \quad + b \sum_{j=k+1}^{n} (\beta(n+1-j)(1+b)^{j-1} \text{ for } 2 \leqslant i \leqslant k, \\ b \sum_{j=i}^{n} (\beta(n+1-j)(1+b)^{j-i} \text{ for } k+1 \leqslant i \leqslant n. \end{cases} \quad (7.27)$$

We also define the following two functions:

$$g(i) := \alpha(i-1) + n\gamma + m_{i+1} \quad \text{for} \quad 1 \leqslant i \leqslant k \quad (7.28)$$

and

$$f(b) := (\alpha + n\gamma)b - \alpha + bm_3. \quad (7.29)$$

Based on definitions (7.27)–(7.29), Cheng et al. [63] proved a few properties, concerning possible relations between $f(b)$, $g(i)$ and m_i.

Property 7.126. (Cheng et al. [63]) (a) If $f(b) = (\alpha+n\gamma)b-\alpha+bm_3 > 0$, then $(\alpha i + n\gamma)b - \alpha + bm_{i+2} > 0$ for $1 \leqslant i \leqslant k-1$. (b) and (c) The implication (a) in which symbol '>' has been replaced by symbol '<' and '=', respectively.

Proof. (a), (b), (c) By mathematical induction with respect to i.   □

Remark 7.127. Cheng et al. [63] proved also a property similar to Property 7.126; see [63, Property 5] for details.   ◇

Property 7.128. (Cheng et al. [63]) If k is defined as in Property 7.124 and if (a) $f(b) \geqslant 0$, then for problem $1|p_j = a_j + bt| \sum(\alpha E_j + \beta T_j + \gamma d)$ there exists an optimal schedule $\sigma^\star$ such that $a_{\sigma_1^\star} \leqslant a_{\sigma_2^\star} \leqslant \cdots \leqslant a_{\sigma_k^\star}$;
(b) $f(b) < 0$, then for problem $1|p_j = a_j + bt| \sum(\alpha E_j + \beta T_j + \gamma d)$ in any optimal schedule $\sigma^\star$ inequalities $a_{\sigma_1^\star} \geqslant a_{\sigma_2^\star} \geqslant \cdots \geqslant a_{\sigma_k^\star}$ hold.

Proof. (a), (b) By the pairwise job interchange argument.   □

Based on the above properties, the authors proved the following result.

Theorem 7.129. (Cheng et al. [63]) *If k is defined as in Property 7.124 and (a) if $f(b) \geqslant 0$, then for problem $1|p_j = a_j + bt| \sum(\alpha E_j + \beta T_j + \gamma d)$ there exists an optimal schedule $\sigma^\star$ in which $a_{\sigma_1^\star} \leqslant a_{\sigma_2^\star} \leqslant \cdots \leqslant a_{\sigma_k^\star}$ and $a_{\sigma_{k+1}^\star} \leqslant \cdots \leqslant a_{\sigma_n^\star}$;
(b) if $f(b) < 0$, then for problem $1|p_j = a_j + bt| \sum(\alpha E_j + \beta T_j + \gamma d)$ there exists an optimal schedule $\sigma^\star$ which is V-shaped with respect to the basic processing times a_j and such that $a_{\sigma_k^\star} = \min_{1 \leqslant j \leqslant n}\{a_j\}$.*

Proof. (a) (b) The results follow from Properties 7.126–7.128.   □

Based on these properties, Cheng et al. [63] proposed an algorithm for the considered problem. Since this algorithm is rather complicated (see [63, Algorithm 1] for details), we only present the following result.

Theorem 7.130. (Cheng et al. [63]) *Problem* $1|p_j = a_j + bt| \sum(\alpha E_j + \beta T_j + \gamma d)$ *is solvable in* $O(n \log n)$ *time.*

Proof. See [63, Properties 9–11, Theorem 12].   ◇

For problem $1|p_j = a_j + bt| \sum(\alpha E_j + \beta T_j + \gamma d)$, Kuo and Yang [70] proposed another $O(n \log n)$ algorithm (see Algorithm 7.18), simpler than the algorithm proposed by Cheng et al. [63]. The pseudo-code of the algorithm is as follows.

Algorithm 7.18. for problem $1|p_j = a_j + bt| \sum(\alpha E_j + \beta T_j + \gamma d)$

1 **Input**   : sequence $(a_1, a_2, \ldots, a_n)$, numbers b, α, β, γ
2 **Output:** an optimal due-date d^*, an optimal schedule σ^*
   ▷ Step 1
3 $N_{\mathcal{J}} \leftarrow \{1, 2, \ldots, n\}$;
4 $W \leftarrow \{1, 2, \ldots, n\}$;
5 $k \leftarrow \left\lceil \frac{n(\beta - \gamma)}{\alpha + \beta} \right\rceil$;
6 Assign due-date d^* at the completion of the kth job;
   ▷ Step 2
7 **for** $j \leftarrow 1$ **to** n **do**
8 $\quad$| Calculate W_j;
   end
   ▷ Step 3
9 Arrange jobs in non-decreasing order of the basic processing times a_j;
10 **while** $(N_{\mathcal{J}} \neq \emptyset)$ **do**
11 $\quad$| Assign job J_k such that $a_k = \max\{a_j : j \in N_{\mathcal{J}}\}$ to the rth
   $\quad$| position in σ^*, where r is such that $W_r = \min\{W_j : j \in N_{\mathcal{J}}\}$;
12 $\quad$| $N_{\mathcal{J}} \leftarrow N_{\mathcal{J}} \setminus \{k\}$;
13 $\quad$| $W \leftarrow W \setminus \{r\}$;
   end
   ▷ Step 4
14 **return** σ^*.

Algorithm 7.18 is based on Lemma 1.2 (a) and the following observation. Given due-date d and sequence σ, we have

$$\sum(\alpha E_j + \beta T_j + \gamma d) = \sum_{j=1}^{n}(\alpha E_{[j]}(\sigma) + \beta T_{[j]}(\sigma) + \gamma d) = \sum_{j=1}^{n} W_j a_{[j]},$$

where the coefficients W_j, $1 \leqslant j \leqslant n$, called *positional weights*, are as follows:

$$
\begin{aligned}
W_1 &= w_1 + w_2 b + w_3 b(1+b) + \cdots + w_n b(1+b)^{n-2}, \\
W_2 &= w_2 + w_3 b + w_4 b(1+b) + \cdots + w_n b(1+b)^{n-3}, \\
&\cdots, \\
W_{n-1} &= w_{n-1} + w_n b, \\
W_n &= w_n
\end{aligned}
$$

(see [70, Section 'Preliminary results'] for details). Algorithm 7.18 works as follows. First, the value of an index, k, is calculated. Next, due-date $d^\star$ is assigned at the completion of the kth job. After that, the algorithm calculates the positional weights W_j for $1 \leqslant j \leqslant n$ and, as long as the set of jobs is not empty, assigns the job with the largest basic processing time to the rth position. The value of index r is indicated by the smallest weight among all yet unscheduled jobs.

Theorem 7.131. (Kuo and Yang [70]) *Problem* $1|p_j = a_j + bt|\sum(\alpha E_j + \beta T_j + \gamma d)$ *is solvable in* $O(n \log n)$ *time by Algorithm 7.18.*

Proof. The result follows from definition of the weights W_j and Lemma 1.2 (a); see [70, An algorithm]. □

Remark 7.132. Scheduling problems with deteriorating jobs, earliness and tardiness penalties and common due-date assignment are not popular in time-dependent scheduling. In the classical scheduling, problems of these types have been studied since early 1970s; see the reviews by Baker and Scudder [59] and Gordon et al. [66, 67].

7.4.4 Simple non-linear deterioration

In this section, we consider polynomially solvable single machine time-dependent scheduling problems with simple non-linear deterioration.

Minimizing the total weighted completion time

Kononov [69] proved that some problems with simple non-linear job deterioration and the $\sum w_j C_j$ criterion are polynomially solvable.

Theorem 7.133. (Kononov [69]) *If* $f(t)$ *is a convex function for* $t \geqslant 0$ *and conditions* (7.6) *and* (7.18) *hold, then*
(a) *if* $f(t) \geqslant 0$ *for all* t, $\lim_{t \to +\infty} \frac{df(t)}{dt} = +\infty$ *and* $w_i \geqslant w_l$ *for all* $J_i, J_l \in \mathcal{J}$ *such that* $b_i < b_l$, *then problem* $1|p_j = b_j f(t)|\sum w_j C_j$ *is solvable in* $O(n \log n)$ *time by scheduling jobs in non-decreasing order of the differences* $b_j - w_j$;
(b) *if* $f(t) \geqslant 0$ *for all* t, $\lim_{t \to +\infty} \frac{df(t)}{dt} = H$ *and* $w_i(b_i^{-1} + H) \geqslant w_l(b_l^{-1} + H)$ *for all* $J_i, J_l \in \mathcal{J}$ *such that* $b_i < b_l$, *then problem* $1|p_j = b_j f(t)|\sum w_j C_j$ *is solvable in* $O(n \log n)$ *time by scheduling jobs in non-decreasing order of the differences* $b_j - w_j(b_j^{-1} + H)$;
(c) *if there exists* $\tau > t_0$ *such that* $f(\tau) = 0$ *and* $f(t) > 0$ *for any* $t \in \langle t_0, \tau \rangle$,

$\lim_{t \to +\infty} \frac{df(t)}{dt} = H$ and $w_i(b_i^{-1} + H) \geqslant w_l(b_l^{-1} + H)$ for all $J_i, J_l \in \mathcal{J}$ such that $b_i < b_l$, then problem $1|p_j = b_j f(t)| \sum w_j C_j$ is solvable in $O(n \log n)$ time by scheduling jobs in non-decreasing order of the differences $b_j - w_j(b_j^{-1} + H)$.

Proof. See [69, Theorem 8].   ◇

Remark 7.134. Kononov proved a similar result for concave functions; see [69, Theorem 10].

Minimizing the total general completion time

Kuo and Yang [71] considered single machine time-dependent scheduling problems with non-linear job processing times and with the $\sum C_j^k$ criterion, where k is a given positive integer.

Theorem 7.135. (Kuo and Yang [71])
(a) *If* $f(t) := \sum_{i=1}^{m} \lambda_i t^{r_i}$ *and* $r_i \in \langle 0, +\infty)$ *for* $1 \leqslant i \leqslant m$, *then there exists an optimal schedule for problem* $1|p_j = a_j + f(t)| \sum C_j^k$ *in which jobs are in non-decreasing order of the basic processing times* a_j.
(b) *If* $f(t) := 1 + \sum_{i=1}^{m} \lambda_i t^{r_i}$ *and* $r_i \in \langle 1, +\infty)$ *for* $1 \leqslant i \leqslant m$, *then there exists an optimal schedule for problem* $1|p_j = b_j f(t)| \sum C_j^k$ *in which jobs are in non-decreasing order of the deterioration rates* b_j.
(c) *If* $f(t) := 1 + \sum_{i=1}^{m} \lambda_i t^{r_i}$ *and* $r_i \in (-\infty, 0\rangle$ *for* $1 \leqslant i \leqslant m$, *then there exists an optimal schedule for problem* $1|p_j = b_j f(t)| \sum C_j^k$ *in which all jobs except the first one are in non-decreasing order of the deterioration rates* b_j.

Proof. (a) By the pairwise job interchange argument; see [71, Proposition 6].
   (b) By the pairwise job interchange argument; see [71, Proposition 7].
   (c) By the pairwise job interchange argument; see [71, Proposition 8].   ◇

Remark 7.136. The $\sum C_j^k$ criterion is nothing else than the kth power of the l_p norm, where $p := k$. In Sect. 19.3, we consider time-dependent scheduling problems with the l_p norm as an optimality criterion.

7.4.5 General non-linear deterioration

In this section, we consider polynomially solvable single machine time-dependent scheduling problems with general non-linear deterioration.

Minimizing the total weighted completion time

Sundararaghavan and Kunnathur [79] considered step deteriorating job processing times given by (6.25). Let $\mathcal{J}$ denote a set of jobs and let the jobs have only two distinct weights, $w_j \in \{w_1, w_2\}$ for $1 \leqslant j \leqslant n$. Let $k := \lfloor \frac{D}{a} \rfloor + 1$ denote the maximum number of jobs that can be scheduled without job processing time increase.

Sundararaghavan and Kunnathur [79] proposed for problem $1|p_j| \sum w_j C_j$, where job processing times $p_j \equiv (6.25)$, the following algorithm.

Algorithm 7.19. for problem $1|p_j| \sum w_j C_j$, $p_j \equiv (6.25)$

---

1 **Input** : $(b_1, b_2, \ldots, b_n)$, numbers a, D, w_1, w_2
2 **Output:** an optimal schedule $\sigma^\star$
  ▷ Step 1
3 $W_1 \leftarrow \{J_k : w_k = w_1\}$; $W_2 \leftarrow \{J_k : w_k = w_2\}$; $k \leftarrow \lfloor \frac{D}{a} \rfloor + 1$;
  ▷ Step 2
4 $E \leftarrow W_1$;
5 Arrange E in non-increasing order of the deterioration rates b_j;
6 **if** $(|E| < k)$ **then**
7 | $E' \leftarrow \{J_{[k]} \in W_2' \subseteq W_2 : |W_2'| = |W_1| - k\}$;
8 | Arrange E' in non-increasing order of the deterioration rates b_j;
9 | $E \leftarrow E \cup E'$; $L \leftarrow \{1, 2, \ldots, n\} \setminus E$; $\sigma^\star \leftarrow (E, L)$;
  end
  ▷ Step 3
10 **repeat**
11 | **if** $(J_{[i]} \in E \cap W_1 \wedge J_{[j]} \in L \cap W_2 \wedge \Delta([i] \leftrightarrow [j]) > 0)$ **then**
12 | | Exchange $J_{[i]}$ with $J_{[j]}$;
13 | | Call the obtained schedule σ';
14 | | $\sigma^\star \leftarrow \sigma'$;
  | **end**
  until (*no more exchange* $[i] \leftrightarrow [j]$ *exists for* $J_{[i]} \in E \cap W_1$ *and*
  $J_{[j]} \in L \cap W_2$);
  ▷ Step 4
15 **repeat**
  | **if** $(J_{[i]} \in E \cap W_2 \wedge J_{[j]} \in L \cap W_1 \wedge \Delta([i] \leftrightarrow [j]) > 0)$ **then**
16 | | Exchange $J_{[i]}$ with $J_{[j]}$;
17 | | Call the obtained schedule σ';
18 | | $\sigma^\star \leftarrow \sigma'$;
  | **end**
  until (*no more exchange* $[i] \leftrightarrow [j]$ *exists for* $J_{[i]} \in E \cap W_2$ *and*
  $J_{[j]} \in L \cap W_1$);
  ▷ Step 5
19 **return** $\sigma^\star$.

The proof of optimality of Algorithm 7.19 is based on the following result.

Lemma 7.137. (Sundararaghavan and Kunnathur [79]) *For a given instance of the problem of minimizing the $\sum w_j C_j$ criterion for a single machine and job processing times in the form of (6.25), let $w_j \in \{w_1, w_2 : w_1 > w_2\}$. Let $E := \{J_k \in \mathcal{J} : C_k \leqslant D\}$, $L := \mathcal{J} - E$, $W_i := \{J_j \in \mathcal{J} : w_j = w_i\}$, $1 \leqslant i \leqslant 2$, and let $J_{[r]}$ denote the job with the greatest starting time among the jobs in $E \cap W_1$. Then the following conditions are necessary for optimality of a* *schedule for the problem:*

(a) $w_{[i]} \geqslant w_{[i+1]}$ *for* $1 \leqslant i \leqslant k-1$ *and* $J_{[i]} \in E$;

(b) $\frac{a+b_{[j]}}{w_{[j]}} \leqslant \frac{a+b_{[j+1]}}{w_{[j+1]}}$ *for* $k+1 \leqslant j \leqslant n-1$ *and* $J_{[j]} \in L$;

(c) $b_{[r]} \geqslant b_i$ *for* $i \in L \cap W_1, J_{[r]} \in E$;

(d) $b_{[r+1]} \geqslant b_i$ *for* $i \in L \cap W_2, J_{[r+1]} \in E$.

Proof. The result follows directly from the properties of an optimal schedule for this problem.   □

Remark 7.138. By Lemma 7.137, we know that in an optimal schedule for the problem with two distinct weights there are k jobs in set E which are arranged in non-increasing order of the deterioration rates b_j, and r jobs in set L which are arranged in non-decreasing order of the ratios $\frac{a+b_j}{w_j}$.

For a given schedule σ, let $i \leftrightarrow j$ denote mutual exchange of job $J_i \in E$ with job $J_j \in L$, i.e., rearrangement of jobs in set $\{J_j\} \cup E \setminus \{J_i\}$ in non-increasing order of the weights w_j and rearrangement of jobs in set $\{J_i\} \cup L \setminus \{J_j\}$ in non-decreasing order of the ratios $\frac{a+b_j}{w_j}$. For a given $J_i \leftrightarrow J_j$, let $\Delta(i \leftrightarrow j)$ denote the difference between total weighted completion time for the schedule before the exchange of J_i with J_j and after it.

Theorem 7.139. (Sundararaghavan and Kunnathur [79]) *The problem of minimizing the $\sum w_j C_j$ criterion for a single machine, job processing times (6.25) and two distinct weights is solvable in $O(n \log n)$ time by Algorithm 7.19.*

Proof. Let σ^* be the final schedule generated by Algorithm 7.19, let $J_{[r]}$ be the job with the greatest starting time among jobs in set $E \cap W_1$ and let $k := \lfloor \frac{D}{a} \rfloor + 1$. Then, by Lemma 7.137, $b_{[1]} \geqslant b_{[2]} \geqslant \cdots \geqslant b_{[r]}$ and $b_{[r+1]} \leqslant b_{[r+2]} \leqslant \cdots \leqslant b_{[k]}$. Since in schedule σ^* there do not exist $J_{[i]} \in E$ and $J_{[j]} \in L$ such that $\Delta([i] \leftrightarrow [j]) < 0$ (otherwise σ^* would not be a final schedule), for any $J_{[r]} \in E \cap W_1$ and any $J_{[i]} \in L \cap W_2$ inequality $\Delta([i] \leftrightarrow [j]) \geqslant 0$ holds.

Since $J_{[r]}$ has the lowest deterioration rate of all jobs in $E \cap W_1$, it follows that the exchange of any subset of jobs in $E \cap W_1$ with any subset of jobs in $L \cap W_2$ cannot improve schedule σ^*. To complete the proof it is sufficient to show that other possible exchanges cannot improve schedule σ^* either. Hence, σ^* is optimal.

Since in the exchanges which are performed in either Step 3 or Step 4 a job is exchanged exactly once, and since Step 1 and Step 2 are performed in at most $O(n \log n)$ time, Algorithm 7.19 runs in $O(n \log n)$ time.   ■

Now, let us assume that there are only two distinct deterioration rates, $b_j \in \{b_1, b_2\}$, $b_1 > b_2$. Let $B_1 := \{J_j \in \mathcal{J} : b_j = b_1\}$ and $B_2 := \{J_j \in \mathcal{J} : b_j = b_2\}$. Let σ, E and L be defined as in Lemma 7.137.

Lemma 7.140. (Sundararaghavan and Kunnathur [79]) *For a given instance of the problem of minimizing criterion $\sum w_j C_j$ for a single machine and job*

processing times in the form of (6.25), *let* $b_j \in \{b_1, b_2\}$, *where* $b_1 > b_2$. *Let*
E *denote the set of jobs which can be scheduled before time* D, $L := \mathcal{J} - E$,
$B_i := \{J_j \in \mathcal{J} : b_j = b_i\}$, $1 \leqslant i \leqslant 2$, *and let* $J_{[r]}$ *denote the job with the*
greatest starting time among the jobs in $E \cap B_1$. *Then the necessary conditions*
for optimality of a schedule for the problem are as follows:
(a) *if* $J_i \in E \cap B_1$ *and* $J_j \in L \cap B_1$, *then* $w_i \geqslant w_j$;
(b) *if* $J_i \in E \cap B_2$ *and* $J_j \in L \cap B_2$, *then* $w_i \geqslant w_j$;
(c) *if* $J_i \in E \cap B_1$ *and* $J_j \in L \cap B_1$, *then* $w_i > w_j$.

Proof. The result follows directly from the properties of an optimal schedule
for this problem. □

For the problem with only two deterioration rates, Sundararaghavan and
Kunnathur [79] proposed the modified Algorithm 7.19 in which $(b_1, b_2, \ldots, b_n)$
and (w_1, w_2) in the input are replaced with (b_1, b_2) and $(w_1, w_2, \ldots, w_n)$, respectively, and W_i is replaced with B_i, $1 \leqslant i \leqslant 2$, in each step of the algorithm.

Theorem 7.141. (Sundararaghavan and Kunnathur [79]) *The problem of*
minimizing criterion $\sum w_j C_j$ *for a single machine, job processing times in*
the form of (6.25) *and two distinct deterioration rates is solvable in* $O(n \log n)$
time by the modified Algorithm 7.19.

Proof. Similar to the proof of Theorem 7.139. □

Sundararaghavan and Kunnathur [79] also considered the case of *agreeable*
job weights and deterioration rates, when

$$w_i \geqslant w_j \Rightarrow b_i \geqslant b_j \tag{7.30}$$

for any $1 \leqslant i, j \leqslant n$.

Lemma 7.142. (Sundararaghavan and Kunnathur [79]) *For a given instance*
of the problem of minimizing criterion $\sum w_j C_j$ *for a single machine and job*
processing times in the form of (6.25), *let inequality* $w_i \geqslant w_j$ *imply inequality*
$b_i \geqslant b_j$ *for any* $1 \leqslant i, j \leqslant n$. *Let* E *denote the set of jobs which can be*
scheduled before time D *and* $L := \mathcal{J} - E$. *Then the condition*

$$if \ w_i < w_j \ and \ J_i \in E, \ then \ J_j \in E$$

is a necessary condition for optimality of a schedule for the problem.

Proof. Let us consider two jobs, J_i and J_j, such that (7.30) is satisfied. Let
us assume that schedule $\sigma^\star$, in which job J_i is followed by J_j, $C_i < D$ and
$C_j > D$, is optimal. Let us consider schedule σ' in which jobs J_i and J_j were
mutually exchanged. Let $R \subseteq L$ denote the set of jobs started after J_j in $\sigma^\star$
(started after J_i in σ'). Since

$$C_j(\sigma') = C_i(\sigma^\star) \tag{7.31}$$

and
$$C_i(\sigma') = C_j(\sigma^\star) - (b_j - b_i), \tag{7.32}$$
we have
$$\sum w_j C_j(\sigma') = \sum w_j C_j(\sigma^\star) - w_i C_i(\sigma^\star) - w_j C_j(\sigma^\star) + w_i C_i(\sigma') + w_j C_j(\sigma') + r,$$

where r is the change in the deterioration of jobs in R. By (7.30) we have $r < 0$. From that and from (7.31) and (7.32) we have $\sum w_j C_j(\sigma') - \sum w_j C_j(\sigma^\star) < 0$. A contradiction. $\qquad\square$

For the problem with agreeable parameters (7.30), Sundararaghavan and Kunnathur [79] proposed the following algorithm.

Algorithm 7.20. for problem $1|p_j| \sum w_j C_j$, $p_j \equiv$ (6.25) $\wedge$ (7.30)

---

1 **Input** : sequences $(b_1, b_2, \ldots, b_n)$, $(w_1, w_2, \ldots, w_n)$, numbers a, D,
2 **Output:** an optimal schedule
  ▷ Step 1
3 $k \leftarrow \lfloor \frac{D}{a} \rfloor + 1$;
4 Arrange k jobs in non-increasing order of the ratios $\frac{1}{w_j}$;
  ▷ If necessary, ties are broken in favour of jobs with
    larger b_j
  ▷ Step 2
5 Arrange the remaining $n - k$ jobs in non-decreasing order of the
  ratios $\frac{a+b_j}{w_j}$;
  ▷ Ties are broken arbitrarily
  ▷ Step 3
6 **return**.

Theorem 7.143. (Sundararaghavan and Kunnathur [79]) *The problem of minimizing criterion $\sum w_j C_j$ for a single machine, job processing times given by (6.25) and agreeable parameters is solvable in $O(n \log n)$ time by Algorithm 7.20.*

Proof. The proof is based on Lemma 7.142 and a lemma concerning the case when $w_i = w_j$ for some $1 \leqslant i, j \leqslant n$; see [79, Lemmas 3–5]. $\qquad\diamond$

For the general problem, Cheng and Ding [65] proposed an enumeration algorithm (the modified Algorithm 13.3; see Sect. 13.2.4 for details) which under some assumptions solves the problem in polynomial time.

Theorem 7.144. (Cheng and Ding [65]) *If an instance of problem $1|p_j \in \{a_j, b_j : a_j \leqslant b_j\}, d_j = D| \sum w_j C_j$ has a fixed number of chains, then it can be solved in polynomial time.*

Proof. Similar to the proof of Theorem 7.77. $\qquad\square$

7.4.6 Proportional-linear shortening

In this section, we consider the polynomially solvable single machine time-dependent scheduling problems with proportional-linearly shortening jobs.

Minimizing the total weighted completion time

Strusevich and Rustogi [78] proved that problem $1|p_j = b_j(1 - bt)|\sum w_j C_j$ is solvable in $O(n\log n)$ time by showing that $\omega(j) = \frac{w_j(1-b_j b)}{b_j b}$ is a 1-priority-generating function for the problem. The proof of the result is similar to the proof of Theorem 7.105; see [78, Sect. 9.1] for details.

In the next two results, $b_{\min}$ is defined as in inequality (6.39).

Minimizing the total number of tardy jobs

Theorem 7.145. (Wang and Xia [80]) *Problem* $1|p_j = b_j(1 - bt), b > 0,$
$$b\left(\sum_{j=1}^{n} b_j - b_{\min}\right) < 1|\sum U_j \text{ is solvable in } O(n\log n) \text{ time by Algorithm 7.15.}$$

Proof. Similar to the proof of Theorem 7.98.   □

Minimizing the maximal cost

Theorem 7.146. (Wang and Xia [80]) *Problem* $1|p_j = b_j(1 - bt), b > 0,$
$$b\left(\sum_{j=1}^{n} b_j - b_{\min}\right) < 1|f_{\max} \text{ is solvable in } O(n^2) \text{ time by Algorithm 7.17.}$$

Proof. Similar to the proof of Theorem 7.111.   □

7.4.7 Linear shortening

In this section, we consider polynomially solvable single machine time-dependent scheduling problem with linear shortening.

Minimizing the total weighted completion time

For job processing times given by (6.43), the following results are known.

Property 7.147. (Bachman et al. [58]) Problem $1|p_j = a - bt|\sum w_j C_j$ is solvable in $O(n\log n)$ time by scheduling jobs in non-increasing order of the weights w_j.

Proof. Since all jobs have the same form of job processing times, the result follows from Lemma 1.2 (b).   ∎

Property 7.148. (Bachman et al. [58]) Problem $1|p_j = a_j - ka_j t| \sum w_j C_j$, where k is a given constant, $k(\sum_{j=1}^{n} a_j - a_{\min}) < 1$ and $a_{\min} = \min_{1 \leqslant j \leqslant n} \{a_j\}$, is solvable in $O(n \log n)$ time by scheduling jobs in non-decreasing order of the ratios $\frac{a_j}{w_j(1-ka_j)}$.

Proof. By the pairwise job interchange argument; see [58, Property 2].   ◇

Property 7.149. (Bachman et al. [58]) Let $\sigma = (\sigma_1, \sigma_2, \ldots, \sigma_n)$ be a schedule for problem $1|p_j = a_j - bt| \sum w_j C_j$, where $w_j = ka_j$ for $1 \leqslant j \leqslant n$, and let $\bar{\sigma} = (\sigma_n, \sigma_{n-1}, \ldots, \sigma_1)$. Then $\sum w_j C_j(\sigma) = \sum w_j C_j(\bar{\sigma})$.

Proof. By direct computation; see [58, Property 3].   ◇

Property 7.150. (Bachman et al. [58]) For problem $1|p_j = a_j - bt| \sum w_j C_j$, where $w_j = ka_j$ for $1 \leqslant j \leqslant n$, there exists an optimal schedule that is V-shaped with respect to the basic processing times a_j.

Proof. Let us consider three schedules σ^*, σ' and σ'', differing only in the order of jobs $J_{\sigma_{i-1}}$, J_{σ_i} and $J_{\sigma_{i+1}}$. Let us assume that $a_{\sigma_i} - a_{\sigma_{i-1}} > 0$, $a_{\sigma_i} - a_{\sigma_{i+1}} > 0$ and that σ^* is optimal. Let us calculate differences $\sum w_j C_j(\sigma^*) - \sum w_j C_j(\sigma')$ and $\sum w_j C_j(\sigma^*) - \sum w_j C_j(\sigma'')$. Since σ^* is optimal, both these differences are positive. This, in turn, leads to a contradiction (see [58, Property 4]).   □

Minimizing the total number of tardy jobs

Theorem 7.49 implies that if jobs have identical ready times and deadlines, the problem of minimizing the number of tardy jobs is $\mathcal{NP}$-complete. Ho et al. [68] stated the problem of minimizing the number of tardy jobs for a set of jobs with a common deadline, D, as an open problem. The problem has been solved independently by two authors: Chen [61] and Woeginger [82]. We first consider the result obtained by Chen.

Lemma 7.151. (Chen [61]) *There exists an optimal schedule for problem $1|r_j = R, p_j = a_j - b_j t, a_j > 0, 0 < b_j < 1, b_j d_j < a_j \leqslant d_j, d_j = D| \sum U_j$, where the on-time jobs are scheduled in non-increasing order of the ratios $\frac{a_j}{b_j}$, and the late jobs are scheduled in an arbitrary order.*

Proof. First, note that without loss of generality we can consider only schedules without idle times. Second, since $d_j = D$, each on-time job is processed before all late jobs. Thus, since by Theorem 7.14 scheduling jobs in non-increasing order of the ratios $\frac{b_j}{a_j}$ minimizes the $C_{\max}$, the result follows.   □

Chen [61] proposed constructing a schedule for problem $1|r_j, p_j = a_j - b_j t| \sum U_j$, where $r_j = R$, $0 < b_j < 1$, $b_j d_j < a_j \leqslant d_j$ and $d_j = D$, using the following Algorithm 7.21 which assigns an unscheduled job with the largest ratio $\frac{a_j}{b_j}$ either to the position immediately following the current last on-time job or to the position following the current last late job.

Algorithm 7.21. for problem $1|r_j = R, p_j = a_j - b_j t, a_j > 0, 0 < b_j < 1,$
$b_j d_j < a_j \leqslant d_j, d_j = D| \sum U_j$

---

1 **Input** : sequences $(a_1, a_2, \ldots, a_n)$, $(b_1, b_2, \ldots, b_n)$, numbers R, D

2 **Output:** the minimum number of late jobs

   ▷ **Step 1**

3 Renumber jobs in non-increasing order of the ratios $\frac{a_j}{b_j}$;

4 $C(1,1) \leftarrow \begin{cases} a_{[1]}, & \text{if } a_{[1]} \leqslant D \\ +\infty, & \text{otherwise} \end{cases}$;

   ▷ **Step 2**

5 **for** $j \leftarrow 1$ **to** n **do**

6 $\quad$ $C(j,0) \leftarrow 0; S(j,0) \leftarrow \emptyset$;

7 $\quad$ $S(1,1) \leftarrow \begin{cases} \{1\}, & \text{if } C(1,1) = a_{[1]} \\ \emptyset, & \text{otherwise} \end{cases}$;

   end

   ▷ **Step 3**

8 **for** $j \leftarrow 2$ **to** n **do**

9 $\quad$ **for** $k \leftarrow 1$ **to** $j-1$ **do**

10 $\quad\quad$ $C(j,k) \leftarrow \min\{C(j-1,k), F(j,k)\}$;

11 $\quad\quad$ $x \leftarrow a_{[j]} + (1 - b_{[j]})C(j-1, j-1)$;

12 $\quad\quad$ $C(j,j) \leftarrow \begin{cases} x, & \text{if } x \leqslant D \\ +\infty, & \text{otherwise} \end{cases}$;

13 $\quad\quad$ $S(j,j) \leftarrow \begin{cases} \{j\} \cup S(j-1, j-1), & \text{if } C(j,j) \leqslant D \\ \emptyset, & \text{otherwise} \end{cases}$;

   $\quad$ **end**

   end

   ▷ **Step 4**

14 Calculate $k^\star \leftarrow \arg\max\{k : C(n,k) \leqslant D\}$;

   ▷ **Step 5**

15 **return** $n - k^\star$.

In the pseudo-code of Algorithm 7.21 symbols $C(j,k)$ and $S(j,k)$ denote the minimum completion time of the on-time jobs in a partial schedule containing the first j jobs, $1 \leqslant j \leqslant n$, among which there are exactly $k \leqslant j$ on-time jobs, and the set of on-time jobs in the schedule corresponding to $C(j,k)$, respectively. Function $F(j,k)$ is defined as follows:

$$F(j,k) := \begin{cases} a_j + (1 - b_j)C(j-1, k-1), & \text{if } a_j + (1 - b_j)C(j-1, k-1) \leqslant D, \\ +\infty, & \text{otherwise.} \end{cases}$$

The set $S(j,k)$ is defined as follows:

$$S(j,k) := \begin{cases} \{j\} \cup S(j-1, k-1), & \text{if } C(j,k) = F(j,k) \leqslant D, \\ S(j-1,k), & \text{if } C(j,k) = C(j-1,k) \leqslant D, \\ \emptyset, & \text{otherwise.} \end{cases}$$

Theorem 7.152. (Chen [61]) *Problem* $1|r_j, p_j = a_j - b_j t| \sum U_j$, *where* $r_j = R$, $0 < b_j < 1, b_j d_j < a_j \leqslant d_j$ *and* $d_j = D$ *is solvable in* $O(n^2)$ *time by Algorithm 7.21.*

Proof. Knowing the minimum number of late jobs, we can construct an optimal schedule by scheduling jobs from set $S(n, k^*)$ in the order of their renumbered indices, and then scheduling the remaining jobs in an arbitrary order. Since in Algorithm 7.21 for each $1 \leqslant j \leqslant n$ there are j possible states, the overall time complexity of the algorithm is $O(n^2)$.   ∎

Problem $1|r_j = R, p_j = a_j - b_j t, a_j > 0, 0 < b_j < 1, b_j d_j < a_j \leqslant d_j$, $d_j = D| \sum U_j$ was also considered by Woeginger [82].

Lemma 7.153. (Woeginger [82]) *A subset of jobs,* $j_1, j_2, \ldots, j_m$, *can be executed during time interval* $\langle t_1, t_2 \rangle$ *if and only if the jobs are processed in non-decreasing order of the ratios* $\frac{b_j}{a_j}$.

Proof. See [82, Lemma 2].   ◇

Applying Lemma 7.153, Woeginger showed how to find the solution using dynamic programming in $O(n^3)$ time (see [82, Theorem 3] for details).

Minimizing the total earliness cost

Zhao and Tang [85] considered job processing times given by (7.20). Let g be a non-decreasing function.

Theorem 7.154. (Zhao and Tang [85]) *Problem* $1|p_j = a_j - bt| \sum g(E_j)$ *is solvable in* $O(n \log n)$ *time by scheduling jobs in non-increasing order of the basic processing times* a_j.

Proof. By the pairwise job interchange argument.   □

Minimizing the total earliness and tardiness

Cheng et al. [64] considered the problem of minimizing the sum of earliness, tardiness and due-date penalties and common due-date assignment for job processing times given by (6.43) with $b_j = b$ for $1 \leqslant i \leqslant n$.

Applying the approach used previously for problem $1|p_j = a_j + bt| \sum(\alpha E_j + \beta T_j + \gamma d)$, Cheng et al. [64] proposed for problem $1|p_j = a_j - bt$, $0 < b < 1, b(\sum_{j=1}^{n} a_j - a_i) < a_i| \sum(\alpha E_j + \beta T_j + \gamma d)$ an algorithm (see [64, Algorithm A]). Since the properties (see [64, Properties 1–8]) on which this algorithm is based are counterparts of the properties from Sect. 7.4.3, we do not present their formulations, giving only the formulation of the main result.

Theorem 7.155. (Cheng et al. [64]) *Problem* $1|p_j = a_j - bt$, $0 < b < 1$, $b(\sum_{j=1}^{n} a_j - a_i) < a_i| \sum(\alpha E_j + \beta T_j + \gamma d)$ *is solvable in* $O(n \log n)$ *time.*

Proof. See [64, Properties 10–13, Theorem 14].   ◇

References

Minimizing the maximum completion time

1. A. Allahverdi, J. N. D. Gupta and T. Aldowaisan, A review of scheduling research involving setup considerations. *Omega* **27** (1999), no. 2, 219–239.
2. M. S. Barketau, T-C. E. Cheng, C-T. Ng, V. Kotov and M. Y. Kovalyov, Batch scheduling of step deteriorating jobs. *Journal of Scheduling* **11** (2008), no. 1, 17–28.
3. S. Browne and U. Yechiali, Scheduling deteriorating jobs on a single processor. *Operations Research* **38** (1990), no. 3, 495–498.
4. J-Y. Cai, P. Cai and Y. Zhu, On a scheduling problem of time deteriorating jobs. *Journal of Complexity* **14** (1998), no. CM980473, 190–209.
5. M-B. Cheng and S-J. Sun, Two scheduling problems in group technology with deteriorating jobs. *Applied Mathematics Journal of Chinese Universities*, Series B, **20** (2005), no. 2, 225–234.
6. T-C. E. Cheng and Q. Ding, Single machine scheduling with deadlines and increasing rates of processing times. *Acta Informatica* **36** (2000), no. 9–10, 673–692.
7. T-C. E. Cheng and Q. Ding, Single machine scheduling with step-deteriorating processing times. *European Journal of Operational Research* **134** (2001), no. 3, 623–630.
8. T-C. E. Cheng and Q. Ding, Scheduling start time dependent tasks with deadlines and identical initial processing times on a single machine. *Computers and Operations Research* **30** (2003), no. 1, 51–62.
9. T-C. E. Cheng and Q. Ding, The complexity of scheduling starting time dependent tasks with release times. *Information Processing Letters* **65** (1998), no. 2, 75–79.
10. T-C. E. Cheng, Q. Ding, M. Y. Kovalyov, A. Bachman and A. Janiak, Scheduling jobs with piecewise linear decreasing processing times. *Naval Research Logistics* **50** (2003), no. 6, 531–554.
11. S. Gawiejnowicz and B. M-T. Lin, Scheduling time-dependent jobs under mixed deterioration. *Applied Mathematics and Computation* **216** (2010), no. 2, 438–447.
12. (a) S. Gawiejnowicz and L. Pankowska, Scheduling jobs with variable processing times. Report 020/1994, Faculty of Mathematics and Computer Science, Adam Mickiewicz University, Poznań, October 1994 (in Polish).
(b) S. Gawiejnowicz and L. Pankowska, Scheduling jobs with variable processing times. Report 027/1995, Faculty of Mathematics and Computer Science, Adam Mickiewicz University, Poznań, January 1995 (in English).
13. S. Gawiejnowicz and L. Pankowska, Scheduling jobs with varying processing times. *Information Processing Letters* **54** (1995), no. 3, 175–178.
14. A-X. Guo and J-B. Wang, Single machine scheduling with deteriorating jobs under the group technology assumption. *International Journal of Pure and Applied Mathematics* **18** (2005), no. 2, 225–231.
15. J. N. D. Gupta and S. K. Gupta, Single facility scheduling with nonlinear processing times. *Computers and Industrial Engineering* **14** (1988), no. 4, 387–393.
16. K. I-J. Ho, J. Y-T. Leung and W-D. Wei, Complexity of scheduling tasks with time-dependent execution times. *Information Processing Letters* **48** (1993), no. 6, 315–320.

17. F. Jaehn and H. A. Sedding, Scheduling with time-dependent discrepancy times. *Journal of Scheduling* **19** (2016), no. 6, 737–757.
18. A. Kononov, Single machine scheduling problems with processing times proportional to an arbitrary function. *Discrete Analysis and Operations Research* **5** (1998), no. 3, 17–37 (in Russian).
19. W. Kubiak and S. L. van de Velde, Scheduling deteriorating jobs to minimize makespan. *Naval Research Logistics* **45** (1998), no. 5, 511–523.
20. A. S. Kunnathur and S. K. Gupta, Minimizing the makespan with late start penalties added to processing times in a single facility scheduling problem. *European Journal of Operational Research* **47** (1990), no. 1, 56–64.
21. W-H. Kuo and D-L. Yang, Single-machine scheduling problems with start-time dependent processing time. *Computers and Mathematics with Applications* **53** (2007), no. 11, 1658–1664.
22. W-H. Kuo and D-L. Yang, Single-machine scheduling with deteriorating jobs. *International Journal of Systems Science* **43** (2012), 132–139.
23. J. Y-T. Leung, C-T. Ng and T. C-E. Cheng, Minimizing sum of completion times for batch scheduling of jobs with deteriorating processing times. *European Journal of Operational Research* **187** (2008), no. 3, 1090–1099.
24. O. I. Melnikov and Y. M. Shafransky, Parametric problem of scheduling theory. *Kibernetika* **6** (1979), no. 3, 53–57 (in Russian).
    English translation: *Cybernetics and System Analysis* **15** (1980), no. 3, 352–357.
25. G. Mosheiov, Scheduling jobs under simple linear deterioration. *Computers and Operations Research* **21** (1994), no. 6, 653–659.
26. C. N. Potts and M. Y. Kovalyov, Scheduling with batching: a review. *European Journal of Operational Research* **120** (2000), no. 2, 228–249.
27. V. A. Strusevich and K. Rustogi, *Scheduling with Time-Changing Effects and Rate-Modifying Activities*. Berlin-Heidelberg: Springer 2017.
28. V. S. Tanaev, V. S. Gordon and Y. M. Shafransky, *Scheduling Theory: Single-stage Systems*. Dordrecht: Kluwer 1994.
29. V. S. Tanaev, M. Y. Kovalyov and Y. M. Shafransky, *Scheduling Theory. Group Technologies*. Minsk: Institute of Technical Cybernetics, National Academy of Sciences of Belarus 1998 (in Russian).
30. W. Wajs, Polynomial algorithm for dynamic sequencing problem. *Archiwum Automatyki i Telemechaniki* **31** (1986), no. 3, 209–213 (in Polish).
31. J-B. Wang, A-X. Guo, F. Shan, B. Jiang and L-Y. Wang, Single machine group scheduling under decreasing linear deterioration. *Journal of Applied Mathematics and Computing* **24** (2007), no. 1–2, 283–293.
32. J-B. Wang, L. Lin and F. Shan, Single-machine group scheduling problems with deteriorating jobs. *International Journal of Advanced Manufacturing Technology* **39** (2008), no. 7–8, 808–812.
33. J-B. Wang and M-Z. Wang, Single-machine scheduling with nonlinear deterioration. *Optimization Letters* **6** (2012), 87–98.
34. J-B. Wang and Z-Q. Xia, Scheduling jobs under decreasing linear deterioration. *Information Processing Letters* **94** (2005), no. 2, 63–69.
35. S. T. Webster and K. R. Baker, Scheduling groups of jobs on a single machine. *Operations Research* **43** (1995), no. 4, 692–703.
36. C-C. Wu, Y-R. Shiau and W-C. Lee, Single-machine group scheduling problems with deterioration consideration. *Computers and Operations Research* **35** (2008), no. 5, 1652–1659.

37. C-L. Zhao, Q-L. Zhang and H-Y. Tang, Scheduling problems under linear deterioration. *Acta Automatica Sinica* **29** (2003), no. 4, 531–535.
38. C-L. Zhao, Q-L. Zhang and H-Y. Tang, Single machine scheduling with linear processing times. *Acta Automatica Sinica* **29** (2003), no. 5, 703–708.

Minimizing the total completion time

39. T-C. E. Cheng, Q. Ding and B. M. T. Lin, A concise survey of scheduling with time-dependent processing times. *European Journal of Operational Research* **152** (2004), no. 1, 1–13.
40. T-C. E. Cheng and Q. Ding, Single machine scheduling with deadlines and increasing rates of processing times. *Acta Informatica* **36** (2000), no. 9–10, 673–692.
41. T-C. E. Cheng and Q. Ding, Single machine scheduling with step-deteriorating processing times. *European Journal of Operational Research* **134** (2001), no. 3, 623–630.
42. T-C. E. Cheng and Q. Ding, The complexity of single machine scheduling with two distinct deadlines and identical decreasing rates of processing times. *Computers and Mathematics with Applications* **35** (1998), no. 12, 95–100.
43. S. Gawiejnowicz, *Scheduling jobs with varying processing times*. Ph. D. dissertation, Poznań University of Technology, Poznań 1997, 102 pp. (in Polish).
44. G. Mosheiov, Scheduling jobs under simple linear deterioration. *Computers and Operations Research* **21** (1994), no. 6, 653–659.
45. C-T. Ng, T-C. E. Cheng, A. Bachman and A. Janiak, Three scheduling problems with deteriorating jobs to minimize the total completion time. *Information Processing Letters* **81** (2002), no. 6, 327–333.
46. K. Ocetkiewicz, Polynomial case of V-shaped policies in scheduling deteriorating jobs. Report ETI-13/06, Gdańsk University of Technology, Gdańsk, Poland, November 2006.
47. W. E. Smith, Various optimizers for single-stage production. *Naval Research Logistics Quarterly* **3** (1956), no. 1–2, 59–66.
48. V. A. Strusevich and K. Rustogi, *Scheduling with Time-Changing Effects and Rate-Modifying Activities*. Berlin-Heidelberg: Springer 2017.
49. J-B. Wang, A-X. Guo, F. Shan, B. Jiang and L-Y. Wang, Single machine group scheduling under decreasing linear deterioration. *Journal of Applied Mathematics and Computing* **24** (2007), no. 1–2, 283–293.
50. J-B. Wang and M-Z. Wang, Single-machine scheduling with nonlinear deterioration. *Optimization Letters* **6** (2012), 87–98.
51. C-C. Wu, Y-R. Shiau and W-C. Lee, Single-machine group scheduling problems with deterioration consideration. *Computers and Operations Research* **35** (2008), no. 5, 1652–1659.

Minimizing the maximum lateness

52. A. Agnetis, J-C. Billaut, S. Gawiejnowicz, D. Pacciarelli and A. Soukhal, *Multi-agent Scheduling: Models and Algorithms*. Berlin-Heidelberg: Springer 2014.
53. T-C. E. Cheng and Q. Ding, Single machine scheduling with deadlines and increasing rates of processing times. *Acta Informatica* **36** (2000), no. 9–10, 673–692.

54. A. Kononov, Single machine scheduling problems with processing times proportional to an arbitrary function. *Discrete Analysis and Operations Research* **5** (1998), no. 3, 17–37 (in Russian).
55. G. Mosheiov, Scheduling jobs under simple linear deterioration. *Computers and Operations Research* **21** (1994), no. 6, 653–659.
56. C-L. Zhao, Q-L. Zhang and H-Y. Tang, Scheduling problems under linear deterioration. *Acta Automatica Sinica* **29** (2003), no. 4, 531–535.

Other criteria

57. B. Alidaee and F. Landram, Scheduling deteriorating jobs on a single machine to minimize the maximum processing times. *International Journal of Systems Science* **27** (1996), no. 5, 507–510.
58. A. Bachman, T-C. E. Cheng, A. Janiak and C-T. Ng, Scheduling start time dependent jobs to minimize the total weighted completion time. *Journal of the Operational Research Society* **53** (2002), no. 6, 688–693.
59. K. R. Baker and G. D. Scudder, Sequencing with earliness and tardiness penalties: a review. *Operations Research* **38** (1990), no. 1, 22–36.
60. S. Browne and U. Yechiali, Scheduling deteriorating jobs on a single processor. *Operations Research* **38** (1990), no. 3, 495–498.
61. Z-L. Chen, A note on single-processor scheduling with time-dependent execution times. *Operations Research Letters* **17** (1995), no. 3, 127–129.
62. M-B. Cheng and S-J. Sun, Two scheduling problems in group technology with deteriorating jobs. *Applied Mathematics Journal of Chinese Universities, Series B*, **20** (2005), 225–234.
63. T-C. E. Cheng, L-Y. Kang and C-T. Ng, Due-date assignment and single machine scheduling with deteriorating jobs. *Journal of the Operational Research Society* **55** (2004), no. 2, 198–203.
64. T-C. E. Cheng, L-Y. Kang and C-T. Ng, Single machine due-date scheduling of jobs with decreasing start-time dependent processing times. *International Transactions in Operational Research* **12** (2005), no. 4, 355–366.
65. T-C. E. Cheng and Q. Ding, Single machine scheduling with step-deteriorating processing times. *European Journal of Operational Research* **134** (2001), no. 3, 623–630.
66. V. S. Gordon, J-M. Proth and C. Chu, A survey of the state-of-the-art of common due date assignment and scheduling research. *European Journal of Operational Research* **139** (2002), no. 1, 1–5.
67. V. S. Gordon, J-M. Proth and C. Chu, Due date assignment and scheduling: SLK, TWK and other due date assignment models. *Production Planning and Control* **13** (2002), no. 2, 117–32.
68. K. I-J. Ho, J. Y-T. Leung and W-D. Wei, Complexity of scheduling tasks with time-dependent execution times. *Information Processing Letters* **48** (1993), no. 6, 315–320.
69. A. Kononov, Single machine scheduling problems with processing times proportional to an arbitrary function. *Discrete Analysis and Operations Research* **5** (1998), no. 3, 17–37 (in Russian).
70. W-H. Kuo and D-L. Yang, A note on due-date assignment and single-machine scheduling with deteriorating jobs. *Journal of the Operational Research Society* **59** (2008), no. 6, 857–859.

71. W-H. Kuo and D-L. Yang, Single-machine scheduling problems with start-time dependent processing time. *Computers and Mathematics with Applications* **53** (2007), no. 11, 1658–1664.

72. E. L. Lawler, Optimal sequencing of a single machine subject to precedence constraints. *Management Science* **19** (1973), no. 5, 544–546.

73. J. Moore, An *n* job, one machine sequencing algorithm for minimizing the number of late jobs. *Management Science* **15** (1968), no. 1, 102–109.

74. G. Mosheiov, Λ-shaped policies to schedule deteriorating jobs. *Journal of the Operational Research Society* **47** (1996), no. 9, 1184–1191.

75. G. Mosheiov, Scheduling jobs under simple linear deterioration. *Computers and Operations Research* **21** (1994), no. 6, 653–659.

76. D. Oron, Single machine scheduling with simple linear deterioration to minimize total absolute deviation of completion times. *Computers and Operations Research* **35** (2008), no. 6, 2071–2078.

77. S. S. Panwalkar, M. L. Smith and A. Seidmann, Common due-date assignment to minimize total penalty for the one machine scheduling problem. *Operations Research* **30** (1982), no. 2, 391–399.

78. V. A. Strusevich and K. Rustogi, *Scheduling with Time-Changing Effects and Rate-Modifying Activities.* Berlin-Heidelberg: Springer 2017.

79. P. S. Sundararaghavan and A. S. Kunnathur, Single machine scheduling with start time dependent processing times: some solvable cases. *European Journal of Operational Research* **78** (1994), no. 3, 394–403.

80. J-B. Wang and Z-Q. Xia, Scheduling jobs under decreasing linear deterioration. *Information Processing Letters* **94** (2005), no. 2, 63–69.

81. J-B. Wang, L. Lin and F. Shan, Single-machine group scheduling problems with deteriorating jobs. *International Journal of Advanced Manufacturing Technology* **39** (2008), no. 7–8, 808–812.

82. G. J. Woeginger, Scheduling with time-dependent execution times. *Information Processing Letters* **54** (1995), no. 3, 155–156.

83. F. Xu, A-X. Guo, J-B. Wang and F. Shan, Single machine scheduling problem with linear deterioration under group technology. *International Journal of Pure and Applied Mathematics* **28** (2006), no. 3, 401–406.

84. C-L. Zhao, Q-L. Zhang and H-Y. Tang, Scheduling problems under linear deterioration. *Acta Automatica Sinica* **29** (2003), 531–535.

85. C-L. Zhao and H-Y. Tang, Single machine scheduling problems with deteriorating jobs. *Applied Mathematics and Computation* **161** (2005), 865–874.

8

Polynomial parallel machine problems

I n the previous chapter, we discussed polynomially solvable single machine time-dependent scheduling problems. In this chapter, we consider similar results concerning time-dependent scheduling on parallel machines.

Chapter 8 is composed of two sections. In Sect. 8.1, we focus on parallel machines and minimization of the $\sum C_j$ criterion. In Sect. 8.2, we discuss the results concerning parallel machines and minimization of other criteria than $\sum C_j$. The chapter is completed with a list of references.

8.1 Minimizing the total completion time

In this section, we consider polynomially solvable parallel-machine time-dependent scheduling problems with the $\sum C_j$ criterion.

8.1.1 Linear deterioration

The result below can be proved using a transformation of the $\sum C_j$ criterion into a sum of products of job processing times and *job positional weights* which express the contribution of the jobs to the overall value of the criterion for a given schedule. Let $b^{[i]}$ denote a common deterioration rate for jobs assigned to machine $M_{[i]}$, where $1 \leqslant i \leqslant m$.

Theorem 8.1. (Strusevich and Rustogi [6]) (*a*) *Problem* $Qm|p_{ij} = \frac{a_j}{s_i} + b^{[i]}t| \sum C_j$ *is solvable in* $O(n \log n)$ *time.* (*b*) *Problem* $Rm|p_{ij} = a_{ij} + b^{[i]}t| \sum C_j$ *is solvable in* $O(n^3)$ *time.*

Proof. (a) An optimal schedule for the considered problem can be found as follows. First, we construct an $n \times m$ matrix, where each column represents all possible job positional weights that can be associated with a particular machine. Next, we choose the n smallest among the weights and assign the jobs with the largest basic processing times to the positions associated with

© Springer-Verlag GmbH Germany, part of Springer Nature 2020
S. Gawiejnowicz, *Models and Algorithms of Time-Dependent Scheduling*,
Monographs in Theoretical Computer Science. An EATCS Series,
https://doi.org/10.1007/978-3-662-59362-2_8

the smallest positional weights. This matching can be done in $O(n \log n)$; see [6, Sect. 11.3.1] for details.   ◇

(b) We proceed similarly as in case (a), using a cost matrix, where rows and columns correspond to jobs and positions of the jobs, respectively. Since the $\sum C_j$ criterion can be expressed as a sum of costs defined by the matrix, the considered problem reduces to a linear assignment problem solvable in $O(n^3)$ time [1]; see [6, Sect. 11.3.2] for details.   ◇

8.1.2 Simple non-linear deterioration

As we will show in Sect. 11.2.2, many time-dependent parallel machine scheduling problems with non-linearly deteriorating jobs and the $\sum C_j$ criterion are intractable. However, some properties of such problems are known.

Theorem 8.2. (Gawiejnowicz [3]) *If $f(t)$ is an arbitrary increasing function such that $f(t) \geqslant 0$ for $t \geqslant 0$, then there exists an optimal schedule for problem $Pm|p_j = a_j + f(t)|\sum C_j$ in which jobs are scheduled on each machine in non-decreasing order of the basic processing times a_j.*

Proof. Let σ be a schedule for problem $Pm|p_j = a_j + f(t)|\sum C_j$ such that not all jobs are scheduled in non-decreasing order of the basic processing times a_j. Then there must exist a machine M_k, $1 \leqslant k \leqslant m$, such that some jobs assigned to the machine are scheduled in non-increasing order of the processing times. By changing this order to a non-decreasing order we obtain, by Theorem 7.74, a new schedule σ' such that $\sum C_j(\sigma') \leqslant \sum C_j(\sigma)$. Repeating, if necessary, the above change for other machines, we obtain an optimal schedule σ^* in which jobs on all machines are scheduled in non-decreasing order of a_j values.   ∎

Remark 8.3. Theorem 8.2 describes a necessary condition of optimality of problem $Pm|p_j = a_j + f(t)|\sum C_j$. Kuo and Yang [5] proved that for $f(t) = bt$ the condition is also a sufficient one; see [5, Sect. 6] for details.

Remark 8.4. Kuo et al. [4] considered a parallel unrelated machine time-dependent scheduling problem with the same form of job processing times as in Remark 8.3. The authors solved the problem using the approach similar to the one used in the proof of Theorem 8.6; see [4, Thm. 1] for details.

8.1.3 Linear shortening

Some time-dependent scheduling problems with shortening job processing times given by (6.49) are polynomially solvable, if the jobs are scheduled on parallel uniform and parallel unrelated machines.

Theorem 8.5. (Cheng et al. [2]) *Problem $Q|p_j = a_j - b(t - y), y = 0$, $Y = +\infty|\sum C_j$ is solvable in $O(n \log n)$ time by scheduling jobs in non-increasing order of the basic processing times a_j.*

Proof. If the machines are uniform, then $a_{lj} = \frac{a_j}{s_l}$. In this case, the total completion time is a weighted sum of a_j values, with weights $b_{lr} = \frac{1-(1-b)^r}{bs_l}$. No weight may be used more than once. Therefore, in order to minimize the value of the $\sum C_j$ criterion we should select the n smallest weights of all mn weights and match the selected weights with the largest a_j values. This can be done in $O(n \log n)$ time.   □

The next problem is solvable by an algorithm which solves a matching problem. We will call the algorithm AMP.

Theorem 8.6. (Cheng et al. [2]) *Problem* $R|p_{ij} = a_{ij} - b(t - y), y = 0,$ $Y = +\infty| \sum C_j$ *is solvable in* $O(n^3)$ *time by algorithm* AMP.

Proof. Let us introduce the variables $x_{(l,r),j}$ such that $x_{(l,r),j} = 1$ if job j is sequenced rth last on machine M_l, and $x_{(l,r),j} = 0$ otherwise. The problem under consideration is equivalent to the following weighted bipartite matching problem:

$$\text{minimize} \sum_{l,r} \sum_j x_{(l,r),j} a_{lj} \frac{1 - (1 - b)^r}{b}$$

subject to

$$\sum_{l,r} x_{(l,r),j} = 1, \quad j = 1, 2, \ldots, n,$$

$$\sum_j x_{(l,r),j} \leqslant 1, \quad l = 1, 2, \ldots, m, r = 1, 2, \ldots, n,$$

$$x_{(l,r),j} \in \{0, 1\} \quad l = 1, 2, \ldots, m, r = 1, 2, \ldots, n,$$

where the summation is taken over all values of l and r or j. This matching problem can be solved in $O(n^3)$ time.   □

The next result is similar to Theorem 8.1.

Theorem 8.7. (Strusevich and Rustogi [6]) *(a) Problem* $Qm|p_{ij} = \frac{a_j}{s_i} - b^{[i]}t| \sum C_j$ *is solvable in* $O(n \log n)$ *time. (b) Problem* $Rm|p_{ij} = a_{ij} - b^{[i]}t| \sum C_j$ *is solvable in* $O(n^3)$ *time.*

Proof. (a) Similar to the proof of Theorem 8.1 (a); see [6, Sect. 11.3.1] for details.

(b) Similar to the proof of Theorem 8.1 (b); see [6, Sect. 11.1.3] for details.
                                                                    ◇

Remark 8.8. Theorem 8.7 is similar to Theorem 8.1, since properties of time-dependent scheduling problems with linearly shortening job processing times are closely related to their counterparts for linearly deteriorating job processing times. These similarities can be explained using the notion of *mutually related scheduling problems*; see Sect. 19.4 and 19.5 for details.

8.2 Minimizing the total weighted earliness and tardiness

In this section, we consider polynomially solvable parallel-machine time-dependent scheduling problems with linear jobs and the criterion of the total weighted earliness and tardiness.

Cheng et al. [7] proved that problem $Pm|p_j = a_j + bt| \sum(\alpha E_j + \beta T_j + \gamma d)$ can be solved in polynomial time, if $\gamma = 0$. Before we state the result, we introduce new notation.

Let $\sigma = (\sigma_1, \sigma_2, \ldots, \sigma_m)$ be a schedule for problem $Pm|p_j = a_j + bt| \sum(\alpha E_j + \beta T_j)$, let n_i be the number of jobs scheduled on machine M_i and let d_i be the optimal due-date for jobs scheduled on machine M_i, $1 \leqslant i \leqslant m$. (By Lemma 7.124 we have $d_i = C_{\sigma_i(K_i)}$, where $K_i = \lceil \frac{n_i \beta}{\alpha + \beta} \rceil$ and $C_{\sigma_i(K_i)}$ denotes the index of the job scheduled as the K_ith in subschedule σ_i, $1 \leqslant i \leqslant m$.)

Let us define

$$m_{i,k_i} := b \sum_{j=k_i}^{K_i} (\alpha(j-1))B^{j-k_i} + b \sum_{j=K_i+1}^{n_i} \beta(n_i+1-j)B^{j-k_i}$$

for $1 \leqslant i \leqslant m, 2 \leqslant k_i \leqslant K_i$ and

$$m_{i,k_i} := b \sum_{j=k_i}^{n_i} (\beta(n_i+1-1)B^{j-k_i}$$

for $1 \leqslant i \leqslant m, K_i + 1 \leqslant k_i \leqslant n_i$, where $B := 1 + b$.

For $1 \leqslant j \leqslant n$ and $1 \leqslant i \leqslant m$, let us define

$$c_{j,(i,k_i)} := (\alpha(k-1) + m_{i,k_i+1})a_j$$

if $1 \leqslant k_i \leqslant K_i$ and

$$c_{j,(i,k_i)} := (\beta(n_i+1-k_i) + m_{i,k_i+1})a_j$$

if $K_i + 1 \leqslant k_i \leqslant n_i$.

Applying the notation, we have $d = \max\{C_{\sigma_i(K_i)} : 1 \leqslant i \leqslant m\}$ and $\sum(\alpha E_j + \beta T_j) \equiv \sum_{i=1}^{m} \sum_{k_i=1}^{n_i} c_{\sigma_i(k_i),(i,k_i)}$.

Let $A := \{(n_1, n_2, \ldots, n_m)\} \in \mathbb{Z}^m$ be the set of all m-elements sequences such that $1 \leqslant n_i \leqslant n - m$ and $\sum_{i=1}^{m} n_i = n$. For any $(n_1, n_2, \ldots, n_m) \in A$, $1 \leqslant i \leqslant m$, $1 \leqslant j \leqslant n$ and $1 \leqslant k_i \leqslant n_i$, let $x_{j,(i,k_i)}$ be a variable such that $x_{j,(i,k_i)} = 1$ if job J_j is scheduled as the k_ith job on machine M_i and $x_{j,(i,k_i)} = 0$ otherwise. Then problem $Pm|p_j = a_j + bt| \sum(\alpha E_j + \beta T_j)$ is equivalent to the following weighted bipartite matching problem:

$$\text{Minimize} \quad \sum_j \sum_{(i,k_i)} c_{j,(i,k_i)} x_{j,(i,k_i)} \tag{8.1}$$

subject to

$$\sum_{(i,k_i)} x_{j,(i,k_i)} = 1 \quad \text{for} \quad j = 1, 2, \ldots, m; \tag{8.2}$$

$$\sum_j x_{j,(i,k_i)} = 1 \quad \text{for} \quad i = 1, 2, \ldots, m; j = 1, 2, \ldots, m; \tag{8.3}$$

$$x_{j,(i,k_i)} \in \{0,1\} \quad \text{for} \quad j = 1, 2, \ldots, n; i = 1, 2 \ldots, m; k_i = 1, 2, \ldots, n. \tag{8.4}$$

Now we can formulate an algorithm (see Algorithm 8.1), proposed by Cheng et al. [7] for problem $Pm|p_j = a_j + bt|\sum(\alpha E_j + \beta T_j)$.

Algorithm 8.1. for problem $Pm|p_j = a_j + bt|\sum(\alpha E_j + \beta T_j)$

---

1 **Input** : sequence $(a_1, a_2, \ldots, a_n)$, numbers b, α, β
2 **Output:** an optimal schedule $\sigma^\star$
▷ Step 1
3 Construct set

$$A := \left\{ (n_1, n_2, \ldots, n_m) \in \mathbb{Z}^m : 1 \leqslant n_i \leqslant n - m \wedge \sum_{i=1}^m n_i = n \right\};$$

▷ Step 2
4 **for all** $(n_1, n_2, \ldots, n_m) \in A$ **do**
  | Solve problem (8.1)–(8.4);
  end
▷ Step 3
5 Compute $\min \{g(n_1, n_2, \ldots, n_m) : (n_1, n_2, \ldots, n_m) \in A\}$;
▷ Step 4
6 $\sigma^\star \leftarrow$ the schedule corresponding to the minimum computed in Step 3;
7 **return** $\sigma^\star$.

Remark 8.9. In pseudo-code of Algorithm 8.1 we use the following notation:

$$g(n_1, n_2, \ldots, n_m) := \min \left\{ \sum_j \sum_{(i,k_i)} c_{j,(i,k_i)} x_{j,(i,k_i)} \right\},$$

$\sigma(n_1, n_2, \ldots, n_m)$ is the schedule corresponding to $(n_1, n_2, \ldots, n_m)$, and

$$d(n_1, n_2, \ldots, n_m) := \max\{C_{\sigma_i(K_i)} : 1 \leqslant i \leqslant m\}.$$

Remark 8.10. If $C_{\sigma_i(K_i)} < d(n_1, n_2, \ldots, n_m)$, then we change the starting time of machine M_i to $d(n_1, n_2, \ldots, n_m) - C_{\sigma_i(K_i)}$.

Theorem 8.11. (Cheng et al. [7]) *Problem* $Pm|p_j = a_j + bt|\sum(\alpha E_j + \beta T_j)$ *is solvable in* $O(n^{m+1} \log n)$ *time by Algorithm 8.1.*

Proof. Algorithm 8.1 generates all possible sequences $(n_1, n_2, \ldots, n_m)$ in which $1 \leqslant n_i \leqslant n - m$ and $\sum_{i=1}^{m} n_i = n$. There are $O(n^m)$ of such sequences. Since each sequence $(n_1, n_2, \ldots, n_m)$ corresponds to a schedule for problem $Pm|p_j = a_j + bt| \sum (\alpha E_j + \beta T_j)$ and since the solution of problem (8.1)–(8.4) needs $O(n \log n)$ time for jobs arranged in the $a_j \nearrow$ order, the result follows.

$\square$

References

Minimizing the total completion time

1. R. Burkard, M. Dell'Amico, S. Martello, *Assignment Problems*, revised reprint, SIAM, Philadelphia, 2012.
2. T-C. E. Cheng, Q. Ding, M. Y. Kovalyov, A. Bachman and A. Janiak, Scheduling jobs with piecewise linear decreasing processing times. *Naval Research Logistics* **50** (2003), no. 6, 531–554.
3. S. Gawiejnowicz, *Scheduling jobs with varying processing times*. Ph. D. dissertation, Poznań University of Technology, Poznań 1997, 102 pp. (in Polish).
4. W-H. Kuo, C-J. Hsu and D-L. Yang, A note on unrelated parallel machine scheduling with time-dependent processing times. *Journal of the Operational Research Society* **60** (2009), 431–434.
5. W-H. Kuo and D-L. Yang, Parallel-machine scheduling with time-dependent processing times. *Theoretical Computer Science* **393** (2008), 204–210.
6. V. A. Strusevich and K. Rustogi, *Scheduling with Time-Changing Effects and Rate-Modifying Activities*. Berlin-Heidelberg: Springer 2017.

Minimizing the total weighted earliness and tardiness

7. T-C. E. Cheng, L-Y. Kang and C-T. Ng, Due-date assignment and parallel-machine scheduling with deteriorating jobs. *Journal of the Operational Research Society* **58** (2007), no. 8, 1103–1108.

9

Polynomial dedicated machine problems

This is the last chapter of the third part of the book, devoted to polynomially solvable time-dependent scheduling problems. In this chapter, we focus on dedicated machine time-dependent scheduling problems.

Chapter 9 is composed of four sections. In Sect. 9.1, we consider polynomially solvable problems on dedicated machines with minimization of the $C_{\max}$ criterion. In Sects. 9.2 and 9.3, we present similar results for minimization of the $\sum C_j$ and the $L_{\max}$ criterion, respectively. In Sect. 9.4, we address the problems of minimization of other criteria than $C_{\max}$, $\sum C_j$ and $L_{\max}$. The chapter is completed with a list of references.

9.1 Minimizing the maximum completion time

In this section, we consider polynomially solvable time-dependent dedicated machine scheduling problems with the $C_{\max}$ criterion.

9.1.1 Proportional deterioration

Unlike parallel machine time-dependent scheduling problems with proportional job processing times (6.7), some dedicated machine time-dependent scheduling problems with such job processing times are solvable in polynomial time.

Flow shop problems

Lemma 9.1. (Mosheiov [8]) *There exists an optimal schedule for problem* $F2|p_{ij} = b_{ij}t|C_{\max}$, *in which the job sequence is identical on both machines.*

Proof. The result is a special case of Lemma 9.10. $\square$

© Springer-Verlag GmbH Germany, part of Springer Nature 2020
S. Gawiejnowicz, *Models and Algorithms of Time-Dependent Scheduling*,
Monographs in Theoretical Computer Science. An EATCS Series,
https://doi.org/10.1007/978-3-662-59362-2_9

Presented below Algorithm 9.1 which solves problem $F2|p_{ij} = b_{ij}t|C_{\max}$ is an adapted version of Johnson's algorithm, [3], for problem $F2||C_{\max}$: instead of the processing times p_{ij} we use the deterioration rates b_{ij}, where $1 \leqslant i \leqslant m$ and $1 \leqslant j \leqslant n$. The pseudo-code of the algorithm is as follows.

Algorithm 9.1. for problem $F2|p_{ij} = b_{ij}t|C_{\max}$

---

1 **Input** : sequence $((b_{11}, b_{21}), (b_{12}, b_{22}), \ldots, (b_{1n}, b_{2n}))$
2 **Output:** an optimal schedule $\sigma^\star$
  ▷ Step 1
3 $\mathcal{J}_1 \leftarrow \{J_j \in \mathcal{J} : b_{1j} \leqslant b_{2j}\}$;
4 $\mathcal{J}_2 \leftarrow \{J_j \in \mathcal{J} : b_{1j} > b_{2j}\}$;
  ▷ Step 2
5 Arrange jobs in $\mathcal{J}_1$ in non-decreasing order of the deterioration rates b_{1j};
6 Call this sequence $\sigma^{(1)}$;
7 Arrange jobs in $\mathcal{J}_2$ in non-increasing order of the deterioration rates b_{2j};
8 Call this sequence $\sigma^{(2)}$;
  ▷ Step 3
9 $\sigma^\star \leftarrow (\sigma^{(1)}|\sigma^{(2)})$;
10 **return** $\sigma^\star$.

Theorem 9.2. (Kononov [4], Mosheiov [8]) *Problem $F2|p_{ij} = b_{ij}t|C_{\max}$ is solvable in $O(n \log n)$ time by Algorithm 9.1.*

Proof. Kononov [4] proves the result applying Lemma 19.72 and Theorem 19.93; see [4, Section 5] for details.

Mosheiov [8] analyses all possible cases in which jobs are scheduled in another order than the one generated by Algorithm 9.1, and proves that the respective schedule cannot be optimal in any of these cases; see [8, Theorem 1] for details.   □

Example 9.3. Let us consider an instance of problem $F2|p_{ij} = b_{ij}t|C_{\max}$, in which $n = 3$, $b_{11} = 2$, $b_{21} = 3$, $b_{12} = 3$, $b_{22} = 1$, $b_{13} = 4$, $b_{23} = 4$ and $t_0 = 1$. Then $\mathcal{J}_1 = \{J_1, J_3\}$, $\mathcal{J}_2 = \{J_2\}$, $\sigma^{(1)} = (1, 3)$, $\sigma^{(2)} = (2)$ and Algorithm 9.1 generates schedule $\sigma^\star = (\sigma^{(1)}|\sigma^{(2)}) = (1, 3, 2)$.   ◆

Remark 9.4. Wang and Xia [15] considered a number of flow shop problems with dominating machines (see Definitions 9.14–9.15). Since the results (see [15, Section 3]) are special cases of the results for job processing times given by (6.9) with $b = 1$, we do not present them here.

Remark 9.5. Polynomially solvable three-machine flow shop problems with dominating machines, proportional jobs and the $C_{\max}$ criterion were considered by Wang [13] and Wang et al. [14].

Open shop problems

We start this section with a lower bound on the optimal value of the maximum completion time in a two-machine open shop.

Lemma 9.6. (Mosheiov [8]) *The optimal value of the maximum completion time for problem* $O2|p_{ij} = b_{ij}t|C_{\max}$ *is not less than*

$$\max \left\{ \prod_{j=1}^{n}(1 + b_{1j}), \prod_{j=1}^{n}(1 + b_{2j}), \max_{1 \leqslant j \leqslant n}\{(1 + b_{1j})(1 + b_{2j})\} \right\}. \qquad (9.1)$$

Proof. The first component of formula (9.1), $\prod_{j=1}^{n}(1 + b_{1j})$, is equal to the machine load of machine M_1. The second component of formula (9.1), $\prod_{j=1}^{n}(1 + b_{2j})$, is equal to the the machine load of machine M_2. The third component of formula (9.1), $\max_{1 \leqslant j \leqslant n}\{(1 + b_{1j})(1 + b_{2j})\}$, is equal to the total processing time of the largest jobs on both machines.

Since the maximum completion time is not less than either of the three components, the result follows. $\qquad \square$

Remark 9.7. Formula (9.1) is a multiplicative counterpart of the following inequality for problem $O2||C_{\max}$ (see, e.g., Pinedo [11]),

$$C_{\max}^{\star}(\sigma) \geqslant \max \left\{ \sum_{j=1}^{n} p_{1j}, \sum_{j=1}^{n} p_{2j}, \max_{1 \leqslant j \leqslant n}\{p_{1j} + p_{2j}\} \right\}. \qquad (9.2)$$

In Sect. 19.6 we explain similarities between formulae (9.1) and (9.2) using the notion of *isomorphic* scheduling problems.

The two-machine open shop problem with proportional job processing times is polynomially solvable (see Algorithm 9.2).

Algorithm 9.2. for problem $O2|p_{ij} = b_{ij}t|C_{\max}$

---

1 **Input** : sequence $((b_{11}, b_{21}), (b_{12}, b_{22}), \ldots, (b_{1n}, b_{2n}))$
2 **Output:** an optimal schedule
   ▷ Step 1
3 $\mathcal{J}_1 \leftarrow \{J_j \in \mathcal{J} : b_{1j} \leqslant b_{2j}\}$; $\mathcal{J}_2 \leftarrow \mathcal{J} \setminus \mathcal{J}_1$;
   ▷ Step 2
4 In set $\mathcal{J}_1$ find job J_p such that $b_{2p} \geqslant \max_{J_j \in \mathcal{J}_1}\{b_{1j}\}$;
5 In set $\mathcal{J}_2$ find job J_q such that $b_{1q} \geqslant \max_{J_j \in \mathcal{J}_2}\{b_{2j}\}$;
   ▷ Step 3
6 $\sigma^1 \leftarrow (\sigma^{(\mathcal{J}_1-p)} \mid \sigma^{(\mathcal{J}_2-q)} \mid q \mid p)$;
7 $\sigma^2 \leftarrow (p \mid \sigma^{(\mathcal{J}_1-p)} \mid \sigma^{(\mathcal{J}_2-q)} \mid q)$;
8 Schedule jobs on machine M_1 according to sequence σ^1 and jobs on
   machine M_2 according to sequence σ^2 in such a way that no two
   operations of the same job overlap;
9 **return.**

Algorithm 9.2 for problem $O2|p_{ij} = b_{ij}t|C_{\max}$ is an adapted version of Gonzalez-Sahni's algorithm for problem $O2||C_{\max}$, [2]. In the pseudo-code of the algorithm symbols $\sigma^{(\mathcal{J}_1-p)}$ and $\sigma^{(\mathcal{J}_2-q)}$ denote an arbitrary sequence of jobs from the sets $\mathcal{J}_1 \setminus \{p\}$ and $\mathcal{J}_1 \setminus \{q\}$, respectively.

Theorem 9.8. (Kononov [4], Mosheiov [8]) *Problem* $O2|p_{ij} = b_{ij}t|C_{\max}$ *is solvable in* $O(n)$ *time by Algorithm 9.2.*

Proof. Kononov [4] proves the result using the notion of isomorphic problems (see Definition 19.71); the proof is similar to the proof of Theorem 9.2.

Mosheiov [8] analyses five possible types of schedules which can be generated by Algorithm 9.2 and proves that in each case the respective schedule is optimal (see [8, Theorem 3]). □

Example 9.9. Let us consider the instance from Example 9.3 as an instance of problem $O2|p_{ij} = b_{ij}t|C_{\max}$. Algorithm 9.2 generates schedule $\sigma^\star = (\sigma^{\star,1}; \sigma^{\star,2}) = (1, 2, 3; 3, 1, 2)$ such that jobs on the first machine are processed in the order $\sigma^{\star,1} = (1, 2, 3)$, while jobs on the second machine are processed in the order $\sigma^{\star,2} = (3, 1, 2)$. ◆

9.1.2 Proportional-linear deterioration

In this section, we consider dedicated machine time-dependent scheduling problems with proportional-linear processing times given by (6.9).

Flow shop problems

We start this section with the following result.

Lemma 9.10. (Kononov and Gawiejnowicz [6]) *There exists an optimal schedule for problem* $Fm|p_{ij} = b_{ij}(a + bt)|C_{\max}$, *in which*
(a) *machines M_1 and M_2 perform jobs in the same order,*
(b) *machines M_{m-1} and M_m perform jobs in the same order.*

Proof. (a) Let us assume that in a schedule σ jobs executed on machine M_1 are processed according to sequence $\sigma^1 = (i_1, \ldots, i_n)$, while jobs executed on machine M_2 are processed according to sequence $\sigma^2 = (j_1 = i_1, j_2 = i_2, \ldots, j_p = i_p, j_{p+1} = i_{r+1}, \ldots, j_{p+q} = i_r, \ldots, j_n)$, $q > 1$, $r > p \geqslant 0$.

By Theorem 7.9, the following equations are satisfied for schedule σ :

$$S_{2,i_{r+1}}(\sigma) = \max\left\{\frac{a}{b}\prod_{k=1}^{r+1}(b_{2,i_k}b + 1) - \frac{a}{b}, C_{2,j_p}(\sigma)\right\},$$

$$S_{2,i_r}(\sigma) = \max\left\{\frac{a}{b}\prod_{k=1}^{r}(b_{2,i_k}b + 1) - \frac{a}{b}, C_{2,j_{p+q-1}}(\sigma)\right\} = C_{2,j_{p+q-1}}(\sigma) \geqslant$$

$$\geqslant C_{2,i_{r+1}}(\sigma) \geqslant C_{1,i_{r+1}}(\sigma) = \frac{a}{b} \prod_{k=1}^{r+1}(b_{2,i_k}b+1) - \frac{a}{b}.$$

Let us consider schedule σ', differing from schedule σ only in that machine M_1 performs jobs according to sequence $\sigma^3 = (i_1,\ldots,i_{r-1},i_{r+1},i_r,i_{r+2},\ldots,i_n)$. For schedule σ', by Theorem 7.9, we obtain the following:

$$S_{2,i_{r+1}}(\sigma') = \max\left\{\frac{a}{b}(b_{2,i_{r+1}}b+1)\prod_{k=1}^{r-1}(b_{2,i_k}b+1) - \frac{a}{b}, C_{2,j_p}(\sigma')\right\},$$

$$S_{2,i_r}(\sigma') = \max\left\{\frac{a}{b}\prod_{k=1}^{r}(b_{2,i_k}a+1) - \frac{a}{b}, C_{2,j_{p+q-1}}(\sigma')\right\}.$$

Since $C_{2,j_p}(\sigma) = C_{2,j_p}(\sigma')$, we have $S_{2,i_{r+1}}(\sigma) \geqslant S_{2,i_{r+1}}(\sigma')$. From this it follows that $C_{2,j_{p+q-1}}(\sigma) \geqslant C_{2,j_{p+q-1}}(\sigma')$ and hence $S_{2,i_r}(\sigma) \geqslant S_{2,i_r}(\sigma')$.

Repeating the above considerations not more than $O(n^2)$ times, we obtain a schedule $\bar{\sigma}$ in which machines M_1 and M_2 perform jobs in the same order, and such that inequality $C_{2j}(\sigma) \geqslant C_{2j}(\bar{\sigma})$ holds for all $J_j \in \mathcal{J}$. From this it follows that $C_{\max}(\sigma) \geqslant C_{\max}(\bar{\sigma})$.

(b) Similar to the proof of (a). ∎

Lemma 9.11. (Kononov and Gawiejnowicz [6]) *If $m \in \{2,3\}$, then there exists an optimal schedule for problem $Fm|p_{ij} = b_{ij}(a+bt)|C_{\max}$, in which jobs are executed on all machines in the same order.*

Proof. Applying Lemma 9.10 for $m = 2,3$, we obtain the result. □

Kononov [5] generalized Theorem 9.2 to the case of proportional-linear job processing times given by (6.8).

Theorem 9.12. Kononov [5]) *If inequalities (7.5) and (7.6) hold, then problem $F2|p_{ij} = b_{ij}(a+bt)|C_{\max}$ is solvable in $O(n \log n)$ time by Algorithm 9.1.*

Proof. Similar to the proof of Theorem 9.2; see [5, Theorem 33]. ◇

Remark 9.13. A special case of Theorem 9.12, without conditions (7.5) and (7.6) but with assumptions $a > 0, b > 0$, $b_{ij} > 0$ for $1 \leqslant i \leqslant 2$ and $1 \leqslant j \leqslant n$, was given by Zhao et al. [18, Theorem 5].

Wang and Xia [16] considered job processing times given by (6.9) with $b = 1$. The authors, assuming that some relations between the job processing times hold, defined *dominating* machines (see Definitions 9.14 – 9.15) and proved a number of results for the multi-machine flow shop problems with machines of this type.

Definition 9.14. (Dominating machines)
Machine M_r is said to be dominated by machine M_k, $1 \leqslant k \neq r \leqslant m$, if and only if $\max\{b_{r,j} : 1 \leqslant j \leqslant n\} \leqslant \min\{b_{k,j} : 1 \leqslant j \leqslant n\}$.

If machine M_r is dominated by machine M_k, we write $M_r \lessdot M_k$ or $M_k \gtrdot M_r$.

Definition 9.15. (Series of dominating machines)
(a) *Machines $M_1, M_2, \ldots, M_m$ form an* increasing series of dominating machines (idm) *if and only if $M_1 \lessdot M_2 \lessdot \ldots \lessdot M_m$;*
(b) *Machines $M_1, M_2, \ldots, M_m$ form a* decreasing series of dominating machines (ddm) *if and only if $M_1 \gtrdot M_2 \gtrdot \ldots \gtrdot M_m$;*
(c) *Machines $M_1, M_2, \ldots, M_m$ form an* increasing-decreasing series of dominating machines (idm-ddm) *if and only if $M_1 \lessdot M_2 \lessdot \ldots \lessdot M_h$ and $M_h \gtrdot M_{h+1} \gtrdot \ldots \gtrdot M_m$ for some $1 \leqslant h \leqslant m$;*
(d) *Machines $M_1, M_2, \ldots, M_m$ form an* decreasing-increasing series of dominating machines (ddm-idm) *if and only if $M_1 \gtrdot M_2 \gtrdot \ldots \gtrdot M_h$ and $M_h \lessdot M_{h+1} \lessdot \ldots \lessdot M_m$ for some $1 \leqslant h \leqslant m$.*

Wang and Xia [16] proposed the following algorithm for the case when a flow shop is of *idm-ddm, no-wait* and *idm-ddm*, or *no-idle* and *idm-ddm* type.

Algorithm 9.3. for problem $Fm|p_{ij} = b_{ij}(a + t), \delta|C_{\max}$, $\delta \in \{idm\text{-}ddm;$ $no\text{-}wait, idm\text{-}ddm; no\text{-}idle, idm\text{-}ddm\}$

---

1 **Input** : sequences $(b_{1j}, b_{2j}, \ldots, b_{mj})$ for $1 \leqslant j \leqslant n$, numbers a, h
2 **Output:** an optimal schedule $\sigma^\star$

▷ Step 1

3 Find job J_u such that $\prod_{i=1}^{h-1}(1 + b_{iu}) = \min\left\{\prod_{i=1}^{h-1}(1 + b_{ij}) : 1 \leqslant j \leqslant n\right\}$;

4 Find job J_v such that $\prod_{i=h+1}^{m}(1 + b_{iv}) = \min\left\{\prod_{i=h+1}^{m}(1 + b_{ij}) : 1 \leqslant j \leqslant n\right\}$;

▷ Step 2

5 **if** $(u = v)$ **then**

6 | Find job J_w such that

$$\prod_{i=1}^{h-1}(1 + b_{iw}) = \min\left\{\prod_{i=1}^{h-1}(1 + b_{ij}) : 1 \leqslant j \neq u \leqslant n\right\};$$

7 | Find job J_z such that

$$\prod_{i=h+1}^{m}(1 + b_{iz}) = \min\left\{\prod_{i=h+1}^{m}(1 + b_{ij}) : 1 \leqslant j \neq v \leqslant n\right\};$$

8 | **if** $\prod_{i=1}^{h-1}(1 + b_{iw}) \prod_{i=h+1}^{m}(1 + b_{iv}) \leqslant \prod_{i=1}^{h-1}(1 + b_{iu}) \prod_{i=h+1}^{m}(1 + b_{iz})$ **then**

| | $u \leftarrow w$;
| **else**
| | $v \leftarrow z$;
| **end**

end

▷ Step 3

9 $\sigma^\star \leftarrow (u|(\mathcal{J} \setminus \{u, v\})|v)$;
10 **return** $\sigma^\star$.

Theorem 9.16. (Wang and Xia [16]) *Problems $Fm|p_{ij} = b_{ij}(a + t), \delta|C_{\max}$, where $\delta \in \{idm\text{-}ddm; no\text{-}wait, idm\text{-}ddm; no\text{-}idle, idm\text{-}ddm\}$, are solvable in $O(mn)$ time by Algorithm 9.3.*

Proof. See [16, Theorem 1].   ◇

Remark 9.17. For other cases of the multi-machine flow shop with dominating machines, Wang and Xia [16, Theorems 2–4] proposed algorithms which are modifications of Algorithm 9.3.

Open shop problems

Kononov [5] generalized Theorem 9.8 to the case of proportional-linear job processing times given by (6.8).

Theorem 9.18. (Kononov [5]) *If inequalities (7.5) and (7.6) hold, then problem $O2|p_{ij} = b_{ij}(a + bt)|C_{\max}$ is solvable in $O(n)$ time by Algorithm 9.2.*

Proof. Similar to the proof of Theorem 9.8; see [5, Theorem 33].   ◇

9.1.3 Linear deterioration

In this section, we consider dedicated machine time-dependent scheduling problems with linear processing times given by (6.12).

Flow shop problems

Cheng et al. [1] and Sun et al. [12] considered a few multi-machine flow shop problems with dominating machines and linear jobs with equal deterioration rates, $b_{ij} = b$ for $1 \leqslant i \leqslant m$ and $1 \leqslant j \leqslant n$. Let $B := 1 + b$.

Lemma 9.19. (Cheng et al. [1]) *Let $\sigma = ([1], [2], \ldots, [n])$ be a schedule for problem $Fm|p_{ij} = a_{ij} + bt, \delta|C_{\max}$, where $\delta \in \{no\text{-}idle, idm; no\text{-}idle, ddm; no\text{-}idle, idm\text{-}ddm; no\text{-}idle, ddm\text{-}idm\}$. Then*

(a) $C_{[j]} = \sum\limits_{i=1}^{m} a_{i,[1]}B^{m+j-i+1} + \sum\limits_{k=2}^{j} a_{m,[k]}B^{j-k}$ *if $\delta \equiv no\text{-}idle, idm$;*

(b) $C_{[j]} = \sum\limits_{k=1}^{n-1} a_{1,[k]}B^{m+j-k+1} + \sum\limits_{i=1}^{m} a_{i,[n]}B^{m-n+j-i} - \sum\limits_{k=j+1}^{n} a_{m,[k]}B^{j-k}$ *if $\delta \equiv no\text{-}idle, ddm$;*

(c) $C_{[j]} = \sum\limits_{i=1}^{h} a_{i,[1]}B^{m-i+j-1} + \sum\limits_{k=2}^{n-1} a_{h,[k]}B^{m-h-k+j} + \sum\limits_{i=h}^{m} a_{i,[n]}B^{m-n+j-i} - \sum\limits_{k=j+1}^{n} a_{m,[k]}B^{j-k}$ *if $\delta \equiv no - idle, idm - ddm$;*

(d) $C_{[j]} = \sum\limits_{k=1}^{n-1} a_{1,[k]}B^{m+j-k-1} + \sum\limits_{i=1}^{h-1} a_{i,[n]}B^{m+j-n-i} - \sum\limits_{k=2}^{n-1} a_{h,[k]}B^{m+j-h-k} + \sum\limits_{i=h+1}^{m} a_{i,[1]}B^{m+j-i-1} + \sum\limits_{k=2}^{j} a_{m,[k]}B^{j-k}$ *if $\delta \equiv no - idle, ddm - idm$.*

Proof. (a) Since machines $M_1, M_2, \ldots, M_m$ are of *no-idle, idm* type, in any feasible schedule σ we have $C_{i,[j]} \leqslant C_{i+1,[j]}$ for $1 \leqslant i \leqslant m$ and $1 \leqslant j \leqslant n$. Therefore, $C_{[j]} = \sum_{i=1}^{m} p_{i,[1]} + \sum_{k=2}^{j} p_{m,[k]}$.

(b), (c) and (d) Similar to (a).   □

Remark 9.20. Cheng et al. [1] formulated two algorithms for problem $Fm|p_{ij} = a_{ij} + bt, \delta|C_{\max}$, where δ describes the type of a series of dominating machines (see [1, Algorithms A-B]). Sun et al. [12] showed by counter-examples that schedules constructed by the algorithms may not be optimal (cf. [12, Sect. 2]). Sun et al. also formulated without proofs (cf. [12, Sect. 2]) several results saying that an optimal schedule for the considered problems is one of n or $n(n-1)$ schedules, where the first job, or the first and the last jobs, are fixed, respectively, while the remaining jobs are scheduled in non-decreasing order of the job basic processing times.

Some dedicated machine time-dependent scheduling problems with linearly deteriorating jobs are polynomially solvable.

The first result shows that problem $Fm|p_{ij} = a_j + b_j t|C_{\max}$ can be solved in a similar way as problem $1|p_j = a_j + b_j t|C_{\max}$ (see Theorem 7.14).

Theorem 9.21. (Mosheiov et al. [9]) *Problem* $Fm|p_{ij} = a_j + b_j t|C_{\max}$ *is solvable in* $O(n \log n)$ *time by scheduling jobs in non-decreasing order of the ratios* $\frac{a_j}{b_j}$.

Proof. We apply the network representation of the flow shop problem (see, e.g., Pinedo [11]). Since in the network the longest path corresponds to the schedule length for a given instance, in order to complete the proof it is sufficient to show that the path corresponding to non-decreasing sequence of the ratios $\frac{a_j}{b_j}$ is the optimal one; see [9, Theorem 1] for details.   □

Example 9.22. (Mosheiov [9, Example 1]) Let us consider the following instance of problem $F3|p_{ij} = a_j + b_j t|C_{\max}$, in which $a_1 = 5$, $a_2 = 1$, $a_3 = 3$, $a_4 = 3$, $a_5 = 5$, $b_1 = 3.5$, $b_2 = 3$, $b_3 = 4$, $b_4 = 3$ and $b_5 = 3$. Then, by Theorem 9.21, optimal schedule is $\sigma^* = (2, 3, 4, 1, 5)$.   ◆

Ng et al. [10, Sect. 3] considered several multi-machine time-dependent flow shop problems with job processing times in the form of $p_{ij} = a_{ij} + bt$ and criterion $C_{\max}$. For the problems, polynomial algorithms were proposed, provided that the number m of machines is fixed.

Sun et al. [28, Sect. 3] considered three multi-machine flow shop problems in the form of $Fm|p_{ij} = a + b_{ij}t, \delta|C_{\max}$, where δ describes the type of series of dominating machines. For the problems, polynomial algorithms were proposed, provided that the number m of machines is fixed.

9.1.4 Simple non-linear deterioration

In this section, we consider polynomially solvable dedicated machine time-dependent scheduling problems with simple non-linearly deteriorating jobs.

Flow shop problems

Melnikov and Shafransky [7] considered a multi-machine flow shop with job processing times given by (6.13), $Fm|p_{ij} = a_{ij} + f(t)|C_{\max}$, where function $f(t)$ is differentiable, it satisfies (6.14) and

$$\frac{df(t)}{dt} \geqslant \lambda > 0 \quad \text{for} \quad t \geqslant 0. \tag{9.3}$$

For the above problem, the authors estimated the difference between the optimal maximum completion time $C_{\max}^\star$ and the maximum completion time $C_{\max}(\sigma^\nearrow)$ for a schedule $\sigma^\nearrow$ obtained by scheduling jobs in non-decreasing order of the basic processing times a_{1j}, $1 \leqslant j \leqslant n$.

Before we formulate the result, we will state an auxiliary result.

Lemma 9.23. (Melnikov and Shafransky [7]) *For any differentiable function $f(t)$ which satisfies conditions (6.14) and (9.3), there exists a finite number N_0 such that for all $n \geqslant N_0$ inequality*

$$\sum_{j=1}^{n} a_{ij} + f\left(\sum_{j=0}^{n-1} C_{1j}\right) \geqslant \sum_{j=l-1}^{n-1} a_{i+1,j} + f\left(\sum_{j=0}^{l-2} C_{1j}\right) \tag{9.4}$$

holds for any i and l, where $1 \leqslant i \leqslant m - 1$, $2 \leqslant l \leqslant n$, $C_{1,0} := 0$ and $C_{1j} = a_{1j} + f\left(\sum_{k=0}^{j-1} C_{1k}\right)$ for $1 \leqslant j \leqslant n$.

Proof. See [7, Theorem 3]. ◇

Assuming that $f(t)$ is a differentiable function which satisfies conditions (6.14) and (9.3), the following result holds.

Theorem 9.24. (Melnikov and Shafransky [7]) *Let $\sigma^\nearrow$ denote a schedule for problem $Fm|p_{ij} = a_{ij} + f(t)|C_{\max}$ in which jobs are scheduled in non-decreasing order of the basic processing times a_{1j}, $1 \leqslant j \leqslant n$. If the number of jobs n satisfies the inequality (9.4), then either $\sigma^\nearrow$ is an optimal schedule or the optimal schedule is one of $k \leqslant n - 1$ schedules π, in which the last job satisfies inequality*

$$\sum_{i=2}^{m} a_{i,\pi_n} < \sum_{i=2}^{m} a_{i,\sigma_n^\nearrow}$$

and the first $n - 1$ jobs are scheduled in non-decreasing order of the basic processing times a_{1j}.

Proof. By Lemma 9.23 and Theorem 7.25, the result follows; see [7, Theorem 4] for details. ◇

We now pass to the two-machine flow shop problem with job processing times given by (6.13), $F2|p_{ij} = a_{ij} + f(t)|C_{\max}$, where $f(t)$ is defined as in Theorem 9.24. For this problem, the following result holds.

Theorem 9.25. (Melnikov and Shafransky [7]) *Let $\sigma^{\nearrow}$ denote a schedule for problem $F2|p_{ij} = a_{ij}+f(t)|C_{\max}$ in which jobs are scheduled in non-decreasing order of the basic processing times a_{1j}, $1 \leqslant j \leqslant n$. Then*

$$C^{\star}_{\max} - C_{\max}(\sigma^{\nearrow}) \leqslant a_{2,\sigma^{\nearrow}_n} - \min_{1 \leqslant j \leqslant n}\{a_{2j}\}.$$

Proof. See [7, Theorem 5]. ◇

9.1.5 Proportional-linear shortening

In this section, we consider polynomially solvable time-dependent dedicated machine scheduling problems with proportionally-linear shortening job processing times given by (6.41) and the $C_{\max}$ criterion.

Flow shop problems

Wang and Xia [17] considered a multi-machine flow shop with job processing times given by (6.41). Let $b_{\min,\min} := \min_{1 \leqslant i \leqslant m, 1 \leqslant j \leqslant n}\{b_{ij}\}$.

Theorem 9.26. (Wang and Xia [17]) *Problem $F2|p_{ij} = b_{ij}(1-bt)$, $b > 0$,*
$$b\left(\sum_{i=1}^{m}\sum_{j=1}^{n} b_{ij} - b_{\min,\min}\right) < 1|C_{\max}$$ *is solvable in $O(n\log n)$ time by Algorithm 9.1.*

Proof. Similar to the proof of Theorem 9.2. □

Assuming that $b_{ij} = b_j$ for all $1 \leqslant i \leqslant m$, the authors obtained the following result.

Theorem 9.27. (Wang and Xia [17]) *For problem $F2|p_{ij} = b_{ij}(1-bt)$, $b_{ij} = b_j$, $b > 0$, $b\left(n\sum_{i=1}^{m} b_i - b_{\min,\min}\right) < 1|C_{\max}$, the maximum completion time does not depend on the schedule of jobs.*

Proof. The result follows from Theorem 7.1. □

9.2 Minimizing the total completion time

In this section, we consider polynomially solvable time-dependent dedicated machine scheduling problems with the $\sum C_j$ criterion.

9.2.1 Proportional deterioration

Problems of scheduling jobs with proportionally deteriorating processing times on dedicated machines are well investigated in the literature.

Flow shop problems

Wang et al. [22] considered problem $F2|p_{ij} = b_{ij}t| \sum C_j$, where deterioration rates $b_{ij} \in (0, 1)$ for $1 \leqslant i \leqslant 2$ and $1 \leqslant j \leqslant n$.

Lemma 9.28. (Wang et al. [22]) *There exists an optimal schedule for problem $F2|p_{ij} = b_{ij}t, 0 < b_{ij} < 1| \sum C_j$ in which the job sequence is identical on both machines.*

Proof. Similar to the proof of Lemma 9.10 for $m = 2$. $\square$

The next two results concern the cases of equal deterioration rates.

Theorem 9.29. (Wang et al. [22]) *If $b_{2j} = b$ for $1 \leqslant j \leqslant n$, then problem $F2|p_{ij} = b_{ij}t, 0 < b_{ij} < 1| \sum C_j$ is solvable in $O(n \log n)$ time by scheduling jobs in non-decreasing order of the deterioration rates b_{1j}.*

Proof. Let in schedule σ jobs be scheduled in non-decreasing order of the deterioration rates b_{1j}, and let τ be an arbitrary other schedule. Since $C_j(\sigma) \leqslant C_{[j]}(\tau)$ for $1 \leqslant j \leqslant n$, the result follows (see [22, Theorem 1]). $\square$

Example 9.30. Let $n = 3$, $t_0 = 1$, $b_{11} = \frac{1}{3}$, $b_{12} = \frac{2}{3}$, $b_{13} = \frac{1}{4}$, $b = \frac{1}{2}$. Then, by Theorem 9.29, schedule $\sigma^* = (3, 1, 2)$ is optimal.   ◆

Theorem 9.31. (Wang et al. [22]) *If $b_{1j} = b_{2j}$ for $1 \leqslant j \leqslant n$, then problem $F2|p_{ij} = b_{ij}t, 0 < b_{ij} < 1| \sum C_j$ is solvable in $O(n \log n)$ time by scheduling jobs in non-decreasing order of the deterioration rates b_{1j}.*

Proof. By the pairwise job interchange argument. $\square$

Example 9.32. Let $n = 3$, $t_0 = 1$, $b_{11} = \frac{1}{4}$, $b_{12} = \frac{2}{3}$, $b_{13} = \frac{1}{3}$, $b_{21} = \frac{1}{4}$, $b_{22} = \frac{2}{3}$, $b_{23} = \frac{1}{3}$. Then, by Theorem 9.31, schedule $\sigma^* = (1, 3, 2)$ is optimal.   ◆

The next two results concern the cases when flow shop machines are dominating (cf. Definitions 9.14–9.15).

Lemma 9.33. (Wang et al. [22]) *If in an instance of problem $F2|p_{ij} = b_{ij}t$, $0 < b_{ij} < 1, M_1 \prec M_2| \sum C_j$, the first scheduled job is fixed, then in an optimal schedule the remaining jobs are scheduled in non-decreasing order of the deterioration rates b_{2j}.*

Proof. By direct calculation. $\square$

By Lemma 9.33, Wang et al. [22] constructed the following algorithm.

Algorithm 9.4. for problem $F2|p_{ij} = b_{ij}t, 0 < b_{ij} < 1, M_1 \lessdot M_2| \sum C_j$

---

1 **Input**   : sequences (b_{1j}, b_{2j}) for $1 \leqslant j \leqslant n$
2 **Output:** an optimal schedule σ^*
  ▷ Step 1
3 **for** $j \leftarrow 1$ **to** n **do**
4 $\quad\mid\quad$ Schedule first job J_j;
5 $\quad\mid\quad$ Schedule remaining jobs in non-decreasing order of the
  $\quad\quad\quad$ deterioration rates b_{2j};
6 $\quad\mid\quad$ Call the obtained schedule σ^j;
  end
  ▷ Step 2
7 $\sigma^* \leftarrow$ the best schedule from schedules $\sigma^1, \sigma^2, \ldots, \sigma^n$;
8 **return** σ^*.

Theorem 9.34. (Wang et al. [22]) *Problem* $F2|p_{ij} = b_{ij}t, 0 < b_{ij} < 1,$
$M_1 \lessdot M_2| \sum C_j$ *is solvable in* $O(n \log n)$ *time.*

Proof. By Lemma 9.33, we can construct n distinct schedules for the problem, with a fixed first job in each of them. Since each such schedule can be constructed in $O(n \log n)$ time, the result follows. $\qquad\square$

Theorem 9.35. (Wang et al. [22]) *Problem* $F2|p_{ij} = b_{ij}t, 0 < b_{ij} < 1,$
$M_1 \gtrdot M_2| \sum C_j$ *is solvable in* $O(n \log n)$ *time by scheduling jobs in non-decreasing order of the ratios* $\frac{b_{1j}}{(1+b_{1j})(1+b_{2j})}$.

Proof. By the pairwise job interchange argument. $\qquad\square$

9.2.2 Linear deterioration

Some dedicated machine time-dependent scheduling problems with linearly deteriorating jobs are polynomially solvable.

Flow shop problems

Cheng et al. [19, Algorithms C-D] formulated two algorithms for problem $Fm|p_{ij} = a_{ij} + bt, \delta|C_{\max}$, where δ describes the type of series of dominating machines. Sun et al. [21, Sect. 2] showed by counter-examples that schedules constructed by the algorithms may not be optimal. Sun et al. also formulated without proofs (see [21, Sect. 2]), several results saying that an optimal schedule for multi-machine flow shop problems with dominating machines, linear job processing times $p_{ij} = a_{ij} + bt$ and the $\sum C_j$ criterion is similar to the optimal schedule for the $C_{\max}$ criterion (cf. Sect. 9.1.3).

Ng et al. [20, Sect. 4] considered several multi-machine time-dependent flow shop problems with linear job processing times in the form of $p_{ij} = a_{ij} + bt$ and criterion $\sum C_j$. For the problems, polynomial algorithms were proposed, provided that the number m of machines is fixed.

9.2.3 Proportional-linear shortening

In this section, we consider polynomially solvable time-dependent dedicated machine scheduling problems with the $\sum C_j$ criterion and proportionally-linear shortening job processing times.

Flow shop problems

Wang and Xia [23] considered the multi-machine flow shop problem with job processing times given by (6.41). Let $b_{\min,\min}$ be defined as in Sect. 9.1.5.

Theorem 9.36. (Wang and Xia [23]) *Problem* $Fm|p_{ij} = b_{ij}(1 - bt), b_{ij} = b_j,$ $b > 0, b\left(\sum\limits_{j=1}^{n} b_j - b_{\min,\min}\right) < 1|\sum C_j,$ *is solvable in* $O(n\log n)$ *time by scheduling jobs in non-decreasing order of the deterioration rates* b_j.

Proof. The result follows from Theorem 7.1.   □

Remark 9.37. Wang and Xia considered a number of flow shop problems with dominating machines and the $L_{\max}$ criterion. Since the results (see [24, Section 5]) are special cases of the results for job processing times given by (6.9) with $b = 1$, we do not present them here.

9.3 Minimizing the maximum lateness

In this section, we consider polynomially solvable time-dependent dedicated machine scheduling problems with the $L_{\max}$ criterion.

9.3.1 Proportional-linear deterioration

Some restricted cases of time-dependent dedicated machine scheduling problems with proportionally-linear deteriorating jobs are polynomially solvable.

Flow shop problems

Wang and Xia [25] considered job processing times given by (6.9) with $b = 1$, assuming that flow shop machines are dominating.

Lemma 9.38. (Wang and Xia [25]) *If in an instance of problem* $Fm|p_{ij} = b_{ij}(a+t), \delta|L_{\max},$ *where* $\delta \in \{idm; no\text{-}wait, idm; no\text{-}idle, idm\},$ *the first scheduled job is fixed, then in an optimal schedule the remaining jobs are in the EDD order.*

Proof. By contradiction; see [25, Theorem 10].   ◇

By Lemma 9.38, Wang and Xia [25] proposed an algorithm for problem $Fm|p_{ij} = b_{ij}(a + t), \delta|L_{\max}$, where $\delta \in \{idm; no\text{-}wait, idm; no\text{-}idle, idm\}$. The algorithm is an appropriately modified Algorithm 9.4 in which in Step 1 all jobs except the first one are scheduled in the EDD order instead of in non-decreasing order of the deterioration rates b_{2j}.

Theorem 9.39. (Wang and Xia [25]) *Problems $Fm|p_{ij} = b_{ij}(a + t), \delta|L_{\max}$, where $\delta \in \{idm; no\text{-}wait, idm; no\text{-}idle, idm\}$, are solvable in $O(n \log n)$ time by the modified Algorithm 9.4.*

Proof. By Lemma 9.38, in order to find an optimal schedule for problems $Fm|p_{ij} = b_{ij}(a + t), \delta|L_{\max}$, where $\delta \in \{idm; no\text{-}wait, idm; no\text{-}idle, idm\}$, it is sufficient to construct n distinct schedules by inserting in the first position in the jth schedule the jth job and scheduling the remaining jobs in the EDD order. Since each such schedule can be obtained in $O(n \log n)$ time, the result follows.   □

Remark 9.40. For other cases of the multi-machine flow shop problem with dominating machines and the $L_{\max}$ criterion, the authors applied a similar approach; see [25, Theorems 11–12] for details.

9.3.2 Proportional-linear shortening

Research on time-dependent dedicated machine scheduling with the $L_{\max}$ criterion and proportional-linearly shortening jobs is limited.

Flow shop problems

Wang and Xia [26] considered the two-machine flow shop problem with job processing times given by (6.41). Let $b_{\min,\min}$ be defined as in Sect. 9.1.5.

Theorem 9.41. (Wang and Xia [26]) *Problem $F2|p_{ij} = b_{ij}(1 - bt), b_{ij} = b_j,$*

$$b > 0, \ b\left(\sum_{j=1}^{n} b_j - b_{\min,\min}\right) < 1|L_{\max} \ \text{is solvable in } O(n \log n) \ \text{time by}$$

scheduling jobs in non-decreasing order of the deadlines d_j.

Proof. The result follows from Theorem 7.84.   □

9.4 Other criteria

In this section, we consider the problems of time-dependent scheduling on dedicated machines with criteria other than $C_{\max}$, $\sum C_j$ or $L_{\max}$.

9.4.1 Proportional deterioration

Dedicated machine time-dependent scheduling problems with proportionally deteriorating jobs have similar time complexity as their counterparts with fixed job processing times (cf. Sect. 19.6).

Flow shop problems

Wang and Xia [30] considered a number of flow shop problems with dominating machines and the $\sum w_j C_j$ criterion. Since the results (see [30, Section 4]) are special cases of the results for job processing times given by (6.9) with $b = 1$, we do not present them here.

9.4.2 Proportional-linear deterioration

Dedicated machine time-dependent scheduling problems with proportional-linearly deteriorating jobs are similar to their counterparts with proportionally deteriorating jobs.

Flow shop problems

Wang and Xia [31] considered the multi-machine flow shop problem with dominating machines and job processing times given by (6.9) with $b = 1$. The authors proved a number of results for the $\sum w_j C_j$ criterion.

Lemma 9.42. (Wang and Xia [31]) *If in an instance of problem $Fm|p_{ij} = b_{ij}(a + t), \delta| \sum w_j C_j$, where $\delta \in \{idm; no\text{-}wait, idm; no\text{-}idle, idm\}$, the first scheduled job is fixed, then in an optimal schedule the remaining jobs are scheduled in non-decreasing order of the ratios $\frac{b_{mj}}{w_j(1+b_{mj})}$.*

Proof. The result follows from the formula for $\sum w_j C_j$ and Theorem 7.90. □

By Lemma 9.42, Wang and Xia [31] proposed an algorithm for problem $Fm|p_{ij} = b_{ij}(a + t), \delta| \sum w_j C_j$, where $\delta \in \{idm; no\text{-}wait, idm; no\text{-}idle, idm\}$. The algorithm is a slightly modified Algorithm 9.4 in which in Step 1 all jobs except the first one are scheduled in non-decreasing order of the ratios $\frac{b_{mj}}{w_j(1+b_{mj})}$ instead of in non-decreasing order of the deterioration rates b_{2j}.

Theorem 9.43. (Wang and Xia [31]) *Problems $Fm|p_{ij} = b_{ij}(a+t), \delta| \sum w_j C_j$, where $\delta \in \{idm; no\text{-}wait, idm; no\text{-}idle, idm\}$, are solvable in $O(n \log n)$ time by the modified Algorithm 9.4.*

Proof. By Lemma 9.42, in order to find an optimal schedule for problems $Fm|p_{ij} = b_{ij}(a + t), \delta| \sum w_j C_j$, where $\delta \in \{idm; no\text{-}wait, idm; no\text{-}idle, idm\}$, it is sufficient to construct n distinct schedules by inserting in the first position in the jth schedule the jth job. Since each such schedule can be obtained in $O(n \log n)$ time, the result follows. □

Remark 9.44. For other cases of the multi-machine flow shop problem with dominating machines and the $\sum w_j C_j$ criterion, the authors applied a similar approach; see [31, Theorems 6–9] for details.

9.4.3 Linear deterioration

Some dedicated machine time-dependent scheduling problems with linearly deteriorating jobs are polynomially solvable.

Flow shop problems

Fiszman and Mosheiov [27] noticed that problem $Fm|p_{ij} = a_j + b_j t|\sum C_{\max}^{(k)}$ can be solved in a similar way as problem $1|p_j = a_j + b_j t|\sum C_{\max}^{(k)}$ (cf. Theorem 7.14).

9.4.4 Proportional-linear shortening

Restricted cases of dedicated machine time-dependent scheduling problems are solvable in polynomial time.

Flow shop problems

Wang and Xia [32] considered the multi-machine flow shop problem with job processing times given by (6.41). For this problem can be used an appropriately modified Algorithm 7.17 (see [32, Modified Algorithm 1]). Let $b_{\min,\min}$ be defined as in Sect. 9.1.5.

Theorem 9.45. (Wang and Xia [32]) *Problem* $Fm|p_{ij} = b_{ij}(1 - bt)$, $b > 0$,
$$b\left(\sum_{j=1}^{n} b_j - b_{\min,\min}\right) < 1, b_{ij} = b_j|f_{\max} \text{ is solvable in } O(n^2) \text{ time by the modified Algorithm 7.17.}$$

Proof. The result follows from Theorem 18.17.   □

To solve the problem with the $\sum U_j$ criterion, Wang and Xia used an appropriately modified Algorithm 7.15 (see [32, Algorithm 2]).

Theorem 9.46. (Wang and Xia [32]) *Problem* $Fm|p_{ij} = b_{ij}(1 - bt)$, $b > 0$,
$$b\left(\sum_{j=1}^{n} b_j - b_{\min,\min}\right) < 1, b_{ij} = b_j|\sum U_j \text{ is solvable in } O(n \log n) \text{ time by the modified Algorithm 7.15.}$$

Proof. The result follows from Theorem 7.98.   □

Remark 9.47. Wang [29] considered a number of multi-machine flow shop problems with job processing times (6.41) and dominating machines. Since all the problems are solved by algorithms which are similar to Algorithm 7.17 (see [29, Theorems 5–9] for details), we do not present the algorithms here.

This remark completes the third part of the book, devoted to polynomially solvable time-dependent scheduling problems. In the next part, we discuss $\mathcal{NP}$-hard time-dependent scheduling problems.

References

Minimizing the maximum completion time

1. M-B. Cheng, S-J. Sun and L-M. He, Flow shop scheduling problems with deteriorating jobs on no-idle dominant machines. *European Journal of Operational Research* **183** (2007), no. 1, 115–24.
2. T. Gonzalez and S. Sahni, Open shop scheduling to minimize finish time. *Journal of the Association for Computing Machinery* **23** (1976), no. 4, 665–679.
3. S. M. Johnson, Optimal two and three stage production schedules with setup times included. *Naval Research Logistics Quarterly* **1** (1954), no. 1, 61–68.
4. A. Kononov, Combinatorial complexity of scheduling jobs with simple linear deterioration. *Discrete Analysis and Operations Research* **3** (1996), no. 2, 15–32 (in Russian).
5. A. Kononov, *On the complexity of the problems of scheduling with time-dependent job processing times.* Ph. D. dissertation, Sobolev Institute of Mathematics, Novosibirsk 1999, 106 pp. (in Russian).
6. A. Kononov and S. Gawiejnowicz, NP-hard cases in scheduling deteriorating jobs on dedicated machines. *Journal of the Operational Research Society* **52** (2001), no. 6, 708–718.
7. O. I. Melnikov and Y. M. Shafransky, Parametric problem of scheduling theory. *Kibernetika* **6** (1979), no. 3, 53–57 (in Russian).
   English translation: *Cybernetics and System Analysis* **15** (1980), no. 3, 352–357.
8. G. Mosheiov, Complexity analysis of job-shop scheduling with deteriorating jobs. *Discrete Applied Mathematics* **117** (2002), no. 1–3, 195–209.
9. G. Mosheiov, A. Sarig and J. Sidney, The Browne-Yechiali single-machine sequence is optimal for flow-shops. *Computers and Operations Research* **37** (2010), no. 11, 1965–1967.
10. C-T. Ng, J-B. Wang, T-C. E. Cheng and S-S. Lam, Flowshop scheduling of deteriorating jobs on dominating machines. *Computers and Industrial Engineering* **61** (2011), no. 3, 647–654.
11. M. L. Pinedo, *Scheduling: Theory, Algorithms, and Systems*, 5th ed. Berlin-Heidelberg: Springer 2016.
12. L-H. Sun, L-Y. Sun, K. Cui and J-B. Wang, A note on flow shop scheduling problems with deteriorating jobs on no-idle dominant machines. *European Journal of Operational Research* **200** (2010), no. 1, 309–311.
13. J-B. Wang, Flow shop scheduling with deteriorating jobs under dominating machines to minimize makespan. *International Journal of Advanced Manufacturing Technology* **48** (2010), no. 5–8, 719–723.

14. L. Wang, L-Y. Sun, L-H. Sun and J-B. Wang, On three-machine flow shop scheduling with deteriorating jobs. *International Journal of Production Economics* **125** (2010), no. 1, 185–189.
15. J-B. Wang and Z-Q. Xia, Flow shop scheduling with deteriorating jobs under dominating machines. *Omega* **34** (2006), no. 4, 327–336.
16. J-B. Wang and Z-Q. Xia, Flow shop scheduling problems with deteriorating jobs under dominating machines. *Journal of the Operational Research Society* **57** (2006), no. 2, 220–226.
17. J-B. Wang and Z-Q. Xia, Scheduling jobs under decreasing linear deterioration. *Information Processing Letters* **94** (2005), no. 2, 63–69.
18. C-L. Zhao, Q-L. Zhang and H-Y. Tang, Scheduling problems under linear deterioration. *Acta Automatica Sinica* **29** (2003), no. 4, 531–535.

Minimizing the total completion time

19. M-B. Cheng, S-J. Sun and L-M. He, Flow shop scheduling problems with deteriorating jobs on no-idle dominant machines. *European Journal of Operational Research* **183** (2007), no. 1, 115–24.
20. C-T. Ng, J-B. Wang, T-C. E. Cheng and S-S. Lam, Flowshop scheduling of deteriorating jobs on dominating machines. *Computers and Industrial Engineering* **61** (2011), no. 3, 647–654.
21. L-H. Sun, L-Y. Sun, K. Cui and J-B. Wang, A note on flow shop scheduling problems with deteriorating jobs on no-idle dominant machines. *European Journal of Operational Research* **200** (2010), no. 1, 309–311.
22. J-B. Wang, C-T. Ng, T-C. E. Cheng and L-L. Liu, Minimizing total completion time in a two-machine flow shop with deteriorating jobs. *Applied Mathematics and Computation* **180** (2006), no. 1, 185–193.
23. J-B. Wang and Z-Q. Xia, Scheduling jobs under decreasing linear deterioration. *Information Processing Letters* **94** (2005), no. 2, 63–69.
24. J-B. Wang and Z-Q. Xia, Flow shop scheduling with deteriorating jobs under dominating machines. *Omega* **34** (2006), no. 4, 327–336.

Minimizing the maximum lateness

25. J-B. Wang and Z-Q. Xia, Flow shop scheduling problems with deteriorating jobs under dominating machines. *Journal of the Operational Research Society* **57** (2006), no. 2, 220–226.
26. J-B. Wang and Z-Q. Xia, Scheduling jobs under decreasing linear deterioration. *Information Processing Letters* **94** (2005), no. 2, 63–69.

Other criteria

27. S. Fiszman and G. Mosheiov, Minimizing total load on a proportionate flowshop with position-dependent processing times and rejection. *Information Processing Letters* **132** (2018), 39–43.
28. L-H. Sun, L-Y. Sun, M-Z. Wang and J-B. Wang, Flow shop makespan minimization scheduling with deteriorating jobs under dominating machines. *International Journal of Production Economics* **138** (2012), no. 1, 195–200.

29. J-B. Wang, Flow shop scheduling problems with decreasing linear deterioration under dominant machines. *Computers and Operations Research* **34** (2007), no. 7, 2043–2058.

30. J-B. Wang and Z-Q. Xia, Flow shop scheduling with deteriorating jobs under dominating machines. *Omega* **34** (2006), no. 4, 327–336.

31. J-B. Wang and Z-Q. Xia, Flow shop scheduling problems with deteriorating jobs under dominating machines. *Journal of the Operational Research Society* **57** (2006), no. 2, 220–226.

32. J-B. Wang and Z-Q. Xia, Scheduling jobs under decreasing linear deterioration. *Information Processing Letters* **94** (2005), no. 2, 63–69.

NP-HARD PROBLEMS

$\mathcal{NP}$-hard single machine problems

A great many time-dependent scheduling problems are intractable. This chapter opens the fourth part of the book, which is devoted to $\mathcal{NP}$-hard time-dependent scheduling problems.

This part of the book is composed of three chapters. In Chap. 10, we focus on $\mathcal{NP}$-hard single machine time-dependent scheduling problems. In Chaps. 11 and 12, we discuss $\mathcal{NP}$-hard parallel machine and dedicated machine time-dependent scheduling problems, respectively.

Chapter 10 is composed of four sections. In Sect. 10.1, we address the problems of minimization of the $C_{\max}$ criterion. In Sect. 10.2, we present the results related to minimization of the $\sum C_j$ criterion. In Sect. 10.3, we focus on minimization of the $L_{\max}$ criterion. In Sect. 10.4, we review the results related to minimization of other criteria than $C_{\max}$, $\sum C_j$ and $L_{\max}$. The chapter is completed with a list of references.

10.1 Minimizing the maximum completion time

In this section, we consider $\mathcal{NP}$-hard single machine time-dependent scheduling problems with the $C_{\max}$ criterion.

10.1.1 Proportional deterioration

The single machine time-dependent scheduling problem with proportional job processing times, distinct ready times and distinct deadlines is intractable.

Theorem 10.1. (Gawiejnowicz [10]) *The decision version of problem $1|p_j = b_j t, r_j, d_j|C_{\max}$ with two distinct ready times and two distinct deadlines is $\mathcal{NP}$-complete in the ordinary sense.*

Proof. We use the following reduction from the SP problem: $n = p + 1$, $t_0 = 1$, $b_j = y_j - 1$, $r_j = 1$ and $d_j = BY$ for $1 \leqslant j \leqslant p$, $b_{p+1} = B - 1$, $r_{p+1} = B$, $d_{p+1} = B^2$ and the threshold value is $G = BY$, where $Y = \prod_{j=1}^{p} y_j$.

© Springer-Verlag GmbH Germany, part of Springer Nature 2020

S. Gawiejnowicz, *Models and Algorithms of Time-Dependent Scheduling*,

Monographs in Theoretical Computer Science. An EATCS Series,

https://doi.org/10.1007/978-3-662-59362-2_10

Notice that the completion time of the jth job in any feasible schedule for problem $1|p_j = b_j t, r_j, d_j|C_{\max}$ is equal to

$$C_{[j]} = S_{[j]}(1 + b_{[j]}) = \max\left\{C_{[j-1]}, r_{[j]}\right\}(1 + b_{[j]}),$$

where $1 \leqslant j \leqslant n$ and $C_{[0]} := t_0$. Hence, the decision version of problem $1|p_j = b_j t, r_j, d_j|C_{\max}$ is in the $\mathcal{NP}$ class.

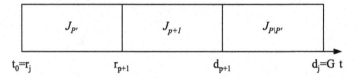

Fig. 10.1: A schedule in the proof of Theorem 10.1

In order to complete the proof it is sufficient to show that the SP problem has a solution if and only if there exists a feasible schedule σ for the above instance of problem $1|p_j = b_j t, r_j, d_j|C_{\max}$ (see Fig. 10.1 and Remark 10.17) such that $C_{\max}(\sigma) \leqslant G$. $\qquad\square$

Remark 10.2. Since in the case of the problems which we consider in the book it is usually easy to show that the decision version of a given problem is in the $\mathcal{NP}$ class or that a given reduction is polynomial (pseudo-polynomial), in most cases we omit these parts of $\mathcal{NP}$-completeness proofs.

10.1.2 Linear deterioration

In this section, we consider single machine time-dependent scheduling problems with linearly deteriorating jobs, distinct ready times and distinct deadlines.

The general problem, with an arbitrary number of distinct ready times and distinct deadlines, is computationally intractable.

Theorem 10.3. *The decision version of problem* $1|p_j = a_j + b_j t, r_j, d_j|C_{\max}$ *is $\mathcal{NP}$-complete in the strong sense.*

Proof. It is sufficient to note that assuming $b_j = 0$ for $1 \leqslant j \leqslant n$, we obtain problem $1|r_j, d_j|C_{\max}$. Since the decision version of the problem is $\mathcal{NP}$-complete in the strong sense (see Lenstra et al. [17]), the result follows. $\qquad\square$

The problem remains computationally intractable if all ready times are equal to zero, i.e., if $r_j = 0$ for $1 \leqslant j \leqslant n$.

Theorem 10.4. (Cheng and Ding [4]) *The decision version of problem* $1|p_j = a_j + b_j t, d_j|C_{\max}$ *is $\mathcal{NP}$-complete in the strong sense.*

Proof. The reduction from the 3-P problem is as follows. Let v denote an integer larger than $2^8 h^3 K^3$. Let us define $n = 4h$ jobs with $a_j = vc_j$ and $b_j = \frac{c_j}{v}$ for $1 \leqslant j \leqslant 3h$, $a_{3h+i} = v$ and $b_{3h+i} = 0$ for $1 \leqslant i \leqslant h$. Deadlines $d_j = D = vh(K+1) + \sum_{i=1}^{n} \sum_{j=1}^{i-1} c_i c_j + \frac{1}{2}h(h-1)K + 1$ for $1 \leqslant j \leqslant 3h$ and $d_{3h+i} = iv + (i-1)(vK + 4hK^2)$ for $1 \leqslant i \leqslant h$. The threshold value is $G = D$.

To complete the proof it is sufficient to show that the 3-P problem has a solution if and only if for the instance of problem $1|p_j = a_j + b_j t, d_j|C_{\max}$ there exists a schedule σ such that $C_{\max}(\sigma) \leqslant G$; see [4, Lemmas 6–7]. □

Problem $1|p_j = a_j + b_j t, r_j, d_j|C_{\max}$ remains computationally intractable, when all basic processing times and all ready times are equal, i.e., when $a_j = 1$ and $r_j = 0$ for $1 \leqslant j \leqslant n$.

Theorem 10.5. (Cheng and Ding [3]) *The problem of whether there exists a feasible schedule for problem $1|p_j = 1 + b_j t, d_j|C_{\max}$ is $\mathcal{NP}$-complete in the strong sense.*

Proof. Given an instance of the 3-P problem, we construct an instance of problem $1|p_j = 1 + b_j t, d_j|C_{\max}$ as follows.

The set of jobs $J = V \cup R \cup Q_1 \cup \cdots \cup Q_{m-1}$, where $q = 32h^2 K$, $v = 16h^2 qK$, $V = \{J_{01}, J_{02}, \ldots, J_{0v}\}$, $R = \{J_1, J_2, \ldots, J_{3h}\}$ and $Q_i = \{J_{i1}, J_{i2}, \ldots, J_{iq}\}$ for $1 \leqslant i \leqslant h-1$. Define $n = v + 3h + (h-1)q$, $E = 4hnK$ and $A = 32n^3 E^2$.

The job deterioration rates and deadlines are the following: $b_{0i} = 0$ and $d_{0i} = v$ for $1 \leqslant i \leqslant v$, $b_{ij} = 0$ and $d_{ij} = D_i$ for $1 \leqslant i \leqslant h-1$ and $1 \leqslant j \leqslant q$, $b_i = \frac{E + c_i}{A}$ and $d_i = G$ for $1 \leqslant i \leqslant 3h$, where the threshold value is

$$G = n + \sum_{k=0}^{h-1} \frac{3E(v + qk + 3k + 1)}{A} + \sum_{k=0}^{h-1} \frac{K(v + qk + 3k + 1)}{A} + \frac{2hK}{A}$$

and the constants $D_i = v + qi + 3i + \sum_{k=0}^{i-1} \frac{3E(v+qk+3k+1)}{A} + \sum_{k=0}^{i-1} \frac{K(v+qk+3k+1)}{A} + \frac{2hK}{A}$ for $1 \leqslant i \leqslant h-1$.

By showing that the 3-P problem has a solution if and only if for the above instance of problem $1|p_j = 1 + b_j t, d_j|C_{\max}$ there exists a feasible schedule σ such that $C_{\max}(\sigma) \leqslant G$, we complete the proof. □

The restricted version of problem $1|p_j = 1 + b_j t, d_j|C_{\max}$, with only two distinct deadlines, is also computationally intractable.

Theorem 10.6. (Cheng and Ding [3]) *The problem whether there exists a feasible schedule for problem $1|p_j = 1 + b_j t, d_j \in \{D_1, D_2\}|C_{\max}$ is $\mathcal{NP}$-complete in the ordinary sense.*

Proof. We use the following reduction from the PP problem. Let us define $n = (k+1)(k+2)$, $E = n^2 2^{2k} k^{2k} A$ and $B = 16n^3 E^2$, where $A = \frac{1}{2} \sum_{j=1}^{k} x_j$. Job deterioration rates are as follows: $b_{00} = b_{01} = \frac{2E}{B}$, $b_{0j} = 0$ for $2 \leqslant j \leqslant k+1$, $b_{i0} = \frac{E + 2^{2k-2i+2}k^{2k-2i+2}A + x_j}{(i+1)B}$, $b_{ij} = \frac{x_i}{(i+1)A}$ for $1 \leqslant i \leqslant k$ and $1 \leqslant j \leqslant k+1$.

The deadlines are the following: $d_{0j} = D_1$ and $d_{ij} = D_2$ for $1 \leqslant i \leqslant k$ and $0 \leqslant j \leqslant k+1$, where $D_1 = 2k + 2 + \frac{4E-2A+1}{2B} + \sum_{i=1}^{k}(i+1)b_{i0}$ and $D_2 = n + \frac{4E+2kA+1}{2B} + \sum_{i=1}^{k}(i+1)b_{i0}\sum_{i=1}^{k}\sum_{j=0}^{k}((i+1)(k+1)+j)\,b_{i,j+1}$. The threshold value is $G = D_2$.

To complete the proof it is sufficient to construct a schedule for the above instance of problem $1|p_j = 1 + b_j t, d_j \in \{D_1, D_2\}|C_{\max}$ and to show that the PP problem has a solution if and only if this schedule is feasible. $\qquad\square$

10.1.3 General non-linear deterioration

Single machine time-dependent scheduling problems with general non-linearly deteriorating jobs are usually intractable, especially when the processing times of the jobs are non-continuous functions.

Theorem 10.7. (Cheng and Ding [5], Mosheiov [18]) *If there hold inequalities (6.27), the decision version of problem $1|p_j \in \{a_j, b_j : a_j \leqslant b_j\}|C_{\max}$ is $\mathcal{NP}$-complete in the ordinary sense, even if $d_j = D$ for $1 \leqslant j \leqslant n$.*

Proof. Mosheiov [18] transformed an integer programming formulation of the considered problem to the KP problem in the following way. Let us introduce 0-1 variables x_j defined as follows: $x_j = 1$ if job J_j starts not later than time $t = d_j$ and $x_j = 0$ otherwise. Then the maximum completion time for a given schedule is equal to

$$\sum_{j=1}^{n} x_j a_j + \sum_{j=1}^{n}(1 - x_j)b_j = \sum_{j=1}^{n} b_j - \sum_{j=1}^{n} x_j(b_j - a_j). \qquad (10.1)$$

Since in (10.1) the value $\sum_{j=1}^{n} b_j$ is a constant, the problem of minimizing the maximum completion time is equal to the problem of maximizing the sum $\sum_{j=1}^{n} x_j(b_j - a_j)$. Therefore, we can reduce the problem of minimizing the $C_{\max}$ criterion to the following problem (P^1):

$$\max \sum_{j=1}^{n} x_j(b_j - a_j)$$

subject to

$$\sum_{j=1}^{i-1} x_j a_j \leqslant d_i + (1 - x_i)L, \quad i = 1, 2, \dots, n, \qquad (10.2)$$

$$x_i \in \{0, 1\} \quad \text{for } i = 1, 2, \dots, n,$$

where $L \geqslant \max\{d_n, \sum_{j=1}^{n} a_j\}$ is sufficiently large.

Let us consider a new problem (P^2) obtained by ignoring constraints (10.2) for $i = 1, 2, \dots, n-1$. The problem (P^2) is as follows:

$$\max \sum_{j=1}^{n} x_j u_j$$

subject to

$$\sum_{j=1}^{i-1} x_j v_j \leqslant D$$

$$x_i \in \{0,1\} \text{ for } i = 1, 2, \dots, n,$$

where $u_j = b_j - a_j$ for $1 \leqslant j \leqslant n$, $c_j = a_j$ for $1 \leqslant j \leqslant n-1$, and $D = d_n + L$. Since the problem (P^2) is equivalent to the KP problem, the result follows.

Cheng and Ding [5] used the following reduction from the PP problem. Let a_0 and b_0 be two numbers larger than A. Let us define $n = k$, $a_j = b_j = x_j$ for $1 \leqslant j \leqslant k$, $D = A$ and $G = a_0 + 3A$.

If the PP problem has a solution, then there exist disjoint subsets of X, X_1 and X_1, such that $X_1 \cup X_2 = X$ and $\sum_{x_i \in X_1} x_i = \sum_{x_i \in X_2} x_i = A$. Let us construct schedule σ in which jobs corresponding to elements of X_1 are scheduled first, the job corresponding to a_0 and b_0 is scheduled next, and it is followed by jobs corresponding to elements of X_2 (see Fig. 10.2 and Remark 10.17). Then we have $C_{\max}(\sigma) = \sum_{x_j \in X_1} a_j + a_0 + \sum_{x_j \in X_2}(a_j + b_j) = G$. Hence, the considered problem has a solution.

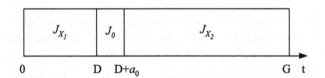

Fig. 10.2: Example schedule in the proof of Theorem 10.7

Let us assume that there exists a schedule $\sigma := (R_1, J_0, R_2)$ for the considered problem, such that $C_{\max} \leqslant G$ and R_1 (R_2) is a set of jobs scheduled before (after) job J_0. By contradiction we can show that neither $C_{\max}(R_1) > A$ nor $C_{\max}(R_1) < A$. Hence, it must be $C_{\max}(R_1) = A$ and by the selection of elements of X which correspond to jobs of R_1 we obtain a solution of the PP problem.   □

Kononov [14] established the time complexity of three cases of problem $1|p_j = a_j + \max\{0, b_j(t - d_j)\}|C_{\max}$.

Theorem 10.8. (Kononov [14]) *The decision version of problem* $1|p_j = a_j + \max\{0, b_j(t - d_j)\}|C_{\max}$ *is*
(a) *$\mathcal{NP}$-complete in the strong sense if deterioration rates b_j are arbitrary;*
(b) *$\mathcal{NP}$-complete in the ordinary sense if $b_j = B$ for $1 \leqslant j \leqslant n$;*
(c) *$\mathcal{NP}$-complete in the ordinary sense if $d_j = D$ for $1 \leqslant j \leqslant n$.*

Proof. (a) We use a reduction from the strongly $\mathcal{NP}$-hard problem $1||\sum w_j T_j$ and to show that given an input for this problem and an arbitrary $\epsilon > 0$ we can construct such an input for problem $1|p_j = a_j + \max\{0, b_j(t - d_j)\}|C_{\max}$ that the solving of the first problem reduces to the solving of the second one, and that an optimal schedule for the second problem is an optimal schedule for the first one for sufficiently small ϵ; see [14, Theorem 1].

(b) By applying a similar reduction from the ordinary $\mathcal{NP}$-hard problem $1||\sum T_j$, the result follows; see [14, Theorem 2].

(c) Given an instance of the SS problem, let us define $u_{\max} := \max_{i \in R}\{u_i\}$, $U := \sum_{i \in R} u_i$, $\epsilon := \frac{1}{u_{\max}^2 n^2}$ and $\mu := \frac{\epsilon}{28 U n}$. Let us construct an instance of problem $1|p_j = a_j + \max\{0, b_j(t - d_j)\}|C_{\max}$ as follows: $n = r + 1$, $a_1 = 1$, $b_1 = \frac{1}{\mu}$, $a_i = \mu u_i$ and $b_i = \epsilon u_i$ for $2 \leqslant i \leqslant r + 1$ and $D = \mu C$. By applying the reasoning similar to (a), the result follows; see [14, Theorem 7].   ◇

Remark 10.9. The ordinary $\mathcal{NP}$-hardness of problem $1||\sum T_j$ was proved by Du and Leung [9]. The strong $\mathcal{NP}$-hardness of problem $1||\sum w_j T_j$ was proved by Lawler [16].

Remark 10.10. Since Theorem 10.8 was originally formulated in the optimization form, the reductions in its proof were made from optimization versions of the problems used in these reductions. The decision form has been used for compatibility with other results presented in the book.

Cai et al. [1] considered a single machine scheduling problem with deteriorating job processing times given by (6.16). The first result proved in [1] concerns the case when $f(t, t_0) := \mathbf{1}_X$ for a set X.

Theorem 10.11. (Cai et al. [1]) *If $X := \{t : t - t_0 > 0\}$, $f(t, t_0) := \mathbf{1}_X$ and $\sum_{j=1}^n a_j > t_0$, then the decision version of problem $1|p_j = a_j + b_j f(t, t_0)|C_{\max}$ is $\mathcal{NP}$-complete in the ordinary sense.*

Proof. Let $X := \{t : t - t_0 > 0\}$ and $f(t, t_0) := \mathbf{1}_X$. Then problem $1|p_j = a_j + b_j f(t, t_0)|C_{\max}$ is equivalent to a version of the KP problem. The version of the KP problem can be formulated as follows. Given $t_0 > 0$ and $k - 1$ pairs of positive integers $\{a_1, b_1), (a_2, b_2), \ldots, (a_{k-1}, b_{k-1})\}$, find a subset $K \subseteq \{1, 2, \ldots, k-1\}$ which maximizes $\sum_{j \in K} b_j$ subject to $\sum_{j \in K} a_j \leqslant t_0$. By letting $a_k > t_0$ and $b_k > \max_{1 \leqslant j \leqslant k-1}\{b_j\}$, we obtain an instance of problem $1|p_j = a_j + b_j f(t, t_0)|C_{\max}$. Since the latter problem has an optimal solution if and only if the KP problem has an optimal solution, and since the KP problem is $\mathcal{NP}$-complete in the ordinary sense, the result follows.   □

Remark 10.12. Cai et al. also proved that a restricted version of problem $1|p_j = a_j + b_j \max\{t - t_0, 0\}|C_{\max}$ is still $\mathcal{NP}$-complete in the ordinary sense; see [1, Sect. 4.1].

Kubiak and van de Velde [15] considered the case of non-linear job processing times given by (6.31).

Theorem 10.13. (Kubiak and van de Velde [15]) *If $d > 0$ and $D = +\infty$, then the decision version of the problem of minimizing the maximum completion time for a single machine and for job processing times given by (6.31) is $\mathcal{NP}$-complete in the ordinary sense.*

Proof. We use the following reduction from the PP problem, provided that $|X| = k = 2l$ for some $l \in \mathbb{N}$. Let $n = k + 1 = 2l + 1$, $H_1 = A^2$, $d = lH_1 + a$, $H_2 = A^5$, $H_3 = A(H_2^{l-1}(H_2 + A + 1) + 1)$ and $H_4 = H_3 \sum_{i=0}^{l} H_2^{l-i} A^i + l^2 H_2^{l-1} d$, where $A = \frac{1}{2} \sum_{j=1}^{k} x_j$.

Job processing times are defined as follows: $a_j = x_j + H_1$ and $b_j = H_2 + x_j - 1$ for $1 \leqslant j \leqslant k$, $a_{k+1} = H_3$ and $b_{k+1} = H_4 + 1$. The threshold value is $G = H_4$.

To complete the proof it is sufficient to show that the PP problem has a solution if and only if for the above instance there exists a feasible schedule σ such that $C_{\max}(\sigma) \leqslant G$ (see [15, Lemmas 2-4]). $\square$

Janiak and Kovalyov [12] considered exponentially deteriorating job processing times given by (6.23).

Theorem 10.14. (Janiak and Kovalyov [12]) *The decision version of problem $1|p_j = a_j 2^{b_j(t-r_j)}|C_{\max}$ is $\mathcal{NP}$-complete in the strong sense.*

Proof. The reduction from the 3-P problem is as follows. There are $n = 4h$ jobs, $r_j = 0, a_j = c_j, b_j = 0$ for $1 \leqslant j \leqslant 3h$, $r_{3h+i} = iK + i - 1$, $a_{3h+i} = 1$, $b_{3h+i} = K$ for $1 \leqslant i \leqslant h$. The threshold value is $G = hK + h$.

In order to complete the proof it is sufficient to show that the 3-P problem has a solution if and only if there exists a schedule σ for the above instance of problem $1|p_j = a_j 2^{b_j(t-r_j)}|C_{\max}$ such that $C_{\max}(\sigma) \leqslant G$. $\square$

If there are only two distinct ready times, the problem remains computationally intractable.

Theorem 10.15. (Janiak and Kovalyov [12]) *The decision version of problem $1|p_j = a_j 2^{b_j(t-r_j)}, r_j \in \{0, R\}|C_{\max}$ is $\mathcal{NP}$-complete in the ordinary sense.*

Proof. The reduction from the PP problem is as follows. There are $n = k + 1$ jobs, $r_j = 0, a_j = x_j, b_j = 0$ for $1 \leqslant j \leqslant k$, $r_{k+1} = A$, $a_{k+1} = 1$ and $b_{k+1} = 1$, where $A = \frac{1}{2} \sum_{j=1}^{k} x_j$. The threshold value is $G = 2A + 1$.

In order to complete the proof it is sufficient to show that the PP problem has a solution if and only if there exists a schedule σ for the above instance of problem $1|p_j = a_j 2^{b_j(t-r_j)}, r_j \in \{0, R\}|C_{\max}$ such that $C_{\max}(\sigma) \leqslant G$ (see Fig. 10.3 and Remark 10.17). $\square$

Remark 10.16. Janiak and Kovalyov stated Theorem 10.15 (see [12, Theorem 2]) without proof. The above reduction comes from the present author.

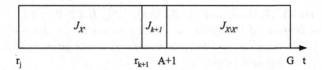

Fig. 10.3: A schedule in the proof of Theorem 10.15

Remark 10.17. In figures included in some $\mathcal{NP}$-completeness proofs, by J_A we denote a set of jobs with deterioration rates corresponding to the elements of set A. For example, in Fig. 10.3, the symbols $J_{X'}$ and $J_{X \setminus X'}$ denote the set of jobs with deterioration rates corresponding to the elements of set X' and $X \setminus X'$, respectively.

10.1.4 Linear shortening

Single machine time-dependent scheduling problems with linearly shortening jobs, similarly to those with linearly deteriorating jobs, are intractable.

Theorem 10.18. (Ho et al. [11]) *If* a_j, b_j *and* d_j *satisfy conditions* (6.44), (6.45) *and* $b_j d_j \leqslant a_j \leqslant d_j$ *for all* $1 \leqslant j \leqslant n$, *then problem* $1|p_j = a_j - b_j t, d_j|-$
(a) *is* $\mathcal{NP}$-complete *in the strong sense if there is an arbitrary number of dead-lines;*
(b) *is* $\mathcal{NP}$-complete *in the ordinary sense if there are only two distinct dead-lines.*

Proof. (a) The reduction from the 3-P problem is as follows: $n = 4h - 1$,
$$d_i = \begin{cases} i(K+1) & \text{for } 1 \leqslant i \leqslant h-1, \\ hK + h - 1 & \text{for } h \leqslant i \leqslant 4h - 1, \end{cases}$$
$$a_i = \begin{cases} iK - 1 & \text{for } 1 \leqslant i \leqslant h-1, \\ c_{i-h+1} & \text{for } h \leqslant i \leqslant 4h - 1, \end{cases}$$
$$b_i = \begin{cases} \frac{iK+i-1}{iK+i-1} & \text{for } 1 \leqslant i \leqslant h-1, \\ \frac{1}{16h^6K^6}, & \text{for } h \leqslant i \leqslant 4h - 1. \end{cases}$$
To complete the proof it is sufficient to show that the 3-P problem has a solution if and only if for the above instance of problem $1|p_j = a_j - b_j t, d_j|C_{\max}$ there exists a non-preemptive feasible schedule.

(b) The reduction from the PP problem is as follows: $n = k + 1$,
$$d_i = \begin{cases} A + 1 & \text{for } i = 1, \\ 2A + 1 & \text{for } 2 \leqslant i \leqslant k + 1, \end{cases}$$
$$a_i = \begin{cases} A & \text{for } i = 1, \\ x_{i-1} & \text{for } 2 \leqslant i \leqslant k + 1. \end{cases}$$
$$b_i = \begin{cases} \frac{A-1}{4} & \text{for } i = 1, \\ \frac{1}{3k^2A}, & \text{for } 2 \leqslant i \leqslant k + 1. \end{cases}$$
To complete the proof it is sufficient to show that the PP problem has a solution if and only if for the above instance of problem $1|p_j = a_j - b_j t, d_j|C_{\max}$ there exists a non-preemptive feasible schedule. $\qquad\square$

The problem of the complexity of minimizing the maximum completion time in the case when there are arbitrary deadlines and equal shortening rates, stated by Ho et al. [11], has been solved by Cheng and Ding [7].

Theorem 10.19. (Cheng and Ding [7]) *The decision version of problem* $1|p_j = a_j - bt, d_j|C_{\max}$ *is $\mathcal{NP}$-complete in the strong sense.*

Proof. The reduction is from the 3-P problem. Let us define $q = 2^6 h^3 K^2$, the identical decreasing processing rate $b = \frac{1}{2^3 q^3 h K}$ and the number of jobs $n = 3h + (h-1)q$. Let us define $D_j = 1 - b((q-1)(jK + j - 1) + \frac{1}{q}\sum_{k=1}^{q-1} k)$ for $1 \leqslant j \leqslant h - 1$. The deadlines are as follows: $d_i = d^0 = hK + \sum_{j=1}^{h-1} D_j$ for $1 \leqslant i \leqslant 3h$, $d_k^j = d^j = jK + j$ for $1 \leqslant j \leqslant h - 1$ and $1 \leqslant k \leqslant q$. The basic processing times are the following: $a_i = c_i$ for $1 \leqslant i \leqslant 3h$, and $a_k^j = a = \frac{1}{q}$ for $1 \leqslant j \leqslant h - 1$ and $1 \leqslant k \leqslant q$. The threshold value is d^0.

To complete the proof it is sufficient to show that an instance of the 3-P problem has a solution if and only if the above constructed instance of problem $1|p_j = a_j - bt, d_j|C_{\max}$ has a solution. The idea which simplifies the proof consists in introducing a special form of a feasible schedule for problem $1|p_j = a_j - bt, d_j|C_{\max}$. Due to regularity of the schedule, further calculations are easier (see [7, Lemmas 1–4]).   □

Theorem 10.20. Cheng and Ding [3]) *The problem of whether there exists a feasible schedule for problem* $1|p_j = 1 - b_j t, d_j|C_{\max}$ *is $\mathcal{NP}$-complete in the strong sense.*

Proof. The reduction from the 3-P problem is as follows. Let us define $q = 2hK$, $v = 32h^2 qK$, $n = v + 3h + (h-1)q$, $A_1 = 3n^3$ and $A_2 = A_3 = 2nhK$.

The deterioration rates and deadlines are the following: $b_{0i} = 0, d_{0i} = v$ for $1 \leqslant i \leqslant v$, $b_{ij} = \frac{1}{A_1 A_3}, d_{ij} = D_i$ for $1 \leqslant i \leqslant h - 1$ and $1 \leqslant j \leqslant q$, $b_i = \frac{c_i}{A_1 A_2 A_3}, d_i = G$ for $1 \leqslant i \leqslant 3h$, where

$$D_i = v + qi + 3i - \sum_{k=1}^{i-1}\sum_{l=1}^{q} \frac{v + qk + 3k - l}{A_1 A_3} - \sum_{k=0}^{i-1} \frac{(v + qk + 3k)K}{A_1 A_2 A_3}$$

for $1 \leqslant i \leqslant h - 1$ and the threshold value is

$$G = n - \sum_{k=1}^{h-1}\sum_{l=1}^{q} \frac{v + qk + 3k - l}{A_1 A_3} - \sum_{k=0}^{h-1} \frac{(v + qk + 3k)K}{A_1 A_2 A_3}.$$

The set of jobs in the above instance of problem $1|p_j = 1 - b_j t, d_j|C_{\max}$ is divided into sets $V = \{J_{01}, J_{02}, \ldots, J_{0v}\}$, $R = \{J_1, J_2, \ldots, J_{3h}\}$ and $Q_i = \{J_{i1}, J_{i2}, \ldots, J_{iq}\}$ for $1 \leqslant i \leqslant h - 1$. Let us construct for this instance a schedule in the form of $(V, R_1, Q_1, R_2, \ldots, Q_{h-1}, R_h)$, where the job order in any of these sets is arbitrary. By showing that the 3-P problem has a solution if and only if the schedule is feasible for problem $1|p_j = 1 - b_j t, d_j|C_{\max}$ (see [3, Lemmas 1–3]), we obtain the result.   □

The restricted version of the above problem, when there are only two distinct deadlines, is also computationally intractable.

Theorem 10.21. (Cheng and Ding [3]) *The problem of whether there exists a feasible schedule for problem* $1|p_j = 1 - b_j t, d_j \in \{D_1, D_2\}|C_{\max}$ *is $\mathcal{NP}$-complete in the ordinary sense.*

Proof. The reduction is from the PP problem. Let us define $n = (k+1)(k+2)$, $A_1 = 4n^3$ and $A_2 = A_3 = 2^{k+1}k^k n^2 A$, where $A = \frac{1}{2}\sum_{j=1}^{k} x_j$.

The job deterioration rates are the following: $b_{00} = b_{01} = 0$, $b_{0j} = \frac{1}{A_2 A_3}$ for $2 \leqslant j \leqslant k+1$, $b_{i0} = \frac{2^i k^i A - x_i}{(i+1)A_1 A_2 A_3}$, $b_{ij} = \frac{2^i k^i A}{(i+1)A_1 A_2 A_3}$ for $1 \leqslant i \leqslant k$ and $1 \leqslant j \leqslant k+1$.

The deadlines are the following: $d_{0j} = D_1$ and $d_{ij} = D_2$ for $1 \leqslant i \leqslant k$ and $0 \leqslant j \leqslant k+1$, where $D_1 = 2k + 2 - \sum_{i=1}^{k}(i+1)b_{i1} - \sum_{j=2}^{k+1}(k+j)b_{0j} +$

$\frac{A}{A_1 A_2 A_3} + \frac{1}{2A_1 A_2 A_3}$, $D_2 = n - \sum_{i=1}^{k}(i+1)b_{i1} - \sum_{j=2}^{k+1}(k+j)b_{0j} - \sum_{i=1}^{k}(i+1)(k+1)b_{i0} -$

$\sum_{i=1}^{k}\sum_{j=1}^{k}((i+1)(k+1)+j)\,b_{i,j+1} - \frac{kA}{A_1 A_2 A_3} + \frac{1}{2A_1 A_2 A_3}$. The threshold value is $G = D_2$.

In order to complete the proof it is sufficient to show that the PP problem has a solution if and only if for the above instance of problem $1|p_j = 1 - b_j t$, $d_j \in \{D_1, D_2\}|C_{\max}$ there exists a feasible schedule. $\square$

Another restricted version of the problem, when basic job processing times are distinct, all deterioration rates are equal and there are only two distinct deadlines, is computationally intractable as well.

Theorem 10.22. (Cheng and Ding [6]) *The decision version of problem* $1|p_j = a_j - bt, d_j \in \{D_1, D_2\}|C_{\max}$ *is $\mathcal{NP}$-complete in the ordinary sense.*

Proof. The reduction is from the PP problem. Let $B = 2^{n+3}n^2 A$ and $v = 2^6 n^3 B$, where $A = \frac{1}{2}\sum_{j=1}^{k} x_j$. Let us define $n = 2k+1$ shortening jobs, where $a_0 = v$, $a_{1i} = v(B + 2^{n-i+}A + x_i)$ and $a_{2i} = v(B + 2^{n-i+1}A)$ for $1 \leqslant i \leqslant n$, $b = \frac{2}{v}$, $d_0 = v(nB + 2^{n+1}A - A + 1)$ and $d_{1i} = d_{2i} = G$ for $1 \leqslant i \leqslant n$. The threshold value is

$$G = \sum_{i=0}^{2n+1}(x_i - b(E - (n+1)Av) + 1),$$

with constant

$$E = \sum_{i=1}^{n}(2n+1)a_{2i} + na_0 + \sum_{i=1}^{n-1}(n-i)a_{1i}.$$

In order to complete the proof it is sufficient to show that the problem PP has a solution if and only if for the above instance of problem $1|p_j = a_j - bt$, $d_j \in \{D_1, D_2\}|C_{\max}$ there exists a feasible schedule σ such that $C_{\max}(\sigma) \leqslant G$ (see [6, Lemmas 1–4]).   □

10.1.5 General non-linear shortening

Single machine time-dependent scheduling problems with non-linearly shortening jobs are intractable.

Theorem 10.23. (Cheng et al. [8]) *The decision version of problem $1|p_j \in \{a_j, a_j - b_j : 0 \leqslant b_j \leqslant a_j\}|C_{\max}$ is $\mathcal{NP}$-complete in the ordinary sense.*

Proof. The reduction from the PP problem is as follows. Let $n = k$, $a_j = 2x_j$ and $b_j = x_j$ for $1 \leqslant j \leqslant n$, $D = 2A$ and the threshold value is $G = 3A$, where $A = \frac{1}{2} \sum_{j=1}^{k} x_j$.

To complete the proof it is sufficient to show that the PP problem has a solution if and only if for the above instance of problem $1|p_j \in \{a_j, a_j - b_j : 0 \leqslant b_j \leqslant a_j\}|C_{\max}$ there exists a schedule σ such that $C_{\max}(\sigma) \leqslant G$.   □

There exists a relationship between the KP problem and problem $1|p_j \in \{a_j, a_j - b_j : 0 \leqslant b_j \leqslant a_j\}|C_{\max}$.

Lemma 10.24. (Cheng et al. [8]) *A version of the KP problem is equivalent to problem $1|p_j \in \{a_j, a_j - b_j : 0 \leqslant b_j \leqslant a_j\}|C_{\max}$.*

Proof. Let us consider an optimal schedule σ for the above scheduling problem. Let $J_E := \{J_k \in \mathcal{J} : S_k < D\}$ and $J_T := \mathcal{J} \setminus J_E = \{J_k \in \mathcal{J} : S_k \geqslant D\}$. Let E and T denote sets of indices of jobs from the set J_E and J_T, respectively. Only two cases are possible: Case 1, when $\sum_{j \in E} a_j \leqslant D - 1$, and Case 2, when $\sum_{j \in E} a_j \geqslant D$.

In Case 1, we have $C_{\max}(\sigma) = D + \sum_{j \in T}(a_j - b_j)$. This, in turn, corresponds to the solution of the following KP problem: $\min \sum_{j \in T}(a_j - b_j)$ subject to $\sum_{j \in T} a_j \geqslant \sum_{j=1}^{n} a_j - D + 1$ for $T \subseteq \{1, 2, \ldots, n\}$.

In Case 2, we have $C_{\max}(\sigma) = \sum_{j \in E}(a_j) + \sum_{j \in T}(a_j - b_j) = \sum_{j \in E} b_j + \sum_{j \in E}(a_j - b_j) + \sum_{j \in T}(a_j - b_j) = \sum_{j=1}^{n}(a_j - b_j) + \sum_{j \in E} b_j$. This, in turn, corresponds to the solution of the following KP problem: $\min \sum_{j \in T} b_j$ subject to $\sum_{j \in E} a_j \geqslant D$ for $E \subseteq \{1, 2, \ldots, n\}$.

The optimal value of the $C_{\max}$ criterion equals

$$\min \left\{ D + z_1, \sum_{j=1}^{n}(a_j - b_j) + z_2 \right\},$$

where z_1 (z_2) is the solution of the first (the second) KP problem.   □

Cheng et al. [2] considered a single machine problem with non-linearly shortening job processing times given by (6.49).

Theorem 10.25. (Cheng et al. [2]) *The decision version of problem $1|p_j = a_j - b_j(t-y), y = 0, 0 < b_j < 1, Y < +\infty|C_{\max}$ is $\mathcal{NP}$-complete in the ordinary sense.*

Proof. The reduction is from the PP problem. Let $V = (k!)(2k)^{3k+6}A^2$, $B = V^4$, $\alpha = \frac{1}{V^{20}}$ and $\beta = \frac{1}{V^{22}}$. Let $2k$ jobs with shortening processing times be as follows: $a_{1j} = B + 2^j A + x_j$ and $a_{2j} = B + 2^j A$ for $1 \leqslant j \leqslant k$, $b_{1j} = \alpha a_{1j} - \beta(2k)^j A - \frac{x_j}{k-j+1}$, $b_{2j} = \alpha a_{2j} - \beta(2k)^j A$ for $1 \leqslant j \leqslant k$, where $A = \frac{1}{2}\sum_{j=1}^{k} x_j$.

The common initial shortening date is $y = 0$ and the common final shortening date is $Y = kB + A(2^{k+1} - 1)$. The threshold value is

$$G = 2Y - \alpha E + \beta BF + 2\alpha V,$$

where the constants E and F are defined as follows:

$$E = \frac{3Y^2 - kB^2}{2} - BA(2^{k+1} - 1)$$

and

$$F = \sum_{j=1}^{k} \left((k+j-1)(2k)^{k+j-1}A + \frac{kx_j}{k-j+1} \right) - A.$$

To complete the proof it is sufficient to show that the problem PP has a solution if and only if for the above instance of problem $1|p_j = a_j - b_j(t-y)$, $y = 0, 0 < b_j < 1, Y < +\infty|C_{\max}$ there exists a feasible schedule σ such that $C_{\max}(\sigma) \leqslant G$ (see [2, Lemmas 1–3]). $\square$

10.1.6 Simple alteration

Jaehn and Sedding [13] proved $\mathcal{NP}$-completness of two single machine time-dependent scheduling problems with alterable job processing times.

Theorem 10.26. (Jaehn and Sedding [13]) (a) *The decision version of problem $1 | p_j = a_j + b|f_j(t) - T| | C_{\max}$ is $\mathcal{NP}$-complete in the ordinary sense.* (b) *The decision version of problem $1 | p_j = a_j + b|f_j(t) - T_j| |C_{\max}$ is $\mathcal{NP}$-complete in the ordinary sense.*

Proof. (a) Let

$$f(x) := \left(\frac{2+b}{2-b} \right)^x$$

and

$$g(x) := \frac{2}{2-b} f(x)$$

for any $x \in \mathbb{R}$. Then, the reduction from the MEOP problem is as follows: $t_0 = 0$, $b = \frac{2q-2}{q+1}$, $a_0 = 0$, $a_j = \frac{x_j}{g(k-\lfloor \frac{j}{2} \rfloor)}$ for $j = 1, 2, \ldots, n$ and $T = \frac{G}{2}$. The threshold value is

$$G = a_0 f(k) + \sum_{j=1}^{k}(a_{2j} + a_{2j-1})g(k-j).$$

To complete the proof it is sufficient to show that the problem MEOP has a solution if and only if for the above instance of the considered problem there exists a feasible schedule σ such that $C_{\max}(\sigma) \leqslant G$ (see [13, Theorem 2]. □

(b) Similar to the proof of case (a); see [13, Theorem 3]. ◇

10.2 Minimizing the total completion time

In this section, we consider $\mathcal{NP}$-hard single machine time-dependent scheduling problems with the $\sum C_j$ criterion.

10.2.1 Linear shortening

Single machine time-dependent scheduling problems with linearly shortening jobs are usually $\mathcal{NP}$-hard.

Theorem 10.27. (Cheng and Ding [24]) *The decision version of problem* $1|p_j = a_j - bt, d_j = D|\sum C_j$ *is $\mathcal{NP}$-complete in the ordinary sense.*

Proof. The authors only give the idea of the proof; see [24, Theorem 1]. ◇

Now we pass to the results concerning job processing times given by (6.49).

Theorem 10.28. (Cheng et al. [23]) *The decision version of problem* $1|p_j = a_j - b_j(t - y), y = 0, 0 < b_j < 1, Y < +\infty|\sum C_j$ *is $\mathcal{NP}$-complete in the ordinary sense.*

Proof. The reduction from the PP problem is as follows. Let $V = (2kA)^6$ and $B = V^3$, where $A = \frac{1}{2}\sum_{j=1}^{k} x_j$. Let us define $2k+1$ jobs with shortening processing times as follows: $a_0 = 4k^2B$, $b_0 = 1$, $a_{1j} = jB + x_j(\frac{1}{2} + (2k - 3j + 2))$ and $a_{2j} = jB$ for $1 \leqslant j \leqslant k$, and $b_{1j} = 0$, $b_{2j} = \frac{x_j}{jB}$ for $1 \leqslant j \leqslant k$.

The common initial shortening date is $y = 0$ and the common final shortening date is

$$Y = \sum_{j=1}^{k}(a_{1j} + a_{2j}) - \sum_{j=1}^{k}(j-1)x_j - A.$$

The threshold value is

$$G = E + a_0 - F + \frac{H}{2} + \frac{1}{V},$$

where constants E and F are defined as follows:

$$E = \sum_{j=1}^{k} \left((4k - 4j + 3)jB + (2k - 2j - 1)(\frac{1}{2} + (2k - 3j + 2))x_j \right)$$

and

$$F = 2 \sum_{j=1}^{k} (k - j + 1)(j - 1)x_j.$$

To complete the proof it is sufficient to show that the PP problem has a solution if and only if for the above instance of problem $1|p_j = a_j - b_j t,$ $0 < b_j < 1, 0 \leqslant t \leqslant Y| \sum C_j$ there exists a schedule σ such that $\sum C_j(\sigma) \leqslant G$. $\qquad \square$

The problem of minimizing the total completion time for a set of jobs which have the same shortening rate, $b_j = b$, and only two distinct deadlines, $d_j \in \{D_1, D_2\}$, is computationally intractable.

Theorem 10.29. (Cheng and Ding [22]) *The decision version of problem* $1|p_j = a_j - bt, d_j \in \{D_1, D_2\}| \sum C_j$ *is $\mathcal{NP}$-complete in the ordinary sense.*

Proof. The result is a corollary from Theorem 10.22. $\qquad \square$

10.3 Minimizing the maximum lateness

In this section, we consider $\mathcal{NP}$-hard single machine time-dependent scheduling problems with the $L_{\max}$ criterion.

10.3.1 Linear deterioration

Minimizing the maximum lateness is intractable, even for linearly deteriorating processing times.

Theorem 10.30. (Kononov [28]) *The decision version of problem* $1|p_j = a_j + b_j t|L_{\max}$ *is $\mathcal{NP}$-complete in the ordinary sense, even if only one $a_k \neq 0$ for some $1 \leqslant k \leqslant n$, and $d_j = D$ for jobs with $a_j = 0$.*

Proof. The reduction from the SP problem is as follows. There are $n = p + 1$ jobs, where $a_0 = 1, b_0 = 0, d_0 = B + 1$ and $a_j = 0, b_j = y_j - 1, d_j = \frac{Y(B+1)}{B}$ for $1 \leqslant j \leqslant p$, with $Y = \prod_{j=1}^{p} y_j$. All jobs start at time $t_0 = 1$. The threshold value is $G = 0$. To prove the result, it is sufficient to apply (7.1) and show that the SP problem has a solution if and only if for the above instance of problem $1|p_j = a_j + b_j t|L_{\max}$ there exists a schedule σ such that $L_{\max}(\sigma) \leqslant G$. $\qquad \square$

Problem $1|p_j = a_j + b_j t|L_{\max}$ was also studied by other authors.

Theorem 10.31. (Bachman and Janiak [25]) *The decision version of problem* $1|p_j = a_j + b_j t|L_{\max}$ *is* $\mathcal{NP}$*-complete in the ordinary sense, even if there are only two distinct deadlines.*

Proof. The reduction from the PP problem is as follows. We have $n = k + 1$ jobs, $d_i = \left(k^{q+2}A + kA + k + A + 1 + \frac{1}{k^q} + \frac{1}{k^{q-1}}\right)\left(1 + \frac{2}{2k^q-1}\right) - k^q A$, $a_i = x_i$ and $b_i = \frac{x_i}{k^q A}$ for $1 \leqslant i \leqslant k$, and $d_{k+1} = k^{q+2}A + kA + A + k + 1 + \frac{(A+2)(k+1)}{2k^q-1}$, $a_{k+1} = k^{q+2}A$, $b_{k+1} = k$, where $q = \lceil \frac{\ln(A+1) - \ln(2)}{\ln k}\rceil + 3$ and $A = \frac{1}{2}\sum_{i=1}^{k} x_i$. All jobs start at time $t_0 = 1$. The threshold value is $G = 0$. In order to complete the proof it is sufficient to show that the PP problem has a solution if and only if for the above instance of problem $1|p_j = a_j + b_j t|L_{\max}$ there exists a schedule σ such that $L_{\max}(\sigma) \leqslant G$. $\qquad\square$

Theorem 10.32. (Cheng and Ding [26]) *The decision version of problem* $1|p_j = a_j + b_j t|L_{\max}$ *is* $\mathcal{NP}$*-complete in the strong sense.*

Proof. The authors give only a sketch of a proof; see [26, Theorem 6]. $\qquad\diamond$

10.3.2 Linear shortening

The problem of minimizing the maximum lateness for a set of jobs which have the same shortening rate, $b_j = b$, and only two distinct deadlines, $d_j \in \{D_1, D_2\}$, is intractable.

Theorem 10.33. (Cheng and Ding [30]) *The decision version of problem* $1|p_j = a_j - bt, d_j \in \{D_1, D_2\}|L_{\max}$ *is* $\mathcal{NP}$*-complete in the ordinary sense.*

Proof. The result is a corollary from Theorem 10.22. $\qquad\square$

10.3.3 General non-linear shortening

Single machine time-dependent scheduling problems with general non-linear shortening job processing times are usually $\mathcal{NP}$-hard.

Theorem 10.34. (Janiak and Kovalyov [27]) *The decision version of problem* $1|p_j = a_j 2^{-b_j t}|L_{\max}$ *is* $\mathcal{NP}$*-complete in the strong sense.*

Proof. The reduction from the 3-P problem is as follows. There are $n = 4h$ jobs, $a_j = c_j, b_j = 0, d_j = hK + \frac{h-1}{2}$ for $1 \leqslant j \leqslant 3h$, $a_{3h+i} = 1$, $b_{3h+i} = (iK + \frac{i-1}{2})^{-1}$ for $1 \leqslant i \leqslant h$. The threshold value is $G = 0$.

To complete the proof it is sufficient to show that the 3-P problem has a solution if and only if for the above instance of problem $1|p_j = a_j 2^{-b_j t}|L_{\max}$ there exists a schedule σ such that $L_{\max}(\sigma) \leqslant G$. $\qquad\square$

The restricted version of the above problem, with only two distinct deadlines, is computationally intractable as well.

Theorem 10.35. (Janiak and Kovalyov [27]) *The decision version of problem* $1|p_j = a_j 2^{-b_j t}, d_j \in \{d, D\}|L_{\max}$ *is $\mathcal{NP}$-complete in the ordinary sense.*

Proof. The reduction from the PP problem is as follows. There are $n = k + 1$ jobs, $a_j = x_j, b_j = 0, d_j = 2A + 1$ for $1 \leqslant j \leqslant k$, $a_{k+1} = 1$, $b_{k+1} = 0$ and $d_{k+1} = A + 1$, where $A = \frac{1}{2}\sum_{j=1}^{k} x_j$. The threshold value is $G = 0$.

   To complete the proof it is sufficient to show that the PP problem has a solution if and only if for the above instance of problem $1|p_j = a_j 2^{-b_j t}$, $d_j \in \{d, D\}|L_{\max}$ there exists a schedule σ such that $L_{\max}(\sigma) \leqslant G$.   □

Remark 10.36. Janiak and Kovalyov stated Theorem 10.35 (see [27, Theorem 4]) without proof. The above reduction comes from the present author.

10.4 Other criteria

In this section, we consider $\mathcal{NP}$-hard single machine time-dependent scheduling problems with other criteria than $C_{\max}, \sum C_j$ and $L_{\max}$.

10.4.1 Linear deterioration

Bachman et al. [20] proved that single machine scheduling with linear jobs, arbitrary deterioration rates and arbitrary weights is $\mathcal{NP}$-hard.

Theorem 10.37. (Bachman et al. [20]) *If $S_1 = 1$, then the decision version of problem* $1|p_j = a_j + b_j t| \sum w_j C_j$ *is $\mathcal{NP}$-complete in the ordinary sense.*

Proof. Let us assume that jobs start at time $S_1 = 1$. The reduction from the N3P problem is as follows. Let $n = 4w$, $a_i = 0$, $b_i = D^{zi} - 1$ and $w_i = 1$ for $1 \leqslant i \leqslant 3w$ and $a_{3w+i} = D^{iZ}$, $b_{3w+i} = 0$, $w_{3w+i} = D^{(w+1-i)Z}$ for $1 \leqslant i \leqslant w$, where $D = 2w^2 + 1$. The threshold value is

$$G = 2w^2 D^{(w+1)Z}.$$

   To complete the proof it is sufficient to show that the N3P problem has a solution if and only if for the above instance of problem $1|p_j = a_j + b_j t| \sum w_j C_j$ there exists a schedule σ such that $\sum w_j C_j(\sigma) \leqslant G$.   □

Remark 10.38. Notice that the assumption $S_1 = 1$ is not essential. If we assume that $S_1 = 0$ and add to the instance described above an additional job J_0 with parameters $a_0 = 1, b_0 = 0, w_0 = G+1$, and if we change the threshold value to $2G + 1$, then it can be shown that the result of Theorem 10.37 also holds in this case; see [20, Sect. 2] for details.

10.4.2 Linear shortening

Single machine time-dependent scheduling problems with linearly shortening jobs, similarly to those with linearly deteriorating jobs, are intractable.

Theorem 10.39. (Bachman et al. [29]) *The decision version of problem* $1|p_j = a_j - b_j t,\ 0 \leqslant b_j < 1,\ b_i(\sum_{j=1}^{n} a_j - a_i) < a_i|\sum w_j C_j$ *is* $\mathcal{NP}$*-complete in the ordinary sense, even if there exists only one non-zero shortening rate.*

Proof. The reduction from the PP problem is as follows. We have $n = k + 1$ jobs with the following parameters: $a_i = x_i$, $b_i = 0$ and $w_i = x_i$ for $1 \leqslant i \leqslant k$, and $a_{k+1} = 2A$, $b_{k+1} = 1 - \frac{1}{A}$, and $w_{k+1} = 2A^2$, where $A = \frac{1}{2}\sum_{i=1}^{k} x_i$. The threshold value is

$$G = \frac{1}{2}\sum_{i=1}^{k} x_i^2 + A^2 + A(2A + 1)^2.$$

To complete the proof it is sufficient to show that the PP problem has a solution if and only if for the above instance of problem $1|p_j = a_j - b_j t,\ 0 \leqslant b_j < 1, b_i(\sum_{j=1}^{n} a_j - a_i) < a_i|\sum w_j C_j$ there exists a schedule σ such that the total weighted completion time $\sum w_j C_j(\sigma) \leqslant G$. The equivalence can be proved using the equalities

$$\frac{1}{2}\sum_{i=1}^{k} x_i^2 = \frac{1}{2}\left(\sum_{i=1}^{k} x_i\right)^2 - \sum_{1 \leqslant i < j \leqslant k} x_i x_j = A^2 - \sum_{1 \leqslant i < j \leqslant m} x_i x_j;$$

see [19, Theorem 1] for details. □

10.4.3 General non-linear shortening

Single machine time-dependent scheduling problems with non-linearly shortening jobs are intractable.

Theorem 10.40. (Janiak and Kovalyov [31]) *The decision version of problem* $1|p_j = a_j 2^{-b_j t}|\sum w_j C_j$ *is* $\mathcal{NP}$*-complete in the ordinary sense.*

Proof. The reduction from the PP problem is as follows. Let us define $n = k + 1$ jobs, where $w_j = x_j$, $a_j = x_j$, $b_j = 0$ for $1 \leqslant j \leqslant k$, $w_{k+1} = A$, $a_{k+1} = \frac{2A}{2\ln 2 + 1}$, $b_{k+1} = \frac{1}{A}$, where $A = \frac{1}{2}\sum_{j=1}^{k} x_j$. The threshold value is

$$G = \frac{1}{2}\sum_{j=1}^{k} x_j^2 + A^2\left(3 + \frac{2}{2\ln 2 + 1}\right).$$

To complete the proof it is sufficient to show that the PP problem has a solution if and only if for the above instance of problem $1|p_j = a_j 2^{-b_j t}|\sum w_j C_j$ there exists a schedule σ such $\sum w_j C_j(\sigma) \leqslant G$. ∎

References

Minimizing the maximum completion time

1. J-Y. Cai, P. Cai and Y. Zhu, On a scheduling problem of time deteriorating jobs. *Journal of Complexity* **14** (1998), no. 2, 190–209.
2. T-C. E. Cheng, Q. Ding, M. Y. Kovalyov, A. Bachman and A. Janiak, Scheduling jobs with piecewise linear decreasing processing times. *Naval Research Logistics* **50** (2003), no. 6, 531–554.
3. T-C. E. Cheng and Q. Ding, Scheduling start time dependent tasks with deadlines and identical initial processing times on a single machine. *Computers and Operations Research* **30** (2003), no. 1, 51–62.
4. T-C. E. Cheng and Q. Ding, Single machine scheduling with deadlines and increasing rates of processing times. *Acta Informatica* **36** (2000), no. 9–10, 673–692.
5. T-C. E. Cheng and Q. Ding, Single machine scheduling with step-deteriorating processing times. *European Journal of Operational Research* **134** (2001), no. 3, 623–630.
6. T-C. E. Cheng and Q. Ding, The complexity of single machine scheduling with two distinct deadlines and identical decreasing rates of processing times. *Computers and Mathematics with Applications* **35** (1998), no. 12, 95–100.
7. T-C. E. Cheng and Q. Ding, The time dependent makespan problem is strongly NP-complete. *Computers and Operations Research* **26** (1999), no. 8, 749–754.
8. T-C. E. Cheng, Y. He, H. Hoogeveen, M. Ji and G. Woeginger, Scheduling with step-improving processing times. *Operations Research Letters* **34** (2006), no. 1, 37–40.
9. J. Du and J. Y-T. Leung, Minimizing total tardiness on one machine is NP-hard. *Mathematics of Operations Research* **15** (1990), no. 3, 483–495.
10. S. Gawiejnowicz, Scheduling deteriorating jobs subject to job or machine availability constraints. *European Journal of Operational Research* **180** (2007), no. 1, 472–478.
11. K. I-J. Ho, J. Y-T. Leung and W-D. Wei, Complexity of scheduling tasks with time-dependent execution times. *Information Processing Letters* **48** (1993), no. 6, 315–320.
12. A. Janiak and M. Y. Kovalyov, Job sequencing with exponential functions of processing times. *Informatica* **17** (2006), no. 1, 13–24.
13. F. Jaehn and H. A. Sedding, Scheduling with time-dependent discrepancy times. *Journal of Scheduling* **19** (2016), no. 6, 737–757.
14. A. Kononov, On schedules of a single machine jobs with processing times non-linear in time. In: A. D. Korshunov (ed.). *Operations Research and Discrete Analysis*, Kluwer 1997, pp. 109–122.
15. W. Kubiak and S. L. van de Velde, Scheduling deteriorating jobs to minimize makespan. *Naval Research Logistics* **45** (1998), no. 5, 511–523.
16. E. L. Lawler, A 'pseudopolynomial' algorithm for sequencing jobs to minimize total tardiness. *Annals of Discrete Mathematics* **1** (1977), 331–342.
17. J. K. Lenstra, A. H. G. Rinnooy Kan and P. Brucker, Complexity of machine scheduling problems. *Annals of Discrete Mathematics* **1** (1977), 343–362.
18. G. Mosheiov, Scheduling jobs with step-deterioration; Minimizing makespan on a single- and multi-machine. *Computers and Industrial Engineering* **28** (1995), no. 4, 869–879.

Minimizing the total completion time

19. A. Bachman, T- C. E. Cheng, A. Janiak and C-T. Ng, Scheduling start time dependent jobs to minimize the total weighted completion time. *Journal of the Operational Research Society* **53** (2002), no. 6, 688–693.

20. A. Bachman, A. Janiak and M.Y. Kovalyov, Minimizing the total weighted completion time of deteriorating jobs. *Information Processing Letters* **81** (2002), no. 2, 81–84.

21. T-C. E. Cheng and Q. Ding, Single machine scheduling with step-deteriorating processing times. *European Journal of Operational Research* **134** (2001), no. 3, 623–630.

22. T-C. E. Cheng and Q. Ding, The complexity of single machine scheduling with two distinct deadlines and identical decreasing rates of processing times. *Computers and Mathematics with Applications* **35** (1998), no. 12, 95–100.

23. T-C. E. Cheng, Q. Ding, M. Y. Kovalyov, A. Bachman and A. Janiak, Scheduling jobs with piecewise linear decreasing processing times. *Naval Research Logistics* **50** (2003), no. 6, 531–554.

24. T-C. E. Cheng, Q. Ding and B. M. T. Lin, A concise survey of scheduling with time-dependent processing times. *European Journal of Operational Research* **152** (2004), no. 1, 1–13.

Minimizing the maximum lateness

25. A. Bachman and A. Janiak, Minimizing maximum lateness under linear deterioration. *European Journal of Operational Research* **126** (2000), no. 3, 557–566.

26. T-C. E. Cheng and Q. Ding, Single machine scheduling with deadlines and increasing rates of processing times. *Acta Informatica* **36** (2000), no. 9–10, 673–692.

27. A. Janiak and M. Y. Kovalyov, Job sequencing with exponential functions of processing times. *Informatica* **17** (2006), no. 1, 13–14.

28. A. Kononov, Scheduling problems with linear increasing processing times. In: U. Zimmermann et al. (eds.). *Operations Research 1996*. Berlin-Heidelberg: Springer 1997, pp. 208–212.

Other criteria

29. A. Bachman, T-C. E. Cheng, A. Janiak and C-T. Ng, Scheduling start time dependent jobs to minimize the total weighted completion time. *Journal of the Operational Research Society* **53** (2002), no. 6, 688–693.

30. T-C. E. Cheng and Q. Ding, The complexity of single machine scheduling with two distinct deadlines and identical decreasing rates of processing times. *Computers and Mathematics with Applications* **35** (1998), no. 12, 95–100.

31. A. Janiak and M. Y. Kovalyov, Job sequencing with exponential functions of processing times. *Informatica* **17** (2006), no. 1, 13–14.

11

$\mathcal{NP}$-hard parallel machine problems

This is the second chapter of the fourth part of the book, devoted to intractable time-dependent scheduling problems. In this chapter, we discuss $\mathcal{NP}$-hard parallel machine time-dependent scheduling problems.

Chapter 11 is composed of three sections. In Sect. 11.1, we study $\mathcal{NP}$-hard parallel machine scheduling problems with the $C_{\max}$ criterion. In Sect. 11.2, we consider $\mathcal{NP}$-hard parallel machine scheduling problems with the $\sum C_j$ criterion. In Sect. 11.3, we focus on parallel machine scheduling problems with other criteria than $C_{\max}$, $\sum C_j$ and $L_{\max}$. The chapter is completed with a list of references.

11.1 Minimizing the maximum completion time

In this section, we consider intractable parallel machine time-dependent scheduling problems with the $C_{\max}$ criterion.

11.1.1 Proportional deterioration

The problem of multi-machine time-dependent scheduling with proportionally deteriorating jobs and the $C_{\max}$ criterion is computationally intractable.

Theorem 11.1. (Kononov [3], Mosheiov [4]) *The decision version of problem* $Pm|p_j = b_j t|C_{\max}$ *is $\mathcal{NP}$-complete in the ordinary sense even if $m = 2$.*

Proof. Kononov [3] uses the following reduction from the SP problem: $n = p + 2$, $t_0 = 1$, $b_j = y_j - 1$ for $1 \leqslant j \leqslant p$, $b_{p+1} = \frac{2Y}{B} - 1$, $b_{p+2} = 2B - 1$, where $Y = \prod_{j=1}^{p} y_j$. The threshold value is $G = 2Y$.

Mosheiov [4] uses the following reduction from the EPP problem: $n = q$, $t_0 = 1$, $b_j = z_j - 1$ for $1 \leqslant j \leqslant q$. The threshold value is

$$ G = \sqrt{\prod_{j=1}^{n} z_j}. $$

© Springer-Verlag GmbH Germany, part of Springer Nature 2020
S. Gawiejnowicz, *Models and Algorithms of Time-Dependent Scheduling*,
Monographs in Theoretical Computer Science. An EATCS Series,
https://doi.org/10.1007/978-3-662-59362-2_11

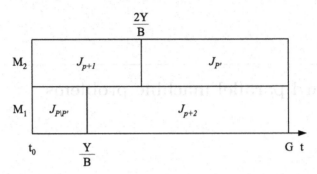

Fig. 11.1: A schedule in the proof of Theorem 11.1

To complete the proof it is sufficient to show that the SP (EPP) problem has a solution if and only if for the above instance of problem $P2|p_j = b_j t|C_{max}$ there exists a schedule σ such that $C_{max}(\sigma) \leqslant G$. (A schedule for the first reduction is given in Fig. 11.1; see also Remark 10.17.)   □

If the number of machines is variable, then the following result holds.

Theorem 11.2.   (Kononov [2])   *The decision version of problem $P|p_j = b_j t|C_{max}$ is $\mathcal{NP}$-complete in the strong sense.*

Proof. The reduction from the 4-P problem is as follows: $n = 4p$, $t_0 = 1$ and $b_j = u_j - 1$ for $1 \leqslant j \leqslant 4p$. The threshold value is $G = D$.

To complete the proof, it is sufficient to show that the 4-P problem has a solution if and only if for the above instance of problem $P|p_j = b_j t|C_{max}$ there exists a schedule σ such that $C_{max}(\sigma) \leqslant G$.   □

11.1.2 Linear deterioration

Since proportional deterioration is a special case of linear deterioration, the results from Sect. 11.1.1 also hold for linear processing times.

Theorem 11.3.
(a) *The decision version of problem $Pm|p_j = a_j + b_j t|C_{max}$ is $\mathcal{NP}$-complete in the ordinary sense, even if $m = 2$.*
(b) *The decision version of problem $P|p_j = a_j + b_j t|C_{max}$ is $\mathcal{NP}$-complete in the strong sense.*

Proof. (a) Since special case of problem $P2|p_j = a_j + b_j t|C_{max}$, when $a_j = 0$ for $1 \leqslant j \leqslant n$, is $\mathcal{NP}$-complete in the ordinary sense by Theorem 11.1, the result follows.

(b) Since special case of problem $P|p_j = a_j + b_j t|C_{max}$, when $a_j = 0$ for $1 \leqslant j \leqslant n$, is $\mathcal{NP}$-complete in the strong sense by Theorem 11.2, the result follows.   □

11.1.3 Linear shortening

Parallel machine scheduling of jobs with decreasing step-linear processing times given by (6.49) is a computationally intractable problem.

Theorem 11.4. (Cheng et al. [1]) *The decision version of problem $Pm|p_j = a_j - b(t - y), y = 0, Y = +\infty|C_{max}$ is $\mathcal{NP}$-complete in the ordinary sense, even if $m = 2$.*

Proof. The main idea is to show that the two-machine problem with variable processing times, $P2|p_j = a_j - b(t - y), y = 0, Y = +\infty|C_{max}$, is equivalent to the problem with fixed procesing times, $P2||C_{max}$, if b is sufficiently small. Let $q_j, 1 \leqslant j \leqslant n$, and G denote job processing times and the threshold value of the C_{max} criterion in the decision version of problem $P2||C_{max}$.

Let us define job processing times and the value of the C_{max} criterion in the problem $P2|p_j = a_j - b(t - y), y = 0, Y = +\infty|C_{max}$ as follows: $a_j = q_j$ for $1 \leqslant j \leqslant n$, $b = 1 - (1 - \frac{1}{a_{max}^2})^{\frac{1}{n}}$, where $a_{max} := \max_{1 \leqslant j \leqslant n}\{a_j\}$.

Let $\mathcal{J}(M_k)$ and C_{M_k} denote a set of jobs assigned to machine M_k in an arbitrary schedule for problem $P2||C_{max}$ and the completion time of the last job from set $\mathcal{J}(M_k), 1 \leqslant k \leqslant m$, respectively. Let us call the schedule σ.

Then

$$\sum_{J_j \in \mathcal{J}(M_k)} a_j - 1 < C_{M_k}(\sigma) < \sum_{J_j \in \mathcal{J}(M_k)} a_j$$

for $1 \leqslant k \leqslant m$. Hence $C'_{max}(\sigma) - 1 < C_{max}(\sigma) < C'_{max}(\sigma)$, where $C_{max}(\sigma)$ and $C'_{max}(\sigma)$ denote the maximum completion time for the schedule σ with fixed and variable job processing times, respectively.

Since all a_j are integers, we have $C_{max}(\sigma) \leqslant G$ if and only if $C'_{max}(\sigma) \leqslant G$. The result follows, because the decision version of problem $P2||C_{max}$ is $\mathcal{NP}$-complete in the ordinary sense. □

Theorem 11.5. (Cheng et al. [1]) *The decision version of problem $P|p_j = a_j - b(t - y), y = 0, Y = +\infty|C_{max}$ is $\mathcal{NP}$-complete in the strong sense.*

Proof. Applying the reasoning from the proof of Theorem 11.4 to problem $P||C_{max}$, we obtain the result. □

11.2 Minimizing the total completion time

In this section, we consider the problems of parallel-machine time-dependent scheduling with the $\sum C_j$ criterion.

11.2.1 Proportional deterioration

Theorem 11.6. (Chen [5], Kononov [8]) *The decision version of problem* $Pm|p_j = b_j t| \sum C_j$ *is $\mathcal{NP}$-complete in the ordinary sense even if* $m = 2$.

Proof. Chen [5] uses the following reduction from the SP problem: $n = p + 4$, $t_0 > 0$ arbitrary, $b_j = y_j - 1$ for $1 \leqslant j \leqslant p$, $b_{p+1} = \frac{Y^2}{B} - 1$, $b_{p+2} = YB - 1$, $b_{p+3} = b_{p+4} = Y^3 - 1$, where $Y = \prod_{j=1}^{p} y_j$. The threshold value is

$$G = (2Y^5 + Y^4)t_0.$$

Kononov [8] uses the reduction from the same problem, SP, but his reduction is slightly different: $n = p + 4$, $t_0 = 1$, $b_j = y_j - 1$ for $1 \leqslant j \leqslant p$, $b_{p+1} = \frac{2Y}{B} - 1$, $b_{p+2} = 2B - 1$, $b_{p+3} = b_{p+4} = 6Y - 1$, where Y is defined as in Chen [5]. The threshold value is

$$G = 24Y^2 + 8Y.$$

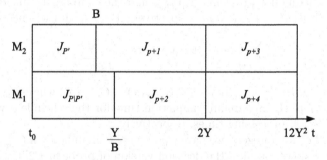

Fig. 11.2: A schedule in the proof of Theorem 11.6

To complete the proof it is sufficient to show that the SP problem has a solution if and only if for the above instance of problem $P2|p_j = b_j t| \sum C_j$ there exists a schedule σ such that $\sum C_j(\sigma) \leqslant G$. (A schedule for the second reduction is presented in Fig. 11.2; see also Remark 10.17.) $\quad\square$

The problem with a variable number of machines is hard to approximate.

Theorem 11.7. (Chen [5]) *If* $\mathcal{P} \neq \mathcal{NP}$, *for problem* $P|p_j = b_j t| \sum C_j$ *there does not exist a polynomial-time heuristic algorithm with a constant worst-case ratio.*

Proof. Assume that for problem $P|p_j = b_j t| \sum C_j$ there exists an algorithm A such that its worst-case ratio $r_A \leqslant V$. Given any instance of the 3-P problem, let us construct an instance of problem $P|p_j = b_j t| \sum C_j$ as follows: there are

h machines, $n = 4h$ jobs, $t_0 = 1$, job deterioration rates $b_j = U^{c_j} - 1$ for $1 \leqslant j \leqslant 3h$, $b_{3h+i} = U^{hK+1} - 1$ for $1 \leqslant i \leqslant h$, where $U = (h + 1)V$.

If we solved the instance by algorithm A with the relative error at most V, we would obtain a pseudo-polynomial algorithm for the 3-P problem, which is strongly $\mathcal{NP}$-complete. A contradiction, since by Lemma 3.19 this is impossible if $\mathcal{P} \neq \mathcal{NP}$.   □

Though from Theorems 11.6–11.7 it follows that $Pm|p_j = b_j t| \sum C_j$ is hard to solve, some properties of an optimal schedule for this problem are known.

Property 11.8. (Gawiejnowicz et al. [7]) In an optimal schedule for problem $Pm|p_j = b_j t| \sum C_j$ jobs assigned to a machine are arranged in non-decreasing order of the deterioration rates b_j and scheduled without idle times.

Proof. Let us assume that σ is a schedule in which jobs assigned to a machine are not in non-decreasing order of deterioration rates b_j, $1 \leqslant j \leqslant n$. By changing the order into non-decreasing order, we obtain a schedule σ' such that $\sum C_j(\sigma') \leqslant \sum C_j(\sigma)$, since by Theorem 7.1 in an optimal schedule for a single machine jobs have to be in non-decreasing order of the deterioration rates b_j. Changing, if necessary, the order of jobs on other machines, we obtain an optimal schedule σ^* such that on each machine jobs are arranged in non-decreasing order of the deterioration rates b_j. Since any idle time increases job completion times, the optimal schedule cannot contain idle times.   ∎

Theorem 11.9. (Gawiejnowicz et al. [7]) *Let* $\sigma^i = (b_1^i, b_2^i, \ldots, b_{n_i}^i)$ *and* $\bar{\sigma}^i = (b_{n_i}^i, b_{n_i-1}^i, \ldots, b_1^i)$ *denote a subsequence of jobs assigned to machine* M_i *and a sequence reversed to* σ^i, *respectively, where* b_j^i *denotes deterioration rate of job* J_j *assigned to machine* M_i, $1 \leqslant j \leqslant n_i$, $1 \leqslant i \leqslant m$ *and* $\sum_{i=1}^m n_i = n$. *Then for every schedule* $\sigma = (\sigma^1, \sigma^2, \ldots, \sigma^m)$ *for problem* $Pm|p_j = b_j t| \sum C_j$ *there exists a corresponding schedule* $\bar{\sigma} = (\bar{\sigma}^1, \bar{\sigma}^2, \ldots, \bar{\sigma}^m)$ *for problem* $Pm|p_j = 1 + b_j t| \sum C_{\max}^{(k)}$ *and for every schedule* $\bar{\sigma}$ *for problem* $Pm|p_j = 1 + b_j t| \sum C_{\max}^{(k)}$ *there exists a corresponding schedule* σ *for problem* $Pm|p_j = b_j t| \sum C_j$ *such that* $\sum C_j(\sigma) = \sum C_{\max}^{(k)}(\bar{\sigma}) - m$, *provided that both these schedules start at time* $t_0 = 1$.

Proof. (Sufficiency) Let us assume that $t_0 = 1$ and let $\sigma = (\sigma^1, \sigma^2, \ldots, \sigma^m)$ be a schedule for problem $Pm|p_j = b_j t| \sum C_j$, where $\sigma^i = (b_1^i, b_2^i, \ldots, b_{n_i}^i)$ for $1 \leqslant i \leqslant m$. Then we have

$$\sum C_j(\sigma) = \sum_{k=1}^m \sum_{i=1}^{n_k} \prod_{j=1}^i (1 + b_j^k) = \sum_{k=1}^m \sum_{i=0}^{n_k} \prod_{j=1}^i (1 + b_j^k) - m =$$

$$\sum_{k=1}^m \sum_{i=0}^{n_k} \prod_{j=1}^i (1 + B_{n_k-j+1}^k) - m = \sum C_{\max}^{(k)}(\bar{\sigma}) - m,$$

where $B_{n_k-j+1}^k = b_j^k$ and $\bar{\sigma}^i = (b_{n_k}^i, b_{n_k-1}^i, \ldots, b_1^i) = (B_1^i, B_2^i, \ldots, B_{n_k}^i)$ for $1 \leqslant i \leqslant m$. The proof of necessity can be made in an analogous way.   ∎

Remark 11.10. Problems which satisfy the conditions of Theorem 11.9 are called *equivalent* problems. We discuss such problems in Chap. 19.

11.2.2 Linear deterioration

Since proportional processing times are a special case of linear processing times, the results from Sect. 11.2.1 also hold for linear processing times.

Theorem 11.11. *The decision version of problem* $Pm|p_j = a_j + b_j t| \sum C_j$ *is* $\mathcal{NP}$-*complete in the ordinary sense even if* $m = 2$.

Proof. The result follows from the fact that a special case of problem $Pm|p_j = a_j + b_j t| \sum C_j$, when $a_j = 0$ for $1 \leqslant j \leqslant n$ and $m = 2$, is $\mathcal{NP}$-complete in the ordinary sense by Theorem 11.6.   □

In the case when $a_j = a > 0$ for $1 \leqslant j \leqslant n$, the following properties of the problem $Pm|p_j = a + b_j t| \sum C_j$ are known.

The first property is a multi-machine counterpart of Property A.2.

Property 11.12. (Gawiejnowicz et al. [6]) Let $k_i = \arg\max\{b_j : j \in U\}$, where $i = 1, 2, \ldots, m$ and U is the set of indices of jobs not yet considered. Then J_{k_i} is the first job on machine M_i in the optimal schedule.

Proof. Let job J_{k_i}, $k_i \in U$, be scheduled as the first one on machine M_i, $i \in \{1, 2, \ldots, m\}$. Then the processing time and the completion time of this job are equal to $p_{k_i} = a + b_{k_i} \times 0 = a$ and $C_{k_i} = a$, respectively. Since C_{k_i} does not depend on b_{k_i}, it is easy to see that the first job in an optimal schedule should be the job with $k_i = \arg\max\{b_j : j \in U\}$.   ∎

The second property does not have a single-machine counterpart.

Property 11.13. (Gawiejnowicz et al. [6]) If $n \geqslant 2m - 1$, then in any optimal schedule at least two jobs are scheduled on each machine.

Proof. Let us assume that there are given $n \geqslant 2m - 1$ jobs and that there exists an optimal schedule, σ^1, such that to some machine, M_l, is assigned only one job. Let M_k be a machine with the largest load in schedule σ^1, j_k be the index of the job assigned to M_k as the last one and let $S_{j_k} > a$ denote the starting time of this job. Then $C_{j_k}(\sigma^1) = a + (1 + b_{j_k})S_{j_k}$ and the total completion time for the schedule σ_1 is $\sum C_j(\sigma^1) = T + a + (1 + b_{j_k})S_{j_k}$, where T denotes the total completion time for jobs other than J_{j_k}.

Now let us construct a new schedule, σ^2, by assigning the job with the index j_k to machine M_l. Then $C_{j_k}(\sigma^2) = a + (1 + b_{j_k})a$ and the total completion time for schedule σ^2 is $\sum C_j(\sigma^2) = T + a + (1 + b_{j_k})a$. Since

$$\sum C_j(\sigma^2) - \sum C_j(\sigma^1) = (1 + b_{j_k})(a - S_{j_k}) < 0,$$

schedule σ^2 is better than σ^1. A contradiction.   ∎

The third property is a multi-machine counterpart of Property A.4.

Property 11.14. (Gawiejnowicz et al. [6]) The total completion time for any sequence of indices of jobs assigned to a given machine and for this sequence in reversed order, starting from the second job, is the same.

Proof. Consider any sequence of indices of jobs assigned to a given machine M_i, $i \in \{1, 2, \dots, m\}$. The result follows, since starting from the second argument the $\sum C_j$ criterion is symmetric with respect to its arguments, i.e.,

$$\sum_{j=0}^{n} \sum_{k=0}^{j} \prod_{l=k+1}^{j} (1 + b_k) = \sum_{j=0}^{n} \sum_{k=0}^{j} \prod_{l=k+1}^{j} (1 + b_{n-k+1}).$$

∎

The last result is a multi-machine counterpart of Theorem A.8.

Theorem 11.15. (Gawiejnowicz et al. [6]) *The optimal schedule for problem $Pm|p_j = a + b_j t| \sum C_j$ is composed of V-shaped subschedules.*

Proof. Let us assume that there exists such an optimal schedule that for some machine, M_k, the sequence of jobs assigned to the machine is not V-shaped. By rearranging the jobs in such a way that their sequence is V-shaped, we obtain a new schedule with a decreased value of the criterion function since, by the V-shape property for a single machine (cf. Theorem A.8), we decreased the total completion time for machine M_k. A contradiction. ∎

11.3 Other criteria

In this section, we present a few results concerning intractable time-dependent parallel machine scheduling problems with criteria other than $C_{\max}$ and $\sum C_j$.

11.3.1 Proportional deterioration

We start with a result concerning the total machine load criterion, $\sum C_{\max}^{(k)}$.

Lemma 11.16. (Mosheiov [11]) *The optimal total machine load for problem $Pm|p_j = b_j t| \sum C_{\max}^{(k)}$ is not less than $m \times \sqrt[n]{\prod_{j=1}^{n}(1 + b_j)}$.*

Proof. Let $A_1, A_2, \dots, A_m$ and $C_{\max}^{(1)}, C_{\max}^{(2)}, \dots, C_{\max}^{(m)}$ denote the sets of jobs assigned to machines $M_1, M_2, \dots, M_m$ and the corresponding total loads, respectively. Then, by Lemma 1.1 (b),

$$\frac{1}{m} \sum_{i=1}^{m} C_{\max}^{(i)} = \frac{1}{m} \sum_{i=1}^{m} \prod_{j \in A_i} (1 + b_j) \geqslant \sqrt[m]{\prod_{j=1}^{n}(1 + b_j)}.$$

□

Theorem 11.17. (Mosheiov [11], Gawiejnowicz et al. [10]) *The decision version of problem* $Pm|p_j = b_jt| \sum C_{\max}^{(k)}$ *is $\mathcal{NP}$-complete in the ordinary sense even if* $m = 2$.

Proof. Mosheiov [11] gives the following idea of the proof. $P2|p_j = b_jt| \sum C_{\max}^{(k)}$ is equivalent to finding $\min\{L_1 + L_2\}$ subject to $L_1 \times L_2 = A^2$ for some positive integer constant A, where L_1 and L_2 denote products of deterioration rates of the jobs assigned to machine M_1 and machine M_2, respectively (cf. Lemma 11.16 for $m = 2$). Since this is equivalent to the EPP problem, the result follows.

Gawiejnowicz et al. [10] prove the result using the notion of equivalent problems (cf. Theorem 11.9; see also Chap. 19). □

Remark 11.18. Mosheiov [11] gives only a sketch of the proof of Theorem 11.17. The formal reduction may be the following. Given an instance of the EPP problem, let us define $n = q$, $b_j = z_j$ for $1 \leqslant j \leqslant q$ and let the threshold value be $G = 2E$, where $E = \sqrt{\prod_{j=1}^q z_j}$. To complete the proof it is sufficient to show that the EPP problem has a solution if and only if for the above instance of problem $P2|p_j = b_jt| \sum C_{\max}^{(k)}$ there exists a schedule σ such that the inequality $\sum C_{\max}^{(k)}(\sigma) \leqslant 2E = G$ holds.

11.3.2 Linear deterioration

Since linear job processing times are a generalization of proportional job processing times, the following result holds.

Theorem 11.19. *The decision version of problem* $Pm|p_j = a_j + b_jt| \sum C_{\max}^{(k)}$ *is $\mathcal{NP}$-complete in the ordinary sense even if* $m = 2$.

Proof. Since the special case of problem $Pm|p_j = a_j + b_jt| \sum C_{\max}^{(k)}$, when $a_j = 0$ for $1 \leqslant j \leqslant n$ and $m = 2$, is $\mathcal{NP}$-complete in the ordinary sense by Theorem 11.17, the result follows. □

Cheng et al. [9] considered problem $Pm|p_j = a_j + bt| \sum(\alpha E_j + \beta T_j + \gamma d)$.

Theorem 11.20. (Cheng et al. [9]) *The decision version of problem* $Pm|p_j = a_j + bt| \sum(\alpha E_j + \beta T_j + \gamma d)$ *is $\mathcal{NP}$-complete in the ordinary sense, even if* $m = 2$.

Proof. The reduction from the PP problem is as follows: $n = k$ jobs, $a_j = x_j$ for $1 \leqslant j \leqslant n$, $b = (1 + \frac{1}{a_n^2})^{\frac{1}{2n^4}}$, $\alpha = 1$, $\beta = \frac{(2M+1)e+1}{b}$ and $\gamma = \frac{2e}{n}$, where $M := \frac{1}{2}\sum_{j=1}^n x_j$ and $e := \sum_{j=1}^n ja_j$. The threshold value is

$$G = (2M + 1)e + 1.$$

To complete the proof it is sufficient to show that the PP problem has a solution if and only if there exists a schedule σ for the above instance of problem $P2|p_j = a_j + bt| \sum(\alpha E_j + \beta T_j + \gamma d)$ such that $\sum_{j=1}^n(\alpha E_j(\sigma) + \beta T_j(\sigma) + \gamma d) \leqslant G$. □

References

Minimizing the maximum completion time

1. T-C. E. Cheng, Q. Ding, M. Y. Kovalyov, A. Bachman and A. Janiak, Scheduling jobs with piecewise linear decreasing processing times. *Naval Research Logistics* **50** (2003), no. 6, 531–554.
2. A. Kononov, Combinatorial complexity of scheduling jobs with simple linear deterioration. *Discrete Analysis and Operations Research* **3** (1996), no. 2, 15–32 (in Russian).
3. A. Kononov, Scheduling problems with linear increasing processing times. In: U. Zimmermann et al. (eds.). *Operations Research 1996*. Berlin-Heidelberg: Springer 1997, pp. 208–212.
4. G. Mosheiov, Multi-machine scheduling with linear deterioration. *Infor* **36** (1998), no. 4, 205–214.

Minimizing the total completion time

5. Z-L. Chen. Parallel machine scheduling with time dependent processing times. *Discrete Applied Mathematics* **70** (1996), no. 1, 81–93. (Erratum: *Discrete Applied Mathematics* **75** (1996), no. 1, 103.)
6. S. Gawiejnowicz, W. Kurc and L. Pankowska, Minimizing time-dependent total completion time on parallel identical machines. *Lecture Notes in Computer Science* **3019** (2004), 89–96.
7. S. Gawiejnowicz, W. Kurc and L. Pankowska, Parallel machine scheduling of deteriorating jobs by modified steepest descent search. *Lecture Notes in Computer Science* **3911** (2006), 116–123.
8. A. Kononov, Scheduling problems with linear increasing processing times. In: U. Zimmermann et al. (eds.). *Operations Research 1996*, Berlin-Heidelberg: Springer 1997, pp. 208–212.

Other criteria

9. T-C. E. Cheng, L-Y. Kang and C-T. Ng, Due-date assignment and parallel-machine scheduling with deteriorating jobs. *Journal of the Operational Research Society* **58** (2007), no. 8, 1103–1108.
10. S. Gawiejnowicz, W. Kurc and L. Pankowska, Parallel machine scheduling of deteriorating jobs by modified steepest descent search. *Lecture Notes in Computer Science* **3911** (2006), 116–123.
11. G. Mosheiov, Multi-machine scheduling with linear deterioration. *Infor* **36** (1998), no. 4, 205–214.

12

$\mathcal{NP}$-hard dedicated machine problems

With this chapter, including the main results on intractable dedicated machine scheduling problems, we complete the fourth part of the book, devoted to $\mathcal{NP}$-hard time-dependent scheduling problems.

Chapter 12 is composed of three sections. In Sect. 12.1, we consider $\mathcal{NP}$-hard dedicated machine time-dependent scheduling problems with the $C_{\max}$ criterion. In Sect. 12.2, we focus on $\mathcal{NP}$-hard dedicated machine time-dependent scheduling problems with the $\sum C_j$ criterion. In Sect. 12.3, we present similar results for the $L_{\max}$ criterion. The chapter is completed with a list of references.

12.1 Minimizing the maximum completion time

In this section, we consider intractable dedicated machine time-dependent scheduling problems with the $C_{\max}$ criterion.

12.1.1 Proportional deterioration

Dedicated machine time-dependent scheduling problems with $m \geqslant 3$ machines usually are not polynomially solvable.

Flow shop problems

The flow shop problem with proportional job processing times and $m \geqslant 3$ machines is computationally intractable.

Theorem 12.1. (Thörnblad and Patriksson [4]) *The decision version of problem* $F3|p_{ij} = b_{ij}t|C_{\max}$ *is $\mathcal{NP}$-complete in the ordinary sense.*

Proof. The reduction from the EPP problem is as follows. Let $n = q + 3$ jobs and $t_0 = \frac{1}{2}$. Job deterioration rates are defined in the following way: $b_{1j} = 0$,

© Springer-Verlag GmbH Germany, part of Springer Nature 2020
S. Gawiejnowicz, *Models and Algorithms of Time-Dependent Scheduling*,
Monographs in Theoretical Computer Science. An EATCS Series,
https://doi.org/10.1007/978-3-662-59362-2_12

$b_{2j} = z_j - 1$, $b_{3j} = 0$ for $1 \leqslant j \leqslant q$, $b_{1,q+1} = 0$, $b_{2,q+1} = 1$, $b_{3,q+1} = E$, $b_{1,q+2} = 2E - 1$, $b_{2,q+2} = \frac{1}{E}$, $b_{3,q+2} = \frac{E^2}{E+1}$, $b_{1,q+3} = E$, $b_{2,q+3} = \frac{1}{E^2+E}$, $b_{3,q+3} = 0$, where $E = \sqrt{\prod_{j=1}^{q} z_j}$. The threshold value is

$$G = E^2 + 3E + 3.$$

To complete the proof it is sufficient to show that the EPP problem has a solution if and only if for the above instance of problem $F3|p_{ij} = b_{ij}t|C_{\max}$ there exists a schedule σ such that $C_{\max}(\sigma) \leqslant G$.   □

Theorem 12.2. (Kononov [1]) *The decision version of problem $F3|p_{i,j} = b_{i,j}t|C_{\max}$ is $\mathcal{NP}$-complete in the ordinary sense.*

Proof. The reduction from the 4-P problem is as follows. Let $n = 5p + 1$ and $t_0 = 1$. Job deterioration rates are defined in the following way: $b_{1j} = 0$, $b_{2j} = u_j - 1$, $b_{3j} = 0$ for $1 \leqslant j \leqslant 4p$, $b_{1,4p+1} = 0$, $b_{2,4p+1} = 0$, $b_{3,4p+1} = 2D-1$, $b_{1,4p+2} = D - 1$, $b_{2,4p+2} = 1$, $b_{3,4p+2} = 2D - 1$, $b_{1,4p+k} = 2D - 1$, $b_{2,4p+k} = 1$, $b_{3,4p+k} = 2D - 1$ for $3 \leqslant k \leqslant p - 1$, $b_{1,5p} = 2D - 1$, $b_{2,5p} = 1$, $b_{3,5p} = D - 1$, $b_{1,5p+1} = 2D - 1$, $b_{2,5p+1} = 0$, $b_{3,5p+1} = 0$. The threshold value is

$$G = 2^{p-1}D^p.$$

To complete the proof it is sufficient to show that the 4-P problem has a solution if and only if for the above instance of problem $F3|p_{ij} = b_{ij}t|C_{\max}$ there exists a schedule σ such that $C_{\max}(\sigma) \leqslant G$.   □

Problem $F3|p_{ij} = b_{ij}t|C_{\max}$ is hard to approximate, even if deterioration rates of the operations executed on machines M_1 and M_3 are equal.

Theorem 12.3. (Kononov and Gawiejnowicz [2]) *If $\mathcal{P} \neq \mathcal{NP}$, then for problem $F3|p_{ij} = b_{ij}t, b_{1i} = b_{3i} = b|C_{\max}$ there does not exist a polynomial-time approximation algorithm with the worst case ratio bounded by a constant.*

Proof. Suppose that there exists a polynomial-time approximation algorithm A for problem $F3|p_{ij} = b_{ij}t, b_{1i} = b_{3i} = b|C_{\max}$ such that its worst-case ratio $r_A < U$, where $U = const$. We will show that this assumption leads to a contradiction.

Let Q be an instance of the 3-P problem. Let us construct instance Q_U of problem $F3|p_{ij} = b_{ij}t, b_{1i} = b_{3i} = b|C_{\max}$ as follows: $t_0 = 1$, $n = 4h + 1$, $b_{1j} = b_{3j} = U^K - 1$ for $1 \leqslant j \leqslant 4h + 1$, $b_{2j} = U^{c_j} - 1$ for $1 \leqslant j \leqslant 3h$, $b_{2,3h+k} = U^{3K} - 1$ for $1 \leqslant k \leqslant h + 1$.

If we applied algorithm A to this instance, we would obtain a pseudo-polynomial algorithm for the strongly $\mathcal{NP}$-complete 3-P problem. A contradiction, since by Lemma 3.19 this is impossible if $\mathcal{P} \neq \mathcal{NP}$.   □

Open shop problems

Three-machine time-dependent open shop problems have similar complexity status as three-machine time-dependent flow shops.

Theorem 12.4. (Kononov [1], Mosheiov [3]) *The decision version of problem* $O3|p_{ij} = b_{ij}t|C_{\max}$ *is $\mathcal{NP}$-complete in the ordinary sense.*

Proof. Kononov [1] uses the following reduction from the SP problem: $n = p + 3$, $t_0 = 1$, $b_{ij} = y_j - 1$, $b_{i,p+1} = \frac{2Y}{B} - 1$, $b_{i,p+2} = 2B - 1$, $b_{i,p+3} = 2Y - 1$ for $1 \leqslant i \leqslant 3$ and $1 \leqslant j \leqslant p$, where $Y = \prod_{j=1}^{p} y_j$. The threshold value is

$$G = 8Y^3.$$

Mosheiov [3] uses the following reduction from the EPP problem: $n = q+1$, $t_0 = 1$, $b_{1j} = b_{2j} = b_{3j} = z_j - 1$ for $1 \leqslant j \leqslant q$, $b_{1,q+1} = b_{2,q+1} = b_{3,q+1} = E - 1$, where $E = \sqrt{\prod_{j=1}^{q} z_j}$. The threshold value is

$$G = E^3.$$

In order to complete the proof it is sufficient to show that the SP (EPP) problem has a solution if and only if for the above instance of problem $O3|p_{ij} = b_{ij}t|C_{\max}$ there exists a schedule σ such that $C_{\max}(\sigma) \leqslant G$. □

Kononov and Gawiejnowicz [2] proved that problem $O3|p_{ij} = b_{ij}t|C_{\max}$ remains computationally intractable, even if all deterioration rates on machine M_3 are equal.

Theorem 12.5. (Kononov and Gawiejnowicz [2]) *The decision version of problem* $O3|p_{ij} = b_{ij}t, b_{3j} = b|C_{\max}$ *is $\mathcal{NP}$-complete in the ordinary sense.*

Proof. The reduction from the SP problem is as follows. Let $\prod_{j=1}^{p} y_j = Y$, $\bar{Y} = 2Y$, $t_0 = 1$ and $n = p + 4$. The job deterioration rates are as follows:
$b_{1j} = 0, b_{2j} = y_j^2 - 1, b_{3j} = \bar{Y}^2 - 1$ for $1 \leqslant j \leqslant p$,
$b_{1,p+1} = 0, b_{2,p+1} = \frac{\bar{Y}^2}{Y^2} - 1, b_{3,p+1} = \bar{Y}^2 - 1$,
$b_{1,p+2} = 0, b_{2,p+2} = 4\bar{Y}^2 - 1, b_{3,p+2} = \bar{Y}^2 - 1$,
$b_{1,p+3} = \bar{Y}^{p+3} - 1, b_{2,p+3} = \bar{Y}^{p+3} - 1, b_{3,p+3} = \bar{Y}^2 - 1$,
$b_{1,p+4} = \bar{Y}^{p+5} - 1, b_{2,p+4} = \bar{Y}^{p+1} - 1, b_{3,p+4} = \bar{Y}^2 - 1$.
The threshold value is

$$G = \bar{Y}^{2p+8}.$$

To complete the proof it is sufficient to show that the SP problem has a solution if and only if for the above instance of problem $O3|p_{ij} = b_{ij}t$, $b_{3j} = b|C_{\max}$ there exists a schedule σ such that $C_{\max}(\sigma) \leqslant G$. □

Job shop problems

Kononov and Gawiejnowicz [2] conjectured that the two-machine job shop problem with proportional processing times is computationally intractable. The conjecture was proved by Mosheiov.

Theorem 12.6. (Mosheiov [3]) *The decision version of problem $J2|p_{ij} = b_{ij}t|C_{\max}$ is $\mathcal{NP}$-complete in the ordinary sense.*

Proof. The reduction from the EPP problem is as follows. Let $n = q + 1$, $t_0 = 1$, $b_{1j} = z_j$ and $b_{2j} = 0$ for $1 \leqslant j \leqslant q$. Job J_{q+1} consists of three operations: $O_{1,q+1}, O_{2,q+1}$ and $O_{3,q+1}$. The operations have to be done on machine M_2, M_1 and M_2, respectively. The deterioration rates for job J_{q+1} are the following: $b_{1,q+1} = \frac{1}{E}$ and $b_{2,q+1} = E - 1$, where $E = \sqrt{\prod_{j=1}^{q} z_j}$. The threshold value is

$$G = (E + 1)E.$$

To complete the proof it is sufficient to show that the EPP problem has a solution if and only if for the above instance of problem $J2|p_{ij} = b_{ij}t|C_{\max}$ there exists a schedule σ such that $C_{\max}(\sigma) \leqslant G$. □

12.1.2 Linear deterioration

In this section, we consider $\mathcal{NP}$-hard dedicated machine time-dependent scheduling problems with linearly deteriorating jobs and the $C_{\max}$ criterion.

Flow shop problems

The two-machine flow shop problem with linearly deteriorating jobs is intractable.

Theorem 12.7. (Kononov and Gawiejnowicz [2]) *The decision version of problem $F2|p_{ij} = a_{ij} + b_{ij}t|C_{\max}$ is $\mathcal{NP}$-complete in the strong sense.*

Proof. The reduction from the 3-P problem is as follows. Let $t_0 = 0$ and $n = 4h$. The job processing times are defined in the following way: $p_{11} = 0$, $p_{21} = 1 + nt$, $p_{1j} = K + 1$, $p_{2j} = \frac{1}{(j-1)(K+1)}t$ for $2 \leqslant j \leqslant h$, $p_{1j} = 0, p_{2j} = c_j$ for $h + 1 \leqslant j \leqslant 4h$. The threshold value is

$$G = hK + h.$$

In order to complete the proof it is sufficient to show that the 3-P problem has a solution if and only if for the above instance of problem $F2|p_{ij} = a_{ij} + b_{ij}t|C_{\max}$ there exists a schedule σ such that $C_{\max}(\sigma) \leqslant G$. □

Open shop problems

As the two-machine flow shop, the two-machine open shop problem with linearly deteriorating jobs is intractable as well.

Theorem 12.8. (Kononov and Gawiejnowicz [2]) *The decision version of problem $O2|p_{ij} = a_{ij} + b_{ij}t|C_{\max}$ is $\mathcal{NP}$-complete in the ordinary sense.*

Proof. The reduction from the PP problem is as follows. Let $t_0 = 1$ and $n = k + 1$. Let us define job processing times as follows: $p_{1j} = p_{2j} = x_j$ for $1 \leqslant j \leqslant k$, $p_{1,k+1} = p_{2,k+1} = At$, where $A = \frac{1}{2}\sum_{j=1}^{k} x_j$. The threshold value is $G = (A + 1)^2 + A$.

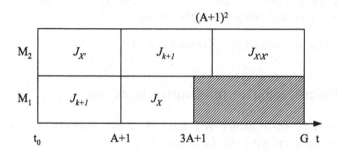

Fig. 12.1: A schedule in the proof of Theorem 12.8

In order to complete the proof it is sufficient to show that the PP problem has a solution if and only if for the above instance of problem $O2|p_{ij} = a_{ij} + b_{ij}t|C_{\max}$ there exists a schedule σ such that $C_{\max}(\sigma) \leqslant G$ (see Fig. 12.1 and Remark 10.17). □

Job shop problems

As in the case of multi-machine flow shop and open shop problems, already the two-machine job shop problem with linearly deteriorating jobs is intractable.

Theorem 12.9. *The decision version of problem $J2|p_{ij} = a_{ij} + b_{ij}t|C_{\max}$ is $\mathcal{NP}$-complete in the ordinary sense.*

Proof. The result is a corollary from Theorem 12.6. □

12.2 Minimizing the total completion time

In this section, we consider intractable time-dependent scheduling problems on dedicated machines with linear jobs and the $\sum C_j$ criterion.

Dedicated machine time-dependent scheduling problems with linearly deteriorating jobs are intractable, even in restricted cases.

Flow shop problems

The two-machine flow shop problem with linear job processing times and the total completion time criterion is intractable.

Theorem 12.10. *The decision version of problem* $F2|p_{ij} = a_{ij} + b_{ij}t| \sum C_j$ *is* $\mathcal{NP}$*-complete in the strong sense.*

Proof. Since the special case of the considered problem is problem $F2|| \sum C_j$ which is $\mathcal{NP}$-complete in the strong sense (Garey et al. [5]), the result follows.
□

Theorem 12.11. (Wu and Lee [6]) *The decision version of problem* $F2|p_{ij} = a_{ij} + bt| \sum C_j$ *is* $\mathcal{NP}$*-complete in the strong sense.*

Proof. Similar to the proof of Theorem 12.10.
□

12.3 Minimizing the maximum lateness

In this section, we consider $\mathcal{NP}$-hard dedicated-machine time-dependent scheduling problems with the $L_{\max}$ criterion.

12.3.1 Proportional deterioration

Dedicated machine time-dependent scheduling problems with proportionally deteriorating jobs and the maximum lateness are intractable.

Flow shop problems

The two-machine flow shop problem with proportional job processing times and the maximum lateness criterion is intractable.

Theorem 12.12. (Kononov [7]) *The decision version of problem* $F2|p_{ij} = b_{ij}t|L_{\max}$ *is* $\mathcal{NP}$*-complete in the strong sense.*

Proof. The reduction from the 4-P problem is as follows: $n = 5p-1$ and $t_0 = 1$, $b_{1j} = 0, b_{2j} = u_j - 1, d_j = 2^{p-1}D^p$ for $1 \leqslant j \leqslant 4p$, $b_{1,4p+1} = D-1, b_{2,4p+1} = 1$, $d_{4p+1} = 2D, b_{1,4p+k} = 2D - 1, b_{2,4p+k} = 1, d_{4p+k} = 2^k D^k$ for $2 \leqslant k \leqslant m - 1$. The threshold value is $G = 0$.

To complete the proof it is sufficient to show that the 4-P problem has a solution if and only if for the above instance of problem $F2|p_{ij} = b_{ij}t|L_{\max}$ there exists a schedule σ such that $L_{\max}(\sigma) \leqslant G$.
□

Open shop problems

The two-machine open shop problem with proportional job processing times and the maximum lateness criterion is intractable.

Theorem 12.13. (Kononov [7]) *The decision version of problem $O2|p_{ij} = b_{ij}t|L_{\max}$ is $\mathcal{NP}$-complete in the ordinary sense.*

Proof. The reduction from the SP problem is as follows: $n = p+3$ and $t_0 = 1$, $b_{ij} = y_j - 1$, $b_{i,p+1} = \frac{2Y}{B} - 1$, $b_{i,p+2} = 2B - 1$, $b_{i,p+3} = 2Y - 1$ for $1 \leqslant i \leqslant 2$, $d_j = 8Y^3$ for $1 \leqslant j \leqslant p+2$ and $d_{p+3} = 4Y^2$, where $Y = \prod_{j=1}^{p} y_j$. The threshold value is $G = 0$.

To complete the proof it is sufficient to show that the SP problem has a solution if and only if for the above instance of problem $O2|p_{ij} = b_{ij}t|L_{\max}$ there exists a schedule σ such that $L_{\max}(\sigma) \leqslant G$. $\qquad\square$

12.3.2 Linear deterioration

In this section, we consider $\mathcal{NP}$-hard dedicated machine time-dependent scheduling problems with linearly deteriorating jobs.

Flow shop problems

The two-machine flow shop problem with linear job processing times and the maximum lateness criterion is intractable.

Theorem 12.14. *The decision version of problem $F2|p_{ij} = a_{ij} + b_{ij}t|L_{\max}$ is $\mathcal{NP}$-complete in the ordinary sense.*

Proof. The result is a corollary from Theorem 12.12. $\qquad\blacksquare$

This result completes the fourth part of the book, devoted to $\mathcal{NP}$-hard time-dependent scheduling problems. In the next part, we discuss various classes of algorithms for intractable time-dependent scheduling problems.

References

Minimizing the maximum completion time

1. A. Kononov, Combinatorial complexity of scheduling jobs with simple linear deterioration. *Discrete Analysis and Operations Research* **3** (1996), no. 2, 15–32 (in Russian).
2. A. Kononov and S. Gawiejnowicz, NP-hard cases in scheduling deteriorating jobs on dedicated machines. *Journal of the Operational Research Society* **52** (2001), no. 6, 708–718.
3. G. Mosheiov, Complexity analysis of job-shop scheduling with deteriorating jobs. *Discrete Applied Mathematics* **117** (2002), no. 1–3, 195–209.
4. K. Thörnblad and M. Patriksson, A note on the complexity of flow-shop scheduling with deteriorating jobs. *Discrete Applied Mathematics* **159** (2011), 251–253.

Minimizing the total completion time

5. M. R. Garey, D. S. Johnson and R. Sethi, The complexity of flowshop and jobshop scheduling. *Mathematics of Operations Research* **1** (1976), no. 2, 117–129.
6. C-C. Wu and W-C. Lee, Two-machine flowshop scheduling to minimize mean flow time under linear deterioration. *International Journal of Production Economics* **103** (2006), no. 2, 572–584.

Minimizing the maximum lateness

7. A. Kononov, Combinatorial complexity of scheduling jobs with simple linear deterioration. *Discrete Analysis and Operations Research* **3** (1996), no. 2, 15–32 (in Russian).

ALGORITHMS

13

Exact algorithms

T his chapter opens the fifth part of the book, where various classes of algorithms for intractable time-dependent scheduling problems are discussed.
Despite the fact that these algorithms generate schedules of a different quality,
in many cases they are capable of finding near-optimal solutions.

This part of the book is composed of five chapters. In Chap. 13, we discuss exact algorithms. In Chap. 14, we focus on approximation algorithms
and approximation schemes. In Chap. 15, we present greedy algorithms based
on functions called *signatures*. In Chap. 16, we review simple heuristic algorithms which construct suboptimal schedules in a step-by-step manner.
Finally, in Chap. 17 we consider local search and meta-heuristic algorithms.

Chapter 13 is composed of four sections. In Sect. 13.1, we describe the
branch-and-bound algorithm which is the most common type of exact algorithm for time-dependent scheduling problems. In Sects. 13.2, 13.3 and 13.4,
we focus on exact algorithms for single, parallel and dedicated machine time-
dependent scheduling problems, respectively. The chapter is completed with
a list of references.

13.1 Preliminaries

Branch-and-bound algorithms (cf. Sect. 2.3.2) are the most common exact algorithms for time-dependent scheduling problems. Therefore, now we describe
the general structure of these algorithms.

A branch-and-bound algorithm consists of two procedures: branching and
bounding. *Branching* is a procedure of partitioning a large problem into a
number of subproblems. The subproblems, in turn, are partitioned into smaller
subproblems. The branching procedure allows us to construct a tree, in which
a node corresponds to a partial solution (called a *child* solution) to a sub-
problem. Two nodes in the tree are connected by an edge if the solutions
corresponding to these nodes are child solutions of a solution. The leaves of
the tree correspond to complete solutions to the problem.

© Springer-Verlag GmbH Germany, part of Springer Nature 2020
S. Gawiejnowicz, *Models and Algorithms of Time-Dependent Scheduling*,
Monographs in Theoretical Computer Science. An EATCS Series,
https://doi.org/10.1007/978-3-662-59362-2_13

Bounding is a procedure which allows us to cut off a certain part of the solution tree. This part includes the partial solutions for which the value of the objective function is not better than the currently best value. The procedure uses an estimation of the optimal value of the objective function, the so-called *lower bound*. (If such an estimation is not known, then the value of the objective function for a known feasible solution is assumed.) Rules which specify the nodes to be cut off are called *dominance rules*, while the cut off (remaining) nodes are called *dominated* (*non-dominated*) *nodes*.

Let $\mathfrak{F}$ denote a set of all possible solutions to an optimization problem. The pseudo-code of a branch-and-bound algorithm (see Algorithm 13.1) can be formulated as follows.

Algorithm 13.1. Branch-and-Bound

---

1 **Input** : initial solution s_0, criterion f
2 **Output:** an optimal solution $s^\star$
   ▷ Step 1: initialization
3 $U \leftarrow f(s_0)$;
4 $s^\star \leftarrow s_0$;
5 $L \leftarrow \mathfrak{F}$;
   ▷ Step 2: the main loop
6 **while** $(L \neq \emptyset)$ **do**
7   Choose a solution $s_{tmp} \in L$;
8     $L \leftarrow L \setminus \{s_{tmp}\}$;
     ▷ Branching
9   Generate all child solutions $s_{n_1}, s_{n_2}, \ldots, s_{n_k}$ of s_{tmp};
     ▷ Bounding
10   **for** $i \leftarrow n_1$ **to** n_k **do**
11   Calculate $LB(s_{n_i})$;
12   **if** $(LB(s_{n_i}) < U)$ **then**
13   **if** (s_{n_i} *is a complete solution*) **then**
14              $U \leftarrow LB(s_{n_i})$;
15              $s^\star \leftarrow s_{n_i}$
          else
16              $L \leftarrow L \cup \{s_{n_i}\}$
          end
       end
    end
  end
   ▷ Step 3: the final solution
17 **return** $s^\star$.

We refer the reader to the literature mentioned in the bibliographic notes in Chap. 2 for more details on branch-and-bound algorithms.

13.2 Exact algorithms for single machine problems

In this section, we consider exact algorithms for intractable single machine time-dependent scheduling problems.

13.2.1 Minimizing the maximum completion time

In this section, we focus on exact algorithms for the $C_{\max}$ criterion.

Linear deterioration

Bosio and Righini [2] proposed an exact dynamic programming algorithm for problem $1|p_j = a_j + b_j t, r_j|C_{\max}$. The algorithm starts from an empty schedule and iteratively adds an unscheduled job to the partial schedule constructed so far. In order to speed up the process of finding the final schedule, the authors used some upper and lower bounds on the optimal value of $C_{\max}$.

Lee et al. [12] considered the above problem in the case when $b_j = b$ for $1 \leqslant j \leqslant n$. The authors established a few properties of an optimal schedule for the latter problem.

Let J_i and J_j, $1 \leqslant i, j \leqslant n$, be two adjacent jobs in a schedule for problem $1|p_j = a_j + bt, r_j|C_{\max}$.

Property 13.1. (Lee et al. [12]) If $S_i < r_i$ and $(1 + b)r_i + a_i < r_j$, then there exists an optimal schedule in which job J_i immediately precedes job J_j.

Proof. Let σ be any feasible schedule for problem $1|p_j = a_j + bt, r_j|C_{\max}$ in which job J_i immediately precedes job J_j. Since $S_i < r_i$ and $(1 + b)r_i + a_i = C_i(\sigma) < r_j$, we have $C_j(\sigma) = (1 + b)r_j + a_j$.

Let σ' denote the schedule σ in which jobs J_i and J_j have been mutually replaced. Since $S_i < r_i < (1 + b)r_i + a_i < r_j$, we have $S_i < r_j$ and $r_i < r_j$. Therefore, $C_j(\sigma') = (1 + b)r_j + a_j$ and $C_i(\sigma') = (1 + b)^2 r_j + (1 + b)a_j + a_i$. Since $C_j(\sigma) < C_i(\sigma')$, we have $C_{\max}(\sigma) < C_{\max}(\sigma')$. ∎

Applying similar reasoning, we can prove the following properties.

Property 13.2. (Lee et al. [12]) If $S_i \geqslant \max\{r_i, r_j\}$ and $a_j > a_i$, then there exists an optimal schedule in which job J_i immediately precedes job J_j.

Property 13.3. (Lee et al. [12]) If $r_i \leqslant S_i \leqslant r_j, (1+b)S_i + a_i \geqslant r_j$ and $a_j > a_i$, then there exists an optimal schedule in which job J_i immediately precedes job J_j.

Property 13.4. (Lee et al. [12]) If $r_i \leqslant S_i$ and $(1 + b)S_i + a_i < r_j$, then there exists an optimal schedule in which job J_i immediately precedes job J_j.

Property 13.5. (Lee et al. [12]) If $r_j \leqslant S_i < r_i, (1 + b)S_i + a_j \geqslant r_i$ and $b(a_j - a_i) + (1 + b)^2(S_i - r_i) > 0$, then there exists an optimal schedule in which job J_i immediately precedes job J_j.

Property 13.6. (Lee et al. [12]) If $r_j > r_i > S_i, (1+b)r_i + a_i \geqslant r_j$ and $a_j > a_i$, then there exists an optimal schedule in which job J_i immediately precedes job J_j.

The authors also proposed the following lower bound on the optimal value of the schedule length for the problem.

Lemma 13.7. (Lee et al. [12]) *Let* $\sigma^{(k)}$ *be a partial schedule for problem* $1|p_j = a_j + bt, r_j|C_{\max}$ *such that* $|\sigma^{(k)}| = k$ *and let* σ *be a complete schedule obtained from* $\sigma^{(k)}$*. Then* $C_{\max}(\sigma) \geqslant \max\{LB_1, LB_2\}$*, where*

$$LB_1 = (1+b)^l C_{[k]}(\sigma) + \sum_{j=1}^{l}(1+b)^{l-j}a_{(k+j)},$$

$$LB_2 = \max\{(1+b)r_{k+j} + a_{k+j} : 1 \leqslant j \leqslant l\}$$

and $a_{(k+1)} \leqslant a_{(k+2)} \leqslant \cdots \leqslant a_{(k+l)}$.

Proof. By direct computation; see [12, Sect. 3.2].   ◇

Based on Properties 13.1–13.6 and Lemma 13.7, Lee et al. [12] proposed a branch-and-bound algorithm for problem $1|p_j = a_j + bt, r_j|C_{\max}$. The algorithm was tested on instances with $12 \leqslant n \leqslant 28$ jobs; see [12, Sect. 5].

General non-linear deterioration

Alidaee [1] and Gupta and Gupta [4] considered polynomially deteriorating job processing times given by (6.19).

Theorem 13.8. (Alidaee [1], Gupta and Gupta [4]) *Let* σ^1 *and* σ^2 *be two different partial schedules for problem* $1|p_j = a_j + b_j t + \cdots + m_j t^m|C_{\max}$*, in which the same subset of set* $\mathcal{J}$ *is scheduled. If* $C_{\max}(\sigma^1) \leqslant C_{\max}(\sigma^2)$*, then* $C_{\max}(\sigma^1, \tau) \leqslant C_{\max}(\sigma^2, \tau)$*, where* τ *is a schedule of remaining jobs.*

Proof. Alidaee [1] gives the proof by the pairwise job interchange argument. Gupta and Gupta [4] state the result without a proof.   ◇

By Theorem 13.8, Gupta and Gupta [4] proposed an exact algorithm for problem $1|p_j = a_j + b_j t + \cdots + m_j t^m|C_{\max}$.

Kunnathur and Gupta [11] proposed two exact algorithms for problem $1|p_j = a_j + \max\{0, b_j(t - d_j)\}|C_{\max}$.

The first exact algorithm is based on dynamic programming. Let S' and S'' be defined as in Lemma 7.36. The algorithm starts with $S'' := \emptyset$ and adds to S'' one job from S' at a time in such a way that the schedule corresponding to the ordered set S'' is always optimal. Let

$$r_j := \max\{0, b_j(t^\star - d_j)\},$$

where

$$t^* = \min_{S''\setminus\{J_j\}} \left\{ \sum_{J_k \in S''\setminus\{J_j\}} a_k + \max\{0, b_k(t - d_k)\} \right\}.$$

An optimal schedule can be found by the recursive equation

$$C_{\max}\left(S''\right) = C_{\max}\left(S'' \setminus \{J_j\}\right) + r_j,$$

with $C_{\max}(\emptyset) := 0$. The time complexity of the dynamic programming algorithm is $O(n2^n)$.

The second exact algorithm proposed by Kunnathur and Gupta [11] is a branch-and-bound algorithm. In this algorithm, the lower bound is computed using Lemma 7.36. The upper bound is obtained by one of the heuristic algorithms and, if possible, improved by the pairwise job interchange. Branching is performed by the standard depth-first search procedure. The branch-and-bound algorithm was tested on instances with up to $n = 15$ jobs. We refer the reader to [11, Sect. 3] for more details on the exact algorithms and to [11, Sect. 6] for the results of a computational experiment, conducted in order to evaluate the quality of schedules generated by the algorithms.

Kubiak and van de Velde [10] established a few properties of the single machine problem with the $C_{\max}$ criterion and job processing times given by (6.31). Before we formulate the properties, we introduce some terminology.

Definition 13.9. (Pivotal, early, tardy and suspended jobs)
(a) *The job that starts by time d and completes after d is called* pivotal.
(b) *The job that starts before d and completes by d is called* early.
(c) *The job that starts after d but completes by D is called* tardy.
(d) *The job that starts after D is called* suspended.

Now we can formulate the above mentioned properties. (We omit simple proofs by the pairwise job interchange argument.)

Property 13.10. (Kubiak and van de Velde [10]) The sequence of the early jobs is immaterial.

Property 13.11. (Kubiak and van de Velde [10]) The tardy jobs are sequenced in non-increasing order of the ratios $\frac{b_j}{a_j}$.

Property 13.12. (Kubiak and van de Velde [10]) The sequence of the suspended jobs is immaterial.

Property 13.13. (Kubiak and van de Velde [10]) The pivotal job has processing time not smaller than any of the early jobs.

Property 13.14. (Kubiak and van de Velde [10]) If $a_k \leqslant a_l$ and $b_k \geqslant b_l$, then job J_k precedes job J_l.

Based on Properties 13.10-13.14, Kubiak and van de Velde [10] proposed a branch-and-bound algorithm. It is reported (see [10, Sect. 5]) that the algorithm solves instances with $n = 50$ jobs within 1 second on a PC, while most of the instances with $n = 100$ jobs are solved within 7 minutes.

Kubiak and van de Velde [10] also proposed three pseudo-polynomial algorithms for the considered problem. The first algorithm, designed for the case of unbounded deterioration, runs in $O(nd\sum_{j=1}^{n} a_j)$ time and $O(nd)$ space. The second algorithm, designed for the case of bounded deterioration, requires $O(n^2d(D - d)\sum_{j=1}^{n} a_j)$ time and $O(nd(D - d))$ space. The third algorithm, also designed for the case of bounded deterioration, requires $O(nd\sum_{j=1}^{n} b_j(\sum_{j=1}^{n} a_j)^2)$ time and $O(nd\sum_{j=1}^{n} b_j \sum_{j=1}^{n} a_j)$ space.

For the general problem, $1|p_j \in \{a_j, b_j : a_j \leqslant b_j\}, d_j|C_{\max}$, Jeng and Lin [8] proposed a pseudo-polynomial exact algorithm, based on dynamic programming.

Lemma 13.15. (Jeng and Lin [8]) *There exists an optimal schedule for problem* $1|p_j \in \{a_j, b_j : a_j \leqslant b_j\}, d_j|C_{\max}$, *in which early jobs precede tardy jobs.*

Proof. Let us assume that in an optimal schedule a tardy job precedes an early job. If we schedule the tardy job as the last one, we do not increase the value of $C_{\max}$, since the processing time of the tardy job will not change. Therefore, repeating this operation for other tardy jobs, we obtain an optimal schedule in which early jobs precede tardy jobs.   □

Lemma 13.16. (Jeng and Lin [8]) *There exists an optimal schedule for problem* $1|p_j \in \{a_j, b_j : a_j \leqslant b_j\}, d_j|C_{\max}$, *in which early jobs are scheduled in non-decreasing order of* $a_j + b_j$ *values.*

Proof. By the pairwise job interchange argument.   □

The dynamic programming algorithm proposed by Jeng and Lin [8] exploits Lemmas 13.15–13.16. This algorithm can be formulated as follows.

Algorithm 13.2. for problem $1|p_j \in \{a_j, b_j : a_j \leqslant b_j\}, d_j|C_{\max}$

Initial conditions: $F(j,t) := \begin{cases} 0, & \text{if } j = 0 \wedge t = 0, \\ +\infty, & \text{otherwise;} \end{cases}$

Recursive formula for $1 \leqslant j \leqslant n, 0 \leqslant t \leqslant a_{\max} + d_{\max}$:

$$F(j,t) := \begin{cases} \min \left\{ \begin{array}{l} F(j-1, t-a_j) + a_j \\ F(j-1, t) + a_j + b_j \end{array} \right\}, & \text{if } t \geqslant a_j, d_j \geqslant t - a_j, \\ F(j-1, t) + a_j + b_j, & \text{otherwise;} \end{cases}$$

Goal: $\min\{F(n,t) : 0 \leqslant t \leqslant a_{\max} + d_{\max}\}$, where $a_{\max}$ is defined as in Sect. 11.1.3 and $d_{\max} := \max_{1 \leqslant j \leqslant n}\{d_j\}$.

The running time of Algorithm 13.2 is $O(n(a_{\max} + d_{\max}))$.

For problem $1|p_j \in \{a_j, b_j : a_j \leqslant b_j\}, d_j|C_{\max}$, Jeng and Lin [8] also proposed a branch-and-bound algorithm. Computational experiments have shown that the algorithm is able to solve instances with $n = 80$ jobs in 10 minutes (see [8, Sect. 'Computational experiments']).

Simple alteration

Jaehn and Sedding [6] proposed a mixed integer programming formulation and a dynamic programming algorithm for problem $1|p_j = a_j + b|f_j(t) - T||C_{\max}$ with fixed starting time of the first job; see [6, Sects. 7.1–7.2] for details.

13.2.2 Minimizing the total completion time

In this section, we consider exact algorithms for single machine time-dependent scheduling problems with the $\sum C_j$ criterion.

General non-linear deterioration

Jeng and Lin [9] formulated a branch-and-bound algorithm for problem $1|p_j \in \{a_j, b_j : a_j \leqslant b_j\}, d_j = D|\sum C_j$ in the case when $d_j = D$ for $1 \leqslant j \leqslant n$. The result given below allows us to estimate the number of early jobs in the so-called *normal* schedule defined as follows.

Definition 13.17. (A normal schedule)
A normal schedule for problem $1|p_j \in \{a_j, b_j : a_j \leqslant b_j\}, d_j = D|\sum C_j$ is a schedule in which the last job in the set of early jobs is started by the common due date D and finished after D.

Lemma 13.18. (Jeng and Lin [9]) *Let $E := \{J_k \in \mathcal{J} : C_k \leqslant d\}$ and let U (L) be the smallest (largest) integer satisfying the inequality $\sum_{j=1}^{U} a_i > D$ $(\sum_{j=n-U}^{U} a_i > D)$. Then for an arbitrary normal schedule for problem $1|p_j \in \{a_j, b_j : a_j \leqslant b_j\}, d_j = D|\sum C_j$ we have $U \leqslant |E| \leqslant L$.*

Proof. Let us notice that without loss of generality we can consider only the schedules in which jobs are indexed in non-decreasing order of the values a_j. Since the time interval $\langle 0, d \rangle$ can be considered as a one-dimensional bin, and job processing times a_j as items to pack into the bin, the result follows. $\square$

Remark 13.19. The terminology used in the proof of Lemma 13.18 is related to the BP problem.

Lemma 13.20. (Jeng and Lin [9]) *Let $b_{(j)}$ be the jth smallest deterioration rate in an instance of problem $1|p_j \in \{a_j, b_j : a_j \leqslant b_j\}, d_j = D|\sum C_j$. Then for any schedule σ for the problem the inequality*

$$\sum C_j(\sigma) \geqslant \sum_{j=1}^{n} (n - j + 1)a_j + \sum_{j=1}^{n-U} (n - U - j + 1)b_{(j)}$$

holds.

Proof. Let $E := \{J_k \in \mathcal{J} : C_k \leqslant D\}$. By definition of problem $1|p_j \in \{a_j, b_j : a_j \leqslant b_j\}, d_j = D|\sum C_j$, for any schedule σ we have $\sum C_j(\sigma) = \sum_{j=1}^{n}(n - j + 1)a_{[j]} + \sum_{j=1}^{n-|E|}(n - |E| - j + 1)b_{[j]}$. Since jobs are indexed in non-decreasing order of the processing times a_j, we have

$$\sum C_j(\sigma) = \sum_{j=1}^{n}(n - j + 1)a_{[j]} \geqslant \sum_{j=1}^{n}(n - j + 1)a_j.$$

By Lemma 13.18, $|E| \leqslant U$ and hence $|L| = |\mathcal{J} \setminus E| \geqslant n - U$. Since $b_{(j)} \leqslant b_{[j]}$ for any $1 \leqslant j \leqslant n$, the result follows. $\qquad\square$

By Lemma 13.20, the value $\sum_{j=1}^{n}(n - j + 1)a_j + \sum_{j=1}^{n-U}(n - U - j + 1)b_{(j)}$ may be used as an initial lower bound of the $\sum C_j$ criterion.

Jeng and Lin [9] also obtained a few other results, concerning the possible dominance relationships between jobs. One of the results is as follows.

Lemma 13.21. (Jeng and Lin [9]) *Let σ and E be an arbitrary subschedule for problem $1|p_j \in \{a_j, b_j : a_j \leqslant b_j\}, d_j = D|\sum C_j$ and the set of early jobs in the subschedule, respectively. If there exists a job J_j not belonging to σ such that $\sum_{j \in E} a_j + \sum_{i=j}^{n} a_i \leqslant d$, then each subtree rooted at $E \cup \{k\}$, $j \leqslant k \leqslant n$, can be eliminated.*

Proof. See [9, Lemma 4].   ◇

Based on Lemmas 13.18–13.21 and some other results (see [9, Sect. 4]) Jeng and Lin [9] constructed the above mentioned branch-and-bound algorithm for problem $1|p_j \in \{a_j, b_j : a_j \leqslant b_j\}, d_j = D|\sum C_j$. The results of computational experiments (see [9, Sect. 5]) suggest that the algorithm is quite effective, since it can solve most instances with $n = 100$ jobs in time not longer than 3 minutes.

Now, let us consider problem $1|p_j \in \{a_j, b_j : a_j \leqslant b_j\}, d_j = D|\sum C_j$ in terms of single machine precedence-constrained scheduling. By Lemma 7.76, we can divide the set of jobs $\mathcal{J}$ into a number of *chains* as follows. Let us renumber jobs in non-decreasing order of a_j, and break ties in non-increasing order of $a_j + b_j$. Let us assume that a job is the head of a chain. Let us put the next job at the end of the chain, if it has agreeable parameters with the last job in the chain. Let us repeat the procedure until all jobs are checked. Let us create other chains from the remaining jobs in the same way.

Let us notice that knowing all chains, we know which jobs are early and tardy in each chain. Since the early jobs have known processing times and in this case the total completion time is minimized by scheduling the jobs in non-decreasing order with respect to the values a_j, we obtain a local optimal schedule. If we enumerate all locally optimal schedules, we are able to find an optimal schedule.

Based on the above reasoning, Cheng and Ding [3] proposed the following enumeration algorithm for the considered problem.

Algorithm 13.3. for problem $1|p_j \in \{a_j, b_j : a_j \leqslant b_j\}, d_j = D| \sum C_j$

---

1 **Input** : sequences $(a_1, a_2, \ldots, a_n)$, $(b_1, b_2, \ldots, b_n)$, number D
2 **Output:** an optimal schedule $\sigma^\star$
  ▷ Step 1
3 Construct all chains of jobs $\mathcal{C}_1, \mathcal{C}_2, \ldots, \mathcal{C}_k$;
  ▷ Step 2
4 **for** *all possible* $(e_1, e_2, \ldots, e_k)$ *such that* $e_i \leqslant |\mathcal{C}_i|$ *for* $1 \leqslant i \leqslant k$ **do**
5   **for** $\mathcal{C} \leftarrow \mathcal{C}_1$ **to** $\mathcal{C}_k$ **do**
6 | Set the number of elements of $\{J_k \in \mathcal{C} : C_k \leqslant D\}$ to $e_{\mathcal{C}}$;
     end
7    $E \leftarrow \{J_k \in \mathcal{J} : C_k \leqslant D\}$; $L \leftarrow \mathcal{J} \setminus E$;
8   Schedule jobs in E in non-decreasing order of the basic processing times a_j;
9   Schedule jobs in L in non-increasing order of the sums $a_j + b_j$;
10   **if** $(C_{\max}(E) \leqslant D)$ **then**
       | Compute $\sum C_j(E)$;
      end
  end
  ▷ Step 3
11 $\sigma^\star \leftarrow$ the best schedule among all schedules generated in Step 2;
12 **return** $\sigma^\star$.

Algorithm 13.3 runs in $O(n^k \log n)$ time, where k is the number of chains created in Step 1.

Remark 13.22. If k is a fixed number, then Algorithm 13.3 is polynomial (cf. Theorem 7.77).

Linear shortening

For problem $1|p_j = a - b_j t| \sum C_j$, Ng et al. [13] proposed a pseudo-polynomial dynamic programming algorithm with running time $O(n^3 h^2)$, where h is the product of denominators of all shortening rates b_j, $1 \leqslant j \leqslant n$. We refer the reader to [13, Sect. 4] for more details.

13.2.3 Minimizing the maximum lateness

In this section, we consider exact algorithms for single machine time-dependent scheduling problems with the $L_{\max}$ criterion.

Linear deterioration

Hsu and Lin [5] proposed a branch-and-bound algorithm for problem $1|p_j = a_j + b_j t|L_{\max}$. The algorithm exploits several properties concerning dominance relations among different schedules.

Property 13.23. (Hsu and Lin [5]) Let $\sigma^{(a)}$ and $\sigma^{(b)}$ be two schedules for a given subset of jobs. If $C_{\max}(\sigma^{(a)}) \leqslant C_{\max}(\sigma^{(b)})$ and $L_{\max}(\sigma^{(a)}) \leqslant L_{\max}(\sigma^{(b)})$, then the subtree rooted at $\sigma^{(b)}$ can be eliminated.

Proof. The result follows from the regularity of the $C_{\max}$ and $L_{\max}$ criteria.
□

Property 13.24. (Hsu and Lin [5]) Let $J_i, J_j \in \mathcal{J}$ be any two jobs scheduled consecutively. If $\frac{a_i}{b_i} \leqslant \frac{a_j}{b_j}$ and $d_i \leqslant d_j$, then there exists an optimal schedule in which job J_i is an immediate predecessor of job J_j.

Proof. Let σ' (σ'') be a schedule in which job J_i (job J_j) precedes job J_j (job J_i). First, note that if $\frac{a_i}{b_i} \leqslant \frac{a_j}{b_j}$ and $d_i \leqslant d_j$, then by Theorem 7.14 we have $C_{\max}(\sigma') \leqslant C_{\max}(\sigma'')$. Second, since then $L_i(\sigma') \leqslant L_i(\sigma'')$ and $L_j(\sigma'') = C_j(\sigma') - d_j \leqslant L_i(\sigma')$, we have

$$\max\{L_i(\sigma'), L_j(\sigma'')\} \leqslant \max\{L_i(\sigma''), L_j(\sigma'')\}.$$

Hence $L_{\max}(\sigma') \leqslant L_{\max}(\sigma'')$.   □

Let $(\sigma^{(a)}|\tau^{(a)})$ denote a schedule composed of partial schedules $\sigma^{(a)}$ and $\tau^{(a)}$, where $|\tau^{(a)}| \geqslant 0$.

Property 13.25. (Hsu and Lin [5]) Given a partial schedule $(\sigma^{(a)}|j)$ and an unscheduled job J_i, if $\frac{a_i}{b_i} \leqslant \frac{a_j}{b_j}$ and $C_{\max}(\sigma^{(a)}|i|j) - d_j \leqslant L_{\max}(\sigma^{(a)}|j)$, then the subtree rooted at $(\sigma^{(a)}|j|i)$ can be eliminated.

Proof. First, inequality $\frac{a_i}{b_i} \leqslant \frac{a_j}{b_j}$ implies that $C_{\max}(\sigma^{(a)}|i|j) \leqslant C_{\max}(\sigma^{(a)}|j|i)$. Second, $L_i(\sigma^{(a)}|i|j) \leqslant L_i(\sigma^{(a)}|j|i)$ and $C_{\max}(\sigma^{(a)}|i|j) - d_j \leqslant L_{\max}(\sigma^{(a)}|j)$ by assumption. Hence,

$$L_{\max}(\sigma^{(a)}|i|j) \leqslant \max\{C_{\max}(\sigma^{(a)}|j|i) - d_i, L_{\max}(\sigma^{(a)}|j)\} = L_{\max}(\sigma^{(a)}|j|i).$$

□

Properties 13.23–13.25 allow us to cut off some subtrees during the process of searching for an optimal schedule in the tree of all possible schedules. In order to estimate the lateness of an optimal schedule from below, we need a lower bound on the value of $L_{\max}$ for the optimal schedule.

Property 13.26. (Hsu and Lin [5]) Let $\sigma^{(a)}$ and $\tau^{(a)}$ denote, respectively, a schedule of a subset of jobs and a schedule with the remaining jobs arranged in non-decreasing order of the ratios $\frac{a_j}{b_j}$. Then

$$L_{\max}(\sigma^{(a)}|\tau^{(a)\prime}) \leqslant C_{\max}(\sigma^{(a)}|\tau^{(a)}) - \max\{d_j : j \in N_{\mathcal{J}} \setminus N(\sigma^{(a)})\},$$

where $\tau^{(a)\prime} \neq \tau^{(a)}$ and $N(\sigma^{(a)})$ denotes the set of indices of jobs from the subschedule $\sigma^{(a)}$.

Proof. See [5, Lemma 4]. ◇

Another lower bound is obtained by a transformation of the initial set of jobs into a new one, called an *ideal* set (see [5, Lemma 5]).

The branch-and-bound algorithm, obtained by implementation of the above properties and using the above lower bounds, appears to be quite effective, since it is reported (see [5, Sect. 5]) that problems of no more than 100 jobs can be solved, on average, within 1 minute.

13.2.4 Other criteria

In this section, we consider exact algorithms for single machine time-dependent scheduling problems with criteria other than $C_{\max}$, $\sum C_j$ and $L_{\max}$.

Linear deterioration

Wu et al. [14] proved a few properties of problem $1|p_j = a_j + b_j t|\sum w_j C_j$.

Property 13.27. (Wu et al. [14]) Let $J_i, J_k \in \mathcal{J}$ be any two jobs. If $a_i < a_k$, $b_i = b_k$ and $w_i \geqslant w_k$, then for problem $1|p_j = a_j + b_j t|\sum w_j C_j$ there exists an optimal schedule in which job J_i precedes job J_k.

Proof. By direct computation; see [14, Property 1]. ◇

Property 13.28. (Wu et al. [14]) Let $J_i, J_k \in \mathcal{J}$ be any two jobs to be scheduled consecutively. If $a_i = a_k$, $b_i > b_k$ and $\frac{b_i}{b_k} \leqslant \frac{w_i}{w_k}$, then for problem $1|p_j = a_j + b_j t|\sum w_j C_j$ there exists an optimal schedule in which job J_i immediately precedes job J_k.

Proof. By direct computation; see [14, Property 2]. ◇

The next three properties are similar to Property 13.28, Wu et al. [14] stated them without proofs.

Property 13.29. (Wu et al. [14]) Let $J_i, J_k \in \mathcal{J}$ be any jobs to be scheduled consecutively and let t_0 be the completion time of the last job scheduled before these two jobs. If $\frac{a_i}{b_i} = \frac{a_k}{b_k}$, $w_i \geqslant w_k$ and $\frac{a_i + b_i t_0}{w_i} < \frac{a_k + b_k t_0}{w_k}$, then for problem $1|p_j = a_j + b_j t|\sum w_j C_j$ there exists an optimal schedule in which job J_i immediately precedes job J_k.

Property 13.30. (Wu et al. [14]) Let $J_i, J_k \in \mathcal{J}$ be any jobs to be scheduled consecutively and let t_0 be the completion time of the last job scheduled before these two jobs. If $\frac{a_i}{b_i} \leqslant \frac{a_k}{b_k}$, $w_i = w_k$ and $a_i + b_i t_0 < a_k + b_k t_0$, then for problem $1|p_j = a_j + b_j t| \sum w_j C_j$ there exists an optimal schedule in which job J_i immediately precedes job J_k.

Property 13.31. (Wu et al. [14]) Let $J_i, J_k \in \mathcal{J}$ be any jobs to be scheduled consecutively and let t_0 be the completion time of the last job scheduled before these two jobs. If $\frac{a_i}{b_i} \leqslant \frac{a_k}{b_k}$ and $\frac{w_k}{w_i} < \frac{1+b_i}{1+b_k} \min\{\frac{a_k}{a_i}, \frac{b_k}{b_i}\}$, then for problem $1|p_j = a_j + b_j t| \sum w_j C_j$ there exists an optimal schedule in which job J_i immediately precedes job J_k.

Wu et al. [14] also proposed a lower bound for the considered problem.

Theorem 13.32. (Wu et al. [14]) *Let* $\sigma = (\sigma^{(1)}, \sigma^{(2)})$ *be a schedule for problem* $1|p_j = a_j + b_j t| \sum w_j C_j$, *where* $\sigma^{(1)}$ $(\sigma^{(2)})$ *denotes the sequence of scheduled (unscheduled) jobs,* $|\sigma^{(1)}| = m$ *and* $|\sigma^{(2)}| = r = n - m$. *Then the optimal weighted completion time* $\sum w_j C_j(\sigma)$ *is not less than* $\max\{LB_1, LB_2\}$, *where*

$$
LB_1 = \sum_{k=1}^{m} w_{[k]} C_{[k]}(\sigma) + C_{[m]}(\sigma) \sum_{k=1}^{r} w_{(m+r+1-k)} \prod_{i=1}^{k} (1 + b_{(m+i)}) +
$$
$$
a_{(m+1)} \sum_{k=1}^{r-1} \left(\prod_{j=1}^{k} (1 + b_{(m+j)}) \right) \left(\sum_{i=1}^{r-k} w_{(m+i)} \right) + a_{(m+1)} \sum_{k=1}^{r} w_{m+k},
$$

and

$$
LB_2 = \sum_{k=1}^{m} w_{[k]} C_{[k]}(\sigma) + C_{[m]}(\sigma) \sum_{k=1}^{r} w_{(m+r+1-k)} \prod_{i=1}^{k} (1 + b_{(m+1)})^k +
$$
$$
\sum_{k=1}^{r-1} a_{(m+k)} \sum_{i=k}^{r} w_{(m+r+1-i)} (1 + b_{(m+1)})^i,
$$

and all parameters of unscheduled jobs are in non-decreasing order, i.e., $a_{(m+1)} \leqslant a_{(m+2)} \leqslant \cdots \leqslant a_{(m+r)}$, $b_{(m+1)} \leqslant b_{(m+2)} \leqslant \cdots \leqslant b_{(m+r)}$ *and* $w_{(m+1)} \leqslant w_{(m+2)} \leqslant \cdots \leqslant w_{(m+r)}$.

Proof. By direct computation; see [14, Sect. 3.1]. ◇

Based on Properties 13.27–13.31 and Theorem 13.32, Wu et al. [14] proposed a branch-and-bound algorithm for problem $1|p_j = a_j + b_j t| \sum w_j C_j$. Computational experiments have shown that the algorithm can solve instances with $n = 16$ jobs in time no longer than 3 hours (see [14, Sect. 5]).

Jafari and Moslehi [7] proposed a branch-and-bound algorithm for problem $1|p_j = a_j + bt| \sum U_j$. This algorithm was tested on instances with $n \leqslant 44$ jobs (cf. [7, Sect. 5]).

General non-linear deterioration

Cheng and Ding [3] proposed for problem $1|p_j \in \{a_j, b_j : a_j \leqslant b_j\}| \sum w_j C_j$ an enumeration algorithm, based on a result similar to Lemma 7.76.

Lemma 13.33. (Cheng and Ding [3]) *If $a_i \leqslant a_j$, $\frac{a_i}{w_i} \leqslant \frac{a_j}{w_j}$ and $\frac{a_i+b_i}{w_i} \leqslant \frac{a_j+b_j}{w_j}$, then job J_i precedes job J_j in any optimal schedule for problem $1|p_j \in \{a_j, b_j : a_j \leqslant b_j\}| \sum w_j C_j$.*

Proof. By the pairwise job interchange argument. □

By Lemma 13.33, we can divide the set $\mathcal{J}$ into a number of chains, using the approach applied in Algorithm 13.3 for problem $1|p_j \in \{a_j, b_j : a_j \leqslant b_j\}$, $d_j = D| \sum C_j$. In order to do that it is sufficient to renumber jobs in non-decreasing order of the values a_j, breaking ties in non-decreasing order of the ratios $\frac{a_j}{w_j}$ if jobs have the same values of a_j, and breaking ties in non-decreasing order of the ratios $\frac{a_j+b_j}{w_j}$ if jobs have the same values of the ratios $\frac{a_j}{w_j}$.

Remark 13.34. Cheng and Ding [3] proved that if an instance of problem $1|p_j \in \{a_j, b_j : a_j \leqslant b_j\}, d_j = D| \sum w_j C_j$ has a fixed number of chains, then the modified Algorithm 13.3 is a polynomial-time algorithm for this instance.

13.3 Exact algorithms for parallel machine problems

In this section, we consider exact algorithms for intractable parallel machine time-dependent scheduling problems.

13.3.1 Minimizing the maximum completion time

The following lower bound on the value of $C_{\max}$ can be used in exact algorithms for problem $Pm|p_j = b_j t|C_{\max}$ with $m \geqslant 2$ machines.

Lemma 13.35. (Hsieh and Bricker [16], Mosheiov [17]) *The optimal value of the maximum completion time for problem $Pm|p_j = b_j t|C_{\max}$ is not less than*
$$LB := \sqrt[m]{\prod_{j=1}^{n}(1 + b_j)}.$$

Proof. Let $\mathcal{N}_{M_k}$, $1 \leqslant k \leqslant m$, denote the set of indices of jobs assigned to machine M_k in an optimal schedule. Then

$$C_{\max}^{\star} = \max_{1 \leqslant k \leqslant m} \left\{ \prod_{j \in \mathcal{N}_{M_k}} (1 + b_j) \right\}.$$

Since, by Lemma 1.1 (a), we have $C_{\max}^{\star} \geqslant \frac{1}{m} \sum_{k=1}^{m} \prod_{j \in \mathcal{N}_{M_k}} (1 + b_j)$ and, by Lemma 1.1 (b), we have

$$C^\star_{\max} \geqslant \sqrt[m]{\prod_{j=1}^{n}(1 + b_j)} = LB,$$

the result follows.   □

The following example shows that the ratio of $C^\star_{\max}$ and bound LB from Lemma 13.35 can be arbitrarily large.

Example 13.36. (Mosheiov [17]) Let $n = 2$, $m = 2$, $t_0 = 1$, $b_1 = B > 0$, $b_2 = \epsilon > 0$. Then $C^\star_{\max} = 1 + B$, while $LB = \sqrt{(1 + B)(1 + \epsilon)} \to \sqrt{(1 + B)}$ if $\epsilon \to 0$. Hence $\frac{C^\star_{\max}}{LB} \to \sqrt{(1 + B)} \to \infty$ if $B \to \infty$ and $\epsilon \to 0$.   ♦

It has been shown, however, that if the deterioration rates b_j, $1 \leqslant j \leqslant n$, are independently drawn from a distribution with a finite second moment and positive density at 0, then $\lim_{n \to \infty} \frac{C^\star_{\max}}{LB} = 1$ almost surely (see Mosheiov [17, Proposition 3]).

Remark 13.37. A sequence (x_n) *converges almost surely* to a limit x if the probability of the event $\lim_{n \to \infty} x_n = x$ is equal to 1.

13.3.2 Minimizing the total completion time

In this section, we consider exact algorithms for the $\sum C_j$ criterion.

Proportional deterioration

Ouazene and Yalaoui [18] proposed an exact dynamic programming algorithm for problem $P2|p_j = b_j t| \sum C_j$. This algorithm is based on a similar algorithm for problem $P2||C_{\max}$ by Ho and Wong [15] and runs in $O(2^n)$ time. The main idea is to combine the *lexicographic search* with some dominance rules similar to those applied in branch-and-bound algorithms. The lexicographic search is a systematic exhaustive search performed on a binary tree. At each level of the tree, a given job is added to the partial schedule from the previous level. The left (right) child node of a parent node corresponds to a schedule in which the job corresponding to the child node is assigned to the first (second) machine, and a path from the root to a leaf node corresponds to a schedule.

Simple non-linear deterioration

Toksarı and Güner [19] solved a parallel machine early/tardy job scheduling problem with linear job processing times in the form of $p_j = a_j + bt$ by using a mathematical programming approach.

13.4 Exact algorithms for dedicated machine problems

In this section, we consider exact algorithms for intractable dedicated machine time-dependent scheduling problems.

13.4.1 Minimizing the maximum completion time

In this section, we consider exact algorithms for the $C_{\max}$ criterion.

Proportional deterioration

Wang et al. [27] proposed a branch-and-bound algorithm for problem $F3|p_{ij} = b_{ij}t|C_{\max}$, where $b_{ij} \in (0,1)$. This algorithm was tested on instances with $n \leqslant 14$ jobs (cf. [27, Sect. 6]).

Remark 13.38. Jafari et al. [21] corrected some of dominance rules proposed by Wang et al. [27].

Linear deterioration

Lee et al. [23] gave a lower bound on the optimal value of the total completion time for problem $F2|p_{ij} = a_{ij} + bt|C_{\max}$. Let $B := 1 + b$.

Lemma 13.39. (Lee et al. [23]) *Let k be the number of already scheduled jobs in a subschedule $\sigma^{(1)}$ and let $\mathcal{J}_{\sigma^{(1)}}$ denote the set of jobs in $\sigma^{(1)}$. Then the maximum completion time $C_{\max}(\sigma)$ for schedule $\sigma = (\mathcal{J}_{\sigma^{(1)}}|\mathcal{J} \setminus \mathcal{J}_{\sigma^{(1)}})$ for problem $F2|p_{ij} = a_{ij} + bt|C_{\max}$ is not less than $\max\{LB_1, LB_2\}$, where*

$$LB_1 := B^{n-k+1}C_{1,[k]} + \sum_{i=1}^{n-k} B^{n-k-i+1}a_{1,(k+i)} + a_{2,(k+1)},$$
$$LB_2 := B^{n-k}C_{2,[k]} + \sum_{i=k+1}^{n} B^{n-i}a_{2,(i)},$$

$a_{1,(k+1)} \leqslant a_{1,(k+2)} \leqslant \cdots \leqslant a_{1,(n)}$ *and* $a_{2,(k+1)} \leqslant a_{2,(k+2)} \leqslant \cdots \leqslant a_{2,(n)}$ *are non-decreasingly ordered basic processing times of unscheduled jobs on machine M_1 and M_2, respectively, and* $a_{2,(k+1)} := \min\{a_{2i} : i \in \mathcal{J} \setminus \mathcal{J}_{\sigma^{(1)}}\}$.

Proof. See [23, Sect. 3].   ◇

Lee et al. [23] also formulated a number of dominance properties (see [23, Properties 1–5]) and proposed a branch-and-bound algorithm for this problem (see [23, Sect. 4.1]). The branch-and-bound algorithm was tested on a number of instances with $8 \leqslant n \leqslant 32$ jobs (cf. [23, Sect. 5]).

Lee et al. [22] proposed a branch-and-bound algorithm for problem $F2|p_{ij} = a_{ij} + bt, block|C_{\max}$, where symbol *block* denotes that a job completed on a machine blocks this machine until the next machine is available for processing.

This algorithm was tested on instances with $n = 16, 24$ and 32 jobs; see [22, Sect. 5].

Wang and Wang [28] proposed a branch-and-bound algorithm for problem $F3|p_{ij} = a_{ij} + bt|C_{max}$. This algorithm was tested on instances with $n = 20, 30, \dots, 60$ jobs; see [28, Sect. 5].

Remark 13.40. Jafari et al. [20] corrected some of dominance rules proposed by Wang et al. [28].

13.4.2 Minimizing the total completion time

In this section, we consider exact algorithms for the $\sum C_j$ criterion.

Proportional deterioration

Wang et al. [26] formulated dominance properties (see [26, Propositions 1–8; Theorems 5–6]) and proposed a branch-and-bound algorithm for problem $F2|p_{ij} = b_{ij}t, 0 < b_{ij} < 1|\sum C_j$ (see [26, Sect. 5.3]). The algorithm is based on the properties and the following lower bound.

Lemma 13.41. (Wang et al. [26]) *Let k be the number of already scheduled jobs in a subschedule $\sigma^{(1)}$ and let $\mathcal{J}_{\sigma^{(1)}}$ denote the set of jobs in $\sigma^{(1)}$. Then the total completion time $\sum C_j(\sigma)$ for schedule $\sigma = (\mathcal{J}_{\sigma^{(1)}}|\mathcal{J} \setminus \mathcal{J}_{\sigma^{(1)}})$ for problem $F2|p_{ij} = b_{ij}t, 0 < b_{ij} < 1|\sum C_j$ is not less than $\max\{LB_1, LB_2\}$, where*

$$LB_1 := \sum_{j=1}^{k} C_{[j]}(\sigma^{(1)}) + C_{[k]}(\sigma^{(1)}) \sum_{j=k+1}^{n} \prod_{i=k+1}^{j} (1 + b_{2,(i)}),$$

$$LB_2 := \sum_{j=1}^{k} C_{[j]}(\sigma^{(1)}) + t_0 \left(1 + \min_{k+1 \leqslant j \leqslant n} \{b_{2j}\} \right) \prod_{j=1}^{k} (1 + b_{1,[j]}) \times$$
$$\times \left(\sum_{i=k+1}^{n} \prod_{j=k+1}^{i} (1 + b_{1,(j)}) \right),$$

and $b_{1,(k+1)} \leqslant b_{1,(k+2)} \leqslant \cdots \leqslant b_{1,(n)}$ and $b_{2,(k+1)} \leqslant b_{2,(k+2)} \leqslant \cdots \leqslant b_{2,(n)}$ are non-decreasingly ordered basic processing times of unscheduled jobs on machine M_1 and machine M_2, respectively.

Proof. See [26, Sect. 5.2].   ◇

The branch-and-bound algorithm proposed by Wang et al. [26] was tested on 100 instances with $n \leqslant 14$ jobs (see [26, Sect. 6] for details).

Shiau et al. [25] gave a lower bound on the optimal value of the total completion time for problem $F2|p_{ij} = b_{ij}t|\sum C_j$.

Lemma 13.42. (Shiau et al. [25]) *Let k be the number of already scheduled jobs in a subschedule $\sigma^{(1)}$ and let $\mathcal{J}_{\sigma^{(1)}}$ denote the set of jobs in $\sigma^{(1)}$.*

Let $B := \prod\limits_{i=1}^{k} (1 + b_{1,[i]})$. Then the total completion time $\sum C_j(\sigma)$ for schedule $\sigma = (\mathcal{J}_{\sigma^{(1)}} | \mathcal{J} \setminus \mathcal{J}_{\sigma^{(1)}})$ for problem $F2|p_{ij} = b_{ij}t| \sum C_j$ is not less than $\max\{LB_1, LB_2, LB_3\}$, where

$$LB_1 := \sum_{j=1}^{k} C_{[j]}(\sigma^{(1)}) + C_{[k]}(\sigma^{(1)}) \sum_{j=k+1}^{n} \prod_{i=k+1}^{j} (1 + b_{2,(i)}),$$

$$LB_2 := \sum_{j=1}^{k} C_{[j]}(\sigma^{(1)}) + B \times \sum_{i=k+1}^{n} (1 + b_{2,(n+k+1-i)}) \prod_{j=k+1}^{i} (1 + b_{1,(j)}),$$

$$LB_3 := \sum_{j=1}^{k} C_{[j]}(\sigma^{(1)}) + (n-k)B \times \sqrt[n-k]{\prod_{i=k+1}^{n} (1 + b_{2i}) \prod_{i=k+1}^{n} (1 + b_{1,(i)})^{n+1-i}},$$

and $b_{1,(k+1)} \leqslant b_{1,(k+2)} \leqslant \cdots \leqslant b_{1,(n)}$ and $b_{2,(k+1)} \leqslant b_{2,(k+2)} \leqslant \cdots \leqslant b_{2,(n)}$ are non-decreasingly ordered basic processing times of unscheduled jobs on machine M_1 and machine M_2, respectively.

Proof. See [25, Sect. 4].   ◊

Shiau et al. [25] also formulated a number of dominance properties (see [25, Propositions 1–11]) and proposed a branch-and-bound algorithm for problem $F2|p_{ij} = b_{ij}t| \sum C_j$ (see [25, Sect. 5.1]). The branch-and-bound algorithm was tested on 450 instances with $6 \leqslant n \leqslant 14$ jobs (see [25, Sect. 7] for details).

Proportional-linear deterioration

Ng et al. [24] proposed a branch-and-bound algorithm for problem $F2|p_{ij} = b_{ij}(a+bt)| \sum C_j$. This algorithm was tested on instances with $n \leqslant 15$ jobs (cf. [24, Sect. 4.3]).

Linear deterioration

Wu and Lee [30] proposed for problem $F2|p_{ij} = a_{ij} + bt| \sum C_j$ a branch-and-bound algorithm, based on a number of dominance properties and a lower bound. First, we briefly describe some of these properties. Let

$$\Theta := C_{2,[n]} - (b+1)C_{1,[n]},$$

where $b > 0$ is a common job deterioration rate.

Property 13.43. (Wu and Lee [30]) If jobs J_i and J_j are scheduled consecutively, $a_{1i} \geqslant \Theta$, $\max\{a_{1i}, a_{2i}\} < a_{1j}$ and $a_{2j} \leqslant \min\{a_{1i}, a_{2i}\}$, then there exists an optimal schedule for problem $F2|p_{ij} = a_{ij} + bt| \sum C_j$ in which job J_i is the immediate predecessor of job J_j.

Proof. By the pairwise job interchange argument; see [30, Appendix A].   ◊

Property 13.44. (Wu and Lee [30]) If jobs J_i and J_j are scheduled consecutively, $a_{1i} \geqslant \Theta$, $a_{1j} \geqslant a_{1i}$ and

$$\min\left\{a_{1j}, a_{2j}, \frac{b+1}{b+2}a_{1j} + \frac{1}{b+2}a_{2j}, \frac{ba_{1j}}{b+1} + \frac{ba_{2j} + a_{2i}}{(b+1)^2}\right\} > a_{1i},$$

then there exists an optimal schedule for problem $F2|p_{ij} = a_{ij} + bt|\sum C_j$ in which job J_i is the immediate predecessor of job J_j.

Proof. The result is given without proof; see [30, Property 2].   ◇

Other properties given by the authors without proof (see [30, Properties 3 – 12]) are similar to Property 13.44.

Wu and Lee [30] also proposed the following lower bound for this problem.

Lemma 13.45. (Wu and Lee [30]) *Let k be the number of already scheduled jobs in a subschedule $\sigma^{(1)}$, let $\mathcal{J}_{\sigma^{(1)}}$ denote the set of jobs in $\sigma^{(1)}$ and let $r := n - k$. Then the total completion time $\sum C_j(\sigma)$ for schedule $\sigma = (\mathcal{J}_{\sigma^{(1)}}|\mathcal{J} \setminus \mathcal{J}_{\sigma^{(1)}})$ for problem $F2|p_{ij} = a_{ij} + b_{ij}t|\sum C_j$ is not less than $\max\{LB_1, LB_2\}$, where*

$$LB_1 := \sum_{j=1}^{k} C_{2,[j]}(\sigma) + C_{1,[k]}(\sigma^{(1)}) \sum_{j=1}^{r}(1+b)^{j+1} + \sum_{j=1}^{r}\sum_{i=1}^{r-j+1} a_{1,(k+j)}(1+b)^j +$$

$$+ \sum_{j=1}^{r} a_{2,k+j},$$

$$LB_2 := \sum_{j=1}^{k} C_{2,[j]}(\sigma) + C_{2,[k]}(\sigma) \sum_{j=1}^{r}(1+b)^j + \sum_{j=1}^{r}\sum_{i=1}^{r-j} a_{2,(k+j)}(1+b)^{i-1},$$

$a_{1,(k+1)} \leqslant a_{1,(k+2)} \leqslant \cdots \leqslant a_{1,(k+r)}$ *and* $a_{2,(k+1)} \leqslant a_{2,(k+2)} \leqslant \cdots \leqslant a_{2,(k+r)}$ *are non-decreasingly ordered basic processing times of unscheduled jobs on machine M_1 and M_2, respectively, and $\sum_{j=1}^{r} a_{2,k+j}$, is the sum of the basic processing times of the unscheduled jobs on machine M_2.*

Proof. See [30, Sect. 4].   ◇

The branch-and-bound algorithm, based on the mentioned properties and Lemma 13.45, was tested on instances with $n \leqslant 27$ jobs (see [30, Sect. 6]).

13.4.3 Other criteria

In this section, we consider exact algorithms for intractable time-dependent scheduling problems with criteria other than $C_{\max}$ and $\sum C_j$.

Proportional deterioration

Yang and Wang [31] proposed a branch-and-bound algorithm for problem $F2|p_{ij} = b_{ij}t|\sum w_j C_j$. This algorithm was tested on instances with $n \leqslant 14$ jobs (cf. [31, Sect. 6]).

Proportional-linear shortening

Wang and Wang [29] proposed a branch-and-bound algorithm for problem $F2|p_{ij} = b_{ij}(1 - bt)| \sum w_j C_j$. This algorithm was tested on instances with $n \leqslant 15$ jobs; see [29, Sects. 5–6].

References

Exact algorithms for single machine problems

1. B. Alidaee, A heuristic solution procedure to minimize makespan on a single machine with non-linear cost functions. *Journal of the Operational Research Society* **41** (1990), no. 11, 1065–1068.
2. A. Bosio and G. Righini, A dynamic programming algorithm for the single-machine scheduling problem with release dates and deteriorating processing times. *Mathematical Methods of Operations Research* **69** (2009), no. 2, 271–280.
3. T-C. E. Cheng and Q. Ding, Single machine scheduling with step-deteriorating processing times. *European Journal of Operational Research* **134** (2001), no. 3, 623–630.
4. J. N. D. Gupta and S. K. Gupta, Single facility scheduling with nonlinear processing times. *Computers and Industrial Engineering* **14** (1988), no. 4, 387–393.
5. Y-S. Hsu and B. M-T. Lin, Minimization of maximum lateness under linear deterioration. *Omega* **31** (2003), no. 6, 459–469.
6. F. Jaehn and H. A. Sedding, Scheduling with time-dependent discrepancy times. *Journal of Scheduling* **19** (2016), no. 6, 737–757.
7. A. Jafari and G. Moslehi, Scheduling linear deteriorating jobs to minimize the number of tardy jobs. *Journal of Global Optimization* **54** (2012), no. 2, 389–404.
8. A. A-K. Jeng and B. M-T. Lin, Makespan minimization in single-machine scheduling with step-deterioration of processing times. *Journal of the Operational Research Society* **55** (2004), no. 3, 247–256.
9. A. A-K. Jeng and B. M-T. Lin, Minimizing the total completion time in single-machine scheduling with step-deteriorating jobs. *Computers and Operations Research* **32** (2005), no. 3, 521–536.
10. W. Kubiak and S. L. van de Velde, Scheduling deteriorating jobs to minimize makespan. *Naval Research Logistics* **45** (1998), no. 5, 511–523.
11. A. S. Kunnathur and S. K. Gupta, Minimizing the makespan with late start penalties added to processing times in a single facility scheduling problem. *European Journal of Operational Research* **47** (1990), no. 1, 56–64.
12. W-C. Lee, C-C. Wu and Y-H. Chung, Scheduling deteriorating jobs on a single machine with release times. *Computers and Industrial Engineering* **54** (2008), no. 3, 441–452.
13. C-T. Ng, T-C. E. Cheng, A. Bachman and A. Janiak, Three scheduling problems with deteriorating jobs to minimize the total completion time. *Information Processing Letters* **81** (2002), no. 6, 327–333.
14. C-C. Wu, W-C. Lee and Y-R. Shiau, Minimizing the total weighted completion time on a single machine under linear deterioration. *International Journal of Advanced Manufacturing Technology* **33** (2007), no. 11–12, 1237–243.

Exact algorithms for parallel machine problems

15. J. Ho and J. Wong, Makespan minimization for m parallel identical processors. *Naval Research Logistics* **42** (1995), no. 6, 935–948.
16. Y-C. Hsieh and D. L. Bricker, Scheduling linearly deteriorating jobs on multiple machines. *Computers and Industrial Engineering* **32** (1997), no. 4, 727–734.
17. G. Mosheiov, Multi-machine scheduling with linear deterioration. *Infor* **36** (1998), no. 4, 205–214.
18. Y. Ouazene and F. Yalaoui, Identical parallel machine scheduling with time-dependent processing times. *Theoretical Computer Science* **721** (2018), 70–77.
19. M. D. Toksarı and E. Güner, The common due-date early/tardy scheduling problem on a parallel machine under the effects of time-dependent learning and linear and nonlinear deterioration. *Expert Systems with Application* **37** (2010), 92–112.

Exact algorithms for dedicated machine problems

20. A-A. Jafari, H. Khademi-Zare, M. M. Lotfi and R. Tavakkoli-Moghaddam, A note on "minimizing makespan in three machine flow shop with deteriorating jobs". *Computers and Operations Research* **72** (2016), 93–98.
21. A-A. Jafari, H. Khademi-Zare, M. M. Lotfi and R. Tavakkoli-Moghaddam, A note on "On three-machine flow shop scheduling with deteriorating jobs". *International Journal of Production Economics* **191** (2017), 250–252.
22. W-C. Lee, Y-R. Shiau, S-K. Chen and C-C. Wu, A two-machine flowshop scheduling problem with deteriorating jobs and blocking. *International Journal of Production Economics* **124** (2010), no. 1, 188–197.
23. W-C. Lee, C-C. Wu, C-C. Wen and Y-H. Chung, A two-machine flowshop makespan scheduling problem with deteriorating jobs. *Computers and Industrial Engineering* **54** (2008), no. 4, 737–749.
24. C-T. Ng, J-B. Wang, T-C. E. Cheng and L-L. Liu, A branch-and-bound algorithm for solving a two-machine flow shop problem with deteriorating jobs. *Computers and Operations Research* **37** (2010), no. 1, 83–90.
25. Y-R. Shiau, W-C. Lee, C-C. Wu and C-M. Chang, Two-machine flowshop scheduling to minimize mean flow time under simple linear deterioration. *International Journal on Advanced Manufacturing Technology* **34** (2007), no. 7–8, 774–782.
26. J-B. Wang, C-T. Ng, T-C. E. Cheng and L-L. Liu, Minimizing total completion time in a two-machine flow shop with deteriorating jobs. *Applied Mathematics and Computation* **180** (2006), no. 1, 185–193.
27. L. Wang, L-Y. Sun, L-H. Sun and J-B. Wang, On three-machine flow shop scheduling with deteriorating jobs. *International Journal of Production Economics* **125** (2010), no. 1, 185–189.
28. J-B. Wang and M-Z. Wang, Minimizing makespan in three-machine flow shops with deteriorating jobs. *Computers and Operations Research* **40** (2013), no. 2, 547–557.
29. J-B. Wang and M-Z. Wang, Solution algorithms for the total weighted completion time minimization flow shop scheduling with decreasing linear deterioration. *International Journal of Advanced Manufacturing Technology* **67** (2013), no. 1–4, 243–253.

30. C-C. Wu and W-C. Lee, Two-machine flowshop scheduling to minimize mean flow time under linear deterioration. *International Journal of Production Economics* **103** (2006), no. 2, 572–584.

31. S-H. Yang and J-B. Wang, Minimizing total weighted completion time in a two-machine flow shop scheduling under simple linear deterioration. *Applied Mathematics and Computation* **217** (2011), no. 9, 4819–4826.

14

Approximation algorithms and schemes

Computationally intractable time-dependent scheduling problems may be solved by various algorithms. In this chapter, we present approximation algorithms and schemes for $\mathcal{NP}$-hard time-dependent scheduling problems, which construct suboptimal schedules in a step-by-step manner.

Chapter 14 is composed of three sections. In Sect. 14.1, we give some preliminary definitions. In Sect. 14.2, we focus on approximation algorithms and schemes for minimizing the $C_{\max}$ criterion. In Sect. 14.3, we present approximation algorithms and schemes for the $\sum C_j$ criterion. The chapter is completed with a list of references.

14.1 Preliminaries

$\mathcal{NP}$-hard time-dependent scheduling problems can be efficiently solved with a known accuracy either by approximation algorithms or approximation schemes (see Sects. 2.4.1 and 2.4.2). We begin this chapter with some remarks on the origins of approximation algorithms and schemes.

The first results on approximation algorithms appeared in the late 1960s: the competitive ratio of algorithm LS for problem $Pm|prec|C_{\max}$, proved by Graham [2], and the worst-case ratio of algorithm LPT for problem $Pm||C_{\max}$, proved by Graham [3]. Next, in the early 1970s, Johnson [6] conducted the worst-case analysis of several approximation algorithms for different optimization problems. The first approximation schemes were developed in the mid 1970s: a PTAS for the KP problem by Horowitz and Sahni [4], an FPTAS for the KP problem by Ibarra and Kim [5] and an FPTAS for problem $Pm||C_{\max}$ by Sahni [7]. Garey and Johnson [1] formalized the term 'FPTAS' and identified conditions (cf. Theorem 3.18) that guarantee that an FPTAS does not exist for a given optimization problem. We refer the reader to Sects. 2.4.1, 2.4.2 and 2.5.4 for more details on approximation algorithms and schemes.

Approximation schemes for time-dependent scheduling problems are constructed using three approaches. One of the approaches was introduced by

© Springer-Verlag GmbH Germany, part of Springer Nature 2020
S. Gawiejnowicz, *Models and Algorithms of Time-Dependent Scheduling*,
Monographs in Theoretical Computer Science. An EATCS Series,
https://doi.org/10.1007/978-3-662-59362-2_14

Kovalyov and Kubiak [27]. First, some auxiliary variables, a set of vectors, and some problem-specific functions of the vectors are defined. Next, one iteratively constructs a sequence of sets $Y_1, Y_2, \ldots, Y_{n-1}$ satisfying some conditions. In each iteration, set Y_j is partitioned into subsets in such a way that for any two vectors from the same subset the values of some functions defined earlier are close enough. After that, from each such subset only the solution with the minimal value of an earlier defined function is chosen and used in the next iteration, while the remaining solutions are discarded. The final solution is a vector from the last constructed set, Y_{n-1}.

The second approach, introduced by Woeginger [36], is as follows. First, a dynamic programming algorithm for the considered problem is formulated. This algorithm goes through n phases, where the kth phase of the algorithm, $1 \leqslant k \leqslant n$, generates a set X_k of states. Any state in X_k is described by a vector x, encoding a partial schedule for the problem. The sets $X_1, X_2, \ldots, X_n$ are constructed iteratively, started from an initial state X_0 and using some functions, $F_1, F_2, \ldots, F_l$. Next, as an optimal schedule we choose the set which corresponds to the minimal value of a function $G(x)$. If some conditions are satisfied (see [36, Lemma 6.1, Thm. 2.5]), then one can prove that the problem is a *DP-benevolent problem*, i.e., the above dynamic programming algorithm may be transformed into an FPTAS.

The last approach, introduced by Halman et al. [18, 19], is based on the properties of *K-approximation sets and functions*. The main idea is as follows. For any pair of integers $A \leqslant B$, let $\{A, \ldots, B\}$ denote the set of integers between A and B. Then for an arbitrary function $\phi : \{A, \ldots, B\} \to \mathbb{R}$ function $\tilde{\phi} : \{A, \ldots, B\} \to \mathbb{R}$ is said to be a *K-approximation* of function ϕ if $\phi(x) \leqslant \tilde{\phi}(x) \leqslant K\phi(x)$ for any $x = A, \ldots, B$. The application of these functions to the considered problems, together with the properties of K-approximation sets and functions (see [18, Prop. 5.1]) and dynamic programming formulations of these problems, allows us to construct efficient approximation schemes, sometimes with better running times than the schemes constructed by the two previously mentioned approaches.

Throughout the chapter we apply the same notation as previously, only adding when necessary some new symbols such as L, the logarithm of the maximal value in input, defined separately for each problem.

14.2 Minimizing the maximum completion time

In this section, we present approximation algorithms and schemes for the maximum completion time criterion, $C_{\max}$.

14.2.1 Proportional deterioration

In this section, we focus on approximation algorithms and schemes for time-dependent scheduling problems with proportionally deteriorating jobs.

Single machine problems

Li et al. [29] proposed an FPTAS for a parallel-batch scheduling problem, $1|p\text{-}batch, p_j = b_j t, r_j, b < n|C_{\max}$, where $p\text{-}batch$ means that the total processing time of jobs assigned to a batch is equal to the maximal processing time among the jobs. This FPTAS runs in $O(n^{2m-1}b^m T^{m-1}\epsilon^{-(2m-2)})$ time, where b is the maximal batch capacity and

$$T = \left(\sum_{j=1}^{n} \max\left\{\log(1 + b_j), 1\right\}\right)\log(1 + b_1).$$

Remark 14.1. Time-dependent batch scheduling problems are also considered in the context of two-agent scheduling (cf. Sect. 21.2).

Remark 14.2. In this chapter, we consider approximation schemes only for time-dependent scheduling problems without additional constraints on machines, such as machine non-availability periods. Approximation schemes for other time-dependent scheduling problems are considered in Chap. 21.

Parallel machine problems

There exist various approaches which allow us to construct polynomial-time approximation algorithms for intractable time-dependent scheduling problems. For example, we can appropriately adapt the classical scheduling algorithms for scheduling problems with fixed job processing times. Several authors analyzed the performance of the algorithms, adapted to the needs of parallel machine time-dependent scheduling with proportional jobs by replacement of the fixed job processing times p_j with job deterioration rates b_j, $1 \leqslant j \leqslant n$. We will show on examples that so modified algorithms usually do not behave very well for time-dependent scheduling problems.

First, we consider time-dependent counterpart $\overline{LS}$ of online algorithm LS proposed by Graham [14] for parallel-machine scheduling with fixed job processing times. In both of these algorithms, $\overline{LS}$ and LS, the first available job is assigned to the first available machine.

Remark 14.3. In time-dependent scheduling, algorithm $\overline{LS}$ is applied not only as an approximation algorithm but also as a heuristic algorithm (cf. Chap. 16), in matrix methods (cf. Chap. 19) and in scheduling on machines with limited availability (cf. Chap. 21).

Let $f(A)$ and $f(OPT)$ denote the value of criterion f for a schedule generated by algorithm A and the value of f for an optimal schedule, respectively.

Cheng and Sun [9] analyzed the performance of algorithm $\overline{LS}$ for problem $P2|p_j = b_j t|C_{\max}$ with bounded job deterioration rates.

Theorem 14.4. (Cheng and Sun [9]) (a) If $t_0 = 1$ and $b_j \in (0,1]$, then for algorithm $\overline{LS}$ applied to problem $P2|p_j = b_j t|C_{\max}$ the inequality

$$\frac{C_{\max}(\overline{LS})}{C_{\max}(OPT)} \leqslant \sqrt{2}$$

holds. (b) If $t_0 = 1$ and $b_j \in (0, \alpha]$, where $0 < \alpha \leqslant 1$, then for algorithm $\overline{LS}$ applied to problem $Pm|p_j = b_j t|C_{\max}$ the inequality $\frac{C_{\max}(\overline{LS})}{C_{\max}(OPT)} \leqslant 2^{\frac{m-1}{m}}$ holds.

Proof. (a) By direct computation; see [9, Theorem 1]. (b) The result was given without proof; see [9, Theorem 2].   ◇

Cheng et al. [10] presented Theorem 14.4 (b) in a slightly another form.

Theorem 14.5. (Cheng et al. [10]) If $t_0 = 1$, then for algorithm $\overline{LS}$ applied to problem $Pm|p_j = b_j t|C_{\max}$ the inequality

$$\frac{\log C_{\max}(\overline{LS})}{\log C_{\max}(OPT)} \leqslant 2 - \frac{1}{m}$$

holds.

Proof. By direct computation; see [10, Theorem 1].   ◇

Cheng and Sun [11] analyzed the performance of a semi-online version of algorithm $\overline{LS}$. Let $b_{\max}$ be defined as in Sect. 6.2.2.

Theorem 14.6. (Cheng and Sun [11]) If $t_0 = 1$ and if only the maximum deterioration rate $b_{\max}$ is known, then the semi-online version of algorithm $\overline{LS}$ applied to problem $Pm|p_j = b_j t|C_{\max}$ is $(1 + b_{\max})^{\frac{m-1}{m}}$-competitive.

Remark 14.7. Liu et al. [31] gave a counterpart of Theorem 14.6, where all job deterioration rates are known.

Yu et al. [37] presented a generalization of Theorem 14.6 for the case $r_j \neq 0$.

Theorem 14.8. (Yu et al. [37]) (a) For problem $P2|r_j, p_j = b_j t|C_{\max}$ no on-line algorithm is better than $(1 + b_{\max})$-competitive. (b) For problem $Pm|r_j, p_j = b_j t|C_{\max}$ algorithm $\overline{LS}$ is $(1 + b_{\max})^{2(\frac{m-1}{m})}$-competitive.

Proof. (a)(b) By calculation of the lower bound on the value of criterion $C_{\max}$ for an optimal schedule and the upper bound on the value of criterion $C_{\max}$ for schedule generated by an online algorithm; see [37, Sect. 3] for details. ◇

Theorems 14.4 and 14.5 suggest that for time-dependent scheduling problems algorithm $\overline{LS}$ can produce schedules which are arbitrarily bad. We show by example that this conjecture is true.

For a given instance I, let $C_{\max}^A(I)$ and $C_{\max}^*(I)$ denote the maximum completion time of the schedule generated by an algorithm A and the maximum completion time of the optimal schedule, respectively.

Example 14.9. (Gawiejnowicz [13]) Let us consider the following instance I of problem $P2|p_j = b_j t|C_{\max}$. Let $p_1 = p_2 = Kt, p_3 = K^2 t$ for some constant $K > 0$. Let both machines start at time $t_0 > 0$. Then $R^a_{\overline{LS}}(I) = \frac{C^{\overline{LS}}_{\max}(I)}{C^\star_{\max}(I)} = \frac{K^2+1}{K+1} \to \infty$ as $K \to \infty$.

Remark 14.10. The ratio $R^a_A(I)$ for an algorithm A and a given instance I was introduced in Remark 2.26; see also Definition 2.25.

The next algorithm is an adaptation of offline algorithm LPT proposed by Graham [15] for scheduling jobs with fixed processing times. Algorithm LPT consists in assigning the first available job with the largest processing time to the first available machine. In the adapted algorithm, proposed by Mosheiov [33], job processing times are replaced by job deterioration rates.

Algorithm 14.1. for problem $Pm|p_j = b_j t|C_{\max}$

1 **Input** : sequence $(b_1, b_2, \ldots, b_n)$
2 **Output:** a suboptimal schedule
   ▷ Step 1
3 Arrange jobs in non-increasing order of the deterioration rates b_j;
   ▷ Step 2
4 **for** $i \leftarrow 1$ **to** n **do**
5 $\quad$| $\quad$ Assign job $J_{[i]}$ to the machine with the smallest machine load;
   end
6 **return.**

Remark 14.11. In the time-dependent scheduling literature Algorithm 14.1 is called *LDR* (*Largest Deterioration Rate first*) or $\overline{LPT}$ (see Chap. 19).

Remark 14.12. We will denote ratios related to a given algorithm using symbol A followed by the number of the pseudo-code of the algorithm. For example, ratio $R^a_A(I)$ for Algorithm 14.1 and instance I will be denoted as $R^a_{A14.1}(I)$.

Algorithm 14.1 can also produce arbitrarily bad schedules, as shown below.

Example 14.13. (Mosheiov [33]) Consider the following instance I of problem $P2|p_j = b_j t|C_{\max} : n = 5$, $b_1 = b_2 = K^{\frac{1}{2}} - 1$, $b_3 = b_4 = b_5 = K^{\frac{1}{3}} - 1$ for some constant $K > 1$. Then $R^a_{A14.1}(I) = \frac{C^{A14.1}_{\max}}{C^\star_{\max}} = \frac{K^{\frac{7}{6}}}{K} = K^{\frac{1}{6}} \to \infty$ as $K \to \infty$.

Examples 14.9 and 14.13 show that well-known scheduling algorithms for jobs with fixed processing times are a risky choice for time-dependent scheduling problems with unbounded deterioration. Hsieh and Bricker [20] proved several results concerning bounded job deterioration rates.

Lemma 14.14. (Hsieh and Bricker [20]) *Let I be an arbitrary instance of problem $Pm|p_j = b_j t|C_{\max}$. Then*

$$C_{\max}^{A14.1}(I \setminus \{J_n\}) \leqslant \left(\prod_{j=1}^{n-1}(1 + b_j) \right)^{\frac{1}{m}}.$$

Proof. See [20, Proposition 1 (b)].   ◇

Theorem 14.15. (Hsieh and Bricker [20]) *Let I be an arbitrary instance of problem $Pm|p_j = b_j t|C_{\max}$. If $b_j \in (0,1)$ for $1 \leqslant j \leqslant n$ and job J_n is assigned to the machine whose maximum completion time determines the overall maximum completion time, then*

$$R_{A14.1}^a(I) \leqslant (1 + b_n)^{1 - \frac{1}{m}}.$$

Proof. Let I be an arbitrary instance of problem $Pm|p_j = b_j t|C_{\max}$. Then, by Lemma 13.35, we have $C_{\max}^{\star}(I) \geqslant \left(\prod_{j=1}^{n}(1 + b_j) \right)^{\frac{1}{m}}$. Next, the equality $C_{\max}^{A14.1}(I) = C_{\max}^{A14.1}(I \setminus \{J_n\})(1 + b_n)$ holds. Finally, by Lemma 14.14,

$$C_{\max}^{A14.1}(I \setminus \{J_n\}) \leqslant \left(\prod_{j=1}^{n-1}(1 + b_j) \right)^{\frac{1}{m}}.$$

Therefore, $R_{A14.1}^a(I)$ is not greater than

$$\frac{C_{\max}^{A14.1}(I \setminus \{J_n\})(1 + b_n)}{\left(\prod_{j=1}^{n}(1 + b_j) \right)^{\frac{1}{m}}} \leqslant \frac{\left(\prod_{j=1}^{n-1}(1 + b_j) \right)^{\frac{1}{m}}(1 + b_n)}{\left(\prod_{j=1}^{n}(1 + b_j) \right)^{\frac{1}{m}}} = (1 + b_n)^{1 - \frac{1}{m}}.$$

∎

Theorem 14.16. (Hsieh and Bricker [20]) *Let I be an arbitrary instance of problem $Pm|p_j = b_j t|C_{\max}$. If $b_j \in (0,1)$ for $1 \leqslant j \leqslant n$, and if job J_k, $1 < k < n$, is assigned to the machine whose maximum completion time determines the overall maximum completion time, then*

$$R_{A14.1}^a(I) \leqslant (1 + b_k)^{1 - \frac{1}{m}}(1 + b_n)^{-\frac{n-k}{m}}.$$

Proof. Let I be an arbitrary instance of problem $Pm|p_j = b_j t|C_{\max}$. Then, the equality $C_{\max}^{A14.1}(I) = C_{\max}^{A14.1}(I \setminus \{J_k\})(1 + b_k)$ holds. Next, by Lemma 14.14, we have $C_{\max}^{A14.1}(I \setminus \{J_n\}) \leqslant \left(\prod_{j=1}^{k-1}(1 + b_j) \right)^{\frac{1}{m}}$. Hence

$$R_{A14.1}^a(I) \leqslant \frac{C_{\max}^{A14.1}(I \setminus \{J_k\})(1 + b_k)}{\left(\prod_{j=1}^{n}(1 + b_j) \right)^{\frac{1}{m}}} \leqslant \frac{\left(\prod_{j=1}^{k-1}(1 + b_j) \right)^{\frac{1}{m}}(1 + b_k)}{\left(\prod_{j=1}^{n}(1 + b_j) \right)^{\frac{1}{m}}} =$$

$$(1+b_k)^{1-\frac{1}{m}} \left(\prod_{k+1}^{n} (1+b_j) \right)^{-\frac{1}{m}} \leqslant (1+b_k)^{1-\frac{1}{m}}(1+b_n)^{-\frac{n-k}{m}}.$$

∎

Remark 14.17. In the formulation of Theorem 14.16 we can assume $1 < k \leqslant n$ instead of $1 < k < n$. The new formulation covers formulations of both Theorem 14.15 and Theorem 14.16.

Remark 14.18. From Theorem 14.16 it follows that if the deterioration rates b_j are uniformly distributed in the $(0,1)$ interval, then $\lim_{n\to\infty} \frac{C_{\max}^{A14.1}}{C_{\max}^{\star}} \overset{=}{=} 1$, i.e., Algorithm 14.1 is asymptotically optimal.

Hsieh and Bricker [20, Sects. 3–5] conducted a computational experiment in which instances with up to $n = 500$ jobs were solved by Algorithm 14.1. The results of the experiment confirmed the observation from Remark 14.18.

Several authors used the approach introduced by Kovalyov and Kubiak [27] to propose approximation schemes for parallel machine time-dependent scheduling problems with proportional jobs.

Ren and Kang [34] and Ji and Cheng [23] proposed an FPTAS for problem $Pm|p_j = b_j t|C_{\max}$, running in $O(n^{2m+1}L^{m+1}\epsilon^{-m})$ time, where

$$L = \log \max \left\{ n, \frac{1}{\epsilon}, \max_{1 \leqslant j \leqslant n} \{1 + b_j\}, t_0 \right\}, \tag{14.1}$$

with t_0 defined as in Sect. 6.2.2. Miao et al. [32] adapted the approximation scheme to an FPTAS for problem $Pm|batch, r_j, p_j = b_j t|C_{\max}$.

Dedicated machine problems

Li [28] proposed an FPTAS for a two-machine time-dependent open shop problem, where one of the machines is a non-bottleneck machine. (A machine is a *non-bottleneck* machine if an arbitrary number of jobs can be processed on that machine simultaneously.)

Remark 14.19. Non-bottleneck machines will be denoted in the second field of the three-field notation by symbol 'NB'.

Li [28] proposed an FPTAS for a special case of this problem, $O2|NB, p_{1j} = a_j t, p_{2j} = b_j t|C_{\max}$, based on a pseudo-polynomial algorithm and running in $O(n^2 \sum_{j=1}^{n} \log(1+b_j)\epsilon^{-1})$ time. The result is the counterpart of a corresponding result for a two-machine open shop problem with a non-bottleneck machine and fixed processing times (see Strusevich and Hall [35] for details).

14.2.2 Linear deterioration

In this section, we consider approximation algorithms and schemes for time-dependent scheduling problems with linearly deteriorating jobs.

Parallel machine problems

Kang and Ng [25] proposed an FPTAS for problem $Pm|p_j = a_j + b_j t|C_{max}$. This scheme, based on the same idea as the scheme proposed by Kovalyov and Kubiak [27], is a counterpart of the scheme proposed by Kang et al. [24] for problem $Pm|p_j = a_j - b_j t|C_{max}$ (cf. Sect. 14.2.4) and runs in $O(n^{2m+1}L^{m+1}\epsilon^{-m})$ time, where

$$L := \log \max \left\{ n, \frac{1}{\epsilon}, a_{max}, \max_{1 \leq j \leq n} \{1 + b_j\} \right\},$$

with a_{max} defined as in Sect. 11.1.3. We refer the reader to [25, Sect. 2] for more details on this FPTAS.

Before we describe two next FPTASes, we introduce new notation. Let us define, for $i \in \{1, 2, \ldots, m\}$ and $j \in \{1, 2, \ldots, n\}$, numbers $\Delta_1 := \min\{\frac{a_j}{s_i}\}$ and $\Delta_2 := \min\{1 + \frac{b_j}{s_i}\}$. Let $k_1, k_2 \in \mathbb{N}$ be arbitrary natural numbers such that $10^{k_1}\Delta_1$ and $10^{k_2}\Delta_2$ are natural numbers. Finally, let $A_{max} := \max\{\frac{a_j}{s_i}\}$, $B_{max} := \max\{\frac{b_j}{s_i}\}$, where s_i denotes the speed of machine M_i, $1 \leq i \leq m$.

Liu et al. [30] adapted the FPTAS by Kang et al. [24] to an FPTAS for problem $Qm|p_j = a_j + b_j t|C_{max}$, running in $O(n^{2(m+1)+1}\log n L^{m+1}\epsilon^{-m})$ time, where

$$L := \log B_{max} + \log A_{max} + k_2 \log 10.$$

Zou et al. [38], applying the same approach, proposed an FPTAS for problem $Qm|p_j = a_j + bt|\sum C_j$, running in $O(n^{2m+1}L^{m+1}\epsilon^{-m})$, where

$$L := \log \max \left\{ n, \frac{1}{\epsilon}, A_{max}, 1 + B_{max}, 10^{k_1}, 10^{k_2} \right\}.$$

14.2.3 General non-linear deterioration

In this section, we consider approximation algorithms and schemes for time-dependent scheduling problems with general non-linearly deteriorating job processing times.

Single machine problems

Cai et al. [8] proposed an FPTAS for the single machine problem $1|p_j = a_j + b_j \max\{t - d_0, 0\}|C_{max}$, where $d_0 > 0$ is the time after which job processing times start to deteriorate. The scheme is based on the observation that finding an optimal schedule is equivalent to finding a set of jobs which are completed before d_0 and the first job, J_{d_0}, that is completed after d_0. (Let us notice that by Theorem 7.41, all jobs which start after time $t = d_0$ should be arranged in non-decreasing order of the ratios $\frac{a_j}{b_j}$.) Iteratively constructing, for each possible choice of J_{d_0}, a polynomial number of sets of schedules which differ by the factor $1 + \frac{\epsilon}{2n}$, and choosing from each set the schedule with the minimal value of a certain function, we obtain an approximate schedule differing from the optimal one by the factor $1 + \epsilon$. For a given $\epsilon > 0$ and for n jobs, the FPTAS by Cai et al. [8] runs in $O(n^6 L^2 \epsilon^{-2})$ time, where

$$L := \log \max \left\{ n, d_0, \frac{1}{\epsilon}, a_{\max}, b_{\max} \right\},$$

with $a_{\max}$ and $b_{\max}$ defined as in Sects. 11.1.3 and 6.2.2, respectively. We refer the reader to [8, Sect. 5] for more details on this FPTAS.

The latter FPTAS was improved by Halman [16, 17] by a factor of $O(\frac{n^2}{L})$. The improved scheme was constructed using the technique of K-approximation sets and functions (cf. Halman et al. [19]) and runs in $O(n^4 L^3 \log \frac{nL}{\epsilon} \epsilon^{-2})$ time.

Another FPTAS, for the single-machine problem with job processing times given by (6.31) and the maximum completion time criterion, was proposed by Kovalyov and Kubiak [27]. The main idea of the scheme is as follows.

Let x_j, for $1 \leqslant j \leqslant n$, be a 0-1 variable such that $x_j := 0$ if job J_j is early, and $x_j := 1$ if job J_j is tardy or suspended (see Sect. 7.1.5 for definitions of early, tardy and suspended jobs). Let X be the set of all 0-1 vectors $x = [x_1, x_2, \ldots, x_{n-1}]$. Let us define functions F_j, G_j and P_j as follows:

$$\begin{aligned} &F_0(x) := s, \; G_0(x) := s - d, \; P_0(x) := 0; \\ &F_j(x) := F_{j-1}(x) + x_j(a_j + b_j G_{j-1}(x)); \\ &G_j(x) := \min\{F_j(x), D\} - d \\ &\text{and} \\ &P_j(x) := \sum_{i=1}^{j} a_j(1 - x_i), \end{aligned}$$

where $x \in X$, $1 \leqslant j \leqslant n-1$, and s is the starting time of the earliest tardy job. Applying these functions, the FPTAS by Kubiak and Kovalyov [27] iteratively constructs a sequence of sets $Y_1, Y_2, \ldots, Y_{n-1}$, where

$$Y_j \subseteq X_j := \{x \in X : x_i = 0, j + 1 \leqslant i \leqslant n - 1\}$$

for $1 \leqslant j \leqslant n-1$. In each iteration of the algorithm, set Y_j is partitioned into subsets in such a way that for any two vectors from the same subset the values of functions F_j and G_j are close enough. Next, from each such subset, only the solution with the minimal value of function P_j is chosen and used in the next iteration, while all remaining solutions are discarded. The final solution is a vector $x^\circ \in Y_{n-1}$ such that $F_{n-1}(x^\circ) = \min\{F_{n-1}(x) : x \in Y_{n-1}\}$. This FPTAS runs in $O(n^5 L^4 \epsilon^{-2})$, where

$$L := \log \max \left\{ n, D, \frac{1}{\epsilon}, a_{\max}, b_{\max} \right\},$$

with $a_{\max}$ and $b_{\max}$ defined as in Sects. 11.1.3 and 6.2.2, respectively. We refer the reader to [27, Sects. 2–3] for more details.

14.2.4 Linear shortening

In this section, we consider approximation algorithms and schemes for time-dependent scheduling problems with linearly shortening job processing times.

Parallel machine problems

Kang et al. [24] proposed an FPTAS for problem $Pm|p_j = a_j - b_j t|C_{\max}$, where coefficients a_j and b_j satisfy inequalities (6.44) and (6.45). This approximation scheme is based on the scheme proposed by Kovalyov and Kubiak [27]. The main idea is as follows.

By Theorem 7.49 (c), we can assume that all jobs have been indexed in such a way that $\frac{a_1}{b_1} \geqslant \frac{a_2}{b_2} \geqslant \ldots \frac{a_n}{b_n}$. Let x_j, for $1 \leqslant j \leqslant n$, be a variable such that $x_j := k$ if job J_j is executed on machine M_k, where $k \in \{1, 2, \ldots, m\}$. Let X be the set of all vectors $x = [x_1, x_2, \ldots, x_n]$ such that $x_j := k$, $1 \leqslant j \leqslant n$, $1 \leqslant k \leqslant m$. Let us define functions F_j and Q as follows:

$$
\begin{aligned}
&F_0^i(x) := 0 \text{ for } 1 \leqslant i \leqslant m; \\
&F_j^k(x) := F_{j-1}^k(x) + a_j - b_j F_{j-1}^k(x) \text{ for } x_j = k; \\
&F_j^i := F_{j-1}^i(x) \text{ for } x_j = k, i \neq k \\
&\text{and} \\
&Q(x) := \max_{1 \leqslant j \leqslant m} \left\{ F_n^j(x) \right\}.
\end{aligned}
$$

Starting from the set $Y_0 := \{(0, 0, \ldots, 0)\}$, the FPTAS iteratively constructs a sequence of sets $Y_1, Y_2, \ldots, Y_n$, where Y_j, $1 \leqslant j \leqslant n$, is obtained from Y_{j-1} by adding k, $k = 1, 2, \ldots, m$, in the jth position of each vector in Y_{j-1} and by applying to all obtained vectors the functions $F_j^k(x)$ and $F_j^i(x)$. Next, Y_j is partitioned into subsets in such a way that any two solutions in the same subset are close enough. From each such subset only the solution with the minimal value of a certain function is chosen as the subset's representative for the next iteration (all remaining solutions are discarded). The final solution is a vector $x^\circ \in Y_n$ such that $Q(x^\circ) = \min\{Q(x) : x \in Y_n\}$. This scheme runs in $O(n^{m+1} L^{m+1} \epsilon^{-m})$ time, where

$$
L := \log \max \left\{ n, \frac{1}{\epsilon}, a_{\max} \right\}, \tag{14.2}
$$

with $a_{\max}$ defined as in Sect. 11.1.3. We refer the reader to [24, Sect. 3] for more details on this FPTAS.

Remark 14.20. Based on the scheme by Kovalyov and Kubiak [27], Kang et al. [24] also proposed an FPTAS for problem $P2|p_j = a_j - b_j t|C_{\max}$, running in $O(n^3 L^3 \epsilon^{-2})$ time, where L is defined by (14.2). We refer the reader to [24, Sect. 2] for details.

14.2.5 General non-linear shortening

In this section, we consider approximation algorithms and schemes for time-dependent scheduling problems with general non-linearly shortening jobs.

Single machine problems

Cheng et al. [12], applying the approach introduced by Woeginger [36], proved the existence of an FPTAS for problem $1|p_j \in \{a_j, a_j - b_j : 0 \leqslant b_j \leqslant a_j\}|C_{\max}$. The proof of this result is based on Lemma 10.24 which states that the problem under consideration is a variation of the KP problem. Since for the latter problem there exists an FPTAS (see, e.g., Kellerer et al. [26]), the running time of the FPTAS for problem $1|p_j \in \{a_j, a_j - b_j : 0 \leqslant b_j \leqslant a_j\}|C_{\max}$ depends on the running time of the FPTAS for the KP problem. We refer the reader to [12, Theorem 3] for more details.

An FPTAS for problem $1|p_j = a_j - b_j \min\{t, D\}|C_{\max}$, where $D > 0$ is the common due date and $0 < b_j \leqslant \frac{a_j}{2D}$ for $1 \leqslant j \leqslant n$, was proposed by Ji and Cheng [21]. This scheme is a special case of an FPTAS proposed by the authors (cf. [21]) for problem $Pm|p_j = a_j - b_j \min\{t, D\}|C_{\max}$ (see below for details) and runs in $O(n^3 L^3 \epsilon^{-2})$ time, where L is defined as in (14.2). We refer the reader to [21, Sect. 2] for more details on this FPTAS.

Cheng et al. [12] applied to problem $1|p_j \in \{a_j, a_j - b_j : 0 \leqslant b_j \leqslant a_j\}|C_{\max}$ algorithm $\overline{LS}$ (cf. Sect. 14.2.1), scheduling every new job J_{j+1} after jobs $J_1, J_2, \ldots, J_j$ without any idle time, $1 \leqslant j \leqslant n - 1$.

Theorem 14.21. (Cheng et al. [12]) *Algorithm $\overline{LS}$ always generates a schedule for problem $1|p_j \in \{a_j, a_j - b_j\} : 0 \leqslant b_j \leqslant a_j\}|C_{\max}$ such that its maximum completion time is at most twice the optimal offline maximum completion time.*

Proof. We start with the following three observations. First, let us notice that if $D > \sum_{j=1}^{n} a_j$, the problem is trivial: we can schedule all jobs in an arbitrary order before time D. Hence without loss of generality we can assume that $D \leqslant \sum_{j=1}^{n} a_j$. Second, from the assumption that $D \leqslant \sum_{j=1}^{n} a_j$ it follows that there exists a unique index k such that $\sum_{j=1}^{k-1} a_j < D \leqslant \sum_{j=1}^{k} a_j$. Third, algorithm $\overline{LS}$ schedules the jobs with indices from set N_1 (N_2) before (after) time D, where $N_1 := \{1, 2, \ldots, k-1\}$ and $N_2 := \{k+1, k+2, \ldots, n\}$.

Let us now consider job J_k. Two cases are possible: Case 1, when J_k is executed in the interval $\langle D, D + a_k - b_k \rangle$ and Case 2, when J_k is executed in the interval $\langle \sum_{j \in N_1} a_j, \sum_{j \in N_1} a_j + a_k \rangle$.

In Case 1, the maximum online completion time is equal to

$$D + \sum_{j \in N_2 \cup \{k\}} (a_j - b_j) \leqslant D + \sum_{j=1}^{n} (a_j - b_j).$$

In Case 2, the maximum online completion time is equal to

$$\sum_{j \in N_1} a_j + a_k + \sum_{j \in N_2}(a_j - b_j) \leqslant D - b_k + a_k + \sum_{j \in N_2}(a_j - b_j)$$
$$\leqslant D + \sum_{j=1}^{n}(a_j - b_j).$$

Since the maximum offline completion time is not less than D and not less than $\sum_{j=1}^{n}(a_j - b_j)$, the result follows.   □

Ji et al. [22] proposed the following offline version of algorithm $\overline{\text{LS}}$.

Algorithm 14.2. for problem $1|p_j \in \{a_j, a_j - b_j : 0 \leqslant b_j \leqslant a_j\}|C_{\max}$

---

1 **Input** : sequences $(a_1, a_2, \ldots, a_n), (b_1, b_2, \ldots, b_n)$, numbers y, Y
2 **Output:** a suboptimal schedule
   ▷ Step 1
3 Arrange jobs in non-decreasing order of the ratios $\frac{b_j}{a_j}$;
4 Find $k := \min\{j : \sum_{i=1}^{j} a_i > D\}$;
   ▷ Step 2
5 Schedule jobs $J_1, J_2, \ldots, J_{k-1}$ before time D;
   ▷ Step 3
6 **if** $(C_{k-1} > D - b_k)$ **then**
7 $\quad$| $\quad$Schedule jobs $J_k, J_{k+1}, \ldots, J_n$ starting from time D;
   else
8 $\quad$| $\quad$Schedule jobs $J_k, J_{k+1}, \ldots, J_n$ starting from time C_{k-1};
   end
   ▷ Step 4
9 $\sigma \leftarrow (1, 2, \ldots, k, k+1, \ldots, n)$;
10 **return** σ.

Theorem 14.22. (Ji et al. [22]) *Let I be an arbitrary instance of problem $1|p_j \in \{a_j, a_j - b_j : 0 \leqslant b_j \leqslant a_j\}|C_{\max}$. Then the inequality $R_{A14.2}^a \leqslant \frac{5}{4}$ holds.*

Proof. See [22, Theorem 2].   ◇

Parallel machine problems

An FPTAS for problem $Pm|p_j = a_j - b_j \min\{t, D\}|C_{\max}$, where $D > 0$ is the common due date and $0 < b_j \leqslant \frac{a_j}{2D}$ for $1 \leqslant j \leqslant n$, was proposed by Ji and Cheng [21]. This scheme is based on the scheme proposed by Kovalyov and Kubiak [27] and runs in $O(n^{2m+1}L^{2m+1}\epsilon^{-2m})$ time, where where L is defined as in (14.2); see [21, Sect. 3] for more details.

14.3 Minimizing the total completion time

In this section, we present approximation algorithms and schemes for time-dependent scheduling problems with proportional jobs and the $\sum C_j$ criterion.

Single machine problems

Some authors analyzed the performance of online algorithms for single machine time-dependent scheduling problems with proportionally deteriorating jobs and non-zero ready times. Let $b_{\max}$ be defined as in Sect. 6.2.2.

Theorem 14.23. (Liu et al. [42]) *Any online algorithm for problem* $1|r_j, p_j = b_j t| \sum C_j$ *is at least* $(1 + b_{\max})$-*competitive.*

Proof. We construct an instance of problem $1|r_j, p_j = b_j t| \sum C_j$ and evaluate the total completion time of a schedule generated by an arbitrary online algorithm for the problem and the total completion time of an optimal offline schedule for the instance; see [42, Theorem 4] for details.   ◇

Parallel machine problems

Since problem $1|p_j = b_j t| \sum C_j$ is solvable by algorithm SDR in $O(n \log n)$ time (cf. Theorem 7.61), a natural question is how algorithm SDR will perform for $m \geqslant 2$ machines. Mosheiov [43] applied this algorithm to problem $P2|p_j = b_j t| \sum C_j$ (see Algorithm 14.3).

Algorithm 14.3. for problem $P2|p_j = b_j t| \sum C_j$

1 **Input**   : sequence $(b_1, b_2, \ldots, b_n)$
2 **Output:** a suboptimal schedule
   ▷ Step 1
3 Arrange jobs in non-increasing order of the deterioration rates b_j;
4 $t_{M_1} \leftarrow t_0;\ \sigma^1 \leftarrow (\phi);\ t_{M_2} \leftarrow t_0;\ \sigma^2 \leftarrow (\phi);\ N_{\mathcal{J}} \leftarrow \{1, 2, \ldots, n\}$;
   ▷ Step 2
5 **while** $(N_{\mathcal{J}} \neq \emptyset)$ **do**
6 $\quad$ $k \leftarrow$ the index of the job with the smallest deterioration rate b_j;
7 $\quad$ **if** $(t_{M_1} \leqslant t_{M_2})$ **then**
8 $\quad\quad$ $\sigma^1 \leftarrow (\sigma^1|k)$;
9 $\quad\quad$ $t_{M_1} \leftarrow (1 + b_k)t_{M_1}$;
   $\quad$ **else**
10 $\quad\quad$ $\sigma^2 \leftarrow (\sigma^2|k)$;
11 $\quad\quad$ $t_{M_2} \leftarrow (1 + b_k)t_{M_2}$;
   $\quad$ **end**
12 $\quad$ $N_{\mathcal{J}} \leftarrow N_{\mathcal{J}} \setminus \{k\}$;
   end
   ▷ Step 3
13 $\sigma \leftarrow (\sigma^1, \sigma^2)$;
14 **return** σ.

The following example shows that in the case of unbounded deterioration rates Algorithm 14.3 can produce arbitrarily bad schedules.

Example 14.24. (Chen [39]) Consider the following instance I of problem $P2|p_j = b_j t| \sum C_j : m = 2, n = 3, S_1 = S_2 = 1, p_1 = p_2 = (B-1)t,$ $p_3 = (B^2 - 1)t$, where $B > 0$ is a constant. Then $R^a_{A14.3}(I) = \frac{\sum C_j^{A14.3}(I)}{\sum C_j^\star(I)} = \frac{B^3 + 2B}{2B^2 + B} \to \infty$ as $B \to \infty$.

Chen [39] showed that even for the two-machine case the absolute ratio of Algorithm 14.3 is unbounded.

Theorem 14.25. (Chen [39]) *Let I be an arbitrary instance of problem $P2|p_j = b_j t| \sum C_j$. Then*

$$R^a_{A14.3}(I) \leqslant \max \left\{ \frac{1 + b_n}{1 + b_1}, \frac{2}{n-1} + \frac{(1 + b_1)(1 + b_n)}{1 + b_2} \right\}.$$

Proof. First, note that without loss of generality we can assume that jobs have been re-indexed so that $b_1 \leqslant b_2 \leqslant \cdots \leqslant b_n$.

Second, if $n = 2k$ for some $k \in \mathbb{N}$, the schedule generated by Algorithm 14.3 is in the form of $(1, 3, \ldots, 2k-1)$ for machine M_1 and $(2, 4, \ldots, 2k)$ for machine M_2. The total completion time of the schedule is equal to

$$\sum_{j=1}^k \left(\prod_{i=1}^j (1 + b_{2i-1}) \right) + \sum_{j=1}^k \left(\prod_{i=1}^j (1 + b_{2i}) \right) \leqslant 2 \sum_{j=1}^k \left(\prod_{i=1}^j (1 + b_{2i}) \right). \quad (14.3)$$

Similarly, if $n = 2k + 1$ for some $k \in \mathbb{N}$, the schedule generated by Algorithm 14.3 is in the form of $(1, 3, \ldots, 2k - 1, 2k + 1)$ for machine M_1 and $(2, 4, \ldots, 2k)$ for machine M_2. The total completion time of the schedule is equal to

$$\sum_{j=0}^k \left(\prod_{i=0}^j (1 + b_{2i+1}) \right) + \sum_{j=1}^k \left(\prod_{i=1}^j (1 + b_{2i}) \right) \leqslant 2 \sum_{j=0}^k \left(\prod_{i=0}^j (1 + b_{2i}) \right). \quad (14.4)$$

Third, by direct computation (see [39, Lemma 4.2] for details) and by Lemma 1.1 (b), we have that if $n = 2k$ for some $k \in \mathbb{N}$, the minimal total completion time is not less than

$$2 \sum_{j=1}^k \left(\prod_{i=1}^j (1 + b_{2i-1}) \right). \quad (14.5)$$

Similarly, if $n = 2k + 1$ for some $k \in \mathbb{N}$, then the minimal total completion time is not less than

$$2 \sum_{j=1}^k \left(\prod_{i=1}^j (1 + b_{2i}) \right). \quad (14.6)$$

By calculating the ratio of the right sides of (14.3) and (14.5), and the ratio of the right sides of (14.4) and (14.6), the result follows. $\square$

Theorem 14.25 only concerns Algorithm 14.3, so we may expect to find a better algorithm for the considered problem, with a bounded absolute worst-case ratio. This, however, is impossible, since problem $Pm|p_j = b_j t| \sum C_j$ with an arbitrary number of machines is difficult to approximate.

Theorem 14.26. (Chen [39]) *There is no polynomial-time approximation algorithm with a constant worst-case bound for problem $Pm|p_j = b_j t| \sum C_j$ with an arbitrary number of machines, unless $\mathcal{P} = \mathcal{NP}$.*

Proof. Assuming that there is a polynomial-time approximation algorithm for problem $Pm|p_j = b_j t| \sum C_j$ with an arbitrary number of machines, we would be able to solve the strongly $\mathcal{NP}$-complete 3-P problem (cf. Section 3.2) in pseudo-polynomial time. A contradiction, since by Lemma 3.19 a strongly $\mathcal{NP}$-complete problem cannot be solved by a pseudo-polynomial algorithm, unless $\mathcal{P} = \mathcal{NP}$. □

Ji and Cheng [40] proposed an FPTAS for problem $Pm|p_j = b_j t| \sum C_j$. This scheme is based on the same idea as the scheme proposed by Kovalyov and Kubiak [41] and runs in $O(n^{2m+3} L^{m+2} \epsilon^{-(m+1)})$ time, where

$$L := \log \max \left\{ n, \frac{1}{\epsilon}, \max_{1 \leqslant j \leqslant n} \{1 + b_j\}, t_0 \right\},$$

with t_0 defined as in Sect. 6.2.2. We refer the reader to [40, Sect. 2] for more details on this FPTAS.

Remark 14.27. Several FPTASes for DP-benevolent (cf. Sect. 14.1) time-dependent scheduling problems on parallel identical or parallel uniform machines with the $\sum C_j$ and $\sum w_j C_j$ criteria, including those for problems $Pm|p_j = b_j t|C_{\max}$ and $Pm|p_j = b_j t| \sum C_j$, are described by Woeginger [44].

14.4 Other criteria

In this section, we present approximation algorithms and schemes for time-dependent scheduling problems with criteria other than $C_{\max}$ and $\sum C_j$.

14.4.1 Proportional deterioration

Yu and Wong [47] generalized Theorem 14.23 for case of the total general completion time criterion. Let $b_{\max}$ be defined as in Sect. 6.2.2.

Theorem 14.28. (Yu and Wong [47]) *No online algorithm for problem $1|r_j, p_j = b_j t| \sum C_j^k$, where $k > 0$ is a constant, is better than $(1 + b_{\max})^k$-competitive.*

Proof. Similar to the proof of Theorem 14.23; see [47, Theorem 3]. ◇

Ji and Cheng [45] proposed an FPTAS for problem $Pm|p_j = b_j t| \sum C_{\max}^{(k)}$, running in $O(n^{2m+1} L^{m+1} \epsilon^{-m})$ time, where L is defined by (14.1).

14.4.2 Proportional-linear deterioration

Ma et al. [46] proved a counterpart of Theorem 14.23 for the $\sum w_j C_j$ criterion. Let b_{max} be defined as in Sect. 6.2.2.

Theorem 14.29. (Ma et al. [46]) *Any online algorithm for problem* $1|r_j, p_j = b_j(a + bt)| \sum w_j C_j$ *is at least* $(1 + \mathrm{sgn}(a) + b_{max}b)$-*competitive.*

Proof. Similar to the proof of Theorem 14.23; see [46, Lemma 2.8].   ◇

References

Preliminaries

1. M. R. Garey and D. S. Johnson, 'Strong' NP-completeness results: motivation, examples, and implications. *Journal of the ACM* **25** (1978), no. 3, 499–508.
2. R. L. Graham, Bounds for certain multiprocessing anomalies. *Bell System Technology Journal* **45** (1966), no. 2, 1563–1581.
3. R. L. Graham, Bounds on multiprocessing timing anomalies. *SIAM Journal on Applied Mathematics* **17** (1969), no. 2, 416–429.
4. E. Horowitz and S. Sahni, Exact and approximate algorithms for scheduling nonidentical processors. *Journal of the ACM* **23** (1976), no. 2, 317–327.
5. O. Ibarra and C. E. Kim, Fast approximation algorithms for the knapsack and sum of subset problems. *Journal of the ACM* **22** (1975), no. 4, 463–468.
6. D. S. Johnson, Approximation algorithms for combinatorial problems. *Journal of Computer and Systems Sciences* **9** (1974), no. 3, 256–278.
7. S. Sahni, Algorithms for scheduling independent tasks. *Journal of the ACM* **23** (1976), no. 1, 116–127.

Minimizing the maximum completion time

8. J-Y. Cai, P. Cai and Y. Zhu, On a scheduling problem of time deteriorating jobs. *Journal of Complexity* **14** (1998), no. 2, 190–209.
9. M-B. Cheng and S-J. Sun, A heuristic MBLS algorithm for the two semi-online parallel machine scheduling problems with deterioration jobs. *Journal of Shanghai University* **11** (2007), 451–456.
10. M-B. Cheng, G-Q. Wang and L-M. He, Parallel machine scheduling problems with proportionally deteriorating jobs. *International Journal of Systems Science* **40** (2009), 53–57.
11. Y-S. Cheng and S-J. Sun, Scheduling linear deteriorating jobs with rejection on a single machine. *European Journal of Operational Research* **194** (2009), 18–27.
12. T-C. E. Cheng, Y. He, H. Hoogeveen, M. Ji and G. Woeginger, Scheduling with step-improving processing times. *Operations Research Letters* **34** (2006), no. 1, 37–40.
13. S. Gawiejnowicz, *Scheduling jobs with varying processing times*. Ph. D. dissertation, Poznań University of Technology, Poznań 1997, 102 pp. (in Polish).
14. R. L. Graham, Bounds for certain multiprocessing anomalies. *Bell System Technology Journal* **45** (1966), no. 2, 1563–1581.

15. R. L. Graham, Bounds on multiprocessing timing anomalies. *SIAM Journal on Applied Mathematics* **17** (1969), no. 2, 416–429.

16. N. Halman, FPTASes for minimizing makespan of deteriorating jobs with nonlinear processing times. *Extended Abstracts of the 2nd International Workshop on Dynamic Scheduling Problems*, 2018, 51–56.

17. N. Halman, A technical note: fully polynomial time approximation schemes for minimizing the makespan of deteriorating jobs with nonlinear processing times. *Journal of Scheduling*, 2019, `https://doi.org/10.1007/s10951-019-00616-8`.

18. N. Halman, D. Klabjan, C.-L. Li, J. Orlin and D. Simchi-Levi, Fully polynomial time approximation schemes for stochastic dynamic programs. *SIAM Journal on Discrete Mathematics* **28** (2014), no. 4, 1725–1796.

19. N. Halman, D. Klabjan, M. Mostagir, J. Orlin and D. Simchi-Levi, A fully polynomial time approximation scheme for single-item stochastic inventory control with discrete demand. *Mathematics of Operations Research* **34** (2009), no. 3, 674–685.

20. Y-C. Hsieh and D. L. Bricker, Scheduling linearly deteriorating jobs on multiple machines. *Computers and Industrial Engineering* **32** (1997), no. 4, 727–734.

21. M. Ji and T-C. E. Cheng, An FPTAS for scheduling jobs with piecewise linear decreasing processing times to minimize makespan. *Information Processing Letters* **102** (2007), no. 2–3, 41–47.

22. M. Ji, Y. He and T-C. E. Cheng, A simple linear time algorithm for scheduling with step-improving processing times. *Computers and Operations Research* **34** (2007), no. 8, 2396–2402.

23. M. Ji and T-C. E. Cheng, Parallel-machine scheduling of simple linear deteriorating jobs. *Theoretical Computer Sciences* **410** (2009), no. 38–40, 3761–3768.

24. L-Y. Kang, T-C. E. Cheng, C-T. Ng and M. Zhao, Scheduling to minimize makespan with time-dependent processing times. *Lecture Notes in Computer Science* **3827** (2005), 925–933.

25. L-Y. Kang and C-T. Ng, A note on a fully polynomial-time approximation scheme for parallel-machine scheduling with deteriorating jobs. *International Journal of Production Economics* **109** (2007), no. 1–2, 180–184.

26. H. Kellerer, U. Pferschy and D. Pisinger, *Knapsack Problems*. Berlin-Heidelberg: Springer 2004.

27. M. Y. Kovalyov and W. Kubiak, A fully polynomial approximation scheme for minimizing makespan of deteriorating jobs. *Journal of Heuristics* **3** (1998), no. 4, 287–297.

28. S-S. Li, Scheduling proportionally deteriorating jobs in two-machine open shop with a non-bottleneck machine. *Asia-Pacific Journal of Operational Research* **28** (2011), no. 5, 623–631.

29. S-S. Li, C-T. Ng, T-C.E. Cheng and J-J. Yuan, Parallel-batch scheduling of deteriorating jobs with release dates to minimize the makespan. *European Journal of Operational Research* **210** (2011), no. 3, 482–488.

30. M. Liu, F-F. Zheng, C-B. Chu and J-T. Zhang, An FPTAS for uniform machine scheduling to minimize makespan with linear deterioration. *Journal of Combinatorial Optimization* **23** (2012), no. 4, 483–492.

31. M. Liu, F-F. Zheng, Y-F. Xu and L. Wang, Heuristics for parallel machine scheduling with deterioration effect. *Lecture Notes in Computer Science* **6831** (2011), 46–51.

32. C-X. Miao, Y-Z. Zhang and Z-G. Cao, Bounded parallel-batch scheduling on single and multi machines for deteriorating jobs. *Information Processing Letters* **111** (2011), 798–803.

33. G. Mosheiov, Multi-machine scheduling with linear deterioration. *Infor* **36** (1998), no. 4, 205–214.

34. C-R. Ren and L-Y. Kang, An approximation algorithm for parallel machine scheduling with simple linear deterioration. *Journal of Shanghai University (English Edition)* **11** (2007), no. 4, 351–354.

35. V. A. Strusevich and L. A. Hall, An open shop scheduling problem with a non-bottleneck machine. *Operations Research Letters* **21** (1997), no. 1, 11–18.

36. G. J. Woeginger, When does a dynamic programming formulation guarantee the existence of a fully polynomial time approximation scheme (FPTAS)? *INFORMS Journal on Computing* **12** (2000), no. 1, 57–74.

37. S. Yu, J-T. Ojiaku, P. W-H. Wong and Y. Xu, Online makespan scheduling of linear deteriorating jobs on parallel machines. *Lecture Notes in Computer Science* **7287** (2012), 260–272.

38. J. Zou, Y-Z. Zhang and C-X. Miao, Uniform parallel-machine scheduling with time dependent processing times. *Journal of the Operations Research Society of China* **1** (2013), no. 2, 239–252.

Minimizing the total completion time

39. Z-L. Chen, Parallel machine scheduling with time dependent processing times. *Discrete Applied Mathematics* **70** (1996), no. 1, 81–93. (Erratum: *Discrete Applied Mathematics* **75** (1996), no. 1, 103.)

40. M. Ji and T-C. E. Cheng, Parallel-machine scheduling with simple linear deterioration to minimize total completion time. *European Journal of Operational Research* **188** (2008), no. 2, 342–347.

41. M. Y. Kovalyov and W. Kubiak, A fully polynomial approximation scheme for minimizing makespan of deteriorating jobs. *Journal of Heuristics* **3** (1998), no. 4, 287–297.

42. M. Liu, F-F. Zheng, S-J. Wang and J-Z. Huo, Optimal algorithms for online single machine scheduling with deteriorating jobs. *Theoretical Computer Science* **445** (2012), 75–81.

43. G. Mosheiov, Multi-machine scheduling with linear deterioration. *Infor* **36** (1998), no. 4, 205–214.

44. G. J. Woeginger, When does a dynamic programming formulation guarantee the existence of a fully polynomial time approximation scheme (FPTAS)? *INFORMS Journal on Computing* **12** (2000), no. 1, 57–74.

Other criteria

45. M. Ji and T-C. E. Cheng, Parallel-machine scheduling of simple linear deteriorating jobs. *Theoretical Computer Science* **410** (2009), no. 38–40, 3761–3768.

46. R. Ma, J-P. Tao and J-J. Yuan, Online scheduling with linear deteriorating jobs to minimize the total weighted completion time. *Applied Mathematics and Computation* **273** (2016), 570–583.

47. S. Yu and P. W-H. Wong, Online scheduling of simple linear deteriorating jobs to minimize the total general completion time. *Theoretical Computer Science* **487** (2013), 95–102.

15

Greedy algorithms based on signatures

$\mathbf{H}$euristic algorithms for intractable time-dependent scheduling problems may be constructed in many ways. In this chapter, we present two greedy time-dependent scheduling heuristic algorithms which exploit certain properties of the so-called *signatures* of job deterioration rates.

Chapter 15 is composed of five sections. In Sect. 15.1, we formulate a single machine time-dependent scheduling problem with the total completion time criterion and introduce the signatures. In Sect. 15.2, we present the basic properties of signatures. In Sect. 15.3, based on the properties of signatures, we formulate the first greedy algorithm for the problem introduced in Sect. 15.1. In Sect. 15.4, we introduce the so-called *regular* sequences. In Sect. 15.5, we formulate for the problem introduced in Sect. 15.1 the second greedy algorithm based on the properties of signatures. We also give arguments for the conjecture that the greedy algorithms find optimal schedules for regular sequences. The chapter is completed with a list of references.

15.1 Preliminaries

In this section, we give the formulation of the problem under consideration and define the notion of signatures of job deterioration rates.

15.1.1 Problem formulation

We will consider the following time-dependent scheduling problem. A set $\mathcal{J}$ of $n+1$ deteriorating jobs, $J_0, J_1, J_2, \ldots, J_n$, is to be processed on a single machine, available from time $t_0 = 0$. The job processing times are in the form of $p_j = 1 + b_j t$, where $b_j > 0$ for $0 \leqslant j \leqslant n$. The criterion of schedule optimality is the total completion time, $\sum C_j = \sum_{j=0}^{n} C_j$, where $C_0 = 1$ and $C_{j-1} + p_j(C_{j-1}) = 1 + (1 + b_j)C_{j-1}$ for $1 \leqslant j \leqslant n$. For simplicity of further presentation, let $\beta_j = 1 + b_j$ for $0 \leqslant j \leqslant n$ and $\hat{\beta} := (\beta_0, \beta_1, \beta_2, \ldots, \beta_n)$.

© Springer-Verlag GmbH Germany, part of Springer Nature 2020
S. Gawiejnowicz, *Models and Algorithms of Time-Dependent Scheduling*,
Monographs in Theoretical Computer Science. An EATCS Series,
https://doi.org/10.1007/978-3-662-59362-2_15

Though the time complexity of the problem is still unknown (see Sect. A.1), there exists a hypothesis that it is at least $\mathcal{NP}$-complete in the ordinary sense (Cheng et al. [1, Sect. 3]). Therefore, in accordance with a recommendation by Garey and Johnson [2, Chapter 4], the consideration of special cases of the problem may lead to finding polynomially solvable cases of the problem and delineating the border between its easy and hard cases. Hence, through the chapter we will assume that job deterioration rates are of a special form, e.g., they are consecutive natural numbers.

15.1.2 Definition of signatures

Lemma 15.1. (Gawiejnowicz et al. [4]) *Let* $C(\hat{\beta}) = [C_0, C_1, C_2, \ldots, C_n]$ *be the vector of job completion times in the form of* (7.9) *for a given sequence* $\hat{\beta} = (\beta_0, \beta_1, \beta_2, \ldots, \beta_n)$. *Then*

(a) $\|C(\hat{\beta})\|_1 := \sum\limits_{j=0}^{n} C_j(\hat{\beta}) = \sum\limits_{j=0}^{n} \left(1 + \sum\limits_{i=1}^{j} \prod\limits_{k=i}^{j} \beta_k\right) = \sum\limits_{j=1}^{n} \sum\limits_{i=1}^{j} \prod\limits_{k=i}^{j} \beta_k + (n+1),$

(b) $\|C(\hat{\beta})\|_\infty := \max\limits_{0 \leqslant j \leqslant n} \{C_j(\hat{\beta})\} = 1 + \sum\limits_{i=1}^{n} \prod\limits_{k=i}^{n} \beta_k.$

Proof. (a) The result follows from equality (7.9) and Definition 1.19 of the norm l_p for $p = 1$.

(b) The result follows from equality (7.9), Definition 1.19 of the norm l_p for $p = \infty$ and the fact that $\max_{0 \leqslant j \leqslant n}\{C_j(\hat{\beta})\} = C_n(\hat{\beta})$.   □

Remark 15.2. By Lemma 15.1, $\|C(\hat{\beta})\|_1 \equiv \sum C_j(\hat{\beta})$ and $\|C(\hat{\beta})\|_\infty \equiv C_{\max}(\hat{\beta})$. Hence, minimizing the norm l_1 (l_∞) is equivalent to minimizing the criterion $\sum C_j$ ($C_{\max}$); see Gawiejnowicz et al. [5] for details.

Remark 15.3. Notice that since $S_{[0]} = t_0 = 0$, the coefficient $\beta_{[0]}$ has no influence on the value of $C_j(\hat{\beta})$ for $0 \leqslant j \leqslant n$. Moreover, $C_j(\hat{\beta})$ depends on β_i in a monotone non-decreasing way for each $1 \leqslant i \leqslant j$. Therefore, given any permutation of sequence $\hat{\beta}$, the best strategy for minimizing $\sum C_j(\hat{\beta})$ is to set as $\beta_{[0]}$ the maximal element in this sequence (cf. Theorem A.2). In other words, if we start at $t_0 = 0$, the subject of our interest is the remaining n-element subsequence $\beta = (\beta_1, \beta_2, \ldots, \beta_n)$ of $\hat{\beta}$, with $\beta_{[0]}$ maximal. Hence, from now on, we will assume that $\beta_{[0]} = \max_{0 \leqslant j \leqslant n}\{\beta_j\}$ and consider mainly sequence β.

Definition 15.4. (Gawiejnowicz et al. [4]) *Let* $C(\beta) = [C_1, C_2, \ldots, C_n]$ *be the vector of job completion times* (7.9) *for a given sequence* $\beta = (\beta_1, \beta_2, \ldots, \beta_n)$. *Functions* $F(\beta)$ *and* $M(\beta)$ *are defined as follows:*

$$F(\beta) := \sum_{j=1}^{n} \sum_{i=1}^{j} \prod_{k=i}^{j} \beta_k$$

and

$$M(\beta) := 1 + \sum_{i=1}^{n} \prod_{k=i}^{n} \beta_k.$$

Remark 15.5. We will refer to the minimizing of $F(\beta)$ and $M(\beta)$ as the *F-problem* and *M-problem*, respectively. Since, by Lemma 15.1, $\|C(\hat{\beta})\|_1 = F(\beta) + (n+1)$ and $\|C(\hat{\beta})\|_\infty = M(\beta)$, the *M*-problem and *F*-problem are closely related to the problems of minimization of the $C_{\max}$ criterion (cf. Sect. 7.1.3) and $\sum C_j$ criterion (cf. Sect. 7.2.3), respectively.

Now we define the basic notion in the chapter (Gawiejnowicz et al. [3, 4]).

Definition 15.6. (Signatures of deterioration rate sequence β)
For a given sequence $\beta = (\beta_1, \beta_2, \ldots, \beta_n)$, signatures $S^-(\beta)$ and $S^+(\beta)$ of sequence β are defined as follows:

$$S^-(\beta) := M(\bar{\beta}) - M(\beta) = \sum_{i=1}^{n}\prod_{j=1}^{i}\beta_j - \sum_{i=1}^{n}\prod_{j=i}^{n}\beta_j, \qquad (15.1)$$

and

$$S^+(\beta) := M(\bar{\beta}) + M(\beta), \qquad (15.2)$$

where $\bar{\beta} := (\beta_n, \beta_{n-1}, \ldots, \beta_1)$ is the reverse permutation of elements of β.

Since the signatures (15.1) and (15.2) are essential in further considerations, we will now prove some of their properties.

15.2 Basic properties of signatures

Let us introduce the following notation. Given a sequence $\beta = (\beta_1, \ldots, \beta_n)$ and any two numbers $\alpha > 1$ and $\gamma > 1$, let $(\alpha|\beta|\gamma)$ and $(\gamma|\beta|\alpha)$ denote concatenations of α, β and γ in the indicated orders, respectively. Let $\mathcal{B} := \prod_{j=1}^{n}\beta_j$.

We start with a lemma which shows how to calculate the values of function $F(\cdot)$ for sequence β extended with the elements α and γ if we know the values of $F(\beta)$, $M(\beta)$ and $M(\bar{\beta})$.

Lemma 15.7. (Gawiejnowicz et al. [4]) *For a given sequence β and any numbers $\alpha > 1$ and $\gamma > 1$, the following equalities hold:*

$$F(\alpha|\beta|\gamma) = F(\beta) + \alpha\, M(\bar{\beta}) + \gamma\, M(\beta) + \alpha\, \mathcal{B}\, \gamma \qquad (15.3)$$

and

$$F(\gamma|\beta|\alpha) = F(\beta) + \gamma\, M(\bar{\beta}) + \alpha\, M(\beta) + \alpha\, \mathcal{B}\, \gamma. \qquad (15.4)$$

Proof. Let $\beta = (\beta_1, \ldots, \beta_n)$, $\beta_0 = \alpha > 1$, $\beta_{n+1} = \gamma > 1$. Then $F(\beta_0|\beta|\beta_{n+1}) = \sum_{j=0}^{n+1}\sum_{i=0}^{j}\beta_i\beta_{i+1}\cdots\beta_j = F(\beta) + \sum_{j=0}^{n}\beta_0\beta_1\cdots\beta_j + \sum_{i=0}^{n+1}\beta_i\beta_{i+1}\cdots\beta_{n+1} = F(\beta) + \beta_0(1 + \sum_{j=1}^{n}\beta_1\beta_2\cdots\beta_j) + \beta_{n+1}(1 + \sum_{i=1}^{n}\beta_i\beta_{i+1}\cdots\beta_n) + \beta_0\beta_1\cdots\beta_{n+1}$, and equality (15.3) follows. To prove (15.4) it is sufficient to exchange α and γ in (15.3) and to note that the last term remains unchanged.   ∎

By Lemma 15.7, we obtain general formulae concerning the difference and the sum of values of $F(\cdot)$ for sequences $(\alpha|\beta|\gamma)$ and $(\gamma|\beta|\alpha)$.

Lemma 15.8. (Gawiejnowicz et al. [4]) *For a given sequence β and any numbers $\alpha > 1$ and $\gamma > 1$, the following equalities hold:*

$$F(\alpha|\beta|\gamma) - F(\gamma|\beta|\alpha) = (\alpha - \gamma)S^-(\beta) \tag{15.5}$$

and

$$F(\alpha|\beta|\gamma) + F(\gamma|\beta|\alpha) = (\alpha + \gamma)S^+(\beta) + 2(F(\beta) + \alpha\,\mathcal{B}\,\gamma). \tag{15.6}$$

Proof. Let $\beta = (\beta_1, \ldots, \beta_n)$, $\alpha > 1$ and $\gamma > 1$ be given. Then by subtracting the left and the right sides of equalities (15.3) and (15.4), respectively, and by applying Definition 15.6, equation (15.5) follows.

Similarly, by adding the left and the right sides of equalities (15.3) and (15.4), respectively, and by applying Definition 15.6, we obtain equation (15.6). ∎

Lemma 15.8 shows the relation which holds between a signature and a change of the value of function $F(\cdot)$ if the first and the last element in sequence β have been mutually exchanged.

From identities (15.5) and (15.6) we can obtain another pair of equalities, expressed in terms of signatures $S^-(\cdot)$ and $S^+(\cdot)$.

Lemma 15.9. (Gawiejnowicz et al. [4]) *For a given sequence β and any numbers $\alpha > 1$ and $\gamma > 1$, the following equalities hold:*

$$F(\alpha|\beta|\gamma) = F(\beta) + \frac{1}{2}((\alpha + \gamma)S^+(\beta) + (\alpha - \gamma)S^-(\beta)) + \alpha\,\mathcal{B}\,\gamma \tag{15.7}$$

and

$$F(\gamma|\beta|\alpha) = F(\beta) + \frac{1}{2}((\alpha + \gamma)S^+(\beta) - (\alpha - \gamma)S^-(\beta)) + \alpha\,\mathcal{B}\,\gamma. \tag{15.8}$$

Proof. Indeed, by adding the left and the right sides of equalities (15.5) and (15.6), respectively, we obtain equality (15.7).

Similarly, by subtracting the left and the right sides of equalities (15.6) and (15.5), respectively, we obtain equality (15.8). ∎

The next result shows how to concatenate new elements α and γ with a given sequence β in order to decrease the value of function $F(\cdot)$.

Theorem 15.10. (Gawiejnowicz et al. [4]) *Let there be given a sequence β related to the F-problem and the numbers $\alpha > 1$ and $\gamma > 1$. Then the equivalence*

$$F(\alpha|\beta|\gamma) \leqslant F(\gamma|\beta|\alpha) \text{ iff } (\alpha - \gamma)S^-(\beta) \leqslant 0 \tag{15.9}$$

holds. Moreover, a similar equivalence holds if in equivalence (15.9) the symbol '$\leqslant$' is replaced with '$\geqslant$'.

Proof. The result follows from identity (15.5) in Lemma 15.8.   ∎

From Theorem 15.10 it follows that in order to decrease the value of $F(\cdot|\beta|\cdot)$ we should choose $(\alpha|\beta|\gamma)$ instead of $(\gamma|\beta|\alpha)$ when $(\alpha - \gamma)S^-(\beta) \leqslant 0$, and $(\gamma|\beta|\alpha)$ instead of $(\alpha|\beta|\gamma)$ in the opposite case. Therefore, the behaviour of function $F(\cdot)$ for such concatenations is determined by the sign of the signature $S^-(\beta)$ of the original sequence β.

In the next theorem we give a greedy strategy for solving the F-problem. This strategy is based on the behaviour of the signature $S^-(\cdot)$ only.

Theorem 15.11. (Gawiejnowicz et al. [4]) *Let* $\beta = (\beta_1, \ldots, \beta_n)$ *be a non-decreasingly ordered sequence for the* F-*problem, let* $u = (u_1, \ldots, u_{k-1})$ *be a* V-*sequence constructed from the first* $k - 1$ *elements of* β, *let* $\alpha = \beta_k > 1$ *and* $\gamma = \beta_{k+1} > 1$, *where* $1 < k < n$, *and let* $\alpha \leqslant \gamma$. *Then the implication*

$$\text{if } S^-(u) \geqslant 0, \text{ then } F(\alpha|u|\gamma) \leqslant F(\gamma|u|\alpha) \tag{15.10}$$

holds. Moreover, a similar implication holds if in implication (15.10) the symbol '$\geqslant$' is replaced by '$\leqslant$' and the symbol '$\leqslant$' is by replaced '$\geqslant$'.

Proof. Assume that the sign of the signature $S^-(u)$ is known. Then it is sufficient to note that by equivalence (15.9) the sign of the difference $F(\alpha|u|\gamma) - F(\gamma|u|\alpha)$ is determined by the sign of the difference $\alpha - \gamma$.   ∎

Theorem 15.11 indicates which one of the two sequences, $(\alpha|u|\gamma)$ or $(\gamma|u|\alpha)$, should be chosen if the sign of the signature $S^-(u)$ is known.

The next result shows a relation between signatures of sequences $(\alpha|\beta|\gamma)$ and $(\gamma|\beta|\alpha)$ and the values of function $M(\cdot)$ for sequences β and $\bar{\beta}$.

Theorem 15.12. (Gawiejnowicz et al. [4]) *For a given sequence* β *and any numbers* $\alpha > 1$ *and* $\gamma > 1$, *the following equalities hold:*

$$S^-(\alpha|\beta|\gamma) = \alpha \, M(\bar{\beta}) - \gamma \, M(\beta) \tag{15.11}$$

and

$$S^-(\gamma|\beta|\alpha) = \gamma \, M(\bar{\beta}) - \alpha \, M(\beta). \tag{15.12}$$

Proof. Let $\beta = (\beta_1, \ldots, \beta_n)$, $\beta_0 = \alpha > 1$ and $\beta_{n+1} = \gamma > 1$. Then we have $S^-(\alpha|\beta|\gamma) = \sum_{i=0}^{n+1} \beta_0\beta_1 \cdots \beta_i - \sum_{i=0}^{n+1} \beta_i \cdots \beta_n\beta_{n+1} = \beta_0(1+\sum_{i=1}^{n} \beta_1 \cdots \beta_i) - \beta_{n+1}(1+\sum_{i=1}^{n} \beta_i \cdots \beta_n)$. Since the expressions in the brackets are nothing else than $M(\bar{\beta})$ and $M(\beta)$, respectively, identity (15.11) follows.

Similarly, by exchanging α and γ in (15.11), we obtain (15.12).   ∎

From Theorem 15.12 there follow identities which determine the behaviour of subsequently calculated signatures $S^-(\cdot)$.

Theorem 15.13. (Gawiejnowicz et al. [4]) *For a given sequence β and any numbers $\alpha > 1$ and $\gamma > 1$, the following identities hold:*

$$S^-(\alpha|\beta|\gamma) + S^-(\gamma|\beta|\alpha) = (\alpha + \gamma)S^-(\beta), \tag{15.13}$$

$$S^-(\alpha|\beta|\gamma) - S^-(\gamma|\beta|\alpha) = (\alpha - \gamma)S^+(\beta) \tag{15.14}$$

and

$$
\begin{aligned}
S^-(\alpha|\beta|\gamma)^2 - S^-(\gamma|\beta|\alpha)^2 &= (\alpha^2 - \gamma^2)\,(M(\bar{\beta})^2 - M(\beta)^2) \\
&= (\alpha^2 - \gamma^2)\,S^-(\beta)\,S^+(\beta). \tag{15.15}
\end{aligned}
$$

Proof. Indeed, by adding the left and right sides of equalities (15.11) and (15.12), respectively, we obtain identity (15.13).

Similarly, by subtracting the left and right sides of equalities (15.11) and (15.12), respectively, we obtain identity (15.14).

Multiplying the left and the right sides of identities (15.13) and (15.14), respectively, we obtain identity (15.15). ∎

Remark 15.14. The above results show that we cannot determine uniquely the sign of signatures $S^-(\alpha|\beta|\gamma)$ and $S^-(\gamma|\beta|\alpha)$ in terms of the sign of the signature $S^-(\beta)$ only, even if we know that $\alpha \leqslant \gamma$ (or $\alpha \geqslant \gamma$). Indeed, if we know the sign of $F(\alpha|\beta|\gamma) - F(\gamma|\beta|\alpha)$ or, equivalently, the sign of $(\alpha - \gamma)S^-(\beta)$, then from identities (15.13), (15.14) and (15.15) it follows that for the consecutive signatures we only know the sign of $|S^-(\alpha|\beta|\gamma)| - |S^-(\gamma|\beta|\alpha)|$.

Finally, by Theorem 15.13 we can prove one more pair of identities.

Theorem 15.15. (Gawiejnowicz et al. [4])

For a given sequence β and any numbers $\alpha > 1$ and $\gamma > 1$, the following identities hold:

$$S^-(\alpha|\beta|\gamma) = \frac{1}{2}((\alpha + \gamma)S^-(\beta) + (\alpha - \gamma)S^+(\beta)) \tag{15.16}$$

and

$$S^-(\gamma|\beta|\alpha) = \frac{1}{2}((\alpha + \gamma)S^-(\beta) - (\alpha - \gamma)S^+(\beta)). \tag{15.17}$$

Proof. Indeed, by adding the left and right sides of identities (15.13) and (15.14), respectively, we obtain identity (15.16).

Similarly, by subtracting the left and right sides of identities (15.14) and (15.13), respectively, we obtain identity (15.17). ∎

Remark 15.16. Considering sequence $\bar{\beta}$ instead of β, we can formulate and prove counterparts of Lemmas 15.7–15.9 and Theorems 15.10–15.13 and 15.15. We omit the formulations of these results, since they do not introduce new insights into the problem. Let us notice only that the equality $S^-(\beta) + S^-(\bar{\beta}) = 0$ holds, i.e., the signatures $S^-(\beta)$ and $S^-(\bar{\beta})$ have opposite signs.

15.3 The first greedy algorithm

In this section, we will introduce the first greedy heuristic algorithm for problem $1|p_j = 1 + b_j t| \sum C_j$, based on the properties presented in Sect. 15.2.

15.3.1 Formulation

Let u denote a V-shaped sequence composed of the first $k \geqslant 1$ elements of sequence β, which have been ordered non-decreasingly. Let $\alpha = \beta_{k+1} > 1$ and $\gamma = \beta_{k+2} > 1$ be two consecutive elements of β, where $\alpha \leqslant \gamma$. Then we can extend sequence u either by concatenating α at the beginning of the left branch and γ at the end of the right branch of the constructed sequence, or conversely. Based on this observation and the results from Sect. 15.2, we can formulate the following algorithm.

Algorithm 15.1. for problem $1|p_j = 1 + b_j t| \sum C_j$

---

1 **Input**   : sequence $\hat{\beta} = (\beta_0, \beta_1, \ldots, \beta_n)$
2 **Output:** suboptimal schedule u
  ▷ Step 1
3 $\beta_{[0]} \leftarrow \max \{\beta_j : 0 \leqslant j \leqslant n\}$;
4 Arrange sequence $\hat{\beta} - \{\beta_{[0]}\}$ in non-decreasing order;
  ▷ Step 2
5 **if** (n is odd) **then**
6 $\quad | \quad u \leftarrow (\beta_{[1]}); i_0 \leftarrow 2$;
  else
7 $\quad | \quad u \leftarrow (\beta_{[1]}, \beta_{[2]}); i_0 \leftarrow 3$;
  end
8 **for** $i \leftarrow i_0$ **to** $n-1$ **step** 2 **do**
9 $\quad$ **if** $(S^-(u) \leqslant 0)$ **then**
10 $\quad \quad | \quad u \leftarrow (\beta_{[i+1]}|u|\beta_{[i]})$;
  $\quad$ **else**
11 $\quad \quad | \quad u \leftarrow (\beta_{[i]}|u|\beta_{[i+1]})$;
  $\quad$ **end**
  end
  ▷ Step 3
12 $u \leftarrow (\beta_{[0]}|u)$;
13 **return** u.

The greedy algorithm, starting from an initial sequence, iteratively constructs a new sequence, concatenating the previous sequence with new elements according to the sign of the signature $S^-(u)$ of this sequence.

Notice that since Step 1 runs in $O(n \log n)$ time, Step 2 is a loop running in $O(n)$ time and Step 3 runs in a constant time, the total running time of Algorithm 15.1 is $O(n \log n)$.

We illustrate the performance of Algorithm 15.1 with two examples.

Example 15.17. Let $\beta = (2,3,4,6,8,16,21)$. The optimal V-shaped sequence is $\beta^\star = (21,8,6,3,2,4,16)$, with $\sum C_j(\beta^\star) = 23,226$. Algorithm 15.1 generates the sequence $u_{A_{15.1}} = (21,8,6,2,3,4,16)$, with $\sum C_j(u_{A_{15.1}}) = 23,240$.

Other algorithms, e.g., Algorithms 16.15 and 16.16 (cf. Sect. 16.3.2), give worse results: $u_{A16.15} = (21,8,4,2,3,6,16)$, $u_{A16.16} = (21,6,3,2,4,8,16)$, with $\sum C_j(u_{A16.15}) = 23,418$ and $\sum C_j(u_{A16.16}) = 24,890$, respectively. ◆

Thus, in general, Algorithm 15.1 is not optimal. The following example shows, however, that this algorithm can be optimal for sequences of consecutive natural numbers.

Example 15.18. Let $\beta = (2,3,4,5,6,7,8)$. Algorithm 15.1 generates the optimal V-sequence $\beta^\star = (8,6,5,2,3,4,7)$ with $\sum C_j(\beta^\star) = 7,386$.

The sequences generated by Algorithms 16.15 and 16.16 are the following: $\beta_{A16.15} = (8,6,4,2,3,5,7)$ and $\beta_{A16.16} = (8,5,4,2,3,6,7)$, with $\sum C_j(\beta_{A16.15}) = 7,403$ and $\sum C_j(\beta_{H_{25}}) = 7,638$, respectively. ◆

15.3.2 Computational experiments

In order to evaluate the quality of schedules generated by Algorithm 15.1 a number of computational experiments have been conducted.

In the first experiment, the schedules generated by Algorithm 15.1 have been compared with schedules obtained by Algorithms 16.15 and 16.16. In this experiment, job deterioration coefficients were consecutive natural numbers, $\beta_j = j + 1$ for $0 \leqslant j \leqslant n$. The results of the experiment are summarized in Table 15.1. The star '$\star$' denotes that for a particular value of n, instance I and algorithm A the measure $R_A^r(I)$ (cf. Remark 2.27) is equal to 0.

The aim of the second computational experiment was to find optimal solutions to the problem $1|p_j = 1 + b_j t| \sum C_j$, where $b_j = j + 1$ for $1 \leqslant j \leqslant 20$. The results of the experiment are given in Table 15.2 (the case $n = 2m$) and Table 15.3 (the case $n = 2m - 1$), where $1 \leqslant m \leqslant 10$.

Remark 15.19. In Table 15.2 and Table 15.3, by Remark 15.3, are not regarded the jobs corresponding to the largest deterioration coefficient $\beta_{[0]}$.

Remark 15.20. The solutions given in Tables 15.2–15.3 have been found by an exact algorithm. Since by Theorem A.8 any optimal sequence for the problem $1|p_j = 1 + b_j t| \sum C_j$ has to be V-shaped, the algorithm for a given n constructed all possible V-shaped sequences and selected the optimal one.

The above results suggest that for certain types of β sequences, which will be called *regular*, Algorithm 15.1 constructs optimal schedules. A regular sequence is, e.g., the sequence composed of consecutive natural numbers or elements of an arithmetic (a geometric) progression. (Two latter sequences will be called *arithmetic* and *geometric* sequences, respectively.) We conjecture that for regular sequences it is possible to construct an optimal schedule knowing only the form of sequence β, *without* calculation of the signature $S^-(u)$. The arguments supporting the conjectures are given in the next section.

Table 15.1: Experiment results for consecutive integer deterioration rates

n	$OPT(I)$	$R^r_{A_{15.1}}(I)$	$R^r_{A16.15}(I)$	$R^r_{A16.16}(I)$
2	8	$\star$	$\star$	$\star$
3	21	$\star$	$\star$	0.142857142857
4	65	$\star$	0.015384615385	0.138461538462
5	250	$\star$	0.008000000000	0.084000000000
6	1,232	$\star$	0.008928571429	0.060876623377
7	7,559	$\star$	0.003571901045	0.053049345151
8	55,689	$\star$	0.002621702670	0.033884609169
9	475,330	$\star$	0.000995098142	0.020871815370
10	4,584,532	$\star$	0.000558835667	0.014906428835
11	49,111,539	$\star$	0.000244423210	0.011506155407
12	577,378,569	$\star$	0.000142137247	0.009070282967
13	7,382,862,790	$\star$	0.000080251254	0.007401067385
14	101,953,106,744	$\star$	0.000052563705	0.006210868342
15	1,511,668,564,323	$\star$	0.000035847160	0.005304460215
16	23,947,091,701,857	$\star$	0.000025936659	0.004588979235
17	403,593,335,602,130	$\star$	0.000019321905	0.004013033262
18	7,209,716,105,574,116	$\star$	0.000014779355	0.003541270022
19	136,066,770,200,782,755	$\star$	0.000011522779	0.003149229584
20	2,705,070,075,537,727,250	$\star$	0.000009131461	0.002819574105

Table 15.2: Solutions of problem $1|p_j = 1 + (1 + j)t| \sum C_j$ for even $n \leqslant 20$

m	$n = 2m$
1	$(1, 2)$
2	$(4, 1, 2, 3)$
3	$(5, 4, 1, 2, 3, 6)$
4	$(8, 5, 4, 1, 2, 3, 6, 7)$
5	$(9, 8, 5, 4, 1, 2, 3, 6, 7, 10)$
6	$(12, 9, 8, 5, 4, 1, 2, 3, 6, 7, 10, 11)$
7	$(13, 12, 9, 8, 5, 4, 1, 2, 3, 6, 7, 10, 11, 14)$
8	$(16, 13, 12, 9, 8, 5, 4, 1, 2, 3, 6, 7, 10, 11, 14, 15)$
9	$(17, 16, 13, 12, 9, 8, 5, 4, 1, 2, 3, 6, 7, 10, 11, 14, 15, 18)$
10	$(20, 17, 16, 13, 12, 9, 8, 5, 4, 1, 2, 3, 6, 7, 10, 11, 14, 15, 18, 19)$

15.4 Signatures of regular sequences

In this section, we present some results which strongly support the conjecture that Algorithm 15.1 is optimal for regular sequences of job deterioration rates. We start with the sequence of consecutive natural numbers.

Table 15.3: Solutions of problem $1|p_j = 1 + (1+j)t| \sum C_j$ for odd $n \leqslant 20$

m	$n = 2m - 1$
1	(1)
2	$(3, 1, 2)$
3	$(4, 3, 1, 2, 5)$
4	$(7, 4, 3, 1, 2, 5, 6)$
5	$(8, 7, 4, 3, 1, 2, 5, 6, 9)$
6	$(11, 8, 7, 4, 3, 1, 2, 5, 6, 9, 10)$
7	$(12, 11, 8, 7, 4, 3, 1, 2, 5, 6, 9, 10, 13)$
8	$(15, 12, 11, 8, 7, 4, 3, 1, 2, 5, 6, 9, 10, 13, 14)$
9	$(16, 15, 12, 11, 8, 7, 4, 3, 1, 2, 5, 6, 9, 10, 13, 14, 17)$
10	$(19, 16, 15, 12, 11, 8, 7, 4, 3, 1, 2, 5, 6, 9, 10, 13, 14, 17, 18)$

Let us define the following two sequences:

$$\beta = (r_m + (-1)^m, \ldots, r_2 + 1, r_1 - 1, r_1, r_2, \ldots, r_m) \text{ for } n = 2m, \quad (15.18)$$

$$\beta = (s_{m-1} + 2, \ldots, s_2 + 2, s_1 + 2, s_1, s_2, \ldots, s_m) \text{ for } n = 2m - 1, \quad (15.19)$$

where

$$r_k = 2k - \tfrac{1}{2}((-1)^k + 3) + 1, \ k = 1, 2, \ldots, m \text{ for } n = 2m, \quad (15.20)$$

$$s_k = 2k - \tfrac{1}{2}((-1)^k + 3), \ k = 1, 2, \ldots, m \text{ for } n = 2m - 1. \quad (15.21)$$

We will refer to sequences r_k and s_k, and to the related sequence β, as to the *even* and *odd* sequence, respectively.

Remark 15.21. Since sequences given in Tables 15.2–15.3 correspond to sequences (15.20) and (15.21) for $1 \leqslant m \leqslant 10$, respectively, formulae (15.18) and (15.19) can be considered as generalizations of these sequences for an arbitrary $m \geqslant 1$.

Now we prove a formula that can be derived from Definition (15.1) of the signature $S^-(\beta)$. For simplicity of notation, if sequence β is fixed, we will write S_n^- instead of $S^-(\beta)$:

$$S_n^- = \sum_{i=1}^{m} \beta_1 \cdots \beta_i - \sum_{i=1}^{n-m} \beta_{n-i+1} \cdots \beta_n + \sum_{i=1}^{n-m} \beta_1 \cdots \beta_{m+i} - \sum_{i=1}^{m} \beta_i \cdots \beta_n, \quad (15.22)$$

where $1 \leqslant m \leqslant n$.

From formula (15.22) we can obtain the following representation of the signature for $n = 2m$ and $n = 2m - 1$, respectively.

Lemma 15.22. (Gawiejnowicz et al. [4]) *Let* $\beta = (\beta_1, \ldots, \beta_n)$. *If* $n = 2m$, *then*

$$S_{2m}^- = \sum_{i=1}^{m} \eta_i(m) \left(\prod_{j=1}^{m-i+1} \beta_j - \prod_{j=m+i}^{2m} \beta_j \right),$$ (15.23)

where $\eta_1(m) = 1$ *and* $\eta_i(m) = 1 + \prod_{j=m-i+2}^{m+i-1} \beta_j$ *for* $i = 2, 3, \ldots, m$.

If $n = 2m - 1$, *then*

$$S_{2m-1}^- = \sum_{i=1}^{m-1} \omega_i(m) \left(\prod_{j=1}^{m-i} \beta_j - \prod_{j=m+i}^{2m-1} \beta_j \right),$$ (15.24)

where $\omega_i(m) = 1 + \prod_{j=m-i+1}^{m+i-1} \beta_j$ *for* $i = 1, 2, \ldots, m - 1$.

Proof. Let $n = 2m$. Then

$$S_{2m}^- = \sum_{i=1}^{m} (\beta_1 \cdots \beta_i - \beta_{2m-i+1} \cdots \beta_{2m}) + \sum_{i=1}^{m} \beta_1 \cdots \beta_{m+i} - \sum_{i=1}^{m} \beta_i \cdots \beta_{2m}.$$

Reducing the last term in the second sum with the first one in the third sum we have $S_{2m}^- = \sum_{i=1}^{m} (\beta_1 \cdots \beta_i - \beta_{2m-i+1} \cdots \beta_{2m}) + \sum_{i=1}^{m-1} \beta_{i+1} \cdots \beta_{2m-i} \times (\beta_1 \cdots \beta_i - \beta_{2m-i+1} \cdots \beta_{2m})$. Next, by joining both sums and by changing the index of summation according to $i := m - i + 1$, we have $S_{2m}^- = \sum_{i=2}^{m} (1 + \beta_{m-i+2} \cdots \beta_{m+i-1}) \times (\beta_1 \cdots \beta_{m-i+1} - \beta_{m+i} \cdots \beta_{2m}) + \beta_1 \cdots \beta_m - \beta_{m+1} \cdots \beta_{2m}$. Hence, taking into account definitions of the coefficients η_i, formula (15.23) follows.

To prove formula (15.24) we proceed in the same way. Let $n = 2m - 1$. Then $S_{2m-1}^- = \sum_{i=1}^{m} \beta_1 \cdots \beta_i - \sum_{i=1}^{m-1} \beta_{m+i} \cdots \beta_{2m-1} + \sum_{i=1}^{m-1} \beta_1 \cdots \beta_{m+i} - \sum_{i=1}^{m} \beta_i \cdots \beta_{2m-1}$. Changing the index of the summation in the first sum according to $i := m - i$, in the third sum according to $i := i - 1$ and in the last sum according to $i := m - i + 1$, we obtain $S_{2m-1}^- = \sum_{i=0}^{m-1} \beta_1 \cdots \beta_{m-i} - \sum_{i=1}^{m-1} \beta_{m+i} \cdots \beta_{2m-1} + \sum_{i=2}^{m-1} (\beta_1 \cdots \beta_{m-i}) \times (\beta_{m-i+1} \cdots \beta_{m+i-1}) - \sum_{i=1}^{m-1} (\beta_{m-i+1} \cdots \beta_{m+i-1}) \times (\beta_{m+i} \cdots \beta_{2m-1})$. By moving in the first sum the term with the index $i = 0$ to the third one under the index $i = 1$ and applying definitions of coefficients ω_i, we obtain formula (15.24). $\blacksquare$

Lemma 15.23. (Gawiejnowicz et al. [4]) *Let* $n = 2m$, *and let* β *be an even sequence. Then for each integer* $m \geqslant 1$ *the equality*

$$S_{2m}^- = \sum_{i=1}^{m} \eta_i \left(\prod_{j=i}^{m} (r_j + (-1)^j) - \prod_{j=i}^{m} r_j \right)$$ (15.25)

holds, where

$$\eta_1 = 1 \ and \ \eta_i = 1 + \prod_{j=1}^{i-1} r_j \prod_{j=1}^{i-1} (r_j + (-1)^j) \tag{15.26}$$

for $i = 2, 3, \ldots, m$.

Proof. Applying Lemma 15.22 to sequence β given by formulae (15.18) and (15.20), we obtain formula (15.25) for the signature S_{2m}^-. ∎

Now, on the basis of (15.25), we can state the following result.

Theorem 15.24. (Gawiejnowicz et al. [4]) *Let* $n = 2m$, *and let* β *be an even sequence. Then for the signatures* S_{2m+2}^- *and* S_{2m}^- *the formula*

$$S_{2m+2}^- = r_{m+1} S_{2m}^- + (-1)^{m+1} \sum_{i=1}^{m+1} \eta_i \prod_{j=i}^{m} (r_j + (-1)^j) \tag{15.27}$$

holds, where the signature S_{2m}^- *and coefficients* η_i *are defined by formulae (15.25) and (15.26), respectively. Moreover, the following identity holds:*

$$S_{2m+2}^- = R_m \left((-1)^{m+1} + \Theta_m \right), \tag{15.28}$$

where $\Theta_m = \frac{S_{2m}^-}{R_m} (r_{m+1} + (-1)^{m+1})$ *and* $R_m = \sum_{i=1}^{m+1} \eta_i \prod_{j=i}^{m} r_j$.

Proof. Applying Lemma 15.23, we obtain

$$S_{2m+2}^- = \sum_{i=1}^{m+1} \eta_i \left((r_i + (-1)^i) \cdots (r_{m+1} + (-1)^{m+1}) - r_i \cdots r_{m+1} \right)$$

$$= \sum_{i=1}^{m} \eta_i \left((r_i + (-1)^i) \cdots (r_{m+1} + (-1)^{m+1}) - r_i \cdots r_{m+1} \right)$$

$$+ \eta_{m+1} \left((r_{m+1} + (-1)^{m+1}) - r_{m+1} \right)$$

$$= r_{m+1} S_{2m}^- + (-1)^{m+1} \sum_{i=1}^{m} \eta_i \left((r_i + (-1)^i) \cdots (r_m + (-1)^m) \right)$$

$$+ \eta_{m+1} \left((r_{m+1} + (-1)^{m+1}) - r_{m+1} \right)$$

$$= r_{m+1} S_{2m}^- + (-1)^{m+1} \sum_{i=1}^{m+1} \eta_i \left((r_i + (-1)^i) \cdots (r_m + (-1)^m) \right),$$

and formula (15.27) follows in view of the definition of coefficients η_i. Formula (15.28) follows from the assumed notation and formula (15.27). ∎

Now we consider the case of an *odd* sequence. Applying Lemma 15.22 to sequence β given by formulae (15.19) and (15.21), we obtain the following formula for the signature S_{2m-1}^-.

Lemma 15.25. (Gawiejnowicz et al. [4]) *Let β be an odd sequence, and let $n = 2m - 1$. Then for every integer $m \geqslant 1$ the equality*

$$S_{2m-1}^- = \sum_{i=1}^{m-1} \omega_i \left(\prod_{j=i}^{m-1} (s_j + 2) - \prod_{j=i+1}^{m} s_j \right) \tag{15.29}$$

holds, where $\omega_i = 1 + \prod_{j=1}^{i} s_j \prod_{j=1}^{i-1} (s_j + 2)$ for $i = 1, \ldots, m - 1$.

Proof. Let $n = 2m - 1$, and let β be an odd sequence. Then we have $\beta_{m-i} = s_i + 2$ for $i = 1, \ldots, m - 1$ and $\beta_{m+i-1} = s_i$ for $i = 1, \ldots, m$. Substituting these values in formula (15.24) and noticing that $\omega_i = 1 + \beta_{m-i+1} \cdots \beta_{m+i-1} = 1 + s_1 \cdots s_i (s_1 + 2) \cdots (s_{i-1} + 2)$, formula (15.29) follows. ∎

On the basis of formula (15.29) we can state the following result, concerning the behaviour of the signature S_n^- for $n = 2m + 1$.

Theorem 15.26. (Gawiejnowicz et al. [4]) *Let $n = 2m + 1$, and let β be an odd sequence. Then for the signatures S_{2m+1}^- and S_{2m-1}^- the formula*

$$S_{2m+1}^- = (s_m + 2) S_{2m-1}^- + (-1)^{m+1} \sum_{i=1}^{m} \omega_i \prod_{j=i+1}^{m} s_j \tag{15.30}$$

holds, where $\omega_i = 1 + \prod_{j=1}^{i} s_j \prod_{j=1}^{i-1} (s_j + 2)$ for $i = 1, \ldots, m$. Moreover, for S_{2m+1}^- the identity

$$S_{2m+1}^- = Q_m \left((-1)^{m+1} + \Gamma_m \right), \tag{15.31}$$

holds, where $\Gamma_m = \frac{S_{2m-1}^-}{Q_m}(s_m + 2)$ and $Q_m = \sum_{i=1}^{m} \omega_i \prod_{j=i+1}^{m} s_j$.

Proof. By Lemma 15.25, for $\omega_i = 1 + (s_1 \cdots s_i)(s_1 + 2) \cdots (s_{i-1} + 2)$, we obtain

$$S_{2m+1}^- = \sum_{i=1}^{m} \omega_i \left((s_i + 2) \cdots (s_m + 2) - s_{i+1} \cdots s_{m+1} \right)$$

$$= \sum_{i=1}^{m-1} \omega_i \left((s_i + 2) \cdots (s_m + 2) - q_i + q_i - s_{i+1} \cdots s_{m+1} \right)$$

$$+ \omega_m \left((s_m + 2) - s_{m+1} \right),$$

where $q_i \equiv (s_{i+1} \cdots s_m)(s_m + 2)$. Hence, by Lemma 15.25, we have

$$S_{2m+1}^- = (s_m + 2) \sum_{i=1}^{m-1} \omega_i \left((s_i + 2) \cdots (s_{m-1} + 2) - s_{i+1} \cdots s_m \right)$$

$$+ \sum_{i=1}^{m-1} \omega_i (s_{i+1} \cdots s_m) \left((s_m + 2) - s_{m+1} \right)$$

$$+ \omega_m \left((s_m + 2) - s_{m+1} \right).$$

Collecting the last terms, applying identity (15.25) and using the equality $(s_m + 2) - s_{m+1} = (-1)^{m+1}$, we obtain formula (15.30).

Formula (15.31) is an immediate consequence of formula (15.30) and the assumed notation.   ∎

We will now prove that for an arbitrary m the sign of signatures S_{2m}^- and S_{2m-1}^- varies periodically. Knowing the behaviour of the signatures we are able to simplify Algorithm 15.1, since we will not have to calculate the signatures in Step 2 of the algorithm.

Theorem 15.27. (Gawiejnowicz et al. [4]) *Let there be given V-sequences* (15.18) *and* (15.19) *of sequence* $\beta := (1, 2, \ldots, n)$. *Then for each integer* $m \geqslant 1$ *the sign of the signatures* S_{2m}^- *and* S_{2m-1}^- *for these sequences varies periodically according to the formulae* $\mathrm{sign}(S_{2m}^-) = (-1)^m$ *and* $\mathrm{sign}(S_{2m-1}^-) = (-1)^m$, *respectively.*

Before we prove Theorem 15.27, we will prove some technical lemmas.

Lemma 15.28. (Gawiejnowicz et al. [4]) *For every integer* $m \geqslant 1$ *the following recurrence relations hold:*

$$\Theta_1 = -\frac{4}{5}, \Theta_{m+1} = D_m(\Theta_m - (-1)^m) \text{ for } n = 2m \tag{15.32}$$

and

$$\Gamma_1 = 0, \Gamma_2 = \frac{8}{11}, \Gamma_{m+1} = E_m(\Gamma_m - (-1)^m) \text{ for } n = 2m - 1, \tag{15.33}$$

where

$$D_m = (r_{m+1} + 2)\frac{R_m}{R_{m+1}} \tag{15.34}$$

and

$$E_m = (s_{m+1} + 2)\frac{Q_m}{Q_{m+1}}. \tag{15.35}$$

Proof. Recurrence relations (15.32) and (15.33) follow from Theorem 15.24 and Theorem 15.26, respectively. In the case of formula (15.34) we apply the equality $r_{m+2} + (-1)^{m+2} = r_{m+1} + 2$. In both formulae, (15.34) and (15.35), it is sufficient to apply definitions of Θ_{m+1} and Γ_{m+1}, and the recurrence formulae (15.25) and (15.29) for S_{2m}^- and S_{2m-1}^-, respectively. Clearly, the definitions of R_m (cf. Theorem 15.24) and Q_m (cf. Theorem 15.26) must be also applied.   ∎

Remark 15.29. Notice that we have $|\Theta_1| < 1$, $\Gamma_1 = 0$ and $|\Gamma_2| < 1$. Moreover, $\Theta_1 < 0$ and $\Gamma_2 > 0$.

The next two lemmas are crucial for proofs of inequalities $0 < D_m < 1$ and $0 < E_m < 1$.

Lemma 15.30. (Gawiejnowicz et al. [4]) *For every integer $m \geqslant 1$ the inequality $R_m < \frac{1}{2}\eta_{m+2}$ holds.*

Proof. We will proceed by mathematical induction. The case $m = 1$ is immediate, since $R_1 = \eta_1 r_1 + \eta_2 = r_1 + (1 + r_1(r_1 - 1))$ and $\eta_3 = 1 + r_1 r_2(r_1 + 2)$, where $r_1 = 2$ and $r_2 = 3$.

Now, assume that $R_{m-1} < \frac{1}{2}\eta_{m+1}$. Hence

$$R_m = r_m R_{m-1} + \eta_{m+1} < \frac{1}{2}(r_m + 2)\eta_{m+1}.$$

Thus, it is sufficient to prove that $(r_m + 2)\eta_{m+1} < \eta_{m+2}$. In order to prove this, note first that $(r_m + 2)\eta_{m+1} = (r_m + 2) + \frac{1}{r_{m+1}}(\eta_{m+2} - 1)$. Consequently, we have to prove that $(r_m + 1) + (1 - \frac{1}{r_{m+1}}) < (1 - \frac{1}{r_{m+1}})\eta_{m+2}$, or that

$$r_m + 1 < (1 - \frac{1}{r_{m+1}})(\eta_{m+2} - 1) = \frac{r_{m+1} - 1}{r_{m+1}}(r_1 \cdots r_m r_{m+1})(r_1 + 2) \cdots (r_m + 2).$$

Since $r_{m+1} = r_m + 2 + (-1)^m$, it is sufficient to check the latter inequality in the expression

$$\frac{r_m + 1}{r_m + 1 + (-1)^m} \leqslant 1 + \frac{1}{r_m} < (r_1 \cdots r_m)(r_1 + 2) \cdots (r_m + 2).$$

Finally, since $r_m \nearrow$, it is sufficient to check the case $m = 1$, which is evident. ∎

Lemma 15.31. (Gawiejnowicz et al. [4]) *For every integer $m \geqslant 1$ the inequality $Q_m < \frac{1}{2}\omega_{m+1}$ holds.*

Proof. First we prove the inequality

$$(s_m + 2)\omega_m < \omega_{m+1}. \tag{15.36}$$

We have

$$(s_m + 2)\omega_m = (s_m + 2)(1 + (s_1 \cdots s_m)(s_1 + 2) \cdots (s_{m-1} + 2))$$

$$= (s_m + 2) + \frac{1}{s_{m+1}}((1 + (s_1 \cdots s_{m+1})(s_1 + 2) \cdots (s_m + 2)) - 1)$$

$$= (s_m + 2) + \frac{1}{s_{m+1}}(\omega_{m+1} - 1).$$

It is sufficient to check that $(s_m + 2) + \frac{1}{s_{m+1}}(\omega_{m+1} - 1) < \omega_{m+1}$ or that

$$\frac{s_m + 1}{s_m + (1 + (-1)^m)} \leqslant 1 + \frac{1}{s_m} < (s_1 \cdots s_m)(s_1 + 2) \cdots (s_m + 2). \tag{15.37}$$

In order to obtain these inequalities we have applied the equality $s_{m+1} - 1 = s_m + (1 + (-1)^m)$. Since $s_i \geqslant 1$, inequalities (15.37) are obviously satisfied, which completes the proof of inequality (15.36).

To prove the lemma we will proceed by mathematical induction. Let $m = 1$. Since $2Q_1 = 2\omega_1 = 2(1 + s_1)$ and $\omega_2 = 1 + (s_1 s_2)(s_1 + 2)$, we obtain $2Q_1 < \omega_2$, since $s_1 = 1$ and $s_2 = 2$. Now, assume that $Q_{m-1} < \frac{1}{2}\omega_m$ holds. Since

$$Q_m = \sum_{i=1}^{m} \omega_i \, s_{i+1} \cdots s_m = s_m Q_{m-1} + \omega_m,$$

we obtain $Q_m < \frac{1}{2}(s_m + 2)\omega_m$. Now, applying inequality (15.36), we obtain

$$Q_m < \frac{1}{2}(s_m + 2)\omega_m < \frac{1}{2}\omega_{m+1}$$

as desired.   ∎

Lemma 15.32. (Gawiejnowicz et al. [4]) *For every integer $m \geqslant 1$ the following inequalities hold:*

$$0 < D_m < 1 \quad \text{and} \quad 0 < E_m < 1.$$

Proof. It is easy to see that $D_m > 0$ and $E_m > 0$. From the definition of D_m, the inequality $D_m < 1$ is equivalent to $R_m < \frac{1}{2}\eta_{m+2}$ which is satisfied in view of Lemma 15.30. To prove that $E_m < 1$, we apply Lemma 15.31. Indeed, in view of the definition of E_m, the inequality $E_m < 1$ is equivalent to the inequality $Q_m < \frac{1}{2}\omega_{m+1}$ from Lemma 15.31, which completes the proof.   ∎

Lemma 15.33. (Gawiejnowicz et al. [4]) *For every integer $m \geqslant 1$ the inequality $|\Theta_m| < 1$ and the equality $\text{sign}(\Theta_m) = (-1)^m$ hold.*

Proof. Taking $m = 2k$ or $m = 2k - 1$, for every integer $k \geqslant 1$ we obtain, by Lemma 15.28, respectively:

$$\Theta_{2k+1} = D_{2k}(\Theta_{2k} - (-1)^{2k}) = D_{2k}(\Theta_{2k} - 1)$$

and

$$\Theta_{2k} = D_{2k-1}(\Theta_{2k-1} - (-1)^{2k-1}) = D_{2k-1}(\Theta_{2k-1} + 1).$$

After substituting the value $2k - 1$ for odd indices we have

$$\Theta_{2k+1} = D_{2k}(D_{2k-1}(\Theta_{2k-1} + 1) - 1).$$

To prove that $|\Theta_{2k-1}| < 1$ and $\Theta_{2k-1} < 0$ for every integer $k \geqslant 1$, we will proceed by mathematical induction.

For $k = 1$ we have $|\Theta_1| < 1$ and $\Theta_1 < 0$, since $\Theta_1 = -\frac{4}{5}$ by definition. Assume that $|\Theta_{2k-1}| < 1$ and $\Theta_{2k-1} < 0$. We will prove that $|\Theta_{2k+1}| < 1$ and $\Theta_{2k+1} < 0$. By induction assumption, $0 < D_{2k-1}(\Theta_{2k-1} + 1) < 1$ and consequently $0 < 1 - D_{2k-1}(\Theta_{2k-1}+1) < 1$. Hence $|\Theta_{2k+1}| < 1$ and $\Theta_{2k+1} < 0$.

Now, consider the case of even indices $2k$. Applying the odd case we obtain

$$\Theta_{2k} = D_{2k-1}(\Theta_{2k-1} + 1) > 0$$

with $D_{2k-1}(\Theta_{2k-1} + 1) < 1$. Consequently, $|\Theta_{2k}| < 1$ with $Q_{2k} > 0$.   ∎

Lemma 15.34. (Gawiejnowicz et al. [4]) *For every integer $m > 1$ the inequality $|\Gamma_m| < 1$ and the equality $\text{sign}(\Gamma_m) = (-1)^m$ hold.*

Proof. Taking $m = 2k$ or $m = 2k - 1$, for every integer $k \geqslant 1$, we obtain, by Lemma 15.28, respectively:

$$\Gamma_{2k+1} = E_{2k}(\Gamma_{2k} - (-1)^{2k}) = E_{2k}(\Gamma_{2k} - 1)$$

and

$$\Gamma_{2k} = E_{2k-1}(\Gamma_{2k-1} - (-1)^{2k-1}) = E_{2k-1}(\Gamma_{2k-1} + 1).$$

After substituting the value $2k - 1$ for odd indices we have

$$\Gamma_{2k+1} = E_{2k}(E_{2k-1}(\Gamma_{2k-1} + 1) - 1).$$

To prove that $|\Gamma_{2k-1}| < 1$ and $\Gamma_{2k-1} < 0$ for every integer $k \geqslant 2$, we will proceed by mathematical induction.

Note that for $k = 1$ we have $\Gamma_1 = 0$. For $k = 2$ we have $|\Gamma_3| < 1$ and $\Gamma_3 < 0$ since $\Gamma_3 = E_2(\Gamma_2 - 1) = E_2(\frac{8}{11} - 1)$ and $0 < E_m < 1$.

Now, let $|\Gamma_{2k-1}| < 1$ and $\Gamma_{2k-1} < 0$ for an arbitrary $k > 2$. Then

$$-1 < E_{2k-1}(\Gamma_{2k-1} + 1) - 1 < 0,$$

since $0 < E_{2k-1}(\Gamma_{2k-1}+1) < 1$. Finally, we obtain $|\Gamma_{2k+1}| < 1$ and $\Gamma_{2k+1} < 0$, which finishes the induction step. This result implies that

$$\Gamma_{2k} = E_{2k-1}(\Gamma_{2k-1} + 1) > 0 \quad \text{and} \quad \Gamma_{2k} < 1$$

for each integer $k \geqslant 2$. Moreover, $\Gamma_2 = \frac{8}{11}$, i.e., $\Gamma_2 > 0$ and $\Gamma_2 < 1$.   ∎

Lemmas 15.33 and 15.34 allow us to prove Theorem 15.27.

Proof of Theorem 15.27. In view of the formula $S_{2m+2}^- = R_m((-1)^{m+1} + \Theta_m)$ for an arbitrary integer $m \geqslant 2$, from the fact that $\text{sign}(S_2^-) = 1$ and from Lemma 15.33, we have that $\text{sign}(S_{2m}^-) = (-1)^m$ for an arbitrary integer $m \geqslant 1$.

Similarly, in view of the formula $S_{2m+1}^- = Q_m((-1)^{m+1} + \Gamma_m)$ for an arbitrary integer $m \geqslant 1$, from the fact that $\text{sign}(S_1^-) = -1$ and from Lemma 15.34, we have that $\text{sign}(S_{2m-1}^-) = (-1)^m$ for an arbitrary integer $m \geqslant 1$.   ∎

Conjecture 15.35. Algorithm 15.1 is optimal for the $1|p_j = 1 + b_j t| \sum C_j$ problem in the case when $b_j = j + 1$ for $j = 0, 1, 2, \ldots, n$.

15.5 The second greedy algorithm

The pseudo-code of the second greedy algorithm for problem $1|p_j = 1 + b_j t| \sum C_j$ is as follows.

Algorithm 15.2. for problem $1|p_j = 1 + b_j t| \sum C_j$

---

1 **Input** : sequence $\hat{\beta} = (\beta_0, \beta_1, \ldots, \beta_n)$
2 **Output:** suboptimal schedule u
  ▷ **Step 1**
3 $\beta_{[0]} \leftarrow \max \{\beta_j : 0 \leqslant j \leqslant n\}$;
4 Arrange sequence $\hat{\beta} - \{\beta_{[0]}\}$ in non-decreasing order;
  ▷ **Step 2**
5 **if** (n is odd) **then**
6 $\quad$ $u \leftarrow (\beta_{[1]})$;
7 $\quad$ $sign \leftarrow (-1)$;
8 $\quad$ $i_0 \leftarrow 2$;
  else
9 $\quad$ $u \leftarrow (\beta_{[1]}, \beta_{[2]})$;
10 $\quad$ $sign \leftarrow 1$;
11 $\quad$ $i_0 \leftarrow 3$;
  end
12 **for** $i \leftarrow i_0$ **to** $n - 1$ **step** 2 **do**
13 $\quad$ $sign \leftarrow sign \times (-1)$;
14 $\quad$ **if** ($sign < 0$) **then**
15 $\quad\quad$ $u \leftarrow (\beta_{[i+1]}|u|\beta_{[i]})$;
  $\quad$ **else**
16 $\quad\quad$ $u \leftarrow (\beta_{[i]}|u|\beta_{[i+1]})$;
  $\quad$ **end**
  end
  ▷ **Step 3**
17 $u \leftarrow (\beta_{[0]}|u)$;
18 **return** u.

Algorithm 15.2 is based on the following observation. If Conjecture 15.35 is true, then in Step 2 of Algorithm 15.1 it is not necessary to check the sign of the signature $S^-(u)$, since the sign varies periodically. Hence, we can construct sequence u iteratively, since by Theorem 15.27 we know which order to choose, $(\alpha|u|\gamma)$ or $(\gamma|u|\alpha)$.

In the remaining part of the section, we extend the results of Sect. 15.4 to cover arithmetic and geometric sequences (cf. Gawiejnowicz et al. [6]).

15.5.1 Arithmetic sequences

We start the section with two examples which illustrate the behaviour of Algorithms 15.1 and 15.2 for arithmetic sequences.

Let α_A and ρ_A denote, respectively, the *first term* and the *common difference* in arithmetic sequence $\beta_j = \alpha_A + j\rho_A$, where $0 \leqslant j \leqslant n$.

Example 15.36. Let $\beta = (1.5, 2.0, \ldots, 9.0)$ be an arithmetic sequence in which $n = 15$, $\alpha_A = 1.5$ and $\rho_A = 0.5$. Then the optimal V-sequence is

$$\beta^\star = (9.0, 8.5, 7.0, 6.5, 5.0, 4.5, 3.0, 2.5, 1.5, 2.0, 3.5, 4.0, 5.5, 6.0, 7.5, 8.0),$$

with $\sum C_j(\beta^\star) = 7{,}071{,}220{,}899.8750$.   ♦

Example 15.37. Let $\beta = (1.5, 1.8, \ldots, 6.3)$ be an arithmetic sequence in which $n = 16$, $\alpha_A = 1.5$ and $\rho_A = 0.3$, Then the optimal V-sequence is

$$\beta^\star = (6.3, 6.0, 5.1, 4.8, 3.9, 3.6, 2.7, 2.4, 1.5, 1.8, 2.1, 3.0, 3.3, 4.2, 4.5, 5.4, 5.7),$$

with $\sum C_j(\beta^\star) = 642{,}302{,}077.7853$.   ♦

Since in both cases Algorithms 15.1 and 15.2 generated the optimal schedules, the examples suggest that in the case of arithmetic sequences the algorithms behave similarly to the case of consecutive natural numbers.

Let us now introduce V-sequences of arithmetic sequences by the formulae

$$\beta = (u_m + (-1)^m \alpha_A, \ldots, u_2 + \alpha_A, u_1 - \alpha_A, u_1, u_2, \ldots, u_m) \qquad (15.38)$$
$$\beta = (v_{m-1} + 2\alpha_A, \ldots, v_2 + 2\alpha_A, v_1 + 2\alpha_A, v_1, v_2, \ldots, v_m) \qquad (15.39)$$

for $n = 2m$ and $n = 2m - 1$, respectively, where sequences $u_k = \alpha_A r_k + \rho_A$, $v_k = \alpha_A s_k + \rho_A$ are such that $u_k \geqslant 1$ and $v_k \geqslant 1$ for $1 \leqslant k \leqslant m$, and sequences (r_k) and (s_k) are defined by (15.20) and (15.21), respectively.

In this case, the following counterpart of Theorem 15.27 holds.

Theorem 15.38. (Gawiejnowicz et al. [6]) *Let $\rho_A \geqslant 0$ and $\alpha_A + \rho_A \geqslant 1$. Then the sign of signatures S^- for the arithmetic sequences (15.38) with $\alpha_A \geqslant 0.11$ and the sequences (15.39) with $\alpha_A \geqslant 0.50$ varies according to formulae* $\mathrm{sign}(S_{2m}^-) = (-1)^m$ *and* $\mathrm{sign}(S_{2m-1}^-) = (-1)^m$, *respectively, where $m \geqslant 1$.*

Proof. Let $n = 2m$, $\alpha_A \geqslant 0.11$, $\rho_A \geqslant 0$ and $\alpha_A + \rho_A \geqslant 1$. Then the recurrence relation $\Theta_{m+1} = D_m(\Theta_m + (-1)^{m+1}\alpha_A)$ with

$$\Theta_1 = -\frac{\alpha_A(4\alpha_A + \rho_A)}{(2\alpha_A + \rho_A)(\alpha_A + \rho_A + 1) + 1}$$

holds, where $D_m = (u_{m+1} + 2\alpha_A)\frac{R_m}{R_{m+1}}$.

Similarly, if $n = 2m - 1$, $\alpha_A \geqslant 0.50$, $\rho_A \geqslant 0$ and $\alpha_A + \rho_A \geqslant 1$, then the recurrence relation $\Gamma_{m+1} = F_m(\Gamma_m + (-1)^{m+1}\alpha_A)$, with $\Gamma_1 = 0$ and

$$\Gamma_2 = \frac{\alpha_A(\alpha_A + \rho_A + 1)(4\alpha_A + \rho_A)}{(2\alpha_A + \rho_A)(1 + (\alpha_A + \rho_A)(3\alpha_A + \rho_A + 1)) + 1}$$

holds, where $F_m = (v_{m+1} + 2\alpha_A)\frac{Q_m}{Q_{m+1}}$.

To end the proof it is sufficient to show that $0 < D_m < 1$ and $\Theta_m < \alpha_A$ if $n = 2m$ and that $0 < F_m < 1$ and $\Gamma_n < \alpha_A$ if $n = 2m - 1$.   □

15.5.2 Geometric sequences

We start the section with two examples which illustrate the behaviour of Algorithms 15.1 and 15.2 for geometric sequences.

Let ρ_G denote the *ratio* in geometric sequence $\beta_j = \rho_G^j$, where $1 \leqslant j \leqslant n$.

Example 15.39. Let $\beta = (3, 9, \ldots, 19683)$ be a geometric sequence for $n = 8$ and $\rho_G = 3$. Then the optimal V-sequence is

$$\beta^\star = (19683, 6561, 243, 81, 3, 9, 27, 729, 2187),$$

with $\sum C_j(\beta^\star) = 150, 186, 346, 871, 598, 597$.   ◆

Example 15.40. Let $\beta = (2, 4, \ldots, 2048)$ be a geometric sequence for $n = 10$ and $\rho_G = 2$. Then the optimal V-sequence is

$$\beta^\star = (2048, 256, 128, 16, 8, 2, 4, 32, 64, 512, 1024)$$

with $\sum C_j(\beta^\star) = 36, 134, 983, 945, 485, 585$.   ◆

As previously, Algorithms 15.1 and 15.2 generated the optimal schedules.

We now define two sequences which are counterparts of sequences (15.38) and (15.39) for geometric sequences. We will distinguish between the case $n = 2m$ and the case $n = 2m - 1$. Let for some $\rho_G > 1$

$$\beta = (\rho_G^{r_m + (-1)^m}, \ldots, \rho_G^{r_2 + 1}, \rho_G^{r_1 - 1}, \rho_G^{r_1}, \rho_G^{r_2}, \ldots, \rho_G^{r_m}) \tag{15.40}$$

and

$$\beta = (\rho_G^{s_m - 1 + 2}, \ldots, \rho_G^{s_2 + 2}, \rho_G^{s_1 + 2}, \rho_G^{s_1}, \rho_G^{s_2}, \ldots, \rho_G^{s_m}) \tag{15.41}$$

for $n = 2m$ and $n = 2m - 1$, respectively.

In this case, the following counterpart of Theorem 15.38 holds.

Theorem 15.41. (Gawiejnowicz et al. [6]) *The sign of signatures $S^-(\beta)$ of the geometric sequences (15.40) and (15.41) varies according to formulae* $\text{sign}(S_{2m}^-) = (-1)^m$ *and* $\text{sign}(S_{2m-1}^-) = (-1)^m$, *respectively, where $m \geqslant 1$.*

A computational experiment has been conducted in order to evaluate the quality of schedules generated by Algorithms 15.1 and 15.2 for arithmetic and geometric sequences (see Gawiejnowicz et al. [6] for details). In the experiment, random instances of arithmetic and geometric sequences were generated. Fifty instances were generated for each value of n, where $n = 10, 15, 20$. Algorithms 15.1 and 15.2 found an optimal schedule for all 150 instances.

Hence, the performance of Algorithms 15.1 and 15.2 for arithmetic and geometric sequences of average size is outstanding.

15.5.3 Arbitrary sequences

From Example 15.17 we know that Algorithm 15.1 (and hence 15.2) are not optimal for arbitrary β. A computational experiment has been conducted in order to evaluate the quality of schedules generated by Algorithms 15.1 and 15.2 for arbitrary sequences (see Gawiejnowicz et al. [6] for details).

In the experiment random instances of β sequence were generated. Fifty instances were generated for each value of $n = 10, 15, 20$. The average ratio $R_{A_{15.1}}(I)$, calculated for 50 instances, was equal to 6654×10^{-8}, 5428×10^{-8} and 1695×10^{-8} for $n = 10$, $n = 15$ and $n = 20$, respectively. The average ratio $R^r_{A_{15.2}}(I)$ was equal to 26988×10^{-8}, 12927×10^{-8} and 2698×10^{-8} for $n = 10$, $n = 15$ and $n = 20$, respectively.

Hence, the performance of Algorithms 15.1 and 15.2 for random sequences of average size is quite satisfactory.

In general, Algorithm 15.1 is better than Algorithm 15.2. The first algorithm is recommended for arbitrary sequences, while the second one is recommended for regular sequences. The formal proof of optimality of these algorithms still remains a challenge.

References

1. T-C. E. Cheng, Q. Ding and B. M-T. Lin, A concise survey of scheduling with time-dependent processing times. *European Journal of Operational Research* **152** (2004), no. 1, 1–13.
2. M. R. Garey and D. S. Johnson, *Computers and Intractability: A Guide to the Theory of NP-Completeness.* San Francisco: Freeman 1979.
3. S. Gawiejnowicz, W. Kurc and L. Pankowska, A greedy approach for a time-dependent scheduling problem. *Lecture Notes in Computer Science* **2328** (2002), 79–86.
4. S. Gawiejnowicz, W. Kurc and L. Pankowska, Analysis of a time-dependent scheduling problem by signatures of deterioration rate sequences. *Discrete Applied Mathematics* **154** (2006), no. 15, 2150–2166.
5. S. Gawiejnowicz, W. Kurc, L. Pankowska and C. Suwalski, Approximate solution of a time-dependent scheduling problem for l_p−norm-based criteria. In: B. Fleischmann et al. (eds.), *Operations Research OR2000*, Berlin: Springer 2001, pp. 372–377.
6. S. Gawiejnowicz, W. Kurc and L. Pankowska, Greedy scheduling of time-dependent jobs with arithmetic or geometric deterioration rates. In: S. Domek and R. Kaszyński (eds.), *Proceedings of the 12th IEEE International Conference on Methods and Models in Automation and Robotics*, 2006, pp. 1119–1124.

16

Heuristic algorithms

In the previous chapter, we considered two greedy algorithms for a single machine time-dependent scheduling problem. In this chapter, we focus on heuristic algorithms for time-dependent scheduling problems on a single, parallel and dedicated machines.

Chapter 16 is composed of five sections. In Sect. 16.1, we give some preliminaries. In Sect. 16.2, we consider heuristic algorithms for minimizing the $C_{\max}$ criterion. In Sect. 16.3 and Sect. 16.4, we present heuristic algorithms for the $\sum C_j$ and $L_{\max}$ criteria, respectively. In Sect. 16.5, we discuss heuristic algorithms for criteria other than $C_{\max}$, $\sum C_j$ and $L_{\max}$. The chapter is completed with a list of references.

16.1 Preliminaries

Heuristic algorithms, in short called *heuristics,* similarly to approximation algorithms and approximation schemes, construct suboptimal solutions for $\mathcal{NP}$-hard problems. In this chapter, we consider heuristic algorithms for $\mathcal{NP}$-hard time-dependent scheduling problems.

Heuristic algorithms generate solutions using rules that are only loosely defined. This means that we apply such algorithms at our own risk, since they may generate infeasible solutions as no proofs of their correctness are known. Moreover, it is possible that for some instances the solutions generated by the algorithms are very poor, because no estimations of their worst-case behaviour are available. These disadvantages distinguish heuristic algorithms from approximation algorithms and approximation schemes (cf. Chap. 14), for which such estimations are known (cf. Sect. 2.4). Nevertheless, for many problems the performance of heuristic algorithms is good enough.

In the literature, there are known several groups of heuristic algorithms. In this chapter, we consider the most popular group comprising *constructive heuristics* which start with an empty solution and repeatedly extend the cur-

© Springer-Verlag GmbH Germany, part of Springer Nature 2020
S. Gawiejnowicz, *Models and Algorithms of Time-Dependent Scheduling,*
Monographs in Theoretical Computer Science. An EATCS Series,
https://doi.org/10.1007/978-3-662-59362-2_16

rent solution until a complete one is obtained. Though so constructed solutions are of different quality, they are obtained in a reasonable time.

Remark 16.1. In Chap. 17, we consider two other groups of heuristic algorithms, called *local search algorithms* and *meta-heuristic algorithms*, respectively. Heuristic algorithms from these groups start with a complete solution and then try to improve the current solution iteratively.

A good heuristic algorithm generates a suboptimal solution in polynomial time. This solution is of good quality, sometimes comparable to the quality of the optimal solution. The quality of solutions generated by a heuristic usually is evaluated using a *computational experiment*. The larger the size of instances tested in such an experiment, the more justified is the claim that the evaluated heuristic can be recommended for use in practice.

16.2 Minimizing the maximum completion time

In this section, we consider heuristic algorithms for the maximum completion time criterion, $C_{\max}$.

16.2.1 Linear deterioration

In this section, we focus on heuristic algorithms for time-dependent scheduling problems with linearly deteriorating job processing times. We begin with the case when all jobs have equal ready times and there are no deadlines.

Parallel machine problems

Hsieh and Bricker [6] proposed three heuristic algorithms for problem $Pm|p_j = a_j + b_j t|C_{\max}$, where $b_j \in (0, 1)$ for $1 \leqslant j \leqslant n$. All of these algorithms are adaptations of Algorithm 14.1 for linear job processing times.

The first heuristic algorithm exploits Theorem 7.14 concerning problem $1|p_j = a_j + b_j t|C_{\max}$.

Algorithm 16.1. for problem $Pm|p_j = a_j + b_j t|C_{\max}$

1 **Input** : sequences $(a_1, a_2, \ldots, a_n)$, $(b_1, b_2, \ldots, b_n)$
2 **Output:** a suboptimal schedule
  ▷ Step 1
3 Arrange jobs in non-increasing order of the ratios $\frac{a_j}{b_j}$;
  ▷ Step 2
4 **for** $i \leftarrow 1$ **to** n **do**
5 $\quad \big|\quad$ Assign job $J_{[i]}$ to the machine with the smallest completion time;
  end
6 **return.**

The second heuristic algorithm uses another rule for job ordering.

Algorithm 16.2. for problem $Pm|p_j = a_j + b_j t|C_{\max}$

---

1 **Input**   : sequences $(a_1, a_2, \ldots, a_n)$, $(b_1, b_2, \ldots, b_n)$
2 **Output:** a suboptimal schedule
▷ Step 1
3 Arrange jobs in non-increasing order of the ratios $\frac{1-b_j}{a_j}$;
▷ Step 2
4 **for** $i \leftarrow 1$ **to** n **do**
5 $\quad|\quad$ Assign job $J_{[i]}$ to the machine with the smallest machine load;
   end
6 **return.**

The third heuristic algorithm proposed in [6] contains a random step and is based on the observation that optimal schedules for problem $Pm|p_j = a_j + b_j t|C_{\max}$ are often very close to the schedules generated by Algorithm 16.2.

Algorithm 16.3. for problem $Pm|p_j = a_j + b_j t|C_{\max}$

---

1 **Input**   : sequences $(a_1, a_2, \ldots, a_n)$, $(b_1, b_2, \ldots, b_n)$
2 **Output:** a suboptimal schedule
▷ Step 1
3 $N_{\mathcal{J}} \leftarrow \{1, 2, \ldots, n\}$; $Z_0 \leftarrow 0$; $Z_{n+1} \leftarrow 1$; $j \leftarrow 1$;
▷ Step 2
4 **while** $(j \leqslant n)$ **do**
5 $\quad\Big|\quad tmp \leftarrow \sum\limits_{k \in N_{\mathcal{J}}} \frac{b_k}{a_k}$;
6 $\quad\Big|\quad$ **for** $all\ i \in N_{\mathcal{J}}$ **do**
7 $\quad\Big|\quad\Big|\quad z_i \leftarrow \frac{b_i}{a_i * tmp}$;
8 $\quad\Big|\quad\Big|\quad Z_i \leftarrow \sum\limits_{\substack{j \in N_{\mathcal{J}} \\ j \leqslant i}} z_j$;
9 $\quad\Big|\quad\Big|\quad$ Generate a random number $r \in (0, 1)$;
10 $\quad\Big|\quad\Big|\quad$ **if** $((Z_{i-1} < r \leqslant Z_i) \wedge (i \in N_{\mathcal{J}}))$ **then**
11 $\quad\Big|\quad\Big|\quad\Big|\quad T_j \leftarrow i$
        else
12 $\quad\Big|\quad\Big|\quad\Big|\quad$ **if** $((Z_{i-1} < r \leqslant Z_i) \wedge (i = n+1))$ **then**
13 $\quad\Big|\quad\Big|\quad\Big|\quad\Big|\quad T_j \leftarrow n+1$;
          end
       end
14 $\quad\Big|\quad\Big|\quad N_{\mathcal{J}} \leftarrow N_{\mathcal{J}} \setminus \{i\}$; $j \leftarrow j+1$;
     end
   end
▷ Step 3
15 Rearrange jobs according to table T;
16 **return.**

Computational experiments conducted for $n = 10$ and $n = 15$ jobs show that for these values of n, Algorithms 16.1–16.3 generate schedules that are close to optimal ones; see [6, Sect. 5].

16.2.2 General non-linear deterioration

In this section, we consider heuristic algorithms for time-dependent scheduling problems with non-linearly deteriorating job processing times. We begin with the case when all jobs have equal ready times and there are no deadlines.

Single machine problems

Gupta and Gupta [5] proposed two heuristics for a single machine scheduling problem with job processing times in the form of (6.5), where functions $f_j(t)$ are polynomials of $m \geqslant 2$ degree for $1 \leqslant j \leqslant n$. Both heuristics are based on the calculation of a set of ratios which are similar to the ratios applied in Theorem 7.14 for $f_j = a_j + b_j t$, where $1 \leqslant j \leqslant n$.

In the first algorithm, the ratios are independent of time.

Algorithm 16.4. for problem $1|p_j = a_j + b_j t + \cdots + n_j t^m|C_{\max}$

1 **Input**   : sequences $(a_j, b_j, \ldots, n_j), 1 \leqslant j \leqslant n$
2 **Output:** a suboptimal schedule
   ▷ Step 1
3 $N_{\mathcal{J}} \leftarrow \{1, 2, \ldots, n\}$;
   ▷ Step 2
4 **for** each $j \in N_{\mathcal{J}}$ **do**
5 $\quad q_j(a) \leftarrow a_j$;
6 $\quad q_j(b) \leftarrow \frac{a_j}{b_j}$;
   $\quad \vdots \quad \vdots \quad \vdots$
7 $\quad q_j(m) \leftarrow \frac{a_j}{m_j}$;
   end
   ▷ Step 3
8 Generate m schedules by arranging jobs in non-increasing order of the
   ratios $q_j(x)$ for $a \leqslant x \leqslant m$;
   ▷ Step 4
9 Select the schedule with the minimal completion time;
10 **return**.

For $m = 1$, the ratios calculated in Algorithm 16.4 may be transformed to those in Algorithm 7.3. However, this is not the case for $m \geqslant 2$. Therefore, in the second algorithm, the calculated parameters $q_j(x)$ are functions of time. We formulate this algorithm for the case of quadratic job processing times (6.20), i.e., when $p_j = a_j + b_j t + c_j t^2$ for $1 \leqslant j \leqslant n$.

Algorithm 16.5. for problem $1|p_j = a_j + b_j t + c_j t^2|C_{\max}$

---

1 **Input** : sequences $(a_j, b_j, c_j), 1 \leqslant j \leqslant n$
2 **Output:** a suboptimal schedule
   ▷ Step 1
3 $\sigma \leftarrow (\phi)$;
4 $N_{\mathcal{J}} \leftarrow \{1, 2, \ldots, n\}$;
5 $T \leftarrow 0$;
   ▷ Step 2
6 **while** $(N_{\mathcal{J}} \neq \emptyset)$ **do**
7   **for** *each* $j \in N_{\mathcal{J}}$ **do**
8         $q_j(a) \leftarrow \frac{a_j}{b_j + c_j T}$;
9         $q_j(b) \leftarrow \frac{a_j + b_j T + c_j T^2}{b_j + 2c_j T}$;
10         $q_j(c) \leftarrow a_j + b_j T + c_j T^2$;
        end
11   Find $k \in N_{\mathcal{J}}$ such that
           $q_k(x) = \min\{q_j(x) : j \in N_{\mathcal{J}} \wedge x \in \{a, b, c\}\}$;
12     $\sigma \leftarrow (\sigma|k)$;
13     $N_{\mathcal{J}} \leftarrow N_{\mathcal{J}} \setminus \{k\}$;
14     $T \leftarrow T + C_k$;
   end
   ▷ Step 3
15 Generate m schedules by arranging jobs in non-increasing order of the
   ratios $q_j(x)$ for $a \leqslant x \leqslant m$;
   ▷ Step 4
16 Select the schedule with the minimal completion time;
17 **return** σ.

The performance of Algorithms 16.4 and 16.5 was tested on small instances with $4 \leqslant n \leqslant 7$ jobs. Since the heuristics were found to be unpredictable in producing schedules close to the optimal one, a technique for improvement of the final schedule was proposed; see [5, Tables 6–7].

Problem $1|p_j = g_j(t)|C_{\max}$ with general non-linear processing times was also considered by Alidaee [1], who proposed for this problem a heuristic algorithm, based on the following result.

Theorem 16.2. (Alidaee [1]) *Let σ' be a sequence of jobs, let σ'' be sequence σ' with mutually exchanged positions i and $i + 1$, $\sigma'' = \sigma'(i \leftrightarrow i + 1)$, and let T denote the maximum completion time for the first $i - 1$ jobs in the sequence. Then there exist real numbers $z_{ji} \in [T, T + p_{j,i+1}(T)]$ and $z_{j,i+1} \in [T, T + p_{ji}(T)]$ such that if the inequality*

$$\frac{g_{ji}(z_{ji})}{g_{ji}(T)} \geqslant \frac{g_{j,i+1}(z_{j,i+1})}{g_{j,i+1}(T)} \tag{16.1}$$

holds, then $C_{\max}(\sigma') \leqslant C_{\max}(\sigma'')$.

Proof. Consider sequences of jobs σ' and σ'', where $\sigma'' = \sigma'(i \leftrightarrow i+1)$. Let T denote the maximum completion time for the first $i-1$ jobs in a sequence and let

$$\Delta(\sigma'', \sigma') = C_{\max}(\sigma'') - C_{\max}(\sigma')$$

be the difference between criterion values for schedules σ'' and σ'. Then, if the inequality

$$g_{ji}(T + g_{j,i+1}(T)) - g_{ji}(T) \geqslant g_{j,i+1}(T + g_{j,i}(T)) - g_{j,i+1}(T)$$

holds, then $\Delta(\sigma'', \sigma') \geqslant 0$.

By Theorem 1.16 (b), there exist real numbers z_{ji} and $z_{j,i+1}$ such that $z_{ji} \in [T, T+g_{j,i+1}(T)]$, $z_{j,i+1} \in [T, T+g_{ji}(T)]$ and such that (16.1) is satisfied.
$\square$

On the basis of Theorem 16.2, Alidaee [1] proposed the following algorithm.

Algorithm 16.6. for problem $1|p_j = g_j(t)|C_{\max}$

---

1 **Input** : sequence $(g_1, g_2, \ldots, g_n)$
2 **Output:** a suboptimal schedule
  ▷ Step 1
3 $T \leftarrow 0$;
4 $N_{\mathcal{J}} \leftarrow \{1, 2, \ldots, n\}$;
5 $K \leftarrow n$;
6 $\sigma \leftarrow (\phi)$;
  ▷ Step 2
7 **while** $(N_{\mathcal{J}} \neq \emptyset)$ **do**
8 $\quad$ $z \leftarrow T + \frac{1}{2}K \sum_{k \in N_{\mathcal{J}}} g_k(T)$;
9 $\quad$ Choose the smallest $i \in N_{\mathcal{J}}$ such that $\frac{g_i'(z)}{g_i(T)} = \max\limits_{j \in N_{\mathcal{J}}} \left\{ \frac{g_j'(z)}{g_j(T)} \right\}$;
10 $\quad$ $\sigma \leftarrow (\sigma|i)$;
11 $\quad$ $T \leftarrow T + g_i(T)$;
12 $\quad$ $N_{\mathcal{J}} \leftarrow N_{\mathcal{J}} \setminus \{i\}$;
13 $\quad$ $K \leftarrow K - 1$;
  end
  ▷ Step 3
14 **return** σ.

Algorithm 16.6 was tested on two sets of instances. In the first set of instances, the job processing times were quadratic, i.e. $p_j = a_j + b_j t + c_j t^2$ for $1 \leqslant j \leqslant n$, where $5 \leqslant n \leqslant 8$ (see [1, Table 1]). In the second set, the job processing times were exponential, $p_j = e^{a_j t}$ for $1 \leqslant j \leqslant n$, where $5 \leqslant n \leqslant 9$. Twenty instances were generated for each value of n.

For quadratic processing times, Algorithm 16.6 gave better results than Algorithm 16.5. However, the results became worse for a polynomial of degree higher than 2, and for the case when the coefficients $a_j, b_j, \ldots, n_j$ were considerably different in size for $1 \leqslant j \leqslant n$.

For exponential job processing times given by (6.23), Janiak and Kova-lyov [7] proposed three heuristics. The authors claim (see [7, Sect. 4]) that one of these algorithms, which schedules jobs in non-decreasing order of r_j values, gives the best results.

Now, we will pass to heuristic algorithms for time-dependent scheduling problems with general non-linearly deteriorating job processing times, distinct ready times and deadlines.

For step deteriorating processing times given by (6.26), Mosheiov [10] proposed the following heuristic algorithm (see Algorithm 16.7), which is an adaptation of Moore-Hodgson's algorithm for problem $1 || \sum U_j$, [9]. The main idea of the new algorithm is as follows.

First, a list of jobs is created in which the jobs are arranged in non-decreasing order of a_j values. Next, the first available job from this list is assigned to the first available machine. If there are no tardy jobs, the algorithm stops. Otherwise, we proceed as in the Moore-Hodgson algorithm for problem $1 || \sum U_j$, [9], i.e., we identify the first tardy job, reconsider all previously scheduled jobs (including the tardy job) and remove the larger job, identify the next tardy job, etc.

Let E denote the set of jobs which have not been scheduled yet. Then, the pseudo-code of the considered algorithm may be formulated as follows.

Algorithm 16.7. for problem $1|p_j \equiv (6.26)|C_{\max}$

1 **Input**   : sequences $(a_1, a_2, \ldots, a_n)$, $(b_1, b_2, \ldots, b_n)$, $(d_1, d_2, \ldots, d_n)$
2 **Output:** a suboptimal schedule
  ▷ Step 1
3 $J \leftarrow \emptyset$; $E \leftarrow \{1, 2, \ldots, n\} \setminus J$;
4 Arrange jobs in non-increasing order of the deadlines d_j;
5 Schedule jobs in the obtained order;
  ▷ Step 2
6 **while** (*there exist* $j \in E$ *such that* $S_j > d_j$) **do**
7   $\quad k \leftarrow \min\{j \in J : S_j > d_j\}$;
8   $\quad l \leftarrow \arg \min\{\frac{b_j - a_j}{a_j} : j \in E, j \leqslant k, a_j \geqslant S_k - d_k\}$;
9   $\quad J \leftarrow J \cup \{l\}$; $E \leftarrow E \setminus \{l\}$;
  end
  ▷ Step 3
10 $\sigma \leftarrow (E|J)$;
11 **return** σ.

Though Algorithm 16.7 may generate bad schedules (see [10, Example 1]), the results of the computational experiment reported in [10, Sect. 2.3] suggest that on average the performance of Algorithm 16.7 is quite satisfactory.

For multi-step deteriorating processing times given by (6.28), Mosheiov [10] proposed Algorithm 16.8, which is an adaptation of Algorithm 16.7. We formulate Algorithm 16.8 for the case of two-step job deterioration.

Algorithm 16.8. for problem $1|p_j \equiv (6.28)|C_{\max}$

1 **Input** : sequences $(a_1^1, a_2^1, \ldots, a_n^1)$, $(a_1^2, a_2^2, \ldots, a_n^2)$, $(d_1^i, d_2^i, \ldots, d_n^i)$,
$(d_1^2, d_2^2, \ldots, d_n^2)$
2 **Output:** a suboptimal schedule
▷ Step 1
3 $J^1 \leftarrow \emptyset$; $E^1 \leftarrow \{1, 2, \ldots, n\} \setminus J^1$;
4 $J^2 \leftarrow \emptyset$; $E^2 \leftarrow \{1, 2, \ldots, n\} \setminus J^2$;
5 Arrange jobs in non-increasing order of the deadlines d_j^1;
6 $T \leftarrow 0$;
▷ Step 2
7 **while** (*there exist* $j \in E^1$ *such that* $S_j > d_j$) **do**
8      $l_1 \leftarrow \min\{j \in E^1 : S_j > d_j^1\}$;
9      $l_2 \leftarrow \arg \min\{\frac{a_j^2 - a_j^1}{a_j^1} : j \in E^1, j \leqslant k, a_j^1 \geqslant S_{l_1} - d_{l_1}^1\}$;
10     $J^1 \leftarrow J^1 \cup \{l_1\}$; $E^1 \leftarrow E^1 \setminus \{l_1\}$; $T \leftarrow T + C_{l_1}$;
end
▷ Step 3
11 Rearrange jobs in J^1 in non-decreasing order of the processing times a_j^2;
12 Call the obtained sequence τ;
13 Starting at time T, schedule jobs in J^1 according to τ;
14 **while** (*there exist* $j \in E^2$ *such that* $S_j > d_j^2$) **do**
15     $l_1 \leftarrow \min\{j \in E^2 : S_j > d_j^2\}$;
16     $l_2 \leftarrow \arg \min\{\frac{a_j^3 - a_j^2}{a_j^2} : j \in E^2, j \leqslant l_1, a_j^2 \geqslant S_{l_1} - d_{l_1}^2\}$;
17     $J^2 \leftarrow J^2 \cup \{l_1\}$; $E^2 \leftarrow E^2 \setminus \{l_2\}$;
end
▷ Step 4
18 $\sigma \leftarrow (E^1|(J^1 \setminus J^2)|J^2)$;
19 **return** σ.

For arbitrary m, Algorithm 16.8 runs in $O(n(n + m)) \approx O(n^2)$ time, since usually $n \gg m$. Similarly to Algorithm 16.7, also Algorithm 16.8 may generate arbitrarily bad schedules (see [10, Example 2, Sect. 3.2]) However, the reported results of computational experiments (see [10, Sect. 4.2]) suggest that on average the schedules generated by Algorithm 16.8 are about 10% longer than the optimal ones.

For job processing times given by (6.24), Kunnathur and Gupta [8] proposed five heuristic algorithms which construct the final schedule iteratively. The first of the algorithms iteratively finds the job with the minimal value of function L, where $L(\sigma|k) := C_{\max}(\sigma|k) + \sum_{i \in N_J} (a_i + \max\{0, b_i(C_{\max}(\sigma|i) - d_i)\})$

for $\sigma \in \hat{\mathfrak{S}}_n$ and $k \in \{1, 2, \ldots, n\}$.

Algorithm 16.9. for problem $1|p_j \equiv (6.24)|C_{\max}$

---

1 **Input**   : sequences $(a_j, b_j, d_j), 1 \leqslant j \leqslant n$
2 **Output:** a suboptimal schedule
   ▷ Step 1
3 $\sigma \leftarrow (\phi); N_{\mathcal{J}} \leftarrow \{1, 2, \ldots, n\}; T \leftarrow 0;$
   ▷ Step 2
4 **while** $(N_{\mathcal{J}} \neq \emptyset)$ **do**
5 │   Find $j \in N_{\mathcal{J}}$ such that $L(\sigma|j) = \min\{L(\sigma|k) : k \in N_{\mathcal{J}}\};$
6 │ $\sigma \leftarrow (\sigma|j); N_{\mathcal{J}} \leftarrow N_{\mathcal{J}} \setminus \{j\}; T \leftarrow T + p_j;$
   end
   ▷ Step 3
7 **return** σ.

Ties in Step 2 of Algorithm 16.9 are broken by selecting the smallest d_k, then the largest b_k and then the smallest a_k.

The second heuristic algorithm proposed in [8] iteratively finds the job with the minimal value of the ratio $\frac{a_j}{b_j}$.

Algorithm 16.10. for problem $1|p_j \equiv (6.24)|C_{\max}$

---

1 **Input**   : sequences $(a_j, b_j, d_j), 1 \leqslant j \leqslant n$
2 **Output:** a suboptimal schedule
   ▷ Step 1
3 $\sigma \leftarrow (\phi); N_{\mathcal{J}} \leftarrow \{1, 2, \ldots, n\}; T \leftarrow 0;$
   ▷ Step 2
4 **while** $(N_{\mathcal{J}} \neq \emptyset)$ **do**
5 │   **repeat**
6 │   │ $L \leftarrow \{k \in N_{\mathcal{J}} : d_k < T\}; F \leftarrow N_{\mathcal{J}} \setminus L;$
7 │   │   Find $j \in F$ such that $\frac{a_j}{b_j} = \min_{k \in F}\{\frac{a_k}{b_k}\};$
   │   │   ▷ Break ties by selecting the job with the smallest d_k
8 │   │ $\sigma \leftarrow (\sigma|j); N_{\mathcal{J}} \leftarrow N_{\mathcal{J}} \setminus \{j\}; T \leftarrow T + p_j;$
   │   **until** $(F = \emptyset);$
9 │   Arrange L in non-decreasing order of the ratios $\frac{a_j - d_j b_j}{b_j};$
10 │ $\sigma \leftarrow (\sigma|L); T \leftarrow T + \sum_{i \in L} p_i;$
   end
   ▷ Step 3
11 **return** σ.

The remaining heuristics proposed in [8] differ from Algorithm 16.10 only in Step 2 (see [8, Sect. 4] for details). The reported results of computational experiments (see [8, Sect. 6]) suggest that Algorithm 16.9 is the best one.

16.2.3 Linear shortening

In this section, we consider approximation algorithms and schemes for time-dependent scheduling problems with linearly shortening job processing times, equal job ready times and without deadlines.

Single machine problems

For problem $1|p_j = a_j - b_j(t-y), y > 0, 0 < b_j < 1, Y < +\infty|C_{\max}$, Cheng et al. [3] proposed three heuristic algorithms, all running in $O(n \log n)$ time. The first algorithm is based on Property 7.56 and schedules jobs in non-increasing order of the ratios $\frac{a_j}{b_j}$. In the second algorithm, the jobs starting by time $t = y$ are sequenced in non-decreasing order of the ratios $\frac{a_j}{b_j}$ and Property 7.56 is applied only to jobs which start after time y.

Algorithm 16.11. for problem $1|p_j = a_j - b_j(t - y),\ y > 0,$ $0 < b_j < 1, Y < +\infty|C_{\max}$

---

1 **Input** : sequences $(a_1, a_2, \ldots, a_n), (b_1, b_2, \ldots, b_n)$, numbers y, Y
2 **Output:** a suboptimal schedule
  ▷ Step 1
3 Arrange jobs in non-decreasing order of the ratios $\frac{a_j}{b_j}$;
4 $i \leftarrow 1;\ j \leftarrow n;\ \sigma \leftarrow (\phi);\ T \leftarrow 0;$
  ▷ Step 2
5 **while** $(T < y)$ **do**
6 $\quad |\quad \sigma \leftarrow (\sigma|[i]);\ T \leftarrow T + a_{[i]};\ i \leftarrow i + 1;$
  end
  ▷ Step 3
7 **while** $(i < j)$ **do**
8 $\quad |\quad \sigma \leftarrow (\sigma|[j]);$
9 $\quad |\quad$ **if** $(y \leqslant T < Y)$ **then**
  $\quad \quad |\quad T \leftarrow T + a_{[j]} - b_{[j]}(T - y);$
  $\quad \quad$ **end**
10 $\quad |\quad$ **if** $(T \geqslant Y)$ **then**
  $\quad \quad |\quad T \leftarrow T + a_{[j]} - b_{[j]}(Y - y);$
  $\quad \quad$ **end**
11 $\quad |\quad j \leftarrow j - 1;$
  end
  ▷ Step 4
12 **return** σ.

The third algorithm proposed in [3], in which the final job sequence is composed of two subsequences, can be formulated as follows.

Algorithm 16.12. for problem $1|p_j = a_j - b_j(t - y)$, $y > 0$, $0 < b_j < 1$, $Y < +\infty|C_{\max}$

---

1 **Input** : sequences $(a_1, a_2, \ldots, a_n), (b_1, b_2, \ldots, b_n)$, numbers y, Y
2 **Output:** a suboptimal schedule
  ▷ Step 1
3 Arrange jobs in non-decreasing order of the ratios $\frac{a_j}{b_j}$;
4 $\sigma^1 \leftarrow (\phi)$;
5 $\sigma^2 \leftarrow (\phi)$;
6 $N_{\mathcal{J}} \leftarrow \{1, 2, \ldots, n\}$;
  ▷ Step 2
7 **while** $(N_{\mathcal{J}} \neq \emptyset)$ **do**
8   $\quad$ Choose job $J_k \in N_{\mathcal{J}}$ such that $a_k = \max\limits_{j \in N_{\mathcal{J}}} \{a_j\}$;
9   $\quad$ $\sigma^1 \leftarrow (\sigma^1|k)$;
10 $\quad$ $N_{\mathcal{J}} \leftarrow N_{\mathcal{J}} \setminus \{k\}$;
11 $\quad$ Choose job $J_l \in N_{\mathcal{J}}$ such that $b_l = \max\limits_{j \in N_{\mathcal{J}}} \{b_j\}$;
12 $\quad$ $\sigma^2 \leftarrow (l|\sigma^2)$;
13 $\quad$ $N_{\mathcal{J}} \leftarrow N_{\mathcal{J}} \setminus \{l\}$;
  end
  ▷ Step 3
14 $\sigma \leftarrow (\sigma^1|\sigma^2)$;
15 **return** σ.

The results of computational experiments suggest (see [3, Sect. 2.3]) that scheduling jobs in non-increasing order of the ratios $\frac{a_j}{b_j}$ gives the best results.

16.3 Minimizing the total completion time

In this section, we present heuristic algorithms for time-dependent scheduling problems with the $\sum C_j$ criterion.

16.3.1 Proportional deterioration

In this section, we consider heuristic algorithms for time-dependent scheduling problems with proportionally deteriorating job processing times. We begin with the case when all jobs have equal ready times and there are no deadlines.

Parallel machine problems

Jeng and Lin [15] proposed the following $O(n \log n)$ heuristic algorithm for problem $Pm|p_j = b_j t| \sum C_j$.

Algorithm 16.13. for problem $Pm|p_j = b_j t| \sum C_j$

---

1 **Input** : sequence $(b_1, b_2, \ldots, b_n)$
2 **Output:** a suboptimal schedule
  ▷ **Step 1**
3 Arrange jobs in non-increasing order of the deterioration rates b_j;
  ▷ **Step 2**
4 **for** $i \leftarrow 1$ **to** n **do**
5 | Assign job $J_{[i]}$ to the machine with the smallest machine load;
  end
  ▷ **Step 3**
6 Reverse the job sequence on each machine;
7 **return.**

Since in Step 3 Algorithm 16.13 reverses the sequence of jobs on each machine, in the final schedule the jobs are arranged in non-decreasing order of deterioration rates b_j. Algorithm 16.13 runs in $O(n \log n)$ time.

Remark 16.3. The algorithm applied in Step 3 of Algorithm 16.13 is sometimes called *reverse LPT* algorithm.

Example 16.4. Let us consider an instance of problem $Pm|p_j = b_j t| \sum C_j$, where we are given $m = 2$ machines and $n = 7$ jobs. Job processing times are as follows: $p_1 = 2t$, $p_2 = 3t$, $p_3 = t$, $p_4 = 4t$, $p_5 = 3t$, $p_6 = 5t$ and $p_7 = t$. Both machines start at time $t_0 = 1$.

In Step 1 of Algorithm 16.13 jobs are arranged according to sequence $(6, 4, 5, 2, 1, 7, 3)$. Next, in Step 2, the algorithm assigns sequence $(6, 2, 7, 3)$ of jobs J_2, J_6, J_7 and J_3 to machine M_1, and sequence $(4, 5, 2)$ of jobs J_4, J_5 and J_2 to machine M_2. In Step 3, it returns sequences $(3, 7, 2, 6)$ and $(2, 5, 4)$. ♦

Remark 16.5. The construction of a schedule by assigning jobs to machines in the reverse order compared to the order generated by an algorithm appeared in the classical scheduling theory in the mid 1970s (see, e.g., Bruno et al. [2], Coffman [4] and Strusevich and Rustogi [12]). In the modern scheduling theory, this idea is applied to some new scheduling models such as *master-slave model* (see, e.g., Sahni and Vairaktarakis [11]).

Dedicated machine problems

Wang et al. [17] proposed the following two-phase heuristic algorithm for problem $F2|p_{ij} = b_{ij} t| \sum C_j$. The main idea of this algorithm is to construct a schedule in the first phase and next to improve it in the second phase.

Given a schedule σ, let $\sigma(J_{[i]} \rightarrow \sigma_k)$ denote a schedule obtained from σ by moving $J_{[i]}$ forward to position k. The pseudo-code of the considered algorithm is as follows.

Algorithm 16.14. for problem $F2|p_{ij} = b_{ij}t| \sum C_j$

1 **Input** : sequences $(b_{11}, b_{21}, \ldots, b_{n1}), (b_{12}, b_{22}, \ldots, b_{n2})$
2 **Output:** a suboptimal schedule
  ▷ Step 1: the beginning of Phase I
3 $k \leftarrow 1; N_{\mathcal{J}} \leftarrow \{1, 2, \ldots, n\}; \sigma := (\phi);$
4 Find J_i such that $(1 + b_{i1})(1 + b_{i2}) = \min\{(1 + b_{k1})(1 + b_{k2}) : k \in N_{\mathcal{J}}\};$

5 $\sigma_k \leftarrow i; A_k \leftarrow t_0(1 + b_{i1}); C_{[k]} \leftarrow t_0(1 + b_{i1})(1 + b_{i2}); N_{\mathcal{J}} \leftarrow N_{\mathcal{J}} \setminus \{i\};$
6 **while** $(N_{\mathcal{J}} \neq \emptyset)$ **do**
7 $\quad E := \{J_l \in \mathcal{J} : A_k(1 + b_{l1}) \leqslant C_{[k]}\};$
8 $\quad$ **if** $(|E| \geqslant 1)$ **then**
9 $\quad\quad |$ Find $J_i \in E$ such that $b_{i1} = \min\{b_{k1} : J_k \in E\};$
  $\quad$ **else**
10 $\quad\quad |$ Find $J_i \in \mathcal{J}$ such that $b_{i1} = \min\{b_{k1} : k \in N_{\mathcal{J}}\};$
  $\quad$ **end**
11 $\quad \sigma_{k+1} \leftarrow i; C_{[k+1]} := \max\{A_k(1 + b_{i1}), C_{[k]}\}(1 + b_{i2});$
12 $\quad A_{k+1} := A_k(1 + b_{i1}); N_{\mathcal{J}} := N_{\mathcal{J}} \setminus \{i\}; k := k + 1;$
  end
  ▷ Step 2: the beginning of Phase II
13 **repeat**
14 $\quad k \leftarrow 1; i \leftarrow k + 1;$
15 $\quad$ **while** $(k \leqslant n)$ **do**
16 $\quad\quad \sigma' \leftarrow \sigma(J_{[i]} \rightarrow \sigma_k);$
17 $\quad\quad$ **if** $(\sum C_j(\sigma') < \sum C_j(\sigma))$ **then**
18 $\quad\quad\quad | \quad \sigma \leftarrow \sigma';$
  $\quad\quad$ **end**
19 $\quad\quad i \leftarrow i + 1; k \leftarrow k + 1;$
  $\quad$ **end**
  until $(k \geqslant n);$
20 **return** $\sigma.$

Algorithm 16.14 was tested on 100 instances with $n \leqslant 14$ jobs. In the experiments (see [17, Sect. 6] for details) the mean and the maximum error of generated schedules did not exceed 6.9% and 21.4%, respectively.

16.3.2 Linear deterioration

In this section, we consider heuristic algorithms for time-dependent scheduling problems with linearly deteriorating job processing times. We begin with the case when all jobs have equal ready times and there are no deadlines.

Single machine problems

Problem $1|p_j = 1 + b_j t| \sum C_j$ is one of open time-dependent scheduling problems (we discuss this problem in Appendix A). In Chap. 15, we considered

Algorithms 15.1 and 15.2, based on the so-called *signatures* of a job deterioration rate sequence (cf. Gawiejnowicz et al. [14]). In this section, we present two heuristic algorithms proposed for this problem by Mosheiov [16].

The first of these heuristics, acting on a non-increasingly ordered sequence of deterioration rates $b_1, b_2, \ldots, b_n$, adds the job corresponding to a given b_j either to the left or to the right branch of the constructed V-shaped sequence.

Algorithm 16.15. for problem $1|p_j = 1 + b_j t| \sum C_j$

---

1 **Input** : sequences $(b_1, b_2, \ldots, b_n)$
2 **Output:** a suboptimal V-shaped schedule
  ▷ Step 1
3 Arrange jobs in non-increasing order of the deterioration rates b_j;
  ▷ Step 2
4 $l \leftarrow (\phi); r \leftarrow (\phi); i \leftarrow 1;$
5 **while** $(i \leqslant n)$ **do**
6 | **if** $(i \text{ is odd})$ **then**
7 | | $l \leftarrow (l|b_{[i]});$
  | **else**
8 | | $r \leftarrow (b_{[i]}|r);$
  | **end**
9 | $i \leftarrow i + 1;$
  end
  ▷ Step 3
10 $\sigma \leftarrow (l|r);$
11 **return** σ.

Example 16.6. Let us consider an instance of problem $1|p_j = 1 + b_j t| \sum C_j$, where we are given $n = 5$ jobs with the following deterioration rates: $b_1 = 3$, $b_2 = 1$, $b_3 = 5$, $b_4 = 2$ and $b_5 = 4$.

In Step 1 of Algorithm 16.15, jobs are arranged into sequence $(3, 5, 1, 4, 2)$. In Step 2, the algorithm generates sequences $l = (3, 5, 4)$ and $r = (1, 2)$. Finally, in Step 3, it returns final schedule $\sigma = (l|r) = (3, 5, 4, 1, 2)$.   ◆

Algorithm 16.15 assigns a job to the left (right) branch of the constructed V-shaped sequence in a fixed way, without checking the job deterioration rate. The main disadvantage of this procedure consists in the construction of unbalanced V-shaped sequences, i.e., those in which one branch is composed of jobs with large deterioration rates, while the other one is composed of jobs with small deterioration rates.

Therefore, the next heuristic algorithm proposed in [16] is equipped with a kind of balance control of branches of the constructed V-shaped sequence, in which the 'weight' of a branch is measured by the sum of deterioration rates of jobs assigned to this branch (see Algorithm 16.16). In the new algorithm, Step 1 is the same as in Algorithm 16.15. In Step 2, an element is added either to the left or to the right branch of the constructed V-shaped sequence if the sum of deterioration rates of the left (right) branch is lower than the sum of deterioration rates of the right (left) branch.

Algorithm 16.16. for problem $1|p_j = 1 + b_j t| \sum C_j$

1 **Input** : sequences $(b_1, b_2, \ldots, b_n)$
2 **Output:** a suboptimal V-shaped schedule
  ▷ Step 1
3 Arrange jobs in non-increasing order of the deterioration rates b_j;
  ▷ Step 2
4 $l \leftarrow (\phi); sum_l \leftarrow 0; r \leftarrow (\phi); sum_r \leftarrow 0; i \leftarrow 1;$
5 **while** $(i \leqslant n)$ **do**
6   **if** $(sum_l \leqslant sum_r)$ **then**
7 | $l \leftarrow (l|b_{[i]}); sum_l \leftarrow sum_l + b_{[i]};$
    else
8 | $r \leftarrow (b_{[i]}|r); sum_r \leftarrow sum_r + b_{[i]};$
    end
9     $i \leftarrow i + 1;$
  end
  ▷ Step 3
10 $\sigma \leftarrow (l|r);$
11 **return** σ.

Example 16.7. Let us consider the instance from Example 16.6 and apply to it Algorithm 16.16.

Since Step 1 of Algorithm 16.16 is the same as Step 1 in Algorithm 16.15, jobs are arranged into sequence $(3, 5, 1, 4, 2)$. In Step 2, the algorithm generates sequences $l = (3, 4, 2)$ and $r = (5, 1)$. Finally, in Step 3, it returns final schedule $\sigma = (l|r) = (3, 4, 2, 5, 1)$. ♦

Both Algorithms 16.15 and 16.16 run in $O(n \log n)$ time and are easy to implement. Furthermore, it has been shown (see [16, Sect. 4]) that they are asymptotically optimal, under the assumption that all deterioration rates are independent, identically distributed random variables. The reported results of computational experiments (see [16, Sect. 5]) suggest that the heuristics are, on average, quite efficient, with much better behaviour of Algorithm 16.15 than of Algorithm 16.16.

Dedicated machine problems

For problem $F2|p_{ij} = a_{ij} + bt| \sum C_j$, Wu and Lee [18] proposed six heuristics. The first three of them are $O(n \log n)$ algorithms which schedule jobs in non-decreasing order of the basic job processing times a_{1j}, a_{2j} and sums $a_{1j} + a_{2j}$, respectively.

The remaining three $O(n^2)$ algorithms are modified versions of the previous three algorithms. The modification consists in adding a procedure PI which uses the pairwise job interchange to improve a given schedule. The modified algorithms were evaluated by an experiment in which 4,440 instances with $n \leqslant 27$ jobs were tested (see [18, Sect. 6] for details). The experiment showed that the heuristic which schedules jobs in non-decreasing order of a_{1j} and next applies the procedure PI gives the best results.

16.3.3 Linear shortening

In this section, we consider approximation algorithms for time-dependent scheduling problems with linearly shortening job processing times. We begin with the case when all jobs have equal ready times and there are no deadlines.

Single machine problems

For the problem $1|p_j = a_j - b_j(t - y), y > 0, 0 < b_j < 1, Y < +\infty| \sum C_j$ Cheng et al. [13] proposed four heuristic algorithms.

The first algorithm schedules jobs in non-decreasing order of the basic processing times a_j. The pseudo-code of the second algorithm is as follows.

Algorithm 16.17. for problem $1|p_j = a_j - b_j(t - y), y > 0, 0 < b_j < 1, Y < +\infty| \sum C_j$

---

1 **Input** : sequences $(a_1, a_2, \ldots, a_n), (b_1, b_2, \ldots, b_n)$, numbers y, Y
2 **Output:** a suboptimal schedule
  ▷ Step 1
3 Arrange jobs in non-decreasing order of the basic processing times a_j;
4 $C \leftarrow 0; k \leftarrow 1; N_{\mathcal{J}} \leftarrow \{1, 2, \ldots, n\}$;
  ▷ Step 2
5 **while** $(C \leqslant y)$ **do**
6 $\quad$ Schedule job $J_{[k]}$;
7 $\quad C \leftarrow C + p_{[k]}(C); N_{\mathcal{J}} \leftarrow N_{\mathcal{J}} \setminus \{[k]\}; k \leftarrow k + 1$;
  end
  ▷ Step 3
8 **while** $(C \leqslant Y)$ **do**
9 $\quad$ **for** all $j \in N_{\mathcal{J}}$ **do**
10 $\quad\quad r_j \leftarrow \frac{C - a_j}{1 - b_j}$;
11 $\quad\quad$ Choose job $J_{[k]}$ such that $r_{[k]} = \min\{r_j \in N_{\mathcal{J}}\}$; Schedule job $J_{[k]}$;
12 $\quad\quad C \leftarrow C + p_{[k]}(C); N_{\mathcal{J}} \leftarrow N_{\mathcal{J}} \setminus \{[k]\}; k \leftarrow k + 1$;
  $\quad$ **end**
  end
  ▷ Step 4
13 Schedule remaining jobs in non-decreasing order of the differences
  $\quad a_j - b_j(Y - y)$;
14 **return.**

The third algorithm is a modification of Algorithm 16.17.

Algorithm 16.18. for problem $1|p_j = a_j - b_j(t - y), y > 0, 0 < b_j < 1, Y < +\infty| \sum C_j$

1 **Input** : sequences $(a_1, a_2, \ldots, a_n), (b_1, b_2, \ldots, b_n)$, numbers y, Y
2 **Output:** a suboptimal schedule
  ▷ Step 1
3 $\sigma^1 \leftarrow (\phi)$;
4 $\sigma^2 \leftarrow (\phi)$;
5 $N_{\mathcal{J}} \leftarrow \{1, 2, \ldots, n\}$;
  ▷ Step 2
6 **while** $(N_{\mathcal{J}} \neq \emptyset)$ **do**
7 $\quad$ Choose job J_k such that $a_k = \min_{j \in N_{\mathcal{J}}} \{a_j\}$;
8 $\quad \sigma^1 \leftarrow (\sigma^1|k)$;
9 $\quad N_{\mathcal{J}} \leftarrow N_{\mathcal{J}} \setminus \{k\}$;
10 $\quad$ Choose job J_l such that $b_l = \max_{j \in N_{\mathcal{J}}} \{a_j - b_j(Y - y)\}$;
11 $\quad \sigma^2 \leftarrow (l|\sigma^2)$;
12 $\quad N_{\mathcal{J}} \leftarrow N_{\mathcal{J}} \setminus \{l\}$;
  end
  ▷ Step 3
13 Arrange σ^1 in non-decreasing order of the basic processing times a_j;
14 Arrange σ^2 in non-decreasing order of the differences $a_j - b_j(Y - y)$;
  ▷ Step 4
15 $\sigma \leftarrow (\sigma^1|\sigma^2)$;
16 **return** σ.

The fourth algorithm proposed by Cheng et al. [13] for the considered problem is Algorithm 16.11. Algorithm 16.17 runs in $O(n^2)$ time, while Algorithm 16.18 run in $O(n \log n)$ time.

Algorithms proposed by Cheng et al. [13] were tested on instances with $n \leqslant 1,000$ jobs. Results of computational experiments (see [13, Sect. 2.3]) suggest that the algorithm which schedules jobs in non-decreasing order of the basic processing times a_j is the best.

16.4 Minimizing the maximum lateness

In this section, we present heuristic algorithms for the maximum lateness criterion, $L_{\max}$.

16.4.1 Linear deterioration

In this section, we consider heuristic algorithms for time-dependent scheduling problems with linearly deteriorating job processing times. We begin with the case when all jobs have equal ready times and there are no deadlines.

Single machine problems

Bachman and Janiak [19] proposed two heuristics for problem $1|p_j = a_j + b_j t|L_{\max}$. The first one schedules jobs in EDD order. The second one, given below, first arranges jobs in non-decreasing order of the ratios $\frac{a_j}{b_j}$ and then iteratively improves the solution.

Algorithm 16.19. for problem $1|p_j = a_j + b_j t|L_{\max}$

---

1 **Input** : sequences $(a_1, a_2, \ldots, a_n)$, $(b_1, b_2, \ldots, b_n)$, $(d_1, d_2, \ldots, d_n)$
2 **Output:** a suboptimal schedule
  ▷ Step 1
3 Arrange jobs in non-decreasing order of the ratios $\frac{a_j}{b_j}$;
4 Call the obtained schedule σ;
5 $\sigma' \leftarrow \sigma$;
  ▷ Step 2
6 **while** $(L_{\max}(\sigma) \leqslant L_{\max}(\sigma'))$ **do**
7   Find in σ the position k such that $k = \arg\max\{C_j - d_j\}$;
8   **while** $(TRUE)$ **do**
9   Denote the subset of jobs before job J_k by T_k;
10   Find in $\{J_j \in T_k : 0 \leqslant j \leqslant k-1\}$ job J_i such that $d_i > d_k$;
11   **if** $(i = 0)$ **then**
12   **|** return σ;
         end
13   Construct schedule σ' by inserting job J_i after job J_k;
14   **if** $(L_{\max}(\sigma') < L_{\max}(\sigma))$ **then**
15   **|** $\sigma \leftarrow \sigma'$;
         else
16   **if** $(i > 0)$ **then**
17   **|** $k \leftarrow i$;
            end
         end
     end
end
  ▷ Step 3
18 **return** σ.

The results of the computational experiment conducted for instances with $n = 10$ and $n = 50$ jobs (see [19, Sect. 4]) have shown that Algorithm 16.19 in most cases is better than Algorithm 16.8.

For the same problem, $1|p_j = a_j + b_j t|L_{\max}$, Hsu and Lin [20, Sect. 4] proposed two heuristic algorithms. The algorithms combine Algorithm 16.19 with the *Hill Climbing* algorithm (see Remark 17.7). However, the reported results of a computational experiment (see [20, Sect. 5]) do not allow us to formulate a clear conclusion about the performance of these algorithms.

16.4.2 General non-linear deterioration

In this section, we consider heuristic algorithms for time-dependent scheduling problems with general forms of non-linearly deteriorating job processing times.

Single machine problems

Janiak and Kovalyov [21] proposed three heuristic algorithms for exponential job processing times given by (6.23) in the case when all jobs have equal ready times and there are no deadlines. The first of the algorithms schedules jobs in EDD order. The second algorithm schedules jobs in non-decreasing order of the ratios $\frac{a_j}{b_j+d_j}$. The last one schedules jobs according to sequence $(b_j \nearrow |d_j \searrow)$. The authors claim (see [7, Sect. 4]) that the first and the third algorithms are the best.

16.5 Other criteria

In this section, we present heuristic algorithms for time-dependent scheduling problems with other criteria than $C_{\max}$, $\sum C_j$ or $L_{\max}$.

16.5.1 Proportional deterioration

First we consider heuristic algorithms for time-dependent scheduling problems with proportionally deteriorating job processing times. We begin with the case when all jobs have equal ready times and there are no deadlines.

Single machine problems

For problem $1|p_j = b_j t| \sum(C_i - C_j)$, Oron [29] proposed two heuristic algorithms. Both these heuristics run in $O(n \log n)$ time and construct the final schedule in two steps. The main idea is as follows.

In the first step, jobs are arranged in non-increasing order of deterioration rates. In the second step, jobs are assigned to predetermined positions in schedule. The indices of the positions change depending on whether the number of jobs n is even or odd (cf. Properties 7.99–7.103, Theorem 7.104).

The pseudo-code of the first of these heuristic algorithms is given below (see Algorithm 16.20). In the case when number n is even, the final schedule is in the form of

$$\left(\frac{n}{2}+1, \frac{n}{2}, \ldots, 3, 2, 1, \frac{n}{2}+2, \frac{n}{2}+3, \ldots, n\right),$$

otherwise it is in the form of

$$\left(\frac{n+3}{2}, \frac{n+1}{2}, \ldots, 3, 2, 1, \frac{n+5}{2}, \frac{n+7}{2}, \ldots, n\right).$$

The pseudo-code of this algorithm is as follows.

Algorithm 16.20. for problem $1|p_j = b_j t| \sum (C_i - C_j)$

1 **Input** : sequences $(b_1, b_2, \ldots, b_n)$
2 **Output:** a suboptimal schedule
  ▷ Step 1
3 Arrange jobs in non-increasing order of the deterioration rates b_j;
4 $\sigma \leftarrow \phi$;
  ▷ Step 2
5 **if** n *is even* **then**
6   **for** $i \leftarrow 1$ **to** $\frac{n}{2} + 1$ **do**
7 | Schedule job $J_{\frac{n}{2}+2-i}$ in the ith position in σ;
       end
8   **for** $i \leftarrow \frac{n}{2} + 2$ **to** n **do**
9 | Schedule job J_i in the ith position in σ;
       end
       ▷ The final schedule is
         $(\frac{n}{2}+1, \frac{n}{2}, \ldots, 3, 2, 1, \frac{n}{2}+2, \frac{n}{2}+3, \ldots, n)$
  else
       ▷ n is odd
       for $i \leftarrow 1$ **to** $\frac{n+3}{2}$ **do**
10 | Schedule job $J_{\frac{n+5}{2}-i}$ in the ith position in σ;
       end
       for $i \leftarrow \frac{n+5}{2}$ **to** n **step** 2 **do**
11 | Schedule job J_i in the ith position in σ;
       end
       ▷ The final schedule is $(\frac{n+3}{2}, \frac{n+1}{2}, \ldots, 3, 2, 1, \frac{n+5}{2}, \frac{n+7}{2}, \ldots, n)$
  end
  ▷ Step 3
12 **return** σ.

Example 16.8. (a) Let $n = 4$, $b_1 = 3$, $b_2 = 1$, $b_3 = 2$ and $b_4 = 4$. Then the final schedule generated by Algorithm 16.20 is $\sigma = (3, 2, 1, 4)$.

(b) Let $n = 5$, $b_1 = 3$, $b_2 = 1$, $b_3 = 2$, $b_4 = 4$ and $b_5 = 6$. Then the final schedule generated by Algorithm 16.20 is $\sigma = (4, 3, 2, 1, 5)$.   ◆

The second of these heuristic algorithms works similarly to Algorithm 16.20: if n is even, then the final schedule is in the form of

$$(n, n - 2, \ldots, 4, 2, 1, 3, \ldots, n - 3, n),$$

otherwise it is in the form of

$$(n, n - 2, \ldots, 5, 3, 1, 2, 4, \ldots, n - 3, n - 1).$$

The pseudo-code of this algorithm is as follows.

Algorithm 16.21. for problem $1|p_j = b_j t| \sum (C_i - C_j)$

1 **Input** : sequences $(b_1, b_2, \ldots, b_n)$
2 **Output:** a suboptimal schedule
  ▷ Step 1
3 Arrange jobs in non-increasing order of the deterioration rates b_j;
4 $\sigma \leftarrow \phi$;
  ▷ Step 2
5 **if** n *is even* **then**
6   **for** $i \leftarrow 1$ **to** $\frac{n}{2}$ **do**
7   Schedule job $J_{\frac{n}{2}+2-i}$ in the ith position in σ;
8   Schedule job J_{2i-1} in the $(\frac{n}{2} + i)$th position in σ;
      end
      ▷ The final schedule is $(n, n-2, \ldots, 4, 2, 1, 3, \ldots, n-3, n)$
  else
      ▷ n is odd
9   **for** $i \leftarrow 1$ **to** $\frac{n+1}{2}$ **do**
10   Schedule job J_{n+2-2i} in the ith position in σ;
      end
11   **for** $i \leftarrow 1$ **to** $\frac{n-1}{2}$ **do**
12   Schedule job J_{2i} in the $(\frac{n+1}{2} + i)$th position in σ;
      end
  end
  ▷ The final schedule is $(n, n-2, \ldots, 5, 3, 1, 2, 4, \ldots, n-3, n-1)$
  ▷ Step 3
13 **return** σ.

Example 16.9. (a) Let us consider the instance from Example 16.8 (a). Then the final schedule generated by Algorithm 16.21 is $\sigma = (4, 2, 1, 3)$.

(b) Let us consider the instance from Example 16.8 (b). Then the final schedule generated by Algorithm 16.21 is $\sigma = (5, 3, 1, 2, 4)$.   ◆

The performance of Algorithms 16.20 and 16.21 was tested in computational experiments with instances with $n = 20, 50$ or 100 jobs. For each n, 1,000 random instances were generated, with deterioration rates b_j from distribution $U(0.05, 1)$. The results of these experiments (see [29, Sect. 4]) suggest that Algorithm 16.20, on average, is more effective than Algorithm 16.21.

Parallel machine problems

For problem $Pm|p_j = b_j t| \sum C_{\max}^{(k)}$, Mosheiov [27] proposed to apply Algorithm 14.1. Moreover, he proved that if $n \to \infty$, then the absolute worst-case ratio of this algorithm for the $\sum C_{\max}^{(k)}$ criterion is bounded and is asymptotically close to 1.

Remark 16.10. The result concerning the absolute worst-case ratio can be also proved by using the notion of equivalent problems (cf. Remark 11.10 and Sect. 19.4).

16.5.2 Linear deterioration

In this section, we consider heuristic algorithms for time-dependent scheduling problems with linearly deteriorating job processing times. We begin with the case when all jobs have equal ready times and there are no deadlines.

Single machine problems

Alidaee and Landram [22] proposed an $O(n^2)$ heuristic algorithm for problem $1|p_j = g_j(t)|P_{\max}$. The algorithm is based on Lawler's algorithm for problem $1|prec|f_{\max}$, [26]. Let $C_{\max}(\mathcal{J})$ denote the maximum completion time of the last job in the set of jobs $\mathcal{J}$.

Algorithm 16.22. for problem $1|p_j = g_j(t)|P_{\max}$

---

1 **Input** : sequences $(g_1, g_2, \ldots, g_n)$
2 **Output:** a suboptimal schedule
  ▷ Step 1
3 $N_{\mathcal{J}} \leftarrow \{1, 2, \ldots, n\}$;
4 $\sigma \leftarrow (\phi)$;
  ▷ Step 2
5 **for** $i \leftarrow n$ **downto** 1 **do**
6   **for** *all* $j \in N_{\mathcal{J}}$ **do**
7   Schedule job J_j in the ith position in σ;
8       $T \leftarrow C_{\max}(N_{\mathcal{J}} \setminus \{j\})$;
9       $p_j \leftarrow g_j(T)$;
   end
10    $k \leftarrow \arg \min_{j \in N_{\mathcal{J}}} \{p_j\}$;
11    $\sigma \leftarrow (k|\sigma)$;
12    $N_{\mathcal{J}} \leftarrow N_{\mathcal{J}} \setminus \{k\}$;
  end
  ▷ Step 3
13 **return** σ.

In order to evaluate the quality of schedules generated by Algorithm 16.22, Alidaee and Landram [22] conducted a computational experiment in which $g_j(t) = a_j + b_j t$. For each n, where $6 \leqslant n \leqslant 9$, fifty random instances were generated, with $a_j \in (0, 1)$ and $b_j \in (1, 10)$. Since the performance was rather poor (see [22, Sect.3]), Algorithm 16.22 is not recommended for arbitrary a_j's and b_j's. The algorithm, however, is optimal when either $a_j > 0 \wedge b_j \geqslant 1$ or $b_j = b$ for $1 \leqslant j \leqslant n$ (see Sect. 7.4 for details).

For problem $1|p_j = a_j + bt| \sum w_j C_j$, Mosheiov [28] proposed a heuristic (see Algorithm 16.23), based on the following observation. Applying the pairwise job interchange argument, one can show that for a given starting time t, jobs in an optimal schedule will be executed in increasing order of the values of function $\frac{a_j}{w_j} + \frac{b}{w_j}t$. Therefore, for small values of t non-decreasing order of the ratios $\frac{a_j}{w_j}$ seems to be more attractive, while for large t values non-increasing order of the weights w_j is better.

Algorithm 16.23. for problem $1|p_j = a_j + b_j t| \sum w_j C_j$

---

1 **Input** : sequences $(a_1, a_2, \ldots, a_n)$, $(w_1, w_2, \ldots, w_n)$, number b
2 **Output:** a suboptimal schedule
  ▷ Step 1
3 Schedule jobs in non-decreasing order of the ratios $\frac{a_j}{w_j}$;
4 Call the obtained schedule σ_1;
5 $G_1 \leftarrow \sum w_j C_j(\sigma_1)$;
  ▷ Step 2
6 Schedule jobs in non-increasing order of the weights w_j;
7 Call the obtained schedule σ_2;
8 $G_2 \leftarrow \sum w_j C_j(\sigma_2)$;
  ▷ Step 3
9 **if** $(G_1 < G_2)$ **then**
10 $\quad | \quad \sigma \leftarrow \sigma_1$;
  else
11 $\quad | \quad \sigma \leftarrow \sigma_2$;
  end
12 **return** σ.

The results of a computational experiment (see [28, Section 'The case of general weights']) suggest that Algorithm 16.23 is, on average, quite good.

Now, we consider heuristic algorithms for time-dependent scheduling problems with linear jobs, distinct ready times and deadlines.

Parallel machine problems

Cheng et al. [25] proposed a heuristic algorithm for problem $Pm|p_j = a_j + bt| \sum(\alpha E_j + \beta T_j + \gamma d)$. This algorithm exploits the idea of Algorithm 8.1 (see Sect. 8.2) for problem $Pm|p_j = a_j + bt| \sum(\alpha E_j + \beta T_j)$, and was tested on a few instances with $5 \leqslant n \leqslant 8$. We refer the reader to [25] for more details.

16.5.3 General non-linear deterioration

In this section, we present heuristic algorithms for time-dependent scheduling problems with general non-linearly deteriorating jobs. We begin with the case when all jobs have distinct ready times and distinct deadlines.

Single machine problems

Janiak and Kovalyov [21] proposed three heuristic algorithms for processing times given by (6.23) and the $\sum w_j C_j$ criterion. In the opinion of the authors (see [21, Sect. 4]), the first of these algorithms, sorting jobs in non-decreasing order of the ratios $\frac{p_j}{w_j}$, gives, on average, the best results.

Sundararaghavan and Kunnathur [30] proposed a heuristic algorithm for job processing times by (6.25) and the $\sum w_j C_j$ criterion. Applying the notation introduced in Sect. 7.4.5, the algorithm can be formulated as follows.

Algorithm 16.24. for problem $1|p_j \equiv (6.25)| \sum w_j C_j$

1 **Input** : sequences $(b_1, b_2, \ldots, b_n)$, $(w_1, w_2, \ldots, w_n)$, numbers a, D
2 **Output:** a suboptimal schedule
   ▷ Step 1
3 Arrange jobs in non-increasing order of the weights w_j;
4 Call the obtained sequence σ;
   ▷ Step 2
5 $k \leftarrow \lfloor \frac{D}{a} \rfloor + 1$;
6 Schedule k jobs according to the order given by σ;
7 $E \leftarrow \{J_k \in \mathcal{J} : C_k \leqslant D\}$;
8 $L \leftarrow \mathcal{J} \setminus E$;
9 Arrange jobs in L in non-decreasing order of the ratios $\frac{a+b_j}{w_j}$;
   ▷ Step 3
10 **repeat**
11 | **if** $(J_{[i]} \in E \wedge J_{[j]} \in L \wedge \Delta([i] \leftrightarrow [j]) > 0)$ **then**
12 | | Exchange $J_{[i]}$ with $J_{[j]}$;
   | **end**
   until (*no more exchange* $[i] \leftrightarrow [j]$ *exists for* $J_{[i]} \in E$ *and* $J_{[j]} \in L$);
13 **return**.

Sundararaghavan and Kunnatur [30] conjectured that Algorithm 16.24 is optimal. Cheng and Ding [24] showed that this conjecture is false.

Example 16.11. (Cheng and Ding [24]) Given a sufficiently large number $Y > 0$, define $Z = 2Y^4$. Let $n = 4$, $a = 1$, $D = 1$, $w_1 = 2Y + 3 + \frac{1}{3Y}$, $w_2 = 2Y$, $w_3 = Y+1$, $w_4 = Y$, $b_1 = YZ$, $b_2 = (Y+1)Z$, $b_3 = 2YZ$, $b_4 = (2Y+3+\frac{2}{3Y})Z$.

Since Algorithm 16.24 generates schedule $\sigma = (1, 3, 2, 4)$, and for schedule $\sigma' = (2, 4, 1, 3)$ we have $\sum w_j C_j(\sigma') < \sum w_j C_j(\sigma)$, the algorithm cannot be optimal (see [24, Sect. 4]).   ◇

16.5.4 Linear shortening

In this section, we consider heuristic algorithms for time-dependent scheduling problems with linearly shortening job processing times, equal ready times and no deadlines.

Single machine problems

For problem $1|p_j = a_j - b_j t, 0 \leqslant b_j < 1, b_i(\sum_{j=1}^{n} a_j - a_i) < a_i| \sum w_j C_j$, Bachman et al. [23] proposed two heuristic algorithms. In the first of them, which is based on Property 7.148, jobs are scheduled in non-decreasing order of the ratios $\frac{a_j}{w_j(1-b_j)}$. In the second one, which is based on Properties 7.149-7.150, an even-odd V-shaped schedule is constructed. The results of a computational experiment (see [23, Section 'Heuristic algorithms']) suggest that the second algorithm outperforms the first one.

References

Minimizing the maximum completion time

1. B. Alidaee, A heuristic solution procedure to minimize makespan on a single machine with non-linear cost functions. *Journal of the Operational Research Society* **41** (1990), no. 11, 1065–1068.
2. J. Bruno, E. G. Coffman jr and R. Sethi, Scheduling independent tasks to reduce mean finishing time. *Communications of the ACM* **17** (1974), no. 7, 382–387.
3. T-C. E. Cheng, Q. Ding, M. Y. Kovalyov, A. Bachman and A. Janiak, Scheduling jobs with piecewise linear decreasing processing times. *Naval Research Logistics* **50** (2003), no. 6, 531–554.
4. E. G. Coffman jr, *Computer and Job-Shop Scheduling Theory*. New York: Wiley 1976.
5. J. N. D. Gupta and S. K. Gupta, Single facility scheduling with nonlinear processing times. *Computers and Industrial Engineering* **14** (1988), no. 4, 387–393.
6. Y-C. Hsieh and D. L. Bricker, Scheduling linearly deteriorating jobs on multiple machines. *Computers and Industrial Engineering* **32** (1997), no. 4, 727–734.
7. A. Janiak and M. Y. Kovalyov, Job sequencing with exponential functions of processing times. *Informatica* **17** (2006), no. 1, 13–14.
8. A. S. Kunnathur and S. K. Gupta, Minimizing the makespan with late start penalties added to processing times in a single facility scheduling problem. *European Journal of Operational Research* **47** (1990), no. 1, 56–64.
9. J. Moore, An n job, one machine sequencing algorithm for minimizing the number of late jobs. *Management Science* **15** (1968), no. 1, 102–109.
10. G. Mosheiov, Scheduling jobs with step-deterioration; Minimizing makespan on a single- and multi-machine. *Computers and Industrial Engineering* **28** (1995), no. 4, 869–879.
11. S. Sahni and G. Vairaktarakis, The master-slave scheduling model. In: J. Y-T. Leung (ed.), *Handbook of Scheduling: Algorithms, Models, and Performance Analysis*. Boca Raton: Chapman & Hall/CRC 2004, Chap. 17.
12. V. A. Strusevich and K. Rustogi, *Scheduling with Times-Changing Effects and Rate-Modifying Activities*. Berlin-Heidelberg: Springer 2017.

Minimizing the total completion time

13. T-C. E. Cheng, Q. Ding, M. Y. Kovalyov, A. Bachman and A. Janiak, Scheduling jobs with piecewise linear decreasing processing times. *Naval Research Logistics* **50** (2003), no. 6, 531–554.
14. S. Gawiejnowicz, W. Kurc and L. Pankowska, Analysis of a time-dependent scheduling problem by signatures of deterioration rate sequences. *Discrete Applied Mathematics* **154** (2006), no. 15, 2150–2166.
15. A. A-K. Jeng and B. M-T. Lin, A note on parallel-machine scheduling with deteriorating jobs. *Journal of the Operational Research Society* **58** (2007), no. 6, 824–826.
16. G. Mosheiov, V-shaped policies for scheduling deteriorating jobs. *Operations Research* **39** (1991), no. 6, 979–991.
17. J-B. Wang, C-T. Ng, T-C. E. Cheng and L-L. Liu, Minimizing total completion time in a two-machine flow shop with deteriorating jobs. *Applied Mathematics and Computation* **180** (2006), no. 1, 185–193.
18. C-C. Wu and W-C. Lee, Two-machine flowshop scheduling to minimize mean flow time under linear deterioration. *International Journal of Production Economics* **103** (2006), no. 2, 572–584.

Minimizing the maximum lateness

19. A. Bachman and A. Janiak, Minimizing maximum lateness under linear deterioration. *European Journal of Operational Research* **126** (2000), no. 3, 557–566.
20. Y. S. Hsu and B. M-T. Lin, Minimization of maximum lateness under linear deterioration. *Omega* **31** (2003), no. 6, 459–469.
21. A. Janiak and M. Y. Kovalyov, Job sequencing with exponential functions of processing times. *Informatica* **17** (2006), no. 1, 1–11.

Other criteria

22. B. Alidaee and F. Landram, Scheduling deteriorating jobs on a single machine to minimize the maximum processing times. *International Journal of Systems Science* **27** (1996), no. 5, 507–510.
23. A. Bachman, T-C. E. Cheng, A. Janiak and C-T. Ng, Scheduling start time dependent jobs to minimize the total weighted completion time. *Journal of the Operational Research Society* **53** (2002), no. 6, 688–693.
24. T-C. E. Cheng and Q. Ding, Single machine scheduling with step-deteriorating processing times. *European Journal of Operational Research* **134** (2001), no. 3, 623–630.
25. T-C. E. Cheng, L. Kang and C-T. Ng, Due-date assignment and parallel-machine scheduling with deteriorating jobs. *Journal of the Operational Research Society* **58** (2007), no. 8, 1103–1108.
26. E. L. Lawler, Optimal sequencing of a single machine subject to precedence constraints. *Management Science* **19** (1973), no. 5, 544–546.
27. G. Mosheiov, Multi-machine scheduling with linear deterioration. *Infor* **36** (1998), no. 4, 205–214.
28. G. Mosheiov, Λ-shaped policies to schedule deteriorating jobs. *Journal of the Operational Research Society* **47** (1996), no. 9, 1184–1191.

29. D. Oron, Single machine scheduling with simple linear deterioration to minimize total absolute deviation of completion times. *Computers and Operations Research* **35** (2008), no. 6, 2071–2078.

30. P. S. Sundararaghavan and A. S. Kunnathur, Single machine scheduling with start time dependent processing times: some solvable cases. *European Journal of Operational Research* **78** (1994), no. 3, 394–403.

Local search and meta-heuristic algorithms

$\mathbb{S}$uboptimal schedules for intractable time-dependent scheduling problems may be found by using various algorithms. In this chapter, closing the fifth part of the book, we consider local search and meta-heuristic algorithms.

Chapter 17 is composed of five sections. In Sect. 17.1, we recall basic definitions concerning local search and meta-heuristic algorithms. In Sects. 17.2 and 17.3, we briefly review basic types of local search and meta-heuristic algorithms, respectively. In Sects. 17.4 and 17.5, we discuss local search and meta-heuristic algorithms for time-dependent scheduling problems, respectively. The chapter is completed with a list of references.

17.1 Preliminaries

In this section, we introduce basic definitions and general concepts related to local search and meta-heuristic algorithms.

17.1.1 Basic definitions

An optimization problem Π is specified by a collection of instances of this problem. An instance is defined by the pair $(\mathfrak{F}, f)$, where the *solution space* $\mathfrak{F}$ is the set of all feasible solutions and $f : \mathfrak{F} \to \mathbb{R}$ is a criterion function.

Remark 17.1. Without loss of generality, cf. Section 2.1, we restrict further discussion to minimization problems.

Definition 17.2. (Optimal solutions vs. global minima)
(a) *A solution $s^\star \in \mathfrak{F}$ is* optimal *(a* global minimum*) if inequality $f(s^\star) \leqslant f(s)$ is satisfied for all $s \in \mathfrak{F}$.*
(b) *The set $\mathfrak{F}_{opt} := \{s \in \mathfrak{F} : f(s) = f^\star\}$ is called a set of all optimal solutions (global minima).*

© Springer-Verlag GmbH Germany, part of Springer Nature 2020
S. Gawiejnowicz, *Models and Algorithms of Time-Dependent Scheduling*,
Monographs in Theoretical Computer Science. An EATCS Series,
https://doi.org/10.1007/978-3-662-59362-2_17

Throughout the chapter, we will write that problem Π is solved if a solution $s \in \mathfrak{F}_{opt}$ has been found.

For a given optimization problem Π, a *neighbourhood function* $\mathcal{N} : \mathfrak{F} \to 2^{\mathfrak{F}}$ may be defined. For each solution $s \in \mathfrak{F}$, this function specifies a set $\mathcal{N}(s) \subseteq \mathfrak{F}$ of *neighbours* of s. The set $\mathcal{N}(s)$ is called the *neighbourhood* of solution s. A solution $s^{\circ} \in \mathfrak{F}$ is called a *local minimum with respect to* $\mathcal{N}$ if $f(s^{\circ}) \leqslant f(s)$ for all $s \in \mathcal{N}(s^{\circ})$. A neighbourhood function $\mathcal{N}$ is called *exact* if every local minimum with respect to $\mathcal{N}$ is also a global minimum.

For other definitions related to the subject we refer the reader to the literature (see, e.g., Aarts and Korst [1], Aarts and Lenstra [2], Bozorg-Haddad et al. [8], Burke and Kendall [4]).

17.1.2 General concepts in local search

The term *local search* refers to a general approach to solving intractable optimization problems. The main idea is to start from an initial solution, $s_0 \in \mathfrak{F}$, construct its neighbourhood $\mathcal{N}(s_0)$ and look for better solutions there. Basically, it is assumed that the neighbourhood includes only feasible and complete solutions which are 'close', in the problem-specific sense, to the solution s_0. The pseudo-code of the most general form of local search algorithm, called *General Local Search* algorithm (GLS, see Algorithm 17.1), is as follows.

Algorithm 17.1. General Local Search

---

1 **Input** : initial solution s_0, neighbourhood function $\mathcal{N}$, criterion f
2 **Output:** a locally optimal solution s_{act}
  ▷ Step 1
3 $s_{act} \leftarrow s_0$;
4 *Initialization*;
  ▷ Step 2
5 **repeat**
6   Generate $\mathcal{N}(s_{act})$;
7   **for all** $s \in \mathcal{N}(s_{act})$ **do**
8   **if** $(f(s) < f(s_{act}))$ **then**
9             $s_{act} \leftarrow s$;
            end
       end
  until (*stop_condition*);
  ▷ Step 3
10 **return** s_{act}.

The pseudo-code of Algorithm 17.1 is a generic template of any local search algorithm. Therefore, some remarks are necessary.

Remark 17.3. The *Initialization* procedure includes preliminary operations such as initialization of counters, setting parameters used during the construction of the neighbourhood $\mathcal{N}(s_{act})$, etc.

Remark 17.4. The neighbourhood function $\mathcal{N}$ is problem-specific and can be defined in various ways.

Remark 17.5. The form of a *stop_condition* depends on the applied variant of the local search algorithm.

Remark 17.6. Some steps in the template (e.g. the *Initialization* procedure) may be dropped and some (e.g. *stop_condition* $\equiv FALSE$) may be trivial.

The basic assumptions of the GLS algorithm may be modified in various ways. The most often encountered modifications are as follows.

First, the search not need be conducted in the set $\mathfrak{F}$ of all feasible solutions. In some variants of local search, the solutions are searched for in the set $E(\mathfrak{F})$ which is an image of the set $\mathfrak{F}$ under some mapping E.

Second, the neighbourhood $\mathcal{N}(s_{act})$ can be composed of feasible and infeasible solutions if the set $E(\mathfrak{F})$ is considered instead of $\mathfrak{F}$.

Finally, not only complete solutions can be elements of the neighbourhood $\mathcal{N}(s_{act})$. In general, partial solutions may also belong to the neighbourhood.

17.1.3 Pros and cons of local search algorithms

Local search algorithms have both advantages and disadvantages, and it is not easy to say which ones prevail, since the overall performance of the algorithms depends on many problem-specific factors.

The main advantages of local search algorithms are as follows. First, due to its generality, the GLS algorithm can be applied to various optimization problems, unlike the constructive heuristics which use problem-specific properties and therefore, usually, are not versatile.

Second, since local search algorithms search only the set $\mathcal{N}(s_{act})$, they are capable of solving problems of larger sizes.

Third, local search algorithms, in most cases, produce solutions of acceptable quality, even if no special attention has been paid to choosing appropriate values of the control parameters of the algorithms.

Local search algorithms also have some disadvantages. First, the minimal exact neighbourhood may be exponential with respect to the size of the input of a given optimization problem.

Second, exponential time may be needed to find a local minimum.

Third, the solutions obtained by a local search algorithm may deviate arbitrarily far from the elements of the set $\mathfrak{F}_{opt}$.

Despite the above disadvantages, local search algorithms have been applied successfully to many intractable optimization problems (see, e.g., Blum and Roli [3], Chopard and Tomassini [12], Gendreau and Potvin [5], and the references cited in Sect. 2.5.4).

17.2 Main types of local search algorithms

There exist a great number of types and variants of local search algorithms. In this section, we shortly describe only those which have been applied to time-dependent scheduling problems.

17.2.1 Iterative improvement

The simplest local search algorithm is *Iterative Improvement* algorithm (II, see Algorithm 17.2). In this case, starting from an initial solution $s_0 \in \mathfrak{F}$, we iteratively search for a neighbour s_{loc} of the current solution s_{act} which has the best value of the criterion function f. The neighbourhood function $\mathcal{N}$ defines the way in which the neighbour s_{loc} is generated. The set $\mathcal{N}(s_{act})$ is composed of only one solution.

Algorithm 17.2. Iterative Improvement

1 **Input** : initial solution s_0, neighbourhood function $\mathcal{N}$, criterion f
2 **Output:** a locally optimal solution s_{act}
▷ Step 1
3 $s_{act} \leftarrow s_0$;
▷ Step 2
4 **repeat**
5 | Generate a neighbour s_{loc} of s_{act};
6 | **for all** $s \in \mathcal{N}(s_{act})$ **do**
7 | | **if** $(f(s_{loc}) < f(s_{act}))$ **then**
8 | | | $s_{act} \leftarrow s_{loc}$;
| | **end**
| **end**
until (*stop_condition*);
▷ Step 3
9 **return** s_{act}.

Remark 17.7. A few variants of Algorithm 17.2 are known, such as *Hill Climbing* (HC) or *Iterative Improvement with Random Start* (IIRS) algorithms, differing in the choice of the initial solution s_0 or the form of the criterion f (see the references listed at the end of the chapter for details).

Since the solution generated by the II algorithm is the first-encountered local minimum, the quality of this minimum may be arbitrarily bad. Therefore, some variations of the algorithm have been proposed.

17.2.2 Steepest descent search

One of the variations of II is the algorithm called *Steepest Descent Search* (SDS, see Algorithm 17.3). Unlike the II, in the SDS algorithm all possible neighbours of the current solution s_{act} are generated. The best neighbour is accepted as the final solution. This improves the quality of the final solution at the cost of increasing the time complexity of the algorithm. The pseudo-code of the algorithm is as follows.

Algorithm 17.3. Steepest Descent Search

---

1 **Input** : initial solution s_0, neighbourhood function $\mathcal{N}$, criterion f
2 **Output:** a locally optimal solution s_{act}
  ▷ Step 1
3 $s_{act} \leftarrow s_0$;
  ▷ Step 2
4 **repeat**
5   Generate $N(s_{act})$;
6   **for all** $s \in \mathcal{N}(s_{act})$ **do**
7   **if** $(f(s) < f(s_{act}))$ **then**
8          $s_{act} \leftarrow s$;
      end
    end
  until $(stop\_condition)$;
  ▷ Step 3
9 **return** s_{act}.

The II and SDS algorithms are simple local search algorithms which, in many cases, produce solutions which are of poor quality. The main reason for this fact is the so-called *trap of local optimum*: after finding a solution which is locally optimal, the local search heuristics are not able to find better solutions, since they cannot move out of the neighbourhood of a locally optimal solution. Therefore, more powerful local search algorithms, called *meta-heuristics*, have been proposed in the literature.

17.3 Main types of meta-heuristic algorithms

In this section, we shortly describe several types of meta-heuristic algorithms which have been applied in time-dependent scheduling.

The first meta-heuristic algorithm for time-dependent scheduling problems was proposed in 2001 by Hindi and Mhlanga [35]. A few years later, Shiau et al. [44] proposed a meta-heuristic algorithm for a two-machine time-dependent flow shop problem. These two papers initiated the research on meta-heuristic algorithms applied to time-dependent scheduling problems.

Now, we briefly review simulated annealing, tabu search, genetic, evolutionary, ant colony optimization and particle swarm optimization algorithms, which have been often used to solve time-dependent scheduling problems.

17.3.1 Simulated annealing

Simulated Annealing algorithm (SA, see Algorithm 17.4), proposed by Kirkpatrick et al. [29] and Černý [11] in the mid 1980s, is based on a procedure that imitates the *annealing* (slow cooling) of a solid after it has been heated to its melting point. The SA is a non-deterministic algorithm, since the current solution s_{act} is selected and accepted from the neighbourhood $\mathcal{N}(s_{act})$ in a random way. The pseudo-code of the meta-heuristic is as follows.

Algorithm 17.4. Simulated Annealing

---

1 **Input**   : initial solution s_0, neighbourhood function $\mathcal{N}$, criterion f,
                 acceptance probability function $p(i)$
2 **Output:** a locally optimal solution s_{best}
  ▷ Step 1
3 $s_{act} \leftarrow s_0$; $s_{best} \leftarrow s_0$; $f_{best} \leftarrow f(s_{best})$;
4 $i \leftarrow 1$;
  ▷ Step 2
5 **repeat**
6 | Generate $\mathcal{N}(s_{act})$;
7 | Choose at random $s_{loc} \in \mathcal{N}(s_{act})$;
8 | **if** $(f(s_{loc}) \leqslant f(s_{act}))$ **then**
  | | $s_{act} \leftarrow s_{loc}$;
  | **end**
9 | **if** $(f(s_{loc}) < f_{best})$ **then**
10 | | $f_{best} \leftarrow f(s_{loc})$;
11 | | $s_{best} \leftarrow s_{loc}$;
  | **else**
12 | | Choose at random $p \in \langle 0, 1 \rangle$;
  | **end**
13 | **if** $(p \leqslant p(i))$ **then**
14 | | $s_{act} \leftarrow s_{loc}$;
  | **end**
15 | $i \leftarrow i + 1$;
  until (*stop_condition*);
  ▷ Step 3
16 **return** s_{best}.

The behaviour of the SA algorithm is determined by a number of parameters such as the initial *temperature*, the *cooling rate* and the *function of temperature* (*cooling scheme*) mirroring the decrease in temperature during the annealing. The *acceptance probability function* $p(i)$ is in the form of

$$p(i) := \exp\left(-\frac{1}{T(i)}\Delta f_i\right),$$

where $\Delta f_i := f(s_{loc}) - f(s_{act})$ and $T(i)$ is a non-increasing function of time. The function of temperature $T(i)$ is usually defined as follows. Starting from T_0, the temperature is fixed for L consecutive steps, and next it is decreased according to the formula $T(iL) \equiv T_i := c^i T_0$, where $0 < c < 1$ is a constant factor. The parameters T_0, c and L are called the *initial temperature*, *cooling rate* and *length of plateau*, respectively.

We refer the reader to the literature (see, e.g., Aarts and Korst [6], Burke and Kendall [9], and Salamon et al. [32]) for details on the SA meta-heuristic.

17.3.2 Tabu search

Tabu Search algorithm (TS, see Algorithm 17.5), proposed by Glover [19, 20] in the late 1980s, is a deterministic meta-heuristic algorithm which guides the local search, avoiding local optima. This is done by accepting the moves which worsen the present solution but allow us to escape from local optima, and banning the moves which lead to solutions visited earlier.

Algorithm 17.5. Tabu Search

---

1 **Input** : initial solution s_0, neighbourhood function $\mathcal{N}$, aspiration
               function A, criterion function f
2 **Output:** a locally optimal solution s_{act}
   ▷ Step 1
3 $s_{best} \leftarrow s_0$; $s_{loc} \leftarrow s_0$; $v_{best} \leftarrow +\infty$; $k \leftarrow 1$;
4 Set l_k; $RL \leftarrow \emptyset$; $TL \leftarrow \emptyset$;
   ▷ Step 2
5 **repeat**
6   **repeat**
7   Generate a neighbour $s_{loc} \in \mathcal{N}(s_{act})$;
8        $\Delta \leftarrow f(s_{act}) - f(s_{loc})$;
9   Calculate $A(\Delta, k)$;
       until
       $(A(\Delta, k) < v_{best})$ **or** $(\Delta = \max\{f(s_{act}) - f(s_{loc}) : s_{loc} \notin TL\})$;
10   Update RL;
11     $TL \leftarrow \{$the last l_k entries of $RL\}$;
12   **if** $(A(\Delta, k) < v_{best}))$ **then**
13        $v_{best} \leftarrow A(\Delta, k)$;
       end
14     $s_{act} \leftarrow s_{loc}$;
15   **if** $(f(s_{loc}) < f(s_{best}))$ **then**
16        $s_{best} \leftarrow s_{loc}$;
       end
17     $k \leftarrow k + 1$;
   until (*stop_condition*);
   ▷ Step 3
18 **return** s_{act}.

The basic elements of the TS algorithm are *tabu list TL*, *running list RL*, *aspiration function A* and *neighbourhood function $\mathcal{N}$*.

The key element is the list TL which includes the elements of search space whose use is forbidden in a limited number of iterations. The list TL, updated in every iteration, is maintained during the whole search process, which allows the algorithm to move in the search space in different directions. The aim of this list is to prevent the TS algorithm from oscillating between some solutions. The number of elements of list TL, l_k, is usually different in every iteration.

The running list RL is an ordered list of all moves performed throughout the search process. In other words, the list RL represents the *trajectory* of solutions encountered. Usually, the list RL is of limited length.

Given lists TL and RL, in every iteration the TS algorithm updates the TL list based on the actual RL list, selects the best move which is not forbidden (i.e., *is not tabu*) and updates the list RL. The main loop of the TS algorithm is performed until some stop condition is met.

We refer the reader to the literature (see, e.g., Glover et al. [21], Glover and Laguna [22]) for more details on the TS meta-heuristic.

17.3.3 Genetic algorithms

Genetic Algorithm (GA) is a non-deterministic meta-heuristic described by Holland [25] in 1975 and formalized by Goldberg [23] in 1989. This algorithm mirrors the development of an evolving *population* of solutions modified by *genetic operators* which are responsible for the mechanisms of selection, crossover and mutation, and generate new, potentially better, solutions.

Algorithm 17.6. Simple Genetic Algorithm

---

1 **Input** : procedures *Initialization*, *Evaluation*, operators *Selection*,
              Recombination, *Mutation*, *Replacement*
2 **Output:** a locally optimal solution
   ▷ Step 1
3 $i \leftarrow 0$;
4 Perform *Initialization* on $\mathbf{P}_i$;
   ▷ Step 2
5 **repeat**
6 | Perform *Evaluation* on $\mathbf{P}_i$;
7 | Apply *Selection* to $\mathbf{P}_i$;
8 | Improve $\mathbf{P}_i$ by *Recombination*, *Mutation* and *Replacement*;
9 | Create a new population;
10 | $i \leftarrow i + 1$;
   until (*stop_condition*);
   ▷ Step 3
11 **return** the last best solution.

The GA starts with an initialization procedure, where some control parameters are fixed and an *initial population* P_0 is generated. Next, as long as a stop condition is not met, the evaluation of the current population, P_i, selection of the most promising solutions, recombination and mutation of existing solutions and replacement of the previous population by a new one, P_{i+1}, are performed for $i > 0$. The pseudo-code of the simplest form of the GA, called *Simple Genetic Algorithm* (SGA), is presented as Algorithm 17.6.

We refer the reader to the literature (see, e.g., Chopard and Tomassini [12], Du and Swamy [17], Gendreau and Potvin [18]) for more details.

17.3.4 Evolutionary algorithms

The GA has evolved to *Evolutionary Algorithm* (EA) which is a combination of the GA and the so-called *genetic programming* (proposed by Koza [30] in 1992) and its more advanced counterparts such as *evolutionary strategy* and *evolutionary programming* (see, e.g., Calégari et al. [10] for more details).

Algorithm 17.7. Simple Evolutionary Algorithm

1 **Input** : procedures *Initialization, Evaluation*, operators
             Preselection, CrossOver_and_Mutation, Postselection
2 **Output:** a locally optimal solution
  ▷ Step 1
3 $i \leftarrow 0$; *Initialization*(P_0); *Evaluation*(P_0);
  ▷ Step 2
4 **repeat**
5    $T_i \leftarrow Preselection(P_i)$;
6    $O_i \leftarrow CrossOver\_and\_Mutation(T_i)$;
7   *Evaluation*(O_i);
8    $P_{i+1} \leftarrow Postselection(P_i, O_i)$;
9    $i \leftarrow i + 1$;
   until (*stop_condition*);
   ▷ Step 3
10 **return** the last best solution.

The simplest EA, called *Simple Evolutionary Algorithm* (SEA, see Algorithm 17.7), works on the *base, offspring* and *temporary* populations of *individuals*, and produces *generations* of solutions, indexed by the variable i. In the *initialization* step of the SEA, the base population P_0 is created in a random way. Next, the *evaluation* of P_0 is performed, i.e., for each individual from P_0 a *fitness function* is calculated. In the main step of the SEA, the offspring population P_i, $i > 0$, is created. First, in the process of *preselection*, the temporary population T_i is built from the best (with respect to the fitness function) individuals of P_i. Next, individuals from T_i are *crossed* and *mutated*, which leads through the *postselection* to the next offspring population P_{i+1}. The process is continued until a certain *stop condition* is met.

Evolutionary algorithms have many variants, whose description is beyond the scope of the book. We refer the reader to the literature (see, e.g., Hart et al. [24] and the references cited in Sect. 2.5.4) for details.

17.3.5 Ant colony optimization

Ant colony optimization (ACO, see Algorithm 17.8), introduced by Dorigo et al. [13, 14] in the mid 1990s, is inspired by the observation of the behaviour of real *ants* reacting to a chemical substance called *pheromone*. The main idea is as follows. A colony of ants moves along the ground surface in different directions, creating paths called *trails*. Ants choose the direction of the next move based on local pheromone concentration left on the trails by other ants. Moving in a given direction along a trail, the ants increase the level of pheromone. If the direction is more frequently visited, the level is higher compared to other trails which, in turn, encourages more ants to choose this trail, etc.

In order to model the behavior of ants and apply it to solving an optimization problem, one needs to define ants, trails, pheromone, the way in which the move of an ant is performed and the level of the pheromone is increased. Sometimes, also a local search algorithm is included. The pseudo-code of the ACO algorithm is as follows.

Algorithm 17.8. Ant Colony Optimization

---

1 **Input**   : definitions of ants, pheromones and trails, function of
                pheromone updating, function of ant movement
2 **Output:** a locally optimal solution
   ▷ Step 1
3 Set parameters;
4 Initialize pheromone trails;
   ▷ Step 2
5 **repeat**
6 $\quad\big|\quad$ Construct ant solutions;
7 $\quad\big|\quad$ Update pheromone;
   until (*stop_condition*);
   ▷ Step 3
8 **return** the last best solution.

We refer the reader to the literature (see, e.g., Dorigo and Socha [15], Dorigo and Stützle [16], Mullen et al. [31]) for more details on the ACO metaheuristic and its applications.

17.3.6 Particle swarm optimization

Particle Swarm Optimization (PSO, see Algorithm 17.9), introduced in 1995 by Kennedy and Eberhart [27], is inspired by the behaviour of groups of moving animals (e.g. *bird flock*, *fish school* etc.). A solution of an optimization problem is equated here with finding the best position in a search space. The main idea of the PSO algorithm is as follows.

A population of collectively acting *particles*, called *swarm*, wants to find a globally best position in a search space. Every particle tries to improve its own position, looking for the best one, and informing other particles when such a position is reached. Hence, the movement of each particle is influenced by its locally best position and the best known positions in the search space. The latter ones are updated, whenever better positions are found by other particles. Acting in this way, a swarm is moving toward the best positions (i.e. solutions). The pseudo-code of the PSO is as follows.

Algorithm 17.9. Particle Swarm Optimization

---

1 **Input** : definition of particle, location, velocity, criterion function f
2 **Output:** a locally optimal solution
  ▷ Step 1
3 **for** $i \leftarrow 1$ **to** n **do**
4 | Initialize location x_i and velocity v_i of particle x_i;
  end
5 $t \leftarrow 0$;
6 Find the global best g^* from $\min\{f(x_1), f(x_2), \ldots, f(x_n)\}$;
  ▷ Step 2
7 **repeat**
  | **for** $i \leftarrow 1$ **to** n **do**
8 | | Generate new velocity v_i;
9 | | Calculate new location x_i;
10 | | Evaluate $f(x_i)$;
11 | | Find the current best for x_i;
12 | | $t \leftarrow t + 1$;
  | **end**
13 | Find the global best g^*;
14 | $t \leftarrow t + 1$;
  until (*stop_condition*);
  ▷ Step 3
15 **return** $x_1^*, x_2^*, \ldots, x_n^*, g^*$.

We refer the reader to the literature (see, e.g., Bonyady and Michalewicz [7], Kar [26], Kennedy and Eberhart [28], Yang [33]) for details on the PSO metaheuristic algorithm.

17.4 Local search time-dependent scheduling algorithms

In this section, we present a few local search algorithms which have been used in time-dependent scheduling.

17.4.1 Iterative improvement algorithms

For problem $Pm|p_j = 1 + b_j t| \sum C_j$, Gawiejnowicz et al. [34] proposed an iterative improvement Algorithm 17.10, based on the following idea. If a schedule for the problem is known, it can be improved by successively moving jobs between machines in order to find such an assignment of jobs to machines which gives a smaller total completion time than the initial one.

Algorithm 17.10. for problem $Pm|p_j = 1 + b_j t| \sum C_j$

---

1 **Input** : sequence $(b_1, b_2, \ldots, b_n)$, number k
2 **Output:** a locally optimal solution σ_{act}
   ▷ Step 1: Construction of the initial schedule σ_0
3 Apply Algorithm 17.12 to sequence $(b_1, b_2, \ldots, b_n)$;
4 Call the obtained schedule σ_0;
5 $k \leftarrow 0$; $\sigma_{act} \leftarrow \sigma_0$;
   ▷ Step 2: Iterative improvement of schedule σ_{act}
6 **repeat**
7 | $k \leftarrow k + 1$; $\sigma_{last} \leftarrow \sigma_{act}$;
8 | **for** $i \leftarrow n$ **downto 2 do**
9 | | **for** $k \leftarrow i - 1$ **downto 1 do**
   | | | **if** $(ind(\sigma_{last}, J_i) \neq ind(\sigma_{last}, J_k))$ **then**
10 | | | | $\tau \leftarrow \sigma_{last}(J_i \leftrightarrow J_k)$;
   | | | **end**
   | | | **if** $(\sum C_j(\sigma) - \sum C_j(\tau) > 0)$ **then**
11 | | | | $\sigma_{act} \leftarrow \tau$;
   | | | **end**
   | | **end**
   | **end**
   until $((\sum C_j(\sigma) - \sum C_j(\sigma_{last}) = 0) \vee (k > n))$;
   ▷ Step 3
12 **return** σ_{act}.

In the pseudo-code of Algorithm 17.10 the symbols $\sigma(J_i \leftrightarrow J_k)$ and $ind(\sigma, J_i)$ denote a schedule σ in which jobs J_i and J_k have been mutually replaced and the index of a machine to which job J_i has been assigned in schedule σ, respectively.

The time complexity of Algorithm 17.10 is $O(n^3)$, since in the worst case we have to check n times $O(n^2)$ possibilities of a mutual change of two jobs.

17.4.2 Steepest descent search algorithms

For the parallel-machine problem $Pm|p_j = b_j t|C_{\max}$, Hindi and Mhlanga [35] proposed a steepest descent search Algorithm 17.11. Let σ_0 and σ_{act} denote the initial schedule and the current best schedule, respectively. Since in any schedule for this problem, by Theorem 7.1, the order of jobs assigned to a machine is immaterial, every subschedule can be represented by a subset of indices of jobs assigned to the machine. In view of this fact, in Algorithm 17.11 the neighbourhood $\mathcal{N}(s_{act})$ is defined as a set of all partitions of set $N_{\mathcal{J}} := \{1, 2, \ldots, n\}$ into m parts. Every new partition is obtained from another partition by a single move. A single *move* is either a transfer of one job from one subset of a partition to another subset, or a mutual exchange of two jobs belonging to two different subsets of a partition.

Algorithm 17.11. for problem $Pm|p_j = b_j t|C_{\max}$

1 **Input** : initial schedule σ_0, neighbourhood function $\mathcal{N}$
2 **Output:** a locally optimal solution σ_{act}
  ▷ Step 1
3 $\sigma_{act} \leftarrow \sigma_0$;
  ▷ Step 2
4 **repeat**
5 | $T \leftarrow C_{\max}(\sigma_{act})$;
6 | Find machine $\mathcal{M} \in \{M_1, M_2, \ldots, M_m\}$ such that $C_{\max}(\mathcal{M}) = T$;
7 | $\mathcal{N}(\sigma_{act}) \leftarrow$ set of schedules obtained by all possible moves of jobs
  | assigned to machine $\mathcal{M}$;
8 | **for all** $\sigma \in \mathcal{N}(\sigma_{act})$ **do**
9 | | **if** $(C_{\max}(\sigma) < C_{\max}(\sigma_{act}))$ **then**
10 | | | $\sigma_{act} \leftarrow \sigma$;
  | | **end**
  | **end**
  until no improvement of σ is possible;
  ▷ Step 3
11 **return** σ_{act}.

The running time of Algorithm 17.11 is bounded by the maximal number of iterations of loop **for** in Step 2, which equals $|\mathcal{N}(\sigma_{act})| \approx \frac{m^n}{m!} = O(m^n)$. Therefore, this algorithm is rather impractical for large m.

Algorithm 17.11 was tested on a set of 320 instances, in which deterioration rates were randomly generated values from the $(0, 1)$ interval. The reported results of the experiment (see [35, Sect. 5]) suggest that the schedules generated by the algorithm are, on average, quite satisfactory.

For the problem $Pm|p_j = b_j t|\sum C_j$, Gawiejnowicz et al. [34] proposed a steepest descent search Algorithm 17.12. The algorithm checks iteratively if the necessary condition for the optimality of a schedule σ_{act} is satisfied, i.e., if the inequality $\sum C_j(\sigma_{act}) - \sum C_j(\tau) \leqslant 0$ holds for any schedule $\tau \in \mathcal{N}(\sigma_{act})$.

Algorithm 17.12 is based on the following observation. Given a schedule σ, the main issue is to construct a new schedule σ' efficiently. This can be done, since the jobs assigned to each machine should be in non-increasing order of the deterioration rates b_i (cf. Theorem 7.61), so we only need to choose a job and assign it to a machine in the right position. Let J_{b_i} denote the job corresponding to deterioration rate b_i.

Algorithm 17.12. for problem $Pm|p_j = b_j t| \sum C_j$

---

1 **Input** : sequence $(b_1, b_2, \ldots, b_n)$, neighbourhood function $\mathcal{N}$
2 **Output:** a locally optimal solution σ_{act}
   ▷ Step 1: Construction of the initial schedule σ_0
3 Arrange all jobs in non-increasing order of the deterioration rates b_i;
4 Assign $m - 1$ jobs with greatest rate b_i to machines $M_2, M_3, \ldots, M_m$;
5 Assign the remaining $n - m$ jobs to machine M_1;
   ▷ Step 2: Construction of the set $\mathcal{N}(s_0)$
6 $\sigma_{act} \leftarrow \sigma_0$;
7 **repeat**
8 $\quad$ $\sigma_{last} \leftarrow \sigma_{act}$;
9 $\quad$ $\mathcal{N}(\sigma_{act}) \leftarrow \emptyset$;
10 $\quad$ **for all** jobs assigned to machine M_1 in σ_{act} **do**
11 $\quad\quad$ Choose a job J_{b_i};
12 $\quad\quad$ **for** $\mathcal{M} \in \{M_2, M_3, \ldots, M_m\}$ **do**
13 $\quad\quad\quad$ Construct schedule σ' by moving job J_{b_i} to machine $\mathcal{M}$;
14 $\quad\quad\quad$ $\mathcal{N}(\sigma_{act}) \leftarrow \mathcal{N}(\sigma_{act}) \cup \sigma'$;
   $\quad\quad$ **end**
   $\quad$ **end**
   $\quad$ ▷ Step 3: Selection of the best schedule in $\mathcal{N}(\sigma_{act})$
15 $\quad$ Choose $\tau \in \mathcal{N}(\sigma_{act})$ such that
   $\quad\quad$ $\tau = \arg\max\{\sum C_j(\sigma_{last}) - \sum C_j(\sigma') : \sigma' \in \mathcal{N}(\sigma_{act})\}$;
16 $\quad$ **if** $(\sum C_j(\sigma_{last}) - \sum C_j(\tau) > 0)$ **then**
17 $\quad\quad$ $\sigma_{act} \leftarrow \tau$;
   $\quad$ **end**
   until $((\sum C_j(\sigma) - \sum C_j(\sigma_{last}) = 0) \vee (k > n))$;
   ▷ Step 4
18 **return** σ_{act}.

The time complexity of Algorithm 17.12 depends on the number of iterations of the **repeat-until** loop, which is $O(nm)$, and the cost of checking the condition $\sum C_j(\sigma_{last}) - \sum C_j(\tau) > 0$. Since this latter cost is $O(n)$, the algorithm runs in $O(n^2 m) \equiv O(n^2)$ time for fixed m.

Experimental evaluation of Algorithms 17.12 and 17.10

In order to evaluate the quality of schedules generated by Algorithms 17.12 and 17.10, a computational experiment for $m = 3$ machines was conducted. The obtained schedules were compared to those generated by Algorithm 16.13.

The coefficients $\beta_i = b_i + 1$ were randomly generated values from intervals $(2, 99)$ and $(1, 2)$. The results for $\beta_i \in (2, 99)$ and for $\beta_i \in (1, 2)$ are presented in Table 17.1 and Table 17.2, respectively. Each value in the tables is an average of results for 10 instances. In total, 160 instances have been tested.

Table 17.1: Results of experimental comparison of Algorithms 16.13, 17.10 and 17.12, $\beta_i \in (2, 99)$

n	$R^{\text{avg}}_{A_{16.13}}$	$R^{\text{lb}}_{A_{16.13}}$	$R^{\text{avg}}_{A_{17.10}}$	$R^{\text{lb}}_{A_{17.10}}$	$R^{\text{avg}}_{A_{17.12}}$	$R^{\text{lb}}_{A_{17.12}}$
6	0.0	0.186729	0.167711	0.374054	0.0	0.186729
8	0.267105	0.639685	0.167173	0.493273	0.0	0.293706
10	0.366406	0.384822	0.121466	0.127353	0.016173	0.031644
12	0.116080	0.115118	0.459128	0.444614	0.003993	0.014459
14	-	0.476927	-	0.206961	-	0.090330
16	-	0.354809	-	0.237446	-	0.012126
18	-	0.052520	-	0.344081	-	0.054585
20	-	0.475177	-	0.161075	-	0.031898

Columns $R^{\text{avg}}_{A_{16.13}}$, $R^{\text{avg}}_{A_{17.10}}$ and $R^{\text{avg}}_{A_{17.12}}$ include an average of ratios $R^r_H(I)$ of the total completion time ($\sum C_j$) for Algorithms 16.13, 17.10 and 17.12, respectively, calculated with respect to the optimal value of $\sum C_j$.

Columns $R^{\text{lb}}_{A_{16.13}}$, $R^{\text{lb}}_{A_{17.10}}$ and $R^{\text{lb}}_{A_{17.8}}$ include an average of ratios $R^r_H(I)$ of $\sum C_j$ for Algorithms 16.13, 17.10 and 17.12, respectively, calculated with respect to the lower bound of $\sum C_j$.

Table 17.2: Results of experimental comparison of Algorithms 16.13, 17.10 and 17.12, $\beta_i \in (1, 2)$

n	$R^{\text{avg}}_{A_{16.3}}$	$R^{\text{lb}}_{A_{16.3}}$	$R^{\text{avg}}_{A_{17.10}}$	$R^{\text{lb}}_{A_{17.10}}$	$R^{\text{avg}}_{A_{17.12}}$	$R^{\text{lb}}_{A_{17.12}}$
5	0.0	0.003121	0.0	0.003121	0.0	0.003121
6	0.0	0.002508	0.000693	0.003204	0.0	0.002508
8	0.001603	0.005034	0.004348	0.007798	0.0	0.003425
10	0.001520	0.002659	0.014319	0.015473	0.000026	0.001163
12	0.001170	0.001809	0.020410	0.021059	0.003459	0.004098
14	-	0.002801	-	0.025815	-	0.005598
16	-	0.002348	-	0.031094	-	0.001261
18	-	0.001272	-	0.044117	-	0.013159
20	-	0.003101	-	0.049956	-	0.004320

Table 17.1 and Table 17.2 show that Algorithm 17.12 is better than Algorithm 16.13 for $\beta_i \in (2, 99)$, while for $\beta_i \in (1, 2)$ the algorithms are comparable. The quality of schedules generated by Algorithm 17.10, in comparison to Algorithm 16.13, is an open question which needs further research.

Wu et al. [36] proposed three local search algorithms for the single machine problem $1|p_j = a_j + b_j t| \sum w_j C_j$. The first algorithm tried to improve the initial schedule σ_0 in which jobs are in non-decreasing order of the ratios $\frac{a_j}{w_j}$, by making pairwise interchanges in σ_0 until no improvement can be made.

The second algorithm proposed in [36] tried to improve the initial schedule σ_0 in which jobs are in non-decreasing order of the ratios $\frac{b_j}{w_j}$, by making pairwise interchanges in σ_0 until no improvement can be made.

The third algorithm proposed in [36] (see Algorithm 17.13) tried to improve the initial schedule σ_0 in which jobs are in non-decreasing order of the ratios $\frac{a_j+(1+b_j)S_j}{w_j}$.

Algorithm 17.13. for problem $1|p_j = a_j + b_j t| \sum w_j C_j$

1 **Input**   : sequences $(a_1, a_2, \ldots, a_n)$, $(b_1, b_2, \ldots, b_n)$, $(w_1, w_2, \ldots, w_n)$,
              number t_0
2 **Output:** a suboptimal schedule σ
   ▷ Step 1
3 $N_{\mathcal{J}} \leftarrow \{1, 2, \ldots, n\}$;
4 $\sigma \leftarrow (\phi)$;
   ▷ Step 2
5 **for** $k \leftarrow 1$ **to** n **do**
6 | Choose job J_i for which $\min\left\{ \frac{a_j+(1+b_j)t_{k-1}}{w_j} : j \in N_{\mathcal{J}} \right\}$ is achieved;
7 | $\sigma_k \leftarrow i$;
8 | $N \leftarrow N \setminus \{i\}$;
9 | $t_k \leftarrow a_i + (1 + b_i)t_{k-1}$;
   end
10 Call the final schedule σ_0;
   ▷ Step 3
11 Make pairwise interchanges in σ_0 until no improvement can be made;
12 Call the final schedule σ;
   ▷ Step 4
13 **return** σ.

The three algorithms were tested on a set of 600 instances with randomly generated basic processing times, deterioration rates and job weights. The reported results of the experiment (see [36, Sect. 5]) suggest that the best schedules, on average, were generated by the first algorithm.

17.5 Meta-heuristic time-dependent scheduling algorithms

In this section, we present several meta-heuristic algorithms applied to single, parallel and dedicated machine time-dependent scheduling problems.

17.5.1 Simulated annealing algorithms

For parallel-machine problem $Pm|p_j = a_j + b_j t|C_{\max}$, Hindi and Mhlanga [42] proposed an SA algorithm. The schedule generated by Algorithm 17.11 was selected in this algorithm as the initial schedule σ_0. The initial temperature T_0 was given by the formula

$$T_0 := \Delta^+ \left[\ln\left(\frac{m^+}{xm^+ - (1-x)(m-m^+)}\right)\right]^{-1},$$

where m^+ is the number of moves which increased the cost of a schedule and were performed during the execution of Algorithm 17.11, Δ^+ is the average cost increase over these moves and $0 < x < 1$ is the acceptance ratio (the authors assumed $x := 0.95$).

The temperature decreased according to the formula $T_{i+1} := \frac{T_i}{1+\kappa T_i}$, where $\kappa \ll \frac{1}{U}$ with U being the largest absolute move value found during the execution of Algorithm 17.11.

The SA algorithm was tested on a set of 320 instances in which deterioration rates b_j were randomly generated values from the interval $(0,1)$ and the basic processing times a_j were chosen from a normal distribution, with a mean of 50 and standard deviation of 10. The experiment shows that if we use the proposed SA algorithm in order to improve the schedule generated by Algorithm 17.11, the results are satisfactory (see [42, Sect. 5] for details).

Bank et al. [38] proposed an SA algorithm for a time-dependent flow shop problem with the total tardiness criterion, $Fm|p_{ij} = b_{ij}(a + bt)| \sum T_j$. The algorithm was tested on a set of 675 instances with $n \in \{10, 15, 20, 30, 40\}$ jobs and $m \in \{2, 3, 5\}$ machines.

Shiau et al. [44] proposed an SA algorithm for problem $F2|p_{ij} = b_{ij}t| \sum C_j$. In this algorithm, the solution space $\mathfrak{F}$ consists of all schedules that correspond to permutations of sequence $(1, 2, \ldots, n)$. New schedules from a neighbourhood $\mathcal{N}(\sigma_0)$ of a given schedule σ_0 are generated by the pairwise interchange of two randomly selected jobs. The probability $p(\sigma_0)$ of acceptance of a schedule σ_0 is generated from an exponential distribution. The number of iterations of this algorithm is limited to $50n$, where n is the number of jobs. The SA algorithm was tested on a number of instances with $10 \leqslant n \leqslant 100$ jobs, giving satisfactory results; see [44, Sect. 6] for details.

Wu et al. [46] proposed an SA algorithm for a single machine time-dependent two-agent scheduling problem, $1|p_j = b_j(a + bt)| \max w_j^A T_j^A + \sum w_j^B T_j^B$. This algorithm was stopped after $200n$ iterations, where n is the number of jobs, and was tested on small ($n = 12, 14$ jobs) and large ($n = 40, 80$ jobs) instances, giving satisfactory results; see [46, Sect. 5] for details.

Remark 17.8. We consider time-dependent two-agent scheduling in Sect. 21.2.

17.5.2 Tabu search algorithms

Bahalke et al. [37] proposed a TS algorithm for a single machine time-dependent scheduling problem with setups, $1|p_j = a_j + b_j t, s_j|C_{\max}$. The stopping criterion was the stability of a solution in 1,000 iterations. The algorithm was tested on small instances with $n = 8, 10, 12$ jobs and on larger instances with $n = 20, 30, 40, 50$ jobs; see [37, Sect. 7] for details.

Wu et al. [45] proposed a TS algorithm for a single machine time-dependent scheduling problem with ready times, $1|r_j, p_j = b_j(a + b\min\{t, D\})|C_{\max}$. In this algorithm, the maximum number of iterations was equal to $30n$ iterations, where n is the number of jobs. The algorithm was tested on instances with $n = 12, 16, 20$ jobs; see [45, Sect. 4] for details.

17.5.3 Genetic algorithms

Bahalke et al. [37] proposed a GA for a single machine time-dependent scheduling problem with linearly deteriorating jobs and setups, $1|p_j = a_j + b_j t, s_j|C_{\max}$. Three sizes of population were tested: 20, 30 and 50 elements. The stopping criterion was the stability of a solution in $100n$ iterations, where n is the number of jobs; see [37, Sect. 7] for details.

17.5.4 Evolutionary algorithms

The EAs for time-dependent scheduling problems can be constructed in many ways. However, though all EAs share the same template, the main difficulty in developing a new EA is the effort which is needed to implement it in a programming language. There are various approaches which help overcome this difficulty. One of the approaches consists in the use of *libraries of classes*. The libraries allow programmers to construct new classes in an easy way, using mechanisms such as *encapsulation* and *inheritance* which are built-in in object programming languages. Libraries of classes, in particular, can be used for the construction of EAs.

Gawiejnowicz et al. [41] proposed a new library of this kind, developed in $C\#$ and designated for work on the .NET platform. The library, called *TEAC* (*Toolbox for Evolutionary Algorithms in C#*), includes a number of classes which allow us to implement basic genetic operators applied in EAs. It also includes a few classes which implement genetic and evolutionary algorithms, in particular the SEA. We refer the reader to [41, Sect. 2] for more detailed description of the TEAC library.

In order to evaluate the solutions generated by EAs implemented using the TEAC library, a computational experiment has been conducted. Based on the classes defined in the TEAC library, an EA was constructed for the problem $Jm||C_{\max}$ (cf. Sect. 4.3). We will call it Algorithm 17.14.

Algorithm 17.14 was identical to the SEA and it stopped after 2000 generations. The benchmark data files from Beasley's OR-Library [39] were used in the experiment. The solutions obtained by the EA for job shop problems are presented in Table 17.3. The column *Size* gives the size of a particular instance, $n \times m$, where n and m are the numbers of jobs and machines, respectively. Columns *OPT*, *AVR* and *Best* show the optimal, average and best found result, respectively. The column *Time* presents the running time of the EA (in seconds). All results are average values from 10 independent runs of the algorithm.

Table 17.3: Results of computational experiment for Algorithm 17.14

File	Size	OPT	AVR	Best	Time
FT06	6x6	55	55.0	55	40.50
FT10	10x10	930	979.1	955	130.44
FT20	20x5	1,165	1,228.8	1,202	199.44
LA01	10x5	666	666.0	666	66.02
LA16	10x10	945	972.9	956	141.70
LA20	10x10	902	914.7	907	141.08
LA21	15x10	1,046	1,111.6	1,097	288.40
LA25	15x10	977	1,032.5	1,019	253.66
LA28	20x10	1,216	1,302.2	1,286	412.35
LA29	20x10	1,152	1,266.9	1,248	409.20
LA39	15x15	1,233	1,320.3	1,291	384.13
LA40	15x15	1,222	1,312.4	1,288	385.36

The results were satisfactory, since suboptimal schedules were generated in reasonable time. The average error of the obtained solutions was from 0% (files FT06, LA01) to 9.97% (file LA29); see [41, Sect. 3] for details.

Gawiejnowicz and Suwalski [40] used the TEAC library for construction of an evolutionary algorithm for two- and three-machine time-dependent flow shop scheduling problems (cf. Sects. 4.3 and 6.2). We will call it Algorithm 17.15. As previously, the EA was identical to the simple evolutionary algorithm, and exploited basic genetic operators; see [40, Sect. 4] for details.

The behaviour of the new EA has been tested in a number of experiments concerning problems $Fm|p_{ij} = b_{ij}t|C_{\max}$ and $Fm|p_{ij} = b_{ij}(a + bt)|C_{\max}$, where $m \in \{2,3\}$. For each problem, job deterioration rates were randomly generated integer values: $b_{ij} \in \langle 1, n - 1 \rangle$ and $a, b \in (1, \frac{n}{2})$. In total, in the experiments 360 instances were generated.

Main results of these experiments are presented in Tables 17.4–17.5. Each value is an average for five distinct instances. Symbols $R^{\min}_{A_{17.15}}$, $R^{\text{avg}}_{A_{17.15}}$ and $R^{\max}_{A_{17.15}}$ denote the minimum, average and maximum absolute ratio $R^a_{A_{17.15}}(I)$, respectively. Symbol T_{avg} denotes the computation time (in seconds). Symbol $FmPn$ ($FmLn$) denotes an instance of the m-machine flow shop problem with n jobs with proportional (proportional-linear) job processing times.

Table 17.4: EA solutions vs. exact solutions for F2P/F2L datasets

Dataset	$R^{min}_{A_{17.15}}$	$R^{avg}_{A_{17.15}}$	$R^{max}_{A_{17.15}}$	T_{avg}	Dataset	$R^{min}_{A_{17.15}}$	$R^{avg}_{A_{17.15}}$	$R^{max}_{A_{17.15}}$	T_{avg}
F2P05	1,000	1,000	1,000	0,880	F2L05	1,000	1,000	1,000	1,195
F2P06	1,000	1,000	1,000	1,056	F2L06	1,000	1,000	1,000	1,435
F2P07	1,000	1,000	1,000	1,259	F2L07	1,000	1,000	1,000	1,781
F2P08	1,000	1,000	1,000	1,477	F2L08	1,037	1,184	1,233	2,091
F2P09	1,000	1,109	1,175	1,699	F2L09	1,000	1,083	1,217	2,453
F2P10	1,000	1,000	1,000	1,978	F2L10	1,000	1,077	1,159	2,833
F2P11	1,027	1,000	1,000	2,252	F2L11	1,248	1,535	1,984	3,284
F2P12	1,094	1,018	1,024	2,526	F2L12	1,250	1,567	2,224	3,697

Table 17.5: EA solutions vs. exact solutions for F3P/F3L datasets

Dataset	$R^{min}_{A_{17.15}}$	$R^{avg}_{A_{17.15}}$	$R^{max}_{A_{17.15}}$	T_{avg}	Dataset	$R^{min}_{A_{17.15}}$	$R^{avg}_{A_{17.15}}$	$R^{max}_{A_{17.15}}$	T_{avg}
F3P05	1,000	1,000	1,000	1,186	F3L05	1,000	1,000	1,000	1,195
F3P06	1,000	1,000	1,000	1,444	F3L06	1,000	1,000	1,000	1,435
F3P07	1,000	1,000	1,000	1,781	F3L07	1,000	1,000	1,000	1,781
F3P08	1,000	1,000	1,000	2,105	F3L08	1,037	1,184	1,233	2,091
F3P09	1,000	1,000	1,000	2,440	F3L09	1,000	1,083	1,217	2,453
F3P10	1,000	1,003	1,013	2,822	F3L10	1,000	1,077	1,159	2,833
F3P11	1,027	1,122	1,249	3,260	F3L11	1,248	1,535	1,984	3,284
F3P12	1,094	1,227	1,319	3,722	F3L12	1,250	1,567	2,224	3,697

17.5.5 Ant colony optimization algorithms

Wu et al. [46] proposed an ACO algorithm for a single machine time-dependent two-agent scheduling problem (cf. Remark 17.8), $1|p_j = b_j(a+bt)| \max w_j^A T_j^A + \sum w_j^B T_j^B$. In the ACO, 50 ants were generated and the stopping criterion was the stability of generated schedules in 50 iterations; see [46, Sect. 5] for details.

Wu et al. [45] proposed an ACO algorithm for a single machine time-dependent scheduling problem $1|r_j, p_j = b_j(a + b\min\{t, D\})|C_{\max}$. In the algorithm, 30 ants were generated and the stopping criterion was the stability of generated schedules in 30 iterations; see [46, Sect. 5] for details.

17.5.6 Particle swarm optimization algorithms

Bank et al. [38] proposed a PSO and an SA algorithms for problem $Fm|p_{ij} = b_{ij}(a + bt)| \sum T_j$. Both algorithms were testes on instances with $n \leqslant 40$ jobs. The average error in percent of the PSO algorithm was smaller than the one of the SA algorithm; see [38, Sect. 5] for details.

Remark 17.9. Meta-heuristics were also applied to other scheduling problems with time-dependent parameters. For example, Lu et al. [43] considered a meta-heuristic called *Artificial Bee Colony* (ABC) combined with a TS algorithm in order to find a schedule with minimal $C_{\max}$ for a time-dependent scheduling problem with linearly deteriorating jobs, a set of parallel unrelated machines and maintenance activities of variable, time-dependent length.

This chapter ends the fifth part of the book, devoted to different classes of algorithms for intractable time-dependent scheduling problems. In the next part, we will discuss selected advanced topics in time-dependent scheduling.

References

General concepts in local search

1. E. Aarts and J. Korst, *Simulated Annealing and Boltzmann Machines: A Stochastic Approach to Combinatorial Optimization and Neural Computing.* New York: Wiley 1989.
2. E. Aarts and J. K. Lenstra, *Local Search in Combinatorial Optimization.* New York: Wiley 1997.
3. C. Blum and A. Roli, Metaheuristics in combinatorial optimization: overview and conceptual comparison. *ACM Computing Surveys* **35** (2003), no. 3, 268–308.
4. E. K. Burke and G. Kendall (eds.), *Search Methodologies: Introductory Tutorials in Optimization and Decision Support Techniques.* Berlin-Heidelberg: Springer 2005.
5. M. Gendreau and J-Y. Potvin, Metaheuristics in combinatorial optimization. *Annals of Operations Research* **140** (2005), no. 1, 189–213.

General concepts in meta-heuristic algorithms

6. E. Aarts and J. Korst, *Simulated Annealing and Boltzmann Machines: A Stochastic Approach to Combinatorial Optimization and Neural Computing.* New York: Wiley 1989.
7. M. R. Bonyadi and Z. Michalewicz, Particle swarm optimization for single objective continuous space problems: a review. *Evolutionary Computation* **25** (2017), no. 1, 1–54.
8. O. Bozorg-Haddad, M. Solgi and H. A. Loáciga, *Meta-Heuristic and Evolutionary Algorithms for Engineering Optimization.* New York: Wiley 2017.
9. E. K. Burke and G. Kendall (eds.), *Search Methodologies: Introductory Tutorials in Optimization and Decision Support Techniques,* Berlin-Heidelberg: Springer 2005.
10. P. Calégari, G. Coray, A. Hertz, D. Kobler and P. Kuonen, A taxonomy of evolutionary algorithms in combinatorial optimization. *Journal of Heuristics* **5** (1999), no. 2, 145–158.
11. V. Černý, Thermodynamical approach to the traveling salesman problem: an efficient simulation algorithm. *Journal of Optimization Theory and Applications* **45** (1985), 41–51.

12. B. Chopard and M. Tomassini, *An Introduction to Metaheuristics for Optimization*. Cham: Springer Nature, 2018.
13. M. Dorigo, *Optimization, Learning and Natural Algorithms*. Ph.D. thesis, Politecnico di Milano, Italy, 1992.
14. M. Dorigo, V. Maniezzo and A. Colorni, Ant system: optimization by a colony of cooperating agents. *IEEE Transactions on Systems, Man, and Cybernetics – Part B*, **26** (1996), no. 1, 29–41.
15. M. Dorigo and K. Socha, An introduction to Ant Colony Optimization, in: T. F. Gonzalez, *Handbook of Approximation Algorithms and Metaheuristics*, 2nd ed., Vol. I *Methodologies and Traditional Applications*. Boca Raton: CRC Press, 2018.
16. M. Dorigo and T. Stützle, *Ant Colony Optimization*, Cambridge: MIT Press, 2004.
17. K-L. Du and M. N. S. Swamy, *Search and Optimization by Metaheuristics: Techniques and Algorithms Inspired by Nature*. Basel: Birkhäuser, 2016.
18. M. Gendreau nd J-Y. Potvin (eds.), *Handbook of Metaheuristics*. Cham: Springer, 2019.
19. F. Glover, Tabu search – Part 1. *ORSA Journal on Computing*, **1** (1989), no. 2, 190–206.
20. F. Glover, Tabu search – Part 2. *ORSA Journal on Computing*, **2** (1990), no. 1, 4–32.
21. F. Glover, M. Laguna, E. Taillard, D. de Werra (eds), Tabu search, *Annals of Operations Research* **41**. Basel: Baltzer, 1993.
22. F. Glover and M. Laguna, *Tabu Search*. Boston: Kluwer 1997.
23. D. Goldberg, *Genetic Algorithms in Search, Optimization and Machine Learning*. Reading: Addison-Wesley, 1989.
24. E. Hart, P. Ross and D. Corne, Evolutionary scheduling: a review. *Genetic Programming and Evolvable Machines* **6** (2005), no. 2, 191–220.
25. J. H. Holland, *Adaptation in Natural and Artificial Systems*. Ann Arbor: University of Michigan Press, 1975.
26. A. K. Kar, Bio inspired computing – a review of algorithms and scope of applications. *Expert Systems with Applications* **59** (2016), 20–32.
27. J. Kennedy and R. Eberhart, Particle swarm optimization. Proceedings of IEEE International Conference on Neural Networks, 1995, IV, 1942–1948.
28. J. Kennedy and R. C. Eberhart, *Swarm Intelligence*. San Francisco: Morgan Kaufmann, 2001.
29. S. Kirkpatrick, C. D. Gelatt and M. P. Vecchi, Optimization by simulated annealing. *Science* **220** (1983), 671–680.
30. J. R. Koza, *Genetic Programming: On the Programming of Computers by Means of Natural Selection*. Cambridge: MIT Press, 1992.
31. R. J. Mullen, D. Monekosso, S. Barman and P. Remagnino, A review of ant algorithms. *Expert Systems with Applications* **36** (2009), 9608–9617.
32. P. Salamon, P. Sibani and R. Frost, *Facts, Conjectures, and Improvements for Simulated Annealing*. Philadelphia: SIAM 2002.
33. X-S. Yang, *Nature-Inspired Optimization Algorithms*. Amsterdam: Elsevier 2014.

Local search time-dependent scheduling algorithms

34. S. Gawiejnowicz, W. Kurc and L. Pankowska, Parallel machine scheduling of deteriorating jobs by modified steepest descent search. *Lecture Notes in Computer Science* **3911** (2006), 116–123.
35. K. S. Hindi and S. Mhlanga, Scheduling linearly deteriorating jobs on parallel machines: a simulated annealing approach. *Production Planning and Control* **12** (2001), no. 12, 76–80.
36. C-C. Wu, W-C. Lee and Y-R. Shiau, Minimizing the total weighted completion time on a single machine under linear deterioration. *International Journal of Advanced Manufacturing Technology* **33** (2007), no. 11–12, 1237–1243.

Meta-heuristic time-dependent scheduling algorithms

37. U. Bahalke, A. M. Yolmeh, K. Shahanaghi, Meta-heuristics to solve single-machine scheduling problem with sequence-dependent setup time and deteriorating jobs. *The International Journal of Advanced Manufacturing Technology* **50** (2010), 749–759.
38. M. Bank, S. M. T. Fatemi Ghomi, F. Jolai and J. Behnamian, Application of particle swarm optimization and simulated annealing algorithms in flow shop scheduling problem under linear deterioration. *Advances in Engineering Software* **47** (2012), 1–6.
39. J. E. Beasley, OR-Library, see http://msmcga.ms.ic.ac.uk/info.html.
40. S. Gawiejnowicz and C. Suwalski, Solving time-dependent scheduling problems by local search algorithms. Report 133/2006, Faculty of Mathematics and Computer Science, Adam Mickiewicz University, Poznań, November 2006.
41. S. Gawiejnowicz, T. Onak and C. Suwalski, A new library for evolutionary algorithms. *Lecture Notes in Computer Science* **3911** (2006), 414–421.
42. K. S. Hindi and S. Mhlanga, Scheduling linearly deteriorating jobs on parallel machines: a simulated annealing approach. *Production Planning and Control* **12** (2001), no. 12, 76–80.
43. S-J. Lu, X-B. Liu, J. Pei, M. T. Thai and P. M. Pardalos, A hybrid ABC-TS algorithm for the unrelated parallel-batching machine scheduling problem with deteriorating jobs and maintenance activity. *Applied Soft Computing* **66** (2018), 168–182.
44. Y-R. Shiau, W-C. Lee, C-C. Wu and C-M. Chang, Two-machine flow-shop scheduling to minimize mean flow time under simple linear deterioration. *International Journal of Advanced Manufacturing Technology* **34** (2007), no. 7–8, 774–782.
45. C-C. Wu, W-H. Wu, W-H. Wu, P-H. Hsu, Y-Q. Yin and J-Y. Xu, A single-machine scheduling with a truncated linear deterioration and ready times. *Information Sciences* **256** (2014), 109–125.
46. W-H. Wu, Y-Q. Yin, W-H. Wu, C-C. Wu and P-H. Hsu, A time-dependent scheduling problem to minimize the sum of the total weighted tardiness among two agents. *Journal of Industrial and Management Optimization* **10** (2014), 591–611.

ADVANCED TOPICS

Time-dependent scheduling under precedence constraints

In the previous five parts of the book we presented the results of time-dependent scheduling which were obtained using the classical scheduling theory methods. In the sixth part of the book, we discuss the topics in time-dependent scheduling which can be analyzed and solved using other methods.

This part of the book is composed of four chapters. In Chap. 18, we consider single machine time-dependent scheduling problems with various forms of precedence constraints among jobs. In Chap. 19, we show how to formulate single and parallel machine time-dependent scheduling problems in terms of matrices and vectors. In Chap. 20, we analyze single and dedicated machine bi-criteria time-dependent scheduling problems. Finally, in Chap. 21 we discuss selected new topics in time-dependent scheduling, such as scheduling on machines with limited availability, two-agent time-dependent scheduling and time-dependent scheduling with mixed deterioration.

Chapter 18 is composed of five sections. In Sect. 18.1, we consider problems with proportionally deteriorating jobs and the $\sum w_j C_j$, $\sum w_j C_j^2$, $\sum w_j W_j^k$ and $f_{\max}$ criteria. In Sect. 18.2, we focus on proportional-linearly deteriorating jobs and the $f_{\max}$ criterion. In Sect. 18.3, we study problems with linearly deteriorating jobs and the $C_{\max}$ and $\sum C_j$ criteria. In Sect. 18.4, we address problems with proportional-linearly shortening jobs and the $\sum w_j C_j$ criterion. Finally, in Sect. 18.5, we concentrate on linearly shortening jobs and the $C_{\max}$ and $\sum C_j$ criteria. The chapter is completed with a list of references.

18.1 Proportional deterioration

In this section, we consider precedence-constrained time-dependent scheduling problems with proportionally deteriorating jobs.

18.1.1 Chain precedence constraints

First, we recall the definition of two types of chains (cf. Dror et al. [4]).

© Springer-Verlag GmbH Germany, part of Springer Nature 2020
S. Gawiejnowicz, *Models and Algorithms of Time-Dependent Scheduling*,
Monographs in Theoretical Computer Science. An EATCS Series,
https://doi.org/10.1007/978-3-662-59362-2_18

Definition 18.1. (Strong and weak chains)
(a) *A chain of jobs is called a* strong chain *if between jobs of the chain no job from another chain can be inserted.*
(b) *A chain of jobs is called a* weak chain *if between jobs of the chain jobs from another chain can be inserted.*

Remark 18.2. Strong and weak chains will be denoted in the second field of the $\alpha|\beta|\gamma$ notation by symbols *s-chains* and *w-chains*, respectively.

Wang et al. [12] considered single machine time-dependent scheduling problems with proportional jobs, strong and weak chain precedence constraints and the $\sum w_j C_j^2$ criterion.

In the first of the problems, we are given m strong chains $\mathcal{C}_1, \mathcal{C}_2, \ldots, \mathcal{C}_m$, where chain $\mathcal{C}_i$ is composed of n_i jobs $J_{i_1}, J_{i_2}, \ldots, J_{i_{n_i}}$ with weights w_{i_1}, $w_{i_2}, \ldots, w_{i_{n_i}}$, respectively, where $1 \leqslant i \leqslant m$ and $\sum_{i=1}^{m} n_i = n$. The aim is to find a schedule which minimizes the criterion $\sum_{i=1}^{m} \sum_{j=i_1}^{i_{n_i}} w_{ij} C_{ij}^2$. We will denote this problem as $1|p_j = b_j t, s\text{-}chains| \sum w_j C_j^2$.

For a given strong chain $\mathcal{C}_i = (J_{i_1}, J_{i_2}, \ldots, J_{i_{n_i}})$, let

$$A(i_1, i_{n_i}) := \sum_{k=i_1}^{i_{n_i}} w_{i_k} \prod_{j=i_1}^{k} (1 + b_{i_j})^2 \qquad (18.1)$$

and

$$B(i_1, i_{n_i}) := \prod_{k=i_1}^{i_{n_i}} (1 + b_{i_k})^2 - 1. \qquad (18.2)$$

Under the notation, we can define the following ratio.

Definition 18.3. (The ratio $R_s(i_1, i_{n_i})$ for a strong chain)
Given a strong chain $\mathcal{C}_i$ of jobs $J_{i_1}, J_{i_2}, \ldots, J_{i_{n_i}}$ the ratio

$$R_s(i_1, i_{n_i}) := \frac{B(i_1, i_{n_i})}{A(i_1, i_{n_i})}$$

will be called the ratio $R_s(i_1, i_{n_i})$ (ratio R_s, in short) for this chain.

Lemma 18.4. (Wang et al. [12]) *Let there be given two strong chains, $\mathcal{C}_i$ and $\mathcal{C}_k$, and let in schedule σ_1 chain $\mathcal{C}_i$ precede chain $\mathcal{C}_k$, while in schedule σ_2 chain $\mathcal{C}_k$ precede chain $\mathcal{C}_i$. Then $\sum_i \sum_j w_{ij} C_{ij}^2(\sigma_1) \leqslant \sum_i \sum_j w_{ij} C_{ij}^2(\sigma_2)$ if and only if $R_s(i_1, i_{n_i}) \leqslant R_s(i_1, i_{n_k})$.*

Proof. By direct computation; see [12, Sect. 3]. ⬦

Since in the considered problem we deal with chains which cannot be decomposed into smaller subchains, each such chain can be considered as a single 'big' job. Therefore, applying Lemma 18.4, we can establish the order of the jobs as follows.

Theorem 18.5. (Wang et al. [12]) *Problem* $1|p_j = b_j t, s\text{-}chains| \sum w_j C_j^2$ *is solvable in* $O(n \log n)$ *time by scheduling strong chains in non-decreasing order of their* R_s *ratios.*

In the second of the problems considered in [12], we are given m weak chains $\mathcal{C}_1, \mathcal{C}_2, \ldots, \mathcal{C}_m$, while other assumptions are the same as previously. We will denote this problem as $1|p_j = b_j t, w\text{-}chains| \sum w_j C_j^2$.

Ratio R_s of a weak chain in problem $1|p_j = b_j t, w\text{-}chains| \sum w_j C_j^2$ is defined similarly to Definition 18.3.

Definition 18.6. (Ratio $R_w(i_1, i_{n_i})$ for a weak chain)
Given a weak chain $\mathcal{C}_i$ *of jobs* $J_{i_1}, J_{i_2}, \ldots, J_{i_{n_i}}$ *the ratio*

$$R_w(i_1, i_{n_i}) := \frac{B(i_1, i_{n_i})}{A(i_1, i_{n_i})}$$

will be called ratio $R_w(i_1, i_{n_i})$ *(ratio* R_w*, in short) for this chain.*

Though any weak chain can be decomposed by jobs from other chains into a few subchains, which makes the problem more difficult compared to the problem with strong chains, the following result holds.

Lemma 18.7. (Wang et al. [12]) *Let there be given a weak chain* $\mathcal{C}_k$ *and its two weak subchains,* $\mathcal{C}_k^1$ *and* $\mathcal{C}_k^2$*, such that* $\mathcal{C}_k = (J_{n_1}, \ldots, J_{n_i}, J_{n_i+1}, \ldots, J_{n_k})$*,* $\mathcal{C}_k^1 = (J_{n_1}, \ldots, J_{n_i})$ *and* $\mathcal{C}_k^2 = (J_{n_i+1}, \ldots, J_{n_k})$*, where* $1 \leqslant i \leqslant k-1$*. Then the following implication holds:*

$$\text{if } R_w(n_1, n_k) \leqslant R_w(n_1, n_i), \text{ then } R_w(n_{i+1}, n_k) \leqslant R_w(n_1, n_i).$$

Proof. By contradiction; see [12, Sect. 3]. ⬦

Definition 18.8. (ρ-factor of a weak chain)
Let k^* *be the smallest integer such that*

$$R_w(i_1, i_{k^*}) = \max_{1 \leqslant j \leqslant n_i} \{R_w(i_1, i_j)\}$$

for a given weak chain $\mathcal{C}_i = (J_{i_1}, J_{i_2}, \ldots, J_{i_{n_i}})$*. Then* k^* *is called the* ρ*-factor of chain* $\mathcal{C}_i$ *and denoted as* $k^* = \rho(\mathcal{C}_i)$*.*

Applying Lemma 18.7, one can prove the following result concerning the meaning of the ρ factor.

Lemma 18.9. (Wang et al. [12]) *Let* $n_k = \rho(\mathcal{C}_k)$ *for a given weak chain* $\mathcal{C}_k = (J_{n_1}, \ldots, J_{n_k})$*. Then there exists an optimal schedule in which jobs of chain* $\mathcal{C}_k$ *are executed sequentially, without idle times and such that no jobs from other chains are executed between the jobs of this chain.*

Proof. By contradiction; see [12, Sect. 3]. ⬦

Lemma 18.9 leads to the following algorithm for problem $1|p_j = b_j t,$ $w\text{-}chains| \sum w_j C_j^2$.

Algorithm 18.1. for problem $1|p_j = b_j t, w\text{-}chains| \sum w_j C_j^2$

---

1 **Input**   : sequences $(b_1, b_2, \ldots, b_n)$, $(w_1, w_2, \ldots, w_n)$,
              chains $\mathcal{C}_1, \mathcal{C}_2, \ldots, \mathcal{C}_m$ of precedence constraints
2 **Output:** an optimal schedule σ^*
  ▷ Step 1
3 $\sigma^* \leftarrow (\phi)$;
4 $N_\mathcal{C} \leftarrow \{\mathcal{C}_1, \mathcal{C}_2, \ldots, \mathcal{C}_m\}$;
  ▷ Step 2
5 **while** $(N_\mathcal{C} \neq \emptyset)$ **do**
6 | **for all** $\mathcal{C}_i \in N_\mathcal{C}$ **do**
7 | | Find ρ-factor of chain $\mathcal{C}_i$;
    | **end**
8 | $\mathcal{C} \leftarrow$ the chain with minimal ρ-factor found in lines 6–7;
9 | $\sigma^* \leftarrow (\sigma^*|\mathcal{C})$;
10 | $N_\mathcal{C} \leftarrow N_\mathcal{C} \setminus \mathcal{C}$;
   end
   ▷ Step 3
11 **return** σ^*.

Example 18.10. (Wang et al. [12]) We are given chains $\mathcal{C}_1 = (J_{11}, J_{12}, J_{13}, J_{14})$ and $\mathcal{C}_2 = (J_{21}, J_{22}, J_{23})$. Job deterioration rates and job weights are as follows: $b_{11} = 0.4$, $b_{12} = 0.2$, $b_{13} = 0.1$, $b_{14} = 0.3$, $b_{21} = 0.5$, $b_{22} = 0.3$, $b_{23} = 0.2$, $w_{11} = 3$, $w_{12} = 4$, $w_{13} = 7$, $w_{14} = 2$, $w_{21} = 5$, $w_{22} = 6$ and $2_{23} = 1$.

Then, the schedule constructed by Algorithm 18.1 is in the form of $\sigma^* = (J_{11}, J_{12}, J_{13}, J_{21}, J_{22}, J_{14}, J_{23})$, with $\sum_i \sum_j w_{ij} C_{ij}^2(\sigma^*) = 201.3031$.   ◆

Theorem 18.11. (Wang et al. [12]) *Problem* $1|p_j = b_j t, w\text{-}chains| \sum w_j C_j^2$ *is solvable in* $O(n^2)$ *time by Algorithm 18.1.*

Remark 18.12. Wang et al. [12] formulate Theorem 18.11 without proof.

Remark 18.13. Applying the same approach, Duan et al. [5] obtained similar results for problem $1|p_j = b_j t, w\text{-}chains| \sum w_j C_j^k$, where $k \in \mathbb{Z}_+$.

Remark 18.14. Wang et al. [11] generalized the results of Duan et al. [5] for problems $1|p_j = b_j t, w\text{-}chains| \sum w_j W_j^k$ and $1|p_j = b_j t, s\text{-}chains| \sum w_j W_j^k$, where $W_j := C_j - p_j$ denotes the waiting time of the jth job and $k \in \mathbb{Z}_+$.

18.1.2 Series-parallel precedence constraints

Wang et al. [10] applied the same approach as the one for linear jobs and the $C_{\max}$ criterion (see Sect. 18.3) to problem $1|p_j = b_j t, ser\text{-}par| \sum w_j C_j$. First, for a given chain $\mathcal{C}_i = (J_{i_1}, J_{i_2}, \ldots, J_{i_{n_i}})$ of jobs, they defined counterparts of (18.1) and (18.2) in the form of

$$A(i_1, i_{n_i}) := \sum_{k=i_1}^{i_{n_i}} w_{i_k} \prod_{j=i_1}^{k} (1 + b_{i_j}) \tag{18.3}$$

and

$$B(i_1, i_{n_i}) := \prod_{k=i_1}^{i_{n_i}} (1 + b_{i_k}) - 1, \tag{18.4}$$

respectively. Next, applying Definition 18.3 to (18.3) and (18.4), they defined a new ρ-factor. Finally, repeating the reasoning for the linear case, the authors claimed that problem $1|p_j = b_j t, ser\text{-}par| \sum w_j C_j$ is sovable in $O(n \log n)$ time by an appropriately modified Algorithm 18.6.

Wang et al. [11], applied a similar approach to problem $1|p_j = b_j t, ser\text{-}par| \sum w_j W_j^k$, where W_j and k are defined as in Sect. 18.1.1. In conclusion, the authors claimed that problem $1|p_j = b_j t, ser\text{-}par| \sum w_j W_j^k$ is solvable in $O(n \log n)$ time by an appropriately modified Algorithm 18.6.

18.1.3 Arbitrary precedence constraints

In this section, we consider problem $1|p_j = b_j t, prec|f_{\max}$. We assume that for job J_j there is defined a non-decreasing cost function f_j that specifies a cost $f_j(C_j)$ that has to be paid at the completion time of the job, $1 \leqslant j \leqslant n$. We assume that for a given function f_j, $1 \leqslant j \leqslant n$, the cost $f_j(C_j)$ can be computed in a constant time.

The main idea of Algorithm 18.2 which solves the problem is the same as in Lawler's algorithm for problem $1|prec|f_{\max}$ [7], i.e., from jobs without successors we choose the job that will cause the smallest cost in the given position and we schedule it as the last one.

Remark 18.15. Algorithm 18.2, which we consider below, will remain polynomial if the costs $F_j(C_j)$, $1 \leqslant j \leqslant n$, are computed in a polynomial time instead of a constant time. In such a case, however, the running time of this algorithm may be different than the $O(n^2)$ specified in Theorem 18.17.

Remark 18.16. Dębczyński and Gawiejnowicz [3] observed that a single machine scheduling problem with variable job processing times can be solved in polynomial time by Lawler's algorithm, if for the problem two properties hold: *schedule non-idleness* and *the maximum completion time invariantness*. The first property says that any possible optimal schedule does not include idle times, while the second one states that the maximum completion time is the same for all such schedules (cf. [3, Sect. 3]). These two properties hold for single machine time-dependent scheduling problems with proportionally and proportional-linearly deteriorating jobs.

Let $NoSucc(G)$ denote the set of indices of jobs without immediate successors for a given digraph G of precedence constraints. The pseudo-code of the algorithm for the considered problem is as follows.

Algorithm 18.2. for problem $1|p_j = b_j t, prec|f_{max}$

---

1 **Input** : sequences $(b_1, b_2, \ldots, b_n)$, $(f_1, f_2, \ldots, f_n)$, number t_0,
               digraph G of precedence constraints
2 **Output:** an optimal schedule $\sigma^\star$
  ▷ Step 1
3 $\sigma^\star \leftarrow (\phi)$;
4 $N_{\mathcal{J}} \leftarrow \{1, 2, \ldots, n\}$;
5 $T \leftarrow t_0 \prod_{j=1}^{n}(b_j + 1)$;
  ▷ Step 2
6 **while** $(N_{\mathcal{J}} \neq \emptyset)$ **do**
7   Construct the set $NoSucc(G)$;
8   Find $k \in NoSucc(G)$ such that
         $f_k(T) = \min\{f_j(T) : j \in NoSucc(G)\}$;
9      $\sigma^\star \leftarrow (k|\sigma^\star)$;
10      $T \leftarrow \frac{T}{b_k+1}$;
11      $N_{\mathcal{J}} \leftarrow N_{\mathcal{J}} \setminus \{k\}$;
12      $NoSucc(G) \leftarrow NoSucc(G) \setminus \{k\}$;
13      $G \leftarrow G \setminus \{k\}$;
  end
  ▷ Step 3
14 **return** $\sigma^\star$.

Theorem 18.17. (Gawiejnowicz [6, Chap. 13]) *Problem* $1|p_j = b_j t, prec|f_{max}$ *is solvable in* $O(n^2)$ *time by Algorithm 18.2.*

Proof. First, notice that without loss of generality we can consider only schedules without idle times. Second, by Theorem 7.1 the value of the C_{max} criterion for jobs with proportional processing times does not depend on the sequence of the jobs and is given by formula (7.1). Note also that the value of the C_{max} can be calculated in $O(n)$ time.

Let $f_{max}^\star(\mathcal{J})$ denote the value of the criterion f_{max} for an optimal schedule. Then the following two inequalities are satisfied:

$$f_{max}^\star(\mathcal{J}) \geqslant \min_{J_j \in NoSucc(G)} \{f_j(C_n)\} \tag{18.5}$$

and

$$f_{max}^\star(\mathcal{J}) \geqslant f_{max}^\star(\mathcal{J} \setminus \{J_j\}) \tag{18.6}$$

for $j = 1, 2, \ldots, n$. Let job $J_k \in \mathcal{J}$ be such that

$$f_k(C_n) = \min\{f_j(C_n) : J_j \in NoSucc(G)\}$$

and let $f_k(\mathcal{J})$ denote the value of the criterion f_k, provided that job J_k is executed as the last one. Then

$$f_k(\mathcal{J}) = \max \left\{ f_k(C_n), f^*_{\max}(\mathcal{J} \setminus \{J_k\}) \right\}. \tag{18.7}$$

From (18.5), (18.6) and (18.7) it follows that $f^*_{\max}(J) \geqslant f_k(J)$ and there exists an optimal schedule in which job J_k is executed as the last one.

Repeating this procedure for the remaining jobs, we obtain an optimal schedule. Because in each run of the procedure we have to choose a job from $O(n)$ jobs, and there are n jobs, Algorithm 18.2 runs in $O(n^2)$ time. ∎

Example 18.18. Let $n = 4$, $p_1 = t$, $p_2 = 3t$, $p_3 = 2t$, $p_4 = t$, $t_0 = 1$. Cost functions are in the form of $f_1 = 2C_1^2, f_2 = C_2^2 + 1, f_3 = C_3, f_4 = C_4 + 5$. Precedence constraints are given in Figure 18.1.

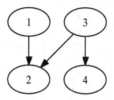

Fig. 18.1: Precedence constraints in Example 18.18

In Step 1 we have $T = 48$, in Step 2 we have $NoSucc(G) = \{2, 4\}$. Because $f_2(48) = 48^2 + 1 > f_4(48) = 48 + 5$, job J_4 is scheduled as the last one. Then $T = \frac{48}{2} = 24$ and $NoSucc(G) = \{2\}$. Therefore, job J_2 is scheduled as the second job from the end. Next $T = \frac{24}{4} = 6$ and $NoSucc(G) = \{1, 2\}$. Because $f_1(6) = 2 \cdot 6^2 > f_3(6) = 6$, job J_3 is scheduled as the third one from the end. After that $T = \frac{6}{2} = 3$ and $NoSucc(G) = \{1\}$. Therefore, job J_1 is scheduled as the first one. The optimal schedule is $\sigma^* = (1, 3, 2, 4)$. ◆

Remark 18.19. Lawler's algorithm is applied not only to the classical and time-dependent scheduling problems. Brauner et al. [1, 2] used this algorithm for some scheduling problems with uncertain job processing times. Other authors applied Lawler's algorithm to a single machine scheduling with two competing agents and job rejection (Mor and Mosheiov [8]; see also Sect. 21.3) and a flow shop problem with the $f_{\max}$ criterion (Mor and Mosheiov [9]).

18.2 Proportional-linear deterioration

In this section, we consider precedence-constrained time-dependent scheduling problems with proportional-linearly deteriorating jobs.

Algorithm 18.2 can be easily modified for proportional-linear job processing times (6.9) by appropriate modification of lines 5 and 10, in which the value of variable T is computed and updated. The pseudo-code of the modified algorithm is as follows.

Algorithm 18.3. for problem $1|p_j = b_j(a + bt), prec|f_{\max}$

---

1 **Input** : sequences $(b_1, b_2, \ldots, b_n)$, $(f_1, f_2, \ldots, f_n)$, numbers a, b, t_0,
   digraph G of precedence constraints
2 **Output:** an optimal schedule σ^*
   ▷ Step 1
3 $\sigma^* \leftarrow (\phi)$;
4 $N_{\mathcal{J}} \leftarrow \{1, 2, \ldots, n\}$;
5 $T \leftarrow (t_0 + \frac{a}{b}) \prod_{j=1}^{n}(1 + b_j b) - \frac{a}{a}$;
   ▷ Step 2
6 **while** $(N_{\mathcal{J}} \neq \emptyset)$ **do**
7   Construct the set $NoSucc(G)$;
8   Find $k \in NoSucc(G)$ such that
         $f_k(T) = \min\{f_j(T) : j \in NoSucc(G)\}$;
9      $\sigma^* \leftarrow (k|\sigma^*)$;
10      $T \leftarrow \frac{T - ab_k}{1 + b_k b}$;
11      $N_{\mathcal{J}} \leftarrow N_{\mathcal{J}} \setminus \{k\}$;
12      $NoSucc(G) \leftarrow NoSucc(G) \setminus \{k\}$;
13      $G \leftarrow G \setminus \{k\}$;
   end
   ▷ Step 3
14 **return** σ^*.

Theorem 18.20. (Kononov [13]) *If inequalities (7.5) and (7.6) hold, then problem $1|p_j = b_j(a + bt), prec|f_{\max}$ is solvable in $O(n^2)$ time by Algorithm 18.3.*

Proof. Similar to proof of Theorem 18.17.   □

Example 18.21. Let $n = 4$, $b_1 = 1$, $b_2 = 3$, $b_3 = 2$, $b_4 = 1$, $a = 2$, $b = 1$ and $t_0 = 0$, i.e., $p_1 = t$, $p_2 = 3(2 + t)$, $p_3 = 2(2 + t)$ and $p_4 = 2 + t$. Cost functions are the same as in Example 18.18, job precedence constraints are the same as those in Figure 18.1.

In Step 1 we have $T = 94$, in Step 2 we have $NoSucc(G) = \{2, 4\}$. Because $f_2(94) = 94^2 + 1 > f_4(94) = 94 + 5$, job J_4 is scheduled as the last one. Then $T = \frac{94-1}{2} = \frac{93}{2}$ and $NoSucc(G) = \{2\}$. Therefore, job J_2 is scheduled as the second job from the end. Next $T = \frac{\frac{93}{2}-3}{4} = \frac{87}{8}$ and $NoSucc(G) = \{1, 2\}$. Because $f_1(\frac{87}{8}) = 2 \cdot (\frac{87}{8})^2 > f_3(\frac{87}{8})$, job J_3 is scheduled as the third one from the end. After that $T = \frac{\frac{87}{8}-2}{3} = \frac{71}{24}$ and $NoSucc(G) = \{1\}$. Therefore, job J_1 is scheduled as the first one. The optimal schedule is $\sigma^* = (1, 3, 2, 4)$.   ♦

18.3 Linear deterioration

In this section, we consider precedence-constrained time-dependent scheduling problems with linearly deteriorating jobs.

18.3.1 Preliminaries

Before we present the main results on single machine scheduling linearly deteriorating jobs, we introduce the following definition (cf. Gawiejnowicz [14, Chap. 13]). For simplicity, we will identify job $J_j \in \mathcal{J}$ and its index j.

Definition 18.22. (A module-chain)
Let G be a graph of precedence constraints in a time-dependent scheduling problem. A chain $(n_1, \ldots, n_k)$ in the precedence graph G is called a module-chain *if for every job $j \in G \setminus \{n_1, \ldots, n_k\}$ one of the following conditions holds:*
(a) there are no precedence constraints between an arbitrary job from the chain $(n_1, \ldots, n_k)$ and job j,
(b) job j precedes all jobs of the chain $(n_1, \ldots, n_k)$,
(c) job j follows all jobs of the chain $(n_1, \ldots, n_k)$.

The following lemma about the total processing time of a chain of jobs is the main tool for results presented in this section.

Lemma 18.23. (Gawiejnowicz [14, Chap. 13]) *Let there be given a chain $(n_1, \ldots, n_k)$ of k jobs with processing times in the form of $p_j = a_j + b_j t, j = n_1, \ldots, n_k$, and let these jobs be processed on a single machine sequentially and without idle times, starting from time $t_0 \geqslant 0$. Then the total processing time $P(n_1, \ldots, n_k)$ of all jobs from this chain is equal to*

$$P(n_1, \ldots, n_k) := \sum_{j=n_1}^{n_k} p_j = \sum_{i=n_1}^{n_k} a_i \prod_{j=i+1}^{n_k} (1+b_j) + \left(\prod_{i=n_1}^{n_k} (1+b_i) - 1 \right) t_0. \quad (18.8)$$

Proof. We prove the lemma by mathematical induction with respect to k, the number of jobs in a chain. Formula (18.8) holds for a single job since $P(n_1) = a_{n_1} + b_{n_1} t_0$. Assume that (18.8) holds for a chain of jobs $(n_1, \ldots, n_k)$. We should prove its validity for the chain $(n_1, \ldots, n_{k+1})$. We have

$$P(n_1, \ldots, n_{k+1}) = P(n_1, \ldots, n_k) + p_{n_{k+1}} =$$

$$P(n_1, \ldots, n_k) + a_{n_{k+1}} + b_{n_{k+1}}(t_0 + P(n_1, \ldots, n_k)) =$$

$$= \sum_{i=n_1}^{n_k} a_i \prod_{j=i+1}^{n_k} (1+b_j) + \left(\prod_{i=n_1}^{n_k} (1+b_i) - 1 \right) t_0 + a_{n_{k+1}} +$$

$$+b_{n_{k+1}} \left(t_0 + \sum_{i=n_1}^{n_k} a_i \prod_{j=i+1}^{n_k} (1+b_j) + \left(\prod_{i=n_1}^{n_k} (1+b_i) - 1 \right) t_0 \right) =$$

$$= \sum_{i=n_1}^{n_{k+1}} a_i \prod_{j=i+1}^{n_{k+1}} (1+b_j) + \left(\prod_{i=n_1}^{n_{k+1}} (1+b_i) - 1 \right) t_0.$$

∎

Remark 18.24. Lemma 18.23 describes the first property of a chain of linearly deteriorating jobs: the processing time of the chain is a linear function of the starting time of the first job from this chain.

By Lemma 18.23, we can express the total processing time $P(n_1, \ldots, n_k)$ as a function of coefficients $A(n_1, n_k)$ and $B(n_1, n_k)$, i.e.,

$$P(n_1, \ldots, n_k) := A(n_1, n_k) + B(n_1, n_k)t_0, \tag{18.9}$$

where

$$A(n_1, n_k) := \sum_{i=n_1}^{n_k} a_i \prod_{j=i+1}^{n_k} (1+b_j) \tag{18.10}$$

and

$$B(n_1, n_k) := \prod_{j=n_1}^{n_k} (1+b_j) - 1. \tag{18.11}$$

Remark 18.25. If a chain is composed of only one job, we will write $P(n_1)$, $A(n_1)$ and $B(n_1)$ instead of $P(n_1, n_1)$, $A(n_1, n_1)$ and $B(n_1, n_1)$, respectively.

The next lemma indicates the importance of the ratio $\frac{B(n_1, n_k)}{A(n_1, n_k)}$ for a given chain of jobs $(n_1, \ldots, n_k)$.

Lemma 18.26. (Gawiejnowicz [14, Chap. 13]) *Let there be given two chains of jobs, $C_1 = (n_1, \ldots, n_k)$ and $C_2 = (n'_1, \ldots, n'_k)$, such that there are no precedence constraints between any job from C_1 and any job from C_2. Let σ' (σ'') denote the schedule in which all jobs from C_1 (C_2) are executed sequentially and without idle times, and are followed by all jobs from C_2 (C_1), and let the execution of the jobs start at the same time t_0 in both schedules. Then $C_{\max}(\sigma') \leqslant C_{\max}(\sigma'')$ if and only if $\frac{B(n_1, n_k)}{A(n_1, n_k)} \geqslant \frac{B(n'_1, n'_k)}{A(n'_1, n_k)}$.*

Proof. Let the execution of jobs start at time t_0. Let us calculate the length of both schedules, $C_{\max}(\sigma')$ and $C_{\max}(\sigma'')$. Using Lemma 18.23 we have

$$C_{\max}(\sigma') = t_0 + A(n_1, n_k) + B(n_1, n_k)t_0 + A(n'_1, n'_k) +$$

$$B(n'_1, n'_k)(t_0 + A(n_1, n_k) + B(n_1, n_k)t_0) = t_0 + A(n_1, n_k) + A(n'_1, n'_k) +$$

$$A(n_1, n_k)B(n'_1, n'_k) + (B(n_1, n_k) + B(n'_1, n'_k) + B(n_1, n_k)B(n'_1, n'_k))t_0$$

and

$$C_{\max}(\sigma'') = t_0 + A(n'_1, n'_k) + B(n'_1, n'_k)t_0 + A(n_1, n_k) +$$

$$B(n_1, n_k)(t_0 + A(n'_1, n'_k) + B(n'_1, n'_k)t_0) = t_0 + A(n_1, n_k) + A(n'_1, n'_k) +$$

$$B(n_1, n_k)A(n'_1, n'_k) + (B(n_1, n_k) + B(n'_1, n'_k) + B(n_1, n_k)B(n'_1, n'_k))t_0.$$

The difference between the length of schedules σ' and σ'' is then equal to

$$C_{\max}(\sigma') - C_{\max}(\sigma'') = A(n_1, n_k)B(n'_1, n'_k) - A(n'_1, n'_k)B(n_1, n_k)$$

and schedule σ' is not worse than σ'' if and only if $\frac{B(n_1,n_k)}{A(n_1,n_k)} \geqslant \frac{B(n'_1,n'_k)}{A(n'_1,n'_k)}$. ∎

Remark 18.27. Lemma 18.26 describes the second property of the considered problem: in the optimal schedule the chain $(n_1, \dots, n_k)$ with the greatest ratio $\frac{B(n_1,n_k)}{A(n_1,n_k)}$ is scheduled as the first one.

Definition 18.28. (Ratio $R(n_1, n_k)$ for a chain $(n_1, n_2, \dots, n_k)$ of jobs)
Given a chain $(n_1, n_2, \dots, n_k)$ of jobs $J_{n_1}, J_{n_2}, \dots, J_{n_k}$, the ratio

$$R(n_1, n_k) := \frac{B(n_1, n_k)}{A(n_1, n_k)}$$

will be called the ratio $R(n_1, n_k)$ *(ratio R, in short) for this chain.*

Remark 18.29. If $A(n_1, n_k) = 0$, we will set $R(n_1, n_k) := \infty$.

Remark 18.30. Note that for chain $(n_1, \dots, n_k)$ the following equations hold:

$$A(n_1, n_1) := a_{n_1} \text{ and } A(n_1, n_{i+1}) = A(n_1, n_i)(1 + b_{i+1}) + a_{i+1} \qquad (18.12)$$

and

$$B(n_1, n_1) := b_{n_1} \text{ and } B(n_1, n_{i+1}) = B(n_1, n_i)(1 + b_{i+1}) + b_{i+1}, \qquad (18.13)$$

where $i = 1, 2, \dots, k - 1$. Hence we can calculate the ratio R for a chain in time that is linear with respect to the length of this chain.

Remark 18.31. Wang et al. [10] give counterparts of Lemma 18.26 (see [10, Lemma 2]) and ratio $R(n_1, n_k)$ (see [10, p. 2687]).

The next lemma describes the monotonicity property of the sequence of ratios $R(n_1, n_j)$ for $j = 1, \dots, k$.

Lemma 18.32. (Gawiejnowicz [14, Chap. 13]) *Let there be given a chain of jobs $(n_1, \dots, n_k)$ and its two subchains, $(n_1, \dots, n_l)$ and $(n_{l+1}, \dots, n_k)$, where $1 \leqslant l \leqslant k - 1$. Then the following implication holds:*

if $R(n_{l+1}, n_k) \geqslant R(n_1, n_l)$, then $R(n_{l+1}, n_k) \geqslant R(n_1, n_k) \geqslant R(n_1, n_l)$.

Moreover, similar implications hold if we replace the sign '$\geqslant$' by '$\leqslant$', respectively.

Proof. We prove the lemma for the case when $R(n_{l+1}, n_k) \geqslant R(n_1, n_l)$; the two remaining cases can be proved by similar reasoning.

Let the chain $(n_1, \ldots, n_k)$ be processed sequentially and without idle times, and let l be any integer from the set $\{1, \ldots, k-1\}$. Then the chain $(n_1, \ldots, n_k)$ is divided by job l into two subchains, $(n_1, \ldots, n_l)$ and $(n_{l+1}, \ldots, n_k)$. For simplicity, let us introduce the following notation: $A(n_1, n_l) = A_1$, $B(n_1, n_l) = B_1$, $A(n_{l+1}, n_k) = A_2$, $B(n_{l+1}, n_k) = B_2$, $A(n_1, n_k) = A$, $B(n_1, n_k) = B$.

From Equations (18.10) and (18.11) we have

$$B = B(n_1, n_k) = \prod_{j=n_1}^{n_k} (1 + b_j) - 1 =$$

$$\prod_{j=n_1}^{n_l} (1 + b_j) \prod_{j=n_{l+1}}^{n_k} (1 + b_j) - 1 = (1 + B_1)(1 + B_2) - 1$$

and

$$A = \sum_{i=n_1}^{n_k} a_i \prod_{j=i+1}^{n_k} (1 + b_j) = \sum_{i=n_1}^{n_l} a_i \prod_{j=i+1}^{n_l} (1 + b_j) \prod_{j=n_{l+1}}^{n_k} (1 + b_j) +$$

$$\sum_{i=n_{l+1}}^{n_k} a_i \prod_{j=i+1}^{n_k} (1 + b_j) = A_1(1 + B_2) + A_2.$$

So we have

$$A = A_1(1 + B_2) + A_2 \quad \text{and} \quad B = (1 + B_1)(1 + B_2) - 1. \qquad (18.14)$$

First, we show that $R(n_{l+1}, n_k) \geqslant R(n_1, n_k)$. Since $\frac{B_2}{A_2} - \frac{B}{A} = \frac{AB_2 - BA_2}{A_2 A}$, by (18.14) we have

$$AB_2 - BA_2 = (1 + B_2)(A_1 B_2 - A_2 B_1). \qquad (18.15)$$

But from assumption of this lemma we know that $\frac{B_2}{A_2} \geqslant \frac{B_1}{A_1}$. This implies that

$$A_1 B_2 - A_2 B_1 \geqslant 0. \qquad (18.16)$$

Hence, by (18.15) and (18.16), we have

$$\frac{B_2}{A_2} = R(n_{l+1}, n_k) \geqslant \frac{B}{A} = R(n_1, n_k). \qquad (18.17)$$

In a similar way we prove the second inequality. Indeed, because

$$\frac{B}{A} - \frac{B_1}{A_1} = \frac{A_1 B - B_1 A}{A A_1} \qquad (18.18)$$

and $A_1B - AB_1 = A_1B_2 - B_1A_2$ we have, by (18.16) and (18.18),

$$\frac{B}{A} = R(n_1, n_k) \geqslant \frac{B_1}{A_1} = R(n_1, n_l). \qquad (18.19)$$

Therefore, by (18.17) and (18.19), $R(n_{l+1}, n_k) \geqslant R(n_1, n_k) \geqslant R(n_1, n_l)$. ∎

On the basis of Lemmas 18.23, 18.26 and 18.32 we get the following result, showing the role played by the element of a chain for which the maximum value of the ratio R is obtained.

Theorem 18.33. (Gawiejnowicz [14, Chap. 13]) *Let* $n_k = \arg \max_{1 \leqslant j \leqslant k} \{R(n_1, n_j)\}$ *for a given module-chain* $(n_1, \ldots, n_k)$. *Then there exists an optimal schedule in which jobs of the chain* $(n_1, \ldots, n_k)$ *are executed sequentially, without idle times and such that no jobs from other chains are executed between the jobs of this chain.*

Proof. Let σ denote a schedule in which some jobs are executed between jobs of chain $(n_1, \ldots, n_k)$. We can represent schedule σ as a set of subchains $\mathcal{CL} = \{\mathcal{C}_1, \mathcal{L}_1, \mathcal{C}_2, \mathcal{L}_2, \ldots, \mathcal{C}_l, \mathcal{L}_l, \mathcal{C}_{l+1}\}$, $l = 1, \ldots, k - 1$, where each subchain $\mathcal{C}_j$ and $\mathcal{L}_j$ contains, respectively, jobs only from and jobs only outside of chain $(n_1, \ldots, n_k)$. Let us denote by $R(\mathcal{C}_j)$, for $j = 1, \ldots, l + 1$ ($R(\mathcal{L}_j)$, for $j = 1, \ldots, l$) the ratio R of subchain $\mathcal{C}_j$ ($\mathcal{L}_j$).

Definition 18.22 implies that there are no precedence constraints between any job from subchain $\mathcal{C}_j$ and any job from subchain $\mathcal{L}_i$. Hence, if we swap two successive subchains from the set $\mathcal{CL}$, then we get a feasible schedule again. We will show that we can always find two successive subchains such that they may be swapped without increasing the length of schedule. Contrary, suppose the opposite. Then, by Lemma 18.26, we have

$$R(\mathcal{C}_1) \geqslant R(\mathcal{L}_1) \geqslant R(\mathcal{C}_2) \geqslant R(\mathcal{L}_2) \geqslant \cdots \geqslant R(\mathcal{C}_l) \geqslant R(\mathcal{L}_l) \geqslant R(\mathcal{C}_{l+1}). \quad (18.20)$$

From (18.20) it follows that $R(\mathcal{C}_1) \geqslant R(\mathcal{C}_2) \geqslant \cdots \geqslant R(\mathcal{C}_l) \geqslant R(\mathcal{C}_{l+1})$. But then Lemma 18.32 implies that $R(\mathcal{C}_1) \geqslant R(n_1, n_k)$. We obtain a contradiction with the definition of n_k. Therefore, we can conclude that in schedule σ there exist two successive subchains of jobs such that they may be swapped without increasing the value of $C_{\max}(\sigma)$. This swap decreases the number l of subchains of jobs that do not belong to the chain $(n_1, \ldots, n_k)$.

Repeating this swapping procedure at most l times, we obtain a schedule σ' in which jobs of the chain $(n_1, \ldots, n_k)$ are executed sequentially, without idle times and such that $C_{\max}(\sigma') \leqslant C_{\max}(\sigma)$. ∎

Remark 18.34. Starting from now, we will assume that if there are two values for which the maximum is obtained, the function 'arg' chooses the job with the larger index.

Remark 18.35. Theorem 18.33 describes the third property of the considered problem: in the set of chains there exist some subchains which are processed in optimal schedule like aggregated jobs, since inserting into these subchains either a separate job (jobs) or a chain (chains) makes the final schedule longer.

Remark 18.36. Wang et al. [10, Lemma 7] give a counterpart of Theorem 18.33.

18.3.2 Chain precedence constraints

We start this section with the pseudo-code of an algorithm for the simplest case of job precedence constraints, i.e., a set of chains.

Algorithm 18.4. for problem $1|p_j = a_j + b_j t, chain|C_{\max}$

1 **Input** : sequences $(a_{n_1}, a_{n_2}, \ldots, a_{n_k})$, $(b_{n_1}, b_{n_2}, \ldots, b_{n_k})$,
                chain $(n_1, n_2, \ldots, n_k)$
2 **Output:** an optimal schedule $\sigma^\star$
   ▷ Step 1
3 $i \leftarrow 1$;
4 $j \leftarrow 1$;
5 $\mathcal{C}_1 \leftarrow (n_1)$;
6 Compute $R(\mathcal{C}_1)$;
   ▷ Step 2
7 **while** $(j + 1 \leqslant k)$ **do**
8 $\quad$ $i \leftarrow i + 1$;
9 $\quad$ $j \leftarrow j + 1$;
10 $\quad$ $\mathcal{C}_i \leftarrow (n_j)$;
11 $\quad$ Compute $R(\mathcal{C}_i)$;
12 $\quad$ **while** $(R(\mathcal{C}_{i-1}) \leqslant R(\mathcal{C}_i))$ **do**
13 $\quad\quad$ $\mathcal{C}_{i-1} \leftarrow \mathcal{C}_{i-1} \cup \mathcal{C}_i$;
14 $\quad\quad$ Compute $R(\mathcal{C}_{i-1})$;
15 $\quad\quad$ $i \leftarrow i - 1$;
   $\quad$ **end**
   end
   ▷ Step 3
16 $\sigma^\star \leftarrow (\mathcal{C}_1, \mathcal{C}_2, \ldots, \mathcal{C}_i)$;
17 **return** $\sigma^\star$.

Algorithm 18.4 constructs a partition $\mathcal{U}$ of chain $\mathcal{C} = (n_1, \ldots, n_k)$ into subchains $\mathcal{C}_i = (n_{i_1}, \ldots, n_{i_{k_i}})$, $i = 1, 2, \ldots, l$. Partition $\mathcal{U}$ has the following properties:

Property 18.37. $\mathcal{C} = \bigcup_{i=1}^{l} \mathcal{C}_i$;

Property 18.38. $n_{i_{k_i}} = \arg \max_{i_1 \leqslant j \leqslant i_{k_i}} \{R(n_{i_1}, n_j)\}$;

Property 18.39. $R(\mathcal{C}_1) \geqslant R(\mathcal{C}_2) \geqslant \ldots \geqslant R(\mathcal{C}_l)$.

Remark 18.40. The subchains generated by Algorithm 18.4 will be called *independent subchains*.

Remark 18.41. Subchains created by Algorithm 18.4 are like 'big' independent jobs: none of them can be divided into smaller subchains and jobs from different subchains cannot be interleaved without an increase in schedule length.

Lemma 18.42. (Gawiejnowicz [14, Chap. 13])
(a) *Algorithm 18.4 constructs a partition $\mathcal{U}$ of chain $\mathcal{C}$ in $O(k)$ time, $k = |\mathcal{C}|$.*
(b) *Partition $\mathcal{U}$ has Properties 18.37–18.39.*

Proof. (a) First, notice that Algorithm 18.4 is composed of three steps and Step 1 is performed in constant time. Second, the algorithm in Step 2 either generates a new subchain from an unconsidered vertex of chain $\mathcal{C}$ or joins two consecutive subchains. Third, each of these procedures can be executed in at most k time units and requires a fixed number of operations. From the considerations above it follows that Algorithm 18.4 runs in $O(k)$ time.

(b) Since in Step 2 we consider iteratively each vertex of chain $\mathcal{C}$, the union of obtained subchains covers $\mathcal{C}$ and Property 18.37 holds.

Now we will prove that Property 18.38 holds. Let $\mathcal{U} = (\mathcal{C}_1, \mathcal{C}_2, \ldots, \mathcal{C}_l)$ be the partition of chain $\mathcal{C}$ and let for a subchain $\mathcal{C}_r = (n_{r_1}, \ldots, n_{r_{k_r}})$ Property 18.38 not hold, i.e., there exists index $p < r_{k_r}$ such that $R(n_{r_1}, n_p) > R(n_{r_1}, n_{r_{k_r}})$. Without loss of generality we can assume that

$$n_p = \arg \max_{1 \leqslant j \leqslant k_r} \{R(n_{r_1}, n_{r_j})\}. \tag{18.21}$$

Let us consider the subset of $\mathcal{U}$, obtained by Algorithm 18.4 before considering element n_{p+1}, and denote it by $\mathcal{U}'$. The algorithm in Step 2 always joins two last elements. Because in the final partition we have subchains $(\mathcal{C}_1, \mathcal{C}_2, \ldots, \mathcal{C}_{r-1})$, they are obtained before considering element n_{r_1} and, in conclusion, they also belong to the set $\mathcal{U}'$. Let $\mathcal{C}_h = (n_{r_1}, n_p)$. Then, by (18.21), we have $\mathcal{U}' = ((\mathcal{C}_1, \mathcal{C}_2, \ldots, \mathcal{C}_{r-1}), \mathcal{C}_h)$. Contrary, assume the opposite. Let $\mathcal{U}' = ((\mathcal{C}_1, \mathcal{C}_2, \ldots, \mathcal{C}_{r-1}), \mathcal{C}_{h_1}, \ldots, \mathcal{C}_{h_p})$. Then, applying Algorithm 18.4, we have $R(\mathcal{C}_{h_1}) \geqslant R(\mathcal{C}_{h_2}) \geqslant \cdots \geqslant R(\mathcal{C}_{h_p})$. Hence, by Lemma 18.32, we have $R(\mathcal{C}_{h_1}) > R(n_{r_1}, n_p)$. A contradiction to equality (18.21).

Let us consider now a subset of $\mathcal{U}$, obtained by Algorithm 18.4 before considering the element n_{r_k+1}, and denote it by $\mathcal{U}''$. From Lemma 18.32 and equality (18.21) it follows that $R(n_{r_1}, n_p) \geqslant R(n_{p+1}, n_j)$ for all $j = p+1, \ldots, r_{k_r}$. Hence, checking the condition in the inner **while** loop for iterations $p+1, \ldots, r_{k_r}$ gives a negative answer. Since $p < r_{k_r}$, it follows that $\mathcal{U}''$ includes subchain $(n_{r_1}, \ldots, n_p)$ and, in conclusion, it does not contain subchain $(n_{r_1}, \ldots, n_{r_{k_r}})$. From the pseudo-code of Algorithm 18.4 it is clear that such a chain cannot be obtained in subsequent iterations, either. A contradiction.

There remains to prove that Property 18.39 also holds. If for some $2 \leqslant i \leqslant l$ we have $R(\mathcal{C}_{i-1}) \leqslant R(\mathcal{C}_i)$, then Algorithm 18.4 joins these subchains in Step 2. This implies that Property 18.39 holds. ∎

Let G, $\mathcal{C}_i$, where $1 \leqslant i \leqslant k$ and $\sum_i^k |\mathcal{C}_i| = n$, and σ^* denote the graph of precedence constraints, the ith module-chain and the optimal job sequence, respectively. Based on Algorithm 18.4, we can formulate the following algorithm for the considered problem with precedence constraints in the form of a set of chains.

Algorithm 18.5. for problem $1|p_j = a_j + b_j t, chains|C_{\max}$

---

1 **Input** : sequences $(a_1, a_2, \ldots, a_n)$, $(b_1, b_2, \ldots, b_n)$,
      chains $\mathcal{C}_1, \mathcal{C}_2, \ldots, \mathcal{C}_k$
2 **Output:** an optimal schedule σ^*
  ▷ Step 1
3 **for** $i \leftarrow 1$ **to** k **do**
4 │ Apply Algorithm 18.4 to chain $\mathcal{C}_i$;
  end
  ▷ Step 2
5 Arrange G in non-increasing order of the ratios R of independent
   chains;
6 Call the obtained sequence σ^*;
  ▷ Step 3
7 **return** σ^*.

Theorem 18.43. (Gawiejnowicz [14, Chap. 13]) *Problem $1|p_j = a_j + b_j t$, chains$|C_{\max}$ is solvable in $O(n \log n)$ time by Algorithm 18.5.*

Proof. Partition $\mathcal{U}$ constructed for each chain satisfies Property 18.39. This implies the feasibility of the schedule generated by Algorithm 18.5. The optimality of the schedule follows from Theorem 18.33 and Lemma 18.26. Algorithm 18.4 runs in linear time with respect to the number of vertices. From that it follows that the running time of Algorithm 18.5 is determined by its Step 2. Since this step needs ordering of at most n elements, Algorithm 18.5 runs in $O(n \log n)$ time. ∎

Example 18.44. We are given $n = 4$ jobs with the following processing times: $p_1 = 2 + t$, $p_2 = 1 + 3t$, $p_3 = 1 + t$, $p_4 = 3 + 2t$. Precedence constraints are as follows: job J_1 precedes job J_2, and job J_3 precedes job J_4 (see Figure 18.2a).

Algorithm 18.4 for chain $(1, 2)$ starts with $i = 1$, $j = 1$, $\mathcal{C}_1 = (1)$ and $R(\mathcal{C}_1) = \frac{1}{2}$. Next we have $i = 2$, $j = 2$, $\mathcal{C}_2 = (2)$ and $R(\mathcal{C}_2) = 3$. Since $R(\mathcal{C}_1) < R(\mathcal{C}_2)$, we join jobs 1 and 2, obtaining subchain $\mathcal{C}_1 = (1, 2)$ with $R(\mathcal{C}_1) = \frac{7}{9}$ (see Figure 18.2b).

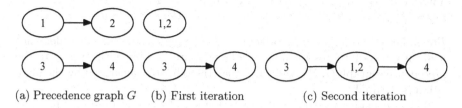

(a) Precedence graph G (b) First iteration   (c) Second iteration

Fig. 18.2: Precedence constraints in Example 18.44

The execution of Algorithm 18.4 for chain $(3,4)$ runs similarly. First we set $i = 1$, $j = 1$, $C_1 = (3)$ and $R(C_1) = 1$. Then $i = 2$, $j = 2$, $C_2 = (4)$ and $R(C_2) = \frac{2}{3}$. Since $R(C_1) \geqslant R(C_2)$, we cannot join vertices 3 and 4. Therefore, we get two subchains, $C_1 = (3)$ and $C_2 = (4)$.

After the completion of Algorithm 18.4 we obtain three independent subchains: $C_1 = (1,2)$, $C_2 = (3)$ and $C_3 = (4)$.

Now we can arrange the independent subchains. Because $R(C_2) \geqslant R(C_1) \geqslant R(C_3)$, the optimal job sequence is $\sigma^* = (3,1,2,4)$; see Figure 18.2c.   ♦

18.3.3 Tree and forest precedence constraints

In this section, we discuss the case when in the considered problem the precedence constraints among jobs are in the form of a tree or a forest.

Let the precedence digraph G be an in-tree. The pseudo-code of an optimal algorithm for this case of precedence constraints is as follows.

Algorithm 18.6. for problem $1|p_j = a_j + b_j t, in\text{-}tree|C_{\max}$

1 **Input** : sequences $(a_1, a_2, \ldots, a_n)$, $(b_1, b_2, \ldots, b_n)$, in-tree G
2 **Output:** an optimal schedule σ^*
  ▷ Step 1
3 **while** (G *is not a single chain*) **do**
4 | Choose $v \in G$ such that $Pred(v)$ is a union of module-chains;
5 | **for** *each module-chain* $C \in Pred(v)$ **do**
6 | | Apply Algorithm 18.4 to C;
  | **end**
7 | Arrange independent subchains of $Pred(v)$ in non-increasing order
  | of the ratios R;
8 | Call the obtained sequence σ;
9 | Replace in G the set $Pred(v)$ by vertices corresponding to its
  | independent subchains in sequence σ;
  end
  ▷ Step 2
10 $\sigma^* \leftarrow$ the order given by G;
11 **return** σ^*.

Theorem 18.45. (Gawiejnowicz [14, Chap. 13]) *Problem* $1|p_j = a_j + b_j t$, *in-tree*$|C_{\max}$ *is solvable in* $O(n \log n)$ *time by Algorithm 18.6.*

Proof. Let us notice that Algorithm 18.6 generates a feasible schedule, since it always looks for subsequences of jobs that are feasible with respect to the precedence digraph G. We will show that the schedule generated by the algorithm is optimal.

Let us consider a vertex $v \in G$ such that set $Pred(v)$ is a union of module-chains, $Pred(v) = \mathcal{C}_1 \cup \mathcal{C}_2 \cup \cdots \cup \mathcal{C}_k$, where $\mathcal{C}_i$ is a module-chain, $1 \leqslant i \leqslant k$. Let us notice that if some job $j \notin Pred(v)$, then either any job from $Pred(v)$ precedes j or there is no precedence between j and any job from $Pred(v)$. Let us apply Algorithm 18.4 to the set $Pred(v)$ and consider the final chain $\mathcal{C} = (\mathcal{C}_{n_1}, \mathcal{C}_{n_2}, \ldots, \mathcal{C}_{n_s})$, where $R(\mathcal{C}_{n_1}) \geqslant R(\mathcal{C}_{n_2}) \geqslant \cdots \geqslant R(\mathcal{C}_{n_s})$. We will show that there exists an optimal schedule in which all subchains are executed in the same order as in $\mathcal{C}$.

Let there exist an optimal schedule σ such that for some $i < j$ the jobs from $\mathcal{C}_{n_j}$ precede the jobs from $\mathcal{C}_{n_i}$. Denote by $\mathcal{L}$ the chain of jobs which are executed in σ, after jobs from $\mathcal{C}_{n_j}$ and before jobs from $\mathcal{C}_{n_i}$. Without loss of generality we can assume that an intersection of $\mathcal{C}$ and $\mathcal{L}$ is empty, $\mathcal{C} \cap \mathcal{L} = \emptyset$. Indeed, if some $\mathcal{C}_{n_k} \in \mathcal{L}$, then either a pair $(\mathcal{C}_{n_k}, \mathcal{C}_{n_j})$ or a pair $(\mathcal{C}_{n_k}, \mathcal{C}_{n_j})$ violates the order of $\mathcal{C}$. Since $\mathcal{L}$ does not contain the jobs from $\mathcal{C}$, then there are no precedence constraints between any job of $\mathcal{L}$ and any job of $\mathcal{C}$.

Let us recall that we have chosen $\mathcal{C}$ in such a way that if some job $j \notin \mathcal{C}$, then either any job from $\mathcal{C}$ precedes j or there are no precedence constraints between j and any job from $\mathcal{C}$. The first case is impossible, because σ is a feasible schedule. The same reasoning implies that $\mathcal{C}_{n_i}$ and $\mathcal{C}_{n_j}$ do not belong to the same module-chain in $\mathcal{C}$. From that it follows that there are no precedence constraints between any job of $\mathcal{C}_{n_i}$ and any job of $\mathcal{C}_{n_j}$.

Let us recall that $R(\mathcal{C}_{n_i}) \geqslant R(\mathcal{C}_{n_j})$. Let us calculate $R(\mathcal{L})$. If $R(\mathcal{L}) \geqslant R(\mathcal{C}_{n_j})$, then schedule $(\mathcal{L}, \mathcal{C}_{n_i}, \mathcal{C}_{n_j})$ has length that is at most equal to length of σ. If $R(\mathcal{L}) < R(\mathcal{C}_{n_j})$, then schedule $(\mathcal{C}_{n_i}, \mathcal{C}_{n_j}, \mathcal{L})$ has length that is at most equal to length of σ. Repeating this reasoning for all $i < j$ such that the jobs of $\mathcal{C}_{n_j}$ precede the jobs of $\mathcal{C}_{n_i}$, we get an optimal schedule σ^* in which all jobs from $\mathcal{L}$ are in the same order as the jobs from $\mathcal{C}$.

We have shown that there exists an optimal schedule σ^* in which all jobs from $\mathcal{L}$ are in the order of jobs from $\mathcal{C}$. Applying this procedure a finite number of times, we obtain from graph G a new graph G^* that is a single chain.

The reasoning for the case of an out-tree is similar.

If we apply 2-3 trees (cf. Remark 1.12), Algorithm 18.6 can be implemented in $O(n \log n)$ time. ∎

Example 18.46. Let $n = 7$, $p_1 = 1 + t$, $p_2 = 2 + 3t$, $p_3 = 1 + 2t$, $p_4 = 2 + t$, $p_5 = 2 + t$, $p_6 = 1 + 3t$, $p_7 = 1 + t$. Precedence constraints are as in Figure 18.3.

In Step 1 of Algorithm 18.6 we choose a set of module-chains that belong to the same vertex. In our case, we can choose chains (1) and (2) that belong to vertex 5, or chains (3) and (4) that belong to vertex 6.

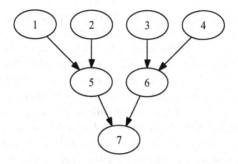

Fig. 18.3: Precedence constraints in Example 18.46

Assume we choose chains (1) and (2). Since $R(2) = \frac{3}{2} > R(1) = 1$, job 2 has to precede job 1. Moreover, (2) and (1) are independent subchains. We transform precedence constraints into the ones given in Figure 18.4a.

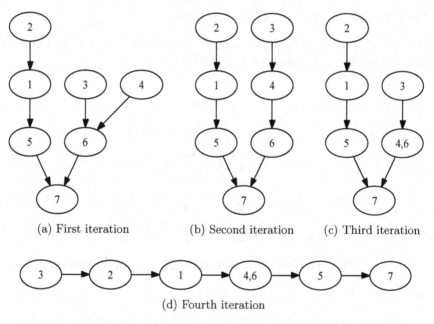

Fig. 18.4: Subsequent iterations of Algorithm 18.4 in Example 18.46

Now consider chains (3) and (4). After execution of Step 1 we obtain two independent subchains, (3) and (4). Since $R(3) = 2 > R(4) = \frac{1}{2}$, job 3 has to precede job 4. Moreover, (3) and (4) are independent subchains. The new form of precedence constraints is given in Figure 18.4b.

Now we have two module-chains $((2),(1),5)$ and $((3),(4),6)$, where by internal brackets we denote the vertices corresponding to independent subchains. Applying Step 1 twice we find two new independent subchains, (5) and (4, 6). Precedence constraints are as in Figure 18.4c.

Finally, we arrange all independent subchains in non-increasing order of the ratios R, obtaining the chain given in Figure 18.4d.

Now graph G is a single chain, the condition in loop **while** is not satisfied and Algorithm 18.6 stops. The optimal job sequence is $\sigma^\star = (3, 2, 1, 4, 6, 5, 7)$.
◆

Note that if for a vertex v the set $Succ(v)$ is a union of module-chains, then by replacing $Pred(v)$ by $Succ(v)$, we can apply Algorithm 18.6 to an out-tree. Let us call the modified Algorithm 18.6 Algorithm 18.6'.

Theorem 18.47. (Gawiejnowicz [14, Chap. 13]) *Problem* $1|p_j = a_j + b_j t, out$-$tree|C_{\max}$ *is solvable in* $O(n \log n)$ *time by Algorithm 18.6'.*

Proof. Similar to the proof of Theorem 18.45.   □

Example 18.48. Let us consider Example 18.46 with reversed orientation of arcs. Precedence constraints are given in Figure 18.5.

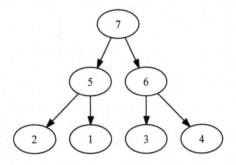

Fig. 18.5: Precedence constraints in Example 18.48

First we choose a set of module-chains that belong to the same vertex. Again, we have two possibilities: we can choose chains (1) and (2) that belong to vertex 5, or chains (3) and (4) that belong to vertex 6. Let us assume we choose chains (1) and (2). Applying Algorithm 18.6', we see that job 2 has to precede job 1, since $R(2) = \frac{3}{2} > R(1) = 1$. Moreover, (2) and (1) are independent subchains. The new form of job precedence constraints is given in Figure 18.6a.

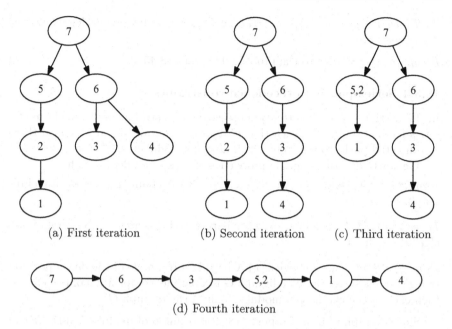

(a) First iteration   (b) Second iteration   (c) Third iteration

(d) Fourth iteration

Fig. 18.6: Subsequent iterations of Algorithm 18.6' in Example 18.48

Now let us consider chains (3) and (4). Since $R(3) = 2 > R(4) = \frac{1}{2}$, job 3 has to precede job 4. Moreover, (3) and (4) are independent subchains. The new digraph of job precedence constraints is given in Figure 18.6b.

Now we have two module-chains: $(5, (2), (1))$ and $(6, (3), (4))$. Applying Algorithm 18.6' we obtain two new independent subchains, $(5, 2)$ and (6), with $R(5, 2) = \frac{7}{10}$ and $R(6) = 3$. The new digraph of job precedence constraints is given in Figure 18.6c.

By arranging all independent subchains in non-increasing order of the ratios R we obtain the chain given in Figure 18.6d.

Now digraph G of job precedence constraints is a single chain and the optimal job sequence is $\sigma^\star = (7, 6, 3, 5, 2, 1, 4)$.   ◆

If job precedence constraints are in the form of a set of trees (a forest), we proceed as follows. Let precedence constraints be in the form of an in-forest. By adding a dummy vertex 0, with processing time $p_0 = \epsilon = const > 0$, and by connecting roots of all in-trees with this vertex, we obtain a new in-tree. For this in-tree we apply Algorithm 18.6 and ignore the dummy job in the final schedule. In a similar way we solve the case of an out-forest.

Theorem 18.49. (Gawiejnowicz [14, Chap. 13])
(a) *Problem* $1|p_j = a_j + b_j t, in\text{-}forest|C_{\max}$ *is solvable in* $O(n \log n)$ *time by Algorithm 18.6.*

(b) *Problem* $1|p_j = a_j + b_j t, out\text{-}forest|C_{\max}$ *is solvable in* $O(n \log n)$ *time by Algorithm 18.6'.*

Proof. (a) (b) Similar to the proof of Theorem 18.45.   □

18.3.4 Series-parallel precedence constraints

In this section, we assume that job precedence constraints are in the form of a series-parallel digraph G such that G does not contain the arc (v, w) if there is a directed path from v to w which does not include (v, w).

We start with three simple properties of series-parallel digraphs.

Property 18.50. (Gawiejnowicz [14, Chap. 13]) A chain is a series-parallel digraph.

Property 18.51. (Gawiejnowicz [14, Chap. 13]) The series composition of two chains is a chain.

Property 18.52. (Gawiejnowicz [14, Chap. 13]) If a node of the decomposition tree $T(G)$ of a series-parallel digraph G is a parallel composition of two chains, then each of the chains is a module-chain in the digraph G.

Since digraph G from Property 18.52 is a union of module-chains, we can apply Algorithm 18.4 to find the optimal sequence of vertices of this digraph. The pseudo-code of an optimal algorithm for series-parallel precedence constraints can be formulated as follows.

Algorithm 18.7. for problem $1|p_j = a_j + b_j t, ser\text{-}par|C_{\max}$

---

1 **Input**   : sequences $(a_1, a_2, \ldots, a_n)$, $(b_1, b_2, \ldots, b_n)$, decomposition
               tree $T(G)$ of series-parallel digraph G
2 **Output:** an optimal schedule σ^*
  ▷ **Step 1**
3 **while** (*there exists* $v \in T(G)$ *such that* $|Succ(v)| = 2$) **do**
4 | **if** (*v has label P*) **then**
5 | | Apply Algorithm 18.5 to chains $\mathcal{C}_1, \mathcal{C}_2 \in Succ(v)$;
6 | | Replace $v, \mathcal{C}_1$ and $\mathcal{C}_2$ in $T(G)$ by the obtained chain;
    | **else**
    | | Replace $v, \mathcal{C}_1$ and $\mathcal{C}_2$ in $T(G)$ by chain $(\mathcal{C}_1, \mathcal{C}_2)$;
    | **end**
7 | Arrange independent subchains of $Pred(v)$ in non-increasing order
    | of the ratios R;
8 | Call the obtained sequence σ;
9 | Replace in G the set $Pred(v)$ by vertices corresponding to its
    | independent subchains in sequence σ;
  end
  ▷ **Step 2**
10 $\sigma^* \leftarrow$ the order given by G;
11 **return** σ^*.

Theorem 18.53. (Gawiejnowicz [14, Chap. 13]) *Problem* $1|p_j = a_j + b_j t$, *ser-par*$|C_{\max}$ *is solvable in* $O(n \log n)$ *time by Algorithm 18.7, provided that decomposition tree* $T(G)$ *of precedence constraints digraph* G *is given.*

Proof. Let us notice that Algorithm 18.7 always generates a feasible job sequence, since it merges vertices of the decomposition tree $T(G)$. In order to show that the final sequence is optimal, we will show how to obtain an optimal job sequence in the case of a parallel or series composition, given the already computed sequence.

We proceed as follows. If we find an optimal sequence for a job precedence digraph (subdigraph), we transform this digraph (subdigraph) into a chain. Since each terminal node (leaf) of the tree $T(G)$ represents a single vertex (job), we will show how to obtain an optimal sequence of the jobs in a parallel or series composition, if both arguments of the composition are chains.

First consider the case when some node of $T(G)$ is a parallel composition of subgraphs G_1 and G_2. Let C_1 and C_2 be two chains that present an optimal sequence of the vertices in subdigraphs G_1 and G_2. Applying Algorithm 18.5 to chains C_1 and C_2, we get an optimal sequence C for digraph G.

Let us assume now that some node of $T(G)$ is a series composition of subdigraphs G_1 and G_2. Without loss of generality we can assume that G_1 precedes G_2. Let C_1 and C_2 be two chains that present an optimal order of vertices in subdigraphs G_1 and G_2. Setting the first vertex of C_2 after the last vertex of C_1, we get an optimal job sequence for digraph G.

By using 2-3 trees, Algorithm 18.7 can be implemented in $O(n \log n)$ time. ∎

Remark 18.54. If the decomposition tree $T(G)$ of a series-parallel digraph G is not given, Algorithm 18.7 must start with the step in which the tree is constructed (cf. Remark 1.14). Since this step needs $O(|V| + |E|) \equiv O(n^2)$ time, in this case the running time of Algorithm 18.7 increases to $O(n^2)$ time.

Remark 18.55. Strusevich and Rustogi [15, Theorem 8.12] proved a counterpart of Theorem 18.53 by showing that the $C_{\max}$ criterion is a priority-generating function for problem $1|p_j = a_j + b_j t$, *ser-par*$|C_{\max}$ and function

$$\omega(\sigma) = \frac{\prod_{j=1}^{n}(1 + b_{\sigma_j}) - 1}{\sum_{j=1}^{n} a_{\sigma_j} \prod_{i=j+1}^{n}(1 + b_{\sigma_i})},$$

where $\sigma = (\sigma_1, \sigma_2, \ldots, \sigma_n)$, is its priority function.

Remark 18.56. A counterpart of Algorithm 18.7 is given by Wang et al. [10, Algorithm 1].

Example 18.57. Let $n = 6, p_1 = 2 + 3t, p_2 = 1 + t, p_3 = 2 + t, p_4 = 3 + 2t$, $p_5 = 3 + 4t, p_6 = 2 + 5t$. The digraph G of precedence constraints is given in Figure 18.7a. The decomposition tree $T(G)$ is given in Figure 18.7b.

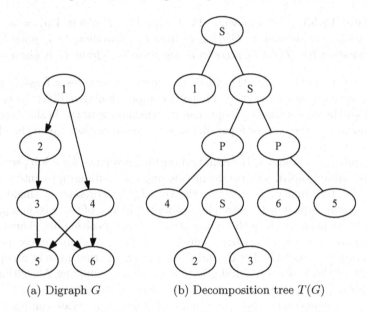

(a) Digraph G (b) Decomposition tree $T(G)$

Fig. 18.7: Precedence constraints in Example 18.57

We start with the vertex labelled S, whose immediate successors are vertices 2 and 3. We can replace these three vertices by chain $(2,3)$. The new decomposition tree is given in Figure 18.8a.

We proceed with the vertex labelled P, whose immediate successors are chain $(2,3)$ and chain (4). Let us calculate ratios R for these vertices. We have $R(2) = 1, R(3) = \frac{1}{2}$ and $R(4) = \frac{2}{3}$. Hence, (2), (3) and (4) are independent chains, and their order is $((2),(4),(3))$. The new form of the decomposition tree is given in Figure 18.8b.

Next, we choose vertices 5 and 6 that are immediate successors of the vertex labelled P. After calculations we have $R(5) = \frac{4}{3} < R(5) = \frac{5}{2}$. Hence, the new form of the decomposition tree is as in Figure 18.8c.

Because now all inner vertices are labelled S, the optimal job sequence is $\sigma^* = (1,2,4,3,6,5)$; see Figure 18.8d.   ♦

Remark 18.58. Scheduling linearly deteriorating jobs was also considered by other authors. The results discussed in Sect. 18.3 are counterparts of the results presented by Tanaev et al. [16, Chapter 3], where two operations on an arbitrary acyclic digraph (the operation of identifying vertices and the operation of including an arc) were introduced. Applying the notion of the priority-generating function (cf. Definition 1.20), Tanaev et al. [16] proved a number of results concerning different scheduling problems with precedence constraints, including the problems considered in this section.

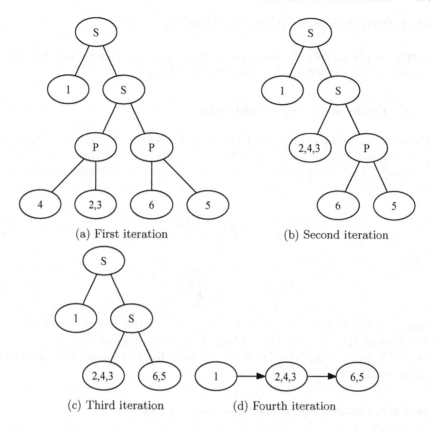

Fig. 18.8: Subsequent iterations of Algorithm 18.7 in Example 18.57

Though problem $1|p_j = a_j + b_jt|\sum C_j$ is much more difficult than its counterpart with the C_{max} criterion and its time complexity is still unknown (see Sect. A.1 for details), for case when $b_j = b$ for $1 \leqslant j \leqslant n$ the following result is known.

Theorem 18.59. (Strusevich and Rustogi [15, Theorem 8.13]) *Problem* $1|p_j = a_j + bt, \, ser\text{-}par|\sum C_j$ *is solvable in* $O(n \log n)$ *time.*

Proof. The result can be proved by showing that the $\sum C_j$ criterion is a priority-generating function for problem $1|p_j = a_j + bt, \, ser\text{-}par|\sum C_j$ and function

$$\omega(\sigma) = \frac{(1+b)^n - 1}{\sum_{j=1}^{n} a_{\sigma_j}(1+b)^{n-j}} \qquad (18.22)$$

is its priority function; see [15, Chap. 8.2] for details.   ◇

18.4 Proportional-linear shortening

In this section, we consider precedence-constrained time-dependent scheduling problems with proportionally-linear shortening jobs.

18.4.1 Chain precedence constraints

Gao et al. [17] applied the approach used for solving problems with linear jobs and the $C_{\max}$ criterion (see Sect. 18.3) to problem $1|p_j = b_j(1 - bt)$, $chains| \sum w_j C_j$. For a given chain $C_i = (J_{i_1}, J_{i_2}, \ldots, J_{i_{n_i}})$ of jobs, they defined

$$A(i_1, i_{n_i}) := \sum_{k=i_1}^{i_{n_i}} w_{i_k} \prod_{j=i_1}^{k} (1 - bb_{i_j}) \tag{18.23}$$

and

$$B(i_1, i_{n_i}) := 1 - \prod_{k=i_1}^{i_{n_i}} (1 - bb_{i_k}), \tag{18.24}$$

and applying Definition 18.3 to (18.23) and (18.24), they obtained a new definition of the ρ-factor. The authors claim that problem $1|p_j = b_j(1 - bt)$, $chains| \sum w_j C_j$ is solvable in $O(n \log n)$ time by an appropriately modified Algorithm 18.6.

18.4.2 Series-parallel precedence constraints

Gao et al. [17] applied the approach from Sect. 18.4.1 to problem $1|p_j = b_j(1 - bt)$, $ser\text{-}par| \sum w_j C_j$. The authors claim that problem $1|p_j = b_j(1 - bt)$, $ser\text{-}par| \sum w_j C_j$ is solvable in $O(n \log n)$ time by an appropriately modified Algorithm 18.6.

18.5 Linear shortening

In this section, we consider precedence-constrained time-dependent scheduling problems with linearly shortening jobs.

18.5.1 Chain precedence constraints

Gawiejnowicz et al. [19] applied the approach used for solving problems with linearly deteriorating jobs, chain precedence constraints and the $C_{\max}$ criterion (see Sect. 18.3.2) to problem $1|p_j = a_j - b_j t, chains|C_{\max}$.

First, for a given chain $C_i = (J_{i_1}, J_{i_2}, \ldots, J_{i_{n_i}})$ of jobs, the authors defined counterparts of (18.1) and (18.2) in the form of

$$A(i_1, i_{n_i}) := \sum_{k=i_1}^{i_{n_i}} a_{i_k} \prod_{j=i_1}^{k} (1 - b_{i_j}) \qquad (18.25)$$

and

$$B(i_1, i_{n_i}) := 1 - \prod_{k=i_1}^{i_{n_i}} (1 - b_{i_k}), \qquad (18.26)$$

respectively.

Next, applying Definition 18.28 to (18.25) and (18.26), they defined a new ratio R for a chain of linearly shortening jobs and, repeating the reasoning from Sect. 18.3.2, they formulated an algorithm for problem $1|p_j = a_j - b_j t, chains|C_{\max}$. This algorithm is a modification of Algorithm 18.5, where in line 5 the phrase 'non-increasing order' is replaced by 'non-decreasing order'. Gawiejnowicz et al. [19] showed that problem $1|p_j = a_j - b_j t, chains|C_{\max}$ is solvable in $O(n \log n)$ time by the new algorithm.

18.5.2 Tree precedence constraints

Gawiejnowicz et al. [19] applied the approach used for solving problems with linearly deteriorating jobs, tree precedence constraints and the $C_{\max}$ criterion (see Sect. 18.3.3) to a few variations of problem $1|p_j = a_j - b_j t, tree|C_{\max}$.

Proceeding as described in Sect. 18.5.1 and repeating the reasoning from Sect. 18.3.3, Gawiejnowicz et al. [19] formulated algorithms for problems $1|p_j = a_j - b_j t, in\text{-}tree|C_{\max}$ and $1|p_j = a_j - b_j t, out\text{-}tree|C_{\max}$.

These algorithms are modifications of Algorithms 18.6 and 18.6', where in line 7 the phrase 'non-increasing order' is replaced by 'non-decreasing order', respectively. Gawiejnowicz et al. [19] showed that these problems are solvable in $O(n \log n)$ time by the new algorithms.

18.5.3 Series-parallel precedence constraints

Gawiejnowicz et al. [19] applied the approach for linearly deteriorating jobs, series-parallel precedence constraints and the $C_{\max}$ criterion (see Sect. 18.3.4) to problem $1|p_j = a_j - b_j t, ser\text{-}par|C_{\max}$.

Applying definitions mentioned in Sect. 18.5.1 and repeating the reasoning from Sect. 18.3.4, Gawiejnowicz et al. [19] showed that $1|p_j = a_j - b_j t, ser\text{-}par|C_{\max}$ is solvable in $O(n \log n)$ time by Algorithm 18.7.

Some polynomially solvable cases are also known for the $\sum C_j$ criterion. Strusevich and Rustogi [21, Chap. 8.2] proved that problem $1|p_j = a_j - bt, ser\text{-}par|\sum C_j$ is solvable in $O(n \log n)$ time by showing that the $\sum Cj$ criterion is a priority-generating function for the problem and function $\omega(\sigma)$ obtained by replacing the deterioration rate b in (18.22) by $-b$ is its priority function.

Remark 18.60. Time-dependent scheduling under precedence constraints is not a fully explored topic. Future research may focus on seeking approximation algorithms constructed by *decomposition methods*, such as those discussed by Buer and Möhring [18] and Muller and Spinrad [20].

References

Proportional deterioration

1. N. Brauner, G. Finke, Y. Shafransky and D. Sledneu, Lawler's minmax cost algorithm: optimality conditions and uncertainty. *Journal of Scheduling* **19** (2016), no. 4, 401–408.
2. N. Brauner, G. Finke and Y. Shafransky, Lawler's minmax cost problem under uncertainty. *Journal of Combinatorial Optimization* **34** (2017), no. 1, 31–46.
3. M. Dębczyński and S. Gawiejnowicz, Scheduling jobs with mixed processing times, arbitrary precedence constraints and maximum cost criterion. *Computers and Industrial Engineering* **64** (2013), no. 1, 273–279.
4. M. Dror, W. Kubiak and P. Dell'Olmo, Scheduling chains to minimize mean flow time. *Information Processing Letters* **61** (1997), no. 6, 297–301.
5. H-L. Duan, W. Wang and Y-B. Wu, Scheduling deteriorating jobs with chain constraints and a power function of job completion times. *Journal of Industrial and Production Engineering* **31** (2014), no. 3, 128–133.
6. S. Gawiejnowicz, *Time-Dependent Scheduling*. Berlin-Heidelberg: Springer 2008.
7. E. L. Lawler, Optimal sequencing of a single machine subject to precedence constraints. *Management Science* **19** (1973), no. 5, 544–546.
8. B. Mor and G. Mosheiov, Minimizing maximum cost on a single machine with two competing agents and job rejection. *Journal of the Operational Research Society* **67** (2016), no. 12, 1524–1531.
9. B. Mor and G. Mosheiov, Polynomial time solutions for scheduling problems on a proportionate flowshop with two competing agents. *Journal of the Operational Research Society* **65** (2014), no. 1, 151–157.
10. J-B. Wang, C-T. Ng and T-C. E. Cheng, Single-machine scheduling with deteriorating jobs under a series-parallel graph constraint. *Computers and Operations Research* **35** (2008), no. 8, 2684–2693.
11. J-B. Wang and J-J. Wang, Single-machine scheduling problems with precedence constraints and simple-linear deterioration. *Applied Mathematical Modelling* **39** (2015), no. 3–4, 1172–1182.
12. J-B. Wang, J-J. Wang and P. Ji, Scheduling jobs with chain precedence constraints and deteriorating jobs. *Journal of the Operations Research Society* **62** (2011), no. 9, 1765–1770.

Proportional-linear deterioration

13. A. Kononov, Single machine scheduling problems with processing times proportional to an arbitrary function. *Discrete Analysis and Operations Research* **5** (1998), no. 3, 17–37 (in Russian).

Linear deterioration

14. S. Gawiejnowicz, *Time-Dependent Scheduling*. Berlin-Heidelberg: Springer 2008.
15. V. A. Strusevich and K. Rustogi, *Scheduling with Time-Changing Effects and Rate-Modifying Activities*. Berlin-Heidelberg: Springer 2017.
16. V. S. Tanaev, V. S. Gordon and Y. M. Shafransky, *Scheduling Theory: Single-Stage Systems*. Dordrecht: Kluwer 1994.

Proportional-linear shortening

17. W-J. Gao, X. Huang and J-B. Wang, Single-machine scheduling with precedence constraints and decreasing start-time dependent processing times. *International Journal of Advanced Manufacturing Technology* **46** (2010), no. 1–4, 291–299.

Linear shortening

18. H. Buer and R. H. Möhring, A fast algorithm for the decomposition of graphs and posets. *Mathematics of Operations Research* **8** (1983), no. 2, 170–184.
19. S. Gawiejnowicz, T-C. Lai and M-H. Chiang, Scheduling linearly shortening jobs under precedence constraints. *Applied Mathematical Modelling* **35** (2011), no. 4, 2005–2015.
20. J. H. Muller and J. Spinrad, Incremental modular decomposition. *Journal of Association for Computing Machinery* **36** (1989), no. 1, 1–19.
21. V. A. Strusevich and K. Rustogi, *Scheduling with Time-Changing Effects and Rate-Modifying Activities*. Berlin-Heidelberg: Springer 2017.

19

Matrix methods in time-dependent scheduling

$\mathbf{M}$atrix methods play an important role in time-dependent scheduling, since they lead to concise formulations of considered problems and simplify the proofs of properties of the problems. In this chapter, we will use these methods in the analysis of some time-dependent scheduling problems.

Chapter 19 is composed of six sections. In Sect. 19.1, we give some preliminaries. In Sect. 19.2, we show how to formulate time-dependent scheduling problems in terms of vectors and matrices. In Sect. 19.3, we consider a single machine time-dependent scheduling problem with the objective to minimize the l_p norm. In the next three sections, we consider pairs of mutually related time-dependent scheduling problems. In Sect. 19.4, we consider equivalent time-dependent scheduling problems, while in Sects. 19.5 and 19.6 we consider conjugate and isomorphic time-dependent scheduling problems, respectively. The chapter is completed with a list of references.

19.1 Preliminaries

Throughout this chapter, we will consider various variations of the following time-dependent scheduling problem.

We are given jobs $J_1, J_2, \ldots, J_n$ to be processed on $m \geqslant 1$ parallel identical machines $M_1, M_2, \ldots, M_m$, which are available at times $t_0^k \geqslant 0$, $1 \leqslant k \leqslant m$. Jobs are independent and neither ready times nor deadlines are given. The processing time p_j of job J_j, $1 \leqslant j \leqslant n$, is in the form of $p_j = a_j + b_j t$, where $a_j \geqslant 0$, $b_j > 0$ and $t \geqslant b_0^k := t_0^k$ for $1 \leqslant k \leqslant m$. The objective will be defined separately for each particular case.

We start with the case when $a_j = 1$ for $1 \leqslant j \leqslant n$ and $m = 1$. Notice that in this case, in view of the form of job processing times, the following recurrence equation holds:

$$C_j = \begin{cases} 1, & j = 0, \\ C_{j-1} + p_j(C_{j-1}) = 1 + \beta_j C_{j-1}, & j = 1, 2, \ldots, n, \end{cases} \tag{19.1}$$

© Springer-Verlag GmbH Germany, part of Springer Nature 2020
S. Gawiejnowicz, *Models and Algorithms of Time-Dependent Scheduling*,
Monographs in Theoretical Computer Science. An EATCS Series,
https://doi.org/10.1007/978-3-662-59362-2_19

where $\beta_j = 1 + b_j$ for $j = 0, 1, \ldots, n$. We will denote the vectors $(\beta_0, \beta_1, \ldots, \beta_n)$ and $(\beta_1, \beta_2, \ldots, \beta_n)$ as $\hat{\beta}$ and β, respectively.

Throughout this section, by $\overline{x}$ we denote the vector with the reverse order of components with respect to vector x. The set of all $(p \times q)$-matrices over $\mathbb{R}$ is denoted by $\mathcal{M}_{p \times q}(\mathbb{R})$. For a given matrix $H \in \mathcal{M}_{p \times q}(\mathbb{R})$, the transposed matrix and the transposed matrix in which rows and columns are in the reverse order are denoted by H^T and $\overline{\overline{H^\mathsf{T}}}$, respectively.

Lemma 19.1. (Gawiejnowicz et al. [4]) *Let $\phi(u, H, v) := u^\mathsf{T} H v$ be a function, where $u \in \mathbb{R}^p$, $v \in \mathbb{R}^q$ and $H \in \mathcal{M}_{p \times q}(\mathbb{R})$. Then the identity*

$$\phi(u, H, v) = \phi(\overline{v}, \overline{\overline{H^\mathsf{T}}}, \overline{u}) \tag{19.2}$$

holds.

Proof. First, let us notice that $u^\mathsf{T} H v = v^\mathsf{T} H^\mathsf{T} u = (Pv)^\mathsf{T} P(H^\mathsf{T}) Q^\mathsf{T} (Qu)$, where $P \in \mathcal{M}_{p \times p}(\mathbb{R})$ and $Q \in \mathcal{M}_{q \times q}(\mathbb{R})$ are arbitrary permutation matrices. To complete the proof it is sufficient to take P and Q such that $Pv = \overline{v}$ and $Qu = \overline{u}$. Then $u^\mathsf{T} H v = \overline{v}^\mathsf{T} \overline{\overline{H^\mathsf{T}}} \, \overline{u}$. ∎

19.2 The matrix approach

In this section, applying an approach introduced by Gawiejnowicz et al. [2, 3], we show how to represent in a matrix form a schedule for the time-dependent scheduling problem formulated in Sect. 19.1.

19.2.1 The matrix form of single machine problems

Expanding formula (19.1) for $j = 0, 1, \ldots, n$ we have $C_0 = 1$, $C_1 = \beta_1 C_0 + 1$, $C_2 = \beta_2 C_1 + 1$, $\ldots$, $C_n = \beta_n C_{n-1} + 1$. These equations can be rewritten in the following form:

$$
\begin{aligned}
C_0 &= 1, \\
-\beta_1 C_0 + C_1 &= 1, \\
-\beta_2 C_1 + C_2 &= 1, \\
\vdots \quad &\vdots \ \vdots \\
-\beta_n C_{n-1} + C_n &= 1.
\end{aligned}
\tag{19.3}
$$

Rewriting equations (19.3) in the matrix form, we have:

$$
\begin{bmatrix}
1 & 0 \ldots & 0\ 0 \\
-\beta_1 & 1 \ldots & 0\ 0 \\
0 & -\beta_2 \ldots & 0\ 0 \\
\vdots & \ldots & \vdots \\
0 & 0 \ldots & -\beta_n\ 1
\end{bmatrix}
\begin{bmatrix}
C_0 \\
C_1 \\
C_2 \\
\vdots \\
C_n
\end{bmatrix}
=
\begin{bmatrix}
1 \\
1 \\
1 \\
\vdots \\
1
\end{bmatrix},
\tag{19.4}
$$

i.e., $A(\beta)C(\beta) = d(1)$, where $A(\beta)$ is the matrix appearing in (19.4) and both vectors, $C(\beta) = [C_0(\beta), C_1(\beta), \ldots, C_n(\beta)]^\top$ and $d(1) = [1, 1, \ldots, 1]^\top$, belong to $\mathbb{R}^{n+1}$.

Since $\det A(\beta) = 1$, there exists an inverse matrix

$$
A^{-1}(\beta) = \begin{bmatrix}
1 & 0 & \ldots 0 & 0 \\
\beta_1 & 1 & \ldots 0 & 0 \\
\beta_1\beta_2 & \beta_2 & \ldots 0 & 0 \\
\beta_1\beta_2\beta_3 & \beta_2\beta_3 & \ldots 0 & 0 \\
\vdots & \vdots & \ldots \vdots & \vdots \\
\beta_1\beta_2\ldots\beta_n & \beta_2\beta_3\ldots\beta_n & \ldots \beta_n & 1
\end{bmatrix}.
\tag{19.5}
$$

Knowing matrix $A^{-1}(\beta)$, given by (19.5), we can calculate the components of vector $C(\beta) = A^{-1}(\beta)d(1)$:

$$
C_k(\beta) = \sum_{i=0}^{k} \prod_{j=i+1}^{k} \beta_j,
\tag{19.6}
$$

where $k = 0, 1, \ldots, n$.

Example 19.2. (Gawiejnowicz [1]) Consider three jobs with the following job processing times: $p_0 = 1 + 10t$, $p_1 = 1 + 2t$, $p_2 = 1 + 3t$.

For this set of jobs we have $\hat{\beta} = (11, 3, 4)$ and $\beta = (3, 4)$. Only two schedules are possible, $\sigma_1 = (1, 2)$ and $\sigma_2 = (2, 1)$. In this case we have

$$
A(\beta_{\sigma_1}) = \begin{bmatrix} 1 & 0 & 0 \\ -3 & 1 & 0 \\ 0 & -4 & 1 \end{bmatrix} \quad \text{and} \quad A(\beta_{\sigma_2}) = \begin{bmatrix} 1 & 0 & 0 \\ -4 & 1 & 0 \\ 0 & -3 & 1 \end{bmatrix},
$$

respectively.

The inverse matrices to matrices $A(\beta_{\sigma_1})$ and $A(\beta_{\sigma_2})$ are as follows:

$$
A^{-1}(\beta_{\sigma_1}) = \begin{bmatrix} 1 & 0 & 0 \\ 3 & 1 & 0 \\ 12 & 4 & 1 \end{bmatrix} \quad \text{and} \quad A^{-1}(\beta_{\sigma_2}) = \begin{bmatrix} 1 & 0 & 0 \\ 4 & 1 & 0 \\ 12 & 3 & 1 \end{bmatrix},
$$

respectively. For the above matrices we have

$$
C(\beta_{\sigma_1}) = A^{-1}(\beta_{\sigma_1})d(1) = \begin{bmatrix} 1 & 0 & 0 \\ 3 & 1 & 0 \\ 12 & 4 & 1 \end{bmatrix} \begin{bmatrix} 1 \\ 1 \\ 1 \end{bmatrix} = [1, 4, 17]^\top
$$

and

$$
C(\beta_{\sigma_2}) = A^{-1}(\beta_{\sigma_1})d(1) = \begin{bmatrix} 1 & 0 & 0 \\ 4 & 1 & 0 \\ 12 & 3 & 1 \end{bmatrix} \begin{bmatrix} 1 \\ 1 \\ 1 \end{bmatrix} = [1, 5, 16]^\top,
$$

respectively. Hence we have $C_0(\sigma_1) = C_0(\sigma_2) = 1$, $C_1(\sigma_1) = 4$, $C_1(\sigma_2) = 5$, $C_2(\sigma_1) = 17$, $C_2(\sigma_2) = 16$. ◆

19.2.2 The matrix form of parallel machine problems

Let us consider a system of linear equations $A(\beta)C(\beta) = D$ in the block form

$$
\begin{bmatrix}
A_1 & O & & O \\
O & A_2 & & O \\
& & \ddots & \\
O & O & & A_m
\end{bmatrix}
\begin{bmatrix}
C^1 \\
C^2 \\
\dots \\
C^m
\end{bmatrix}
=
\begin{bmatrix}
d^1 \\
d^2 \\
\dots \\
d^m
\end{bmatrix},
\tag{19.7}
$$

where

$$
A_i \equiv A(a^i) =
\begin{bmatrix}
1 & 0 & \dots & & 0 & 0 \\
-\beta_1^i & 1 & \dots & & 0 & 0 \\
0 & -\beta_2^i & \dots & & 0 & 0 \\
& \vdots & & \dots & & \vdots \\
0 & 0 & \dots & -\beta_{n_i}^i & 1
\end{bmatrix},
\tag{19.8}
$$

$C(\beta) = [C^1, C^2, \dots, C^m]^T$, where $C^i = [C_0^i, C_1^i, \dots, C_{n_i}^i]$ is a vector of the completion times of the jobs assigned to machine M_i, $1 \leqslant i \leqslant m$, and $D = [d^1, d^2, \dots, d^m]^T$ with $d^i = (d, d, \dots, d) \in R^{n_i}$.

Since $\det(A(\beta)) = 1$, matrix $A(\beta)$ in (19.7) is non-singular. Its inverse, in the block form, is as follows:

$$
A^{-1}(\beta) =
\begin{bmatrix}
A_1^{-1} & O & \dots & O \\
O & A_2^{-1} & \dots & O \\
\vdots & \vdots & & \vdots \\
O & O & \dots & A_m^{-1}
\end{bmatrix},
$$

where O denotes a zero matrix of a suitable size and $A_i^{-1} \in M_{p \times q}(\mathbb{R})$ is in the form of

$$
A^{-1}(a^i) =
\begin{bmatrix}
1 & 0 & \dots & 0 & 0 \\
\beta_1^i & 1 & \dots & 0 & 0 \\
\beta_1^i \beta_2^i & \beta_2^i & \dots & 0 & 0 \\
\vdots & \vdots & & \vdots & \vdots \\
\beta_1^i \dots \beta_{n_i}^i & \beta_2^i \dots \beta_{n_i}^i & \dots & \beta_{n_i}^i & 1
\end{bmatrix}.
$$

Example 19.3. (Gawiejnowicz [1]) Let us consider six jobs with the following job processing times: $p_0 = 1 + 10t$, $p_1 = 1 + 2t$, $p_2 = 1 + 3t$, $p_3 = 1 + 12t$, $p_4 = 1 + 5t$ and $p_5 = 1 + 4t$.

For this set of jobs and schedule $\sigma_1 = ((2, 1)|(4, 5))$ in which jobs J_0, J_2, J_1 and J_3, J_4, J_5 are assigned to machines M_1 and M_2, respectively, we have

$$
A(\beta_{\sigma_2}) =
\begin{bmatrix}
1 & 0 & 0 & 0 & 0 & 0 \\
-4 & 1 & 0 & 0 & 0 & 0 \\
0 & -3 & 1 & 0 & 0 & 0 \\
0 & 0 & 0 & 1 & 0 & 0 \\
0 & 0 & 0 & -5 & 1 & 0 \\
0 & 0 & 0 & 0 & -6 & 1
\end{bmatrix}
$$

$\blacklozenge$

19.3 Minimizing the l_p norm

In this section, based on results by Gawiejnowicz and Kurc [10], we consider the problem defined in Sect. 19.1 with $m = 1$ and the l_p norm criterion (cf. Definition 1.19).

19.3.1 Problem formulation

We are given a set of $n > 1$ independent and non-preemptable jobs that have to be scheduled on a single machine from time 0. The processing time p_j of the jth job is variable and linearly deteriorates in time, i.e. $p_j = 1 + \alpha_j t$, where deterioration rate $\alpha_j > 0$ and the starting time of the job $t \geqslant 0$ for $1 \leqslant j \leqslant n$. The objective is to minimize the l_p norm

$$\|C\|_p := \left(\sum_{j=1}^{n} C_j^p \right)^{\frac{1}{p}}, \tag{19.9}$$

where $C = (C_1, C_2, \ldots, C_n)$ denotes the vector of job completion times and $1 \leqslant p \leqslant +\infty$. Following [10], we will refer to the problem as to *problem* (P_p).

Problem (P_p) can be concisely formulated using the matrix approach introduced in Sect. 19.2. Let $C(a) = (C_0, C_1, \ldots, C_n)^\top$, where $C_0 := 1$, and let $A(a)$ be an $n \times n$ square matrix with 1s on the main diagonal, the components a_i of sequence a multiplied by -1 below the main diagonal and equal to 0 otherwise. Then

$$(P_p) \quad \begin{cases} \min W_{P_p}(a) := \|C(a)\|_p \\ \text{subject to } A(a)C(a) = d, \end{cases} \tag{19.10}$$

where $d = (1, 1, \ldots, 1)^\top \in \mathbb{R}^{n+1}$ and the minimization of the $W_{P_p}(a)$ criterion is taken over all a from set $\mathcal{P}(a^\circ)$ of all permutations of a°.

Remark 19.4. Although the l_p norm is a rarely studied objective, there are known some results on scheduling problems with this objective or its special cases, concerning either fixed job processing times (Bansal and Pruhs [6], Azar et al. [5], Caragiannis [7], Gupta and Sivakumar [11], Liu et al. [13]), or time-dependent job processing times (Chung et al. [8], Wang [17], Kuo and Yang [12], Wei and Wang [18], Yu and Wong [19]).

In the next few sections, we consider the properties of problem (P_p).

19.3.2 Injectivity and convexity

In this section, we consider the properties of mapping $a \mapsto C(a)$, where $a \in \mathbb{R}_+^n \setminus \{0\}$ and $A(a)C(a) = d$, which are related to the uniqueness and convexity of solutions to problem (P_p).

Theorem 19.5. (Gawiejnowicz and Kurc [10]) *The mapping $a \mapsto C(a)$ is an injection.*

Proof. Since matrix $A(a)^{-1}$ exists for any a, the mapping $a \mapsto C(a)$ is a single-valued function. Let $C(a) = C(b)$, where $a, b \in \mathbb{R}^n_+ \setminus \{0\}$. Since $A(a)C(a) = d$ and $A(b)C(b) = d$, $(A(a) - A(b))C(a) = 0$. Thus, $(a_i - b_i) \times C_{i-1}(a) = 0$ for $1 \leqslant i \leqslant n$. Therefore, since $C_i(a) > 1$, $a_i = b_i$ for $1 \leqslant i \leqslant n$, i.e., $a \mapsto C(a)$ is an injection. ∎

The next result is related to the second property showing that solutions to problem (P_p) possess a kind of convexity (cf. Definitions 1.29 and 1.32).

Lemma 19.6. (Gawiejnowicz and Kurc [10]) *If $\alpha, \beta \geqslant 0$ are such that $\alpha + \beta = 1$, $A(a)C(a) = d$, $A(b)C(b) = d$ and $A(\alpha a + \beta b)C(\alpha a + \beta b) = d$ for $a, b \in \mathbb{R}^n_+ \setminus \{0\}$, then*

$$C(\alpha a + \beta b) = \alpha U C(a) + \beta V C(b),$$

where matrices U and V depend only on $\alpha, \beta, A(a)$ and $A(b)$, and satisfy equality $\alpha U + \beta V = I$, where I denotes an identity matrix.

Proof. Let $a, b \in \mathbb{R}^n_+ \setminus \{0\}$, $A(a)C(a) = d$ and $A(b)C(b) = d$. Let, moreover, $A(\alpha a + \beta b)C(\alpha a + \beta b) = d$, where $\alpha, \beta \geqslant 0$ and $\alpha + \beta = 1$.

Since $A(\alpha a + \beta b) = \alpha A(a) + \beta A(b)$, we have $\alpha A(a)C(a) + \beta A(b)C(b) = d$. Then

$$
\begin{aligned}
C(\alpha a + \beta b) &= (\alpha A(a) + \beta A(b))^{-1}(\alpha A(a)C(a) + \beta A(b)C(b)) \\
&= \alpha[(\alpha A(a) + \beta A(b))^{-1}A(a)]C(a) + \beta[(\alpha A(a) \\
&\quad + \beta A(b))^{-1}A(b)]C(b) \\
&= \alpha U C(a) + \beta V C(b),
\end{aligned}
$$

where $U = (\alpha A(a) + \beta A(b))^{-1}A(a)$ and $V = (\alpha A(a) + \beta A(b))^{-1}A(b)$.

Equality $\alpha U + \beta V = I$ holds, since $\alpha U + \beta V = \alpha(\alpha A(a) + \beta A(b))^{-1}A(a) + \beta(\alpha A(a) + \beta A(b))^{-1}A(b) = (\alpha A(a) + \beta A(b))^{-1}(\alpha A(a) + \beta A(b)) = I$. ∎

The next result shows that if $\| \cdot \|$ is an arbitrary vector norm or induced matrix norm on $\mathbb{R}^n$, then near the ends of any interval $[a, b] = \{\alpha a + \beta b : \alpha, \beta \geqslant 0$ and $\alpha + \beta = 1\}$ the function $a \mapsto \|C(a)\|$ behaves asymptotically as a convex function (cf. Definition 1.29).

Theorem 19.7. (Gawiejnowicz and Kurc [10]) *Let $a, b \in \mathbb{R}^n_+ \setminus \{0\}$, $\alpha, \beta \geqslant 0$ be such that $\alpha + \beta = 1$ and $T := A(a)^{-1}A(b)$.*
(a) If β is sufficiently small, then

$$\|C(\alpha a + \beta b)\| \leqslant \frac{K_1}{K_2}\left(\frac{1}{K_1}\|C(a)\| + \frac{K_1 - 1}{K_1}\|C(b)\|\right),$$

where $K_1 = 1 + \frac{\beta}{\alpha}\|T\|$ and $K_2 = 1 - \frac{\beta}{\alpha}\|T\|$.

(b) *If α is sufficiently small, then*

$$\|C(\alpha a + \beta b)\| \leqslant \frac{K_1'}{K_2'} \left(\frac{K_1' - 1}{K_1'} \|C(a)\| + \frac{1}{K_1'} \|C(b)\| \right),$$

where $K_1' = 1 + \frac{\alpha}{\beta}\|T\|$ and $K_2' = 1 - \frac{\alpha}{\beta}\|T\|$.

Moreover, both leading coefficients in (a) and (ii) approach 1 from above for $\beta \searrow 0$ and $\alpha \searrow 0$, respectively.

Proof. (a) By Lemma 19.6, we have $C(\alpha a + \beta b) = \alpha U C(a) + \beta V C(b)$. Hence,

$$\|C(\alpha a + \beta b)\| \leqslant \|\alpha U C(a)\| + \|\beta V C(b)\|. \tag{19.11}$$

Since $U = (\alpha A(a) + \beta A(b))^{-1} A(a)$ and $V = (\alpha A(a) + \beta A(b))^{-1} A(b)$, we have

$$\alpha U = (I + \frac{\beta}{\alpha}T)^{-1} \tag{19.12}$$

and

$$\beta V = (I + \frac{\beta}{\alpha}T)^{-1}(\frac{\beta}{\alpha}T), \tag{19.13}$$

where $T = A(a)^{-1} A(b)$.

If β is sufficiently small, then $\|\frac{\beta}{\alpha}T\| < 1$. Hence, applying the *Neumann series* (see, e.g., Maurin [14, Chap. VIII, Sect. 1]), we have

$$\|(I + \frac{\beta}{\alpha}T)^{-1}\| \leqslant \frac{1}{1 - \frac{\beta}{\alpha}\|T\|} = \frac{1}{K_2} \tag{19.14}$$

and

$$\|(I + \frac{\beta}{\alpha}T)^{-1}(\frac{\beta}{\alpha}T)\| \leqslant \frac{\frac{\beta}{\alpha}\|T\|}{1 - \frac{\beta}{\alpha}\|T\|} = \frac{K_1 - 1}{K_2}, \tag{19.15}$$

where $K_1 = 1 + \frac{\beta}{\alpha}\|T\|$ and $K_2 = 1 - \frac{\beta}{\alpha}\|T\|$.

Combining (19.11) with (19.12), (19.13), (19.14) and (19.15), we have

$$\|\alpha U C(a)\| + \|\beta V C(b)\| \leqslant \frac{1}{K_2}\|C(a)\| + \frac{K_1 - 1}{K_2}\|C(b)\|$$

$$= \frac{K_1}{K_2} \left(\frac{1}{K_1}\|C(a)\| + \frac{K_1 - 1}{K_1}\|C(b)\| \right).$$

(b) Similar reasoning can be used for sufficiently small α. ∎

19.3.3 Bounded logarithmic growth

Let us notice that matrix $A(a)$ is non-singular with the inverse $A^{-1}(a) = (e_{ij})$, where $e_{ij} = \prod_{k=j}^{i-1} a_k$ for $1 \leqslant j \leqslant i$ and $e_{ij} = 0$ otherwise. Thus, since $C(a) = A^{-1}(a)d$, for $0 \leqslant i \leqslant n$, we have

$$C_i = C_i(a) = \sum_{j=0}^{i} \prod_{k=j+1}^{i} a_k. \tag{19.16}$$

The next result explains the role which two boundary cases, $\|C(a)\|_1$ and $\|C(a)\|_\infty$, play in the behaviour of the $\|C(a)\|_p$ criterion for $1 \leqslant p \leqslant +\infty$.

Theorem 19.8. (Gawiejnowicz and Kurc [10]) *If $A(a)C(a) = d$, then*

$$\log \|C(a)\|_p \leqslant \frac{1}{p} \log \|C(a)\|_1 + \left(1 - \frac{1}{p}\right) \log \|C(a)\|_\infty. \tag{19.17}$$

Proof. Let $A = (a_{ij})_{0 \leqslant i,j \leqslant n}$ be any matrix. The l_p-type norm $\|A\|_p$ of the matrix is equal to $\|A\|_p := \left(\sum_{i=0}^{n} \left(\sum_{j=0}^{n} |a_{ij}|\right)^p\right)^{\frac{1}{p}}$ if $1 \leqslant p < +\infty$, and it is equal to $\|A\|_\infty := \max_i \left\{\sum_{j=0}^{n} |a_{ij}|\right\}$ if $p = +\infty$.

Since $A(a)C(a) = d = (1, 1, \dots, 1)^\mathsf{T}$, we have

$$\|C(a)\|_p = \|A^{-1}(a)d\|_p = \|A^{-1}(a)\|_p. \tag{19.18}$$

It follows from (19.18) that problem (P_p) can be formulated equivalently in terms of minimization of the l_p-type norm $\|A^{-1}(a)\|_p$ only, i.e.,

$$(P'_p) \quad \min W_{P_p}(a) := \|A^{-1}(a)\|_p,$$

where the minimization is taken over all permutations $a \in \mathcal{P}(a^\circ)$.

Since

$$W_{P_p}(a) = \left(\sum_{i=0}^{n} C_i^p(a)\right)^{\frac{1}{p}} = \left(\sum_{i=0}^{n} \left(\sum_{j=0}^{i} a_{j+i} \dots a_i\right)^p\right)^{\frac{1}{p}},$$

we can apply to (19.18) the following inequality for the l_p-type norm of matrix A:

$$\|A\|_p^p = \sum_i \left(\sum_j |a_{ij}|\right)^{p-1} \left(\sum_j |a_{ij}|\right)$$
$$\leqslant \max_i \left(\sum_j |a_{ij}|\right)^{p-1} \sum_i \sum_j |a_{ij}|. \tag{19.19}$$

Applying definitions of respective l_p norms and taking into account the structure of $A^{-1}(a)$, we complete the proof of (19.17), since by inequality (19.19) we have $\|C(a)\|_p \leqslant \|C(a)\|_1^{\frac{1}{p}} \|C(a)\|_\infty^{1-\frac{1}{p}}$. □

19.3.4 Asymmetricity

In this section, we show that problems (P_1) and (P_p) for $p > 1$ are different from the point of view of symmetricity of their solutions. Mosheiov [16] proved that for $p = 1$ the symmetry property holds, i.e., $\|C(a)\|_1 = \|C(\bar{a})\|_1$, where $\bar{a}$ denotes the sequence a with elements given in reversed order. We show that the symmetry property for the $W_{P_p}(a) := \|C(a)\|_p$ criterion is no longer satisfied for all $p > 1$, except for a finite number of p.

Lemma 19.9. (Gawiejnowicz and Kurc [10]) *Let $\phi(r) := \sum_{i=1}^n \alpha_i^r$ and $\psi(r) := \sum_{i=1}^n \beta_i^r$, where $r > 1$ and $\alpha_i, \beta_i > 0$ for $1 \leqslant i \leqslant n$. If sequences (α_i) and (β_i) are composed of distinct elements and do not coincide for $1 \leqslant i \leqslant n$, then $\phi(r) = \psi(r)$ only for a finite number of $r > 1$.*

Proof. Let us assume that the assertion of the lemma is not satisfied, i.e., $f(r_k) := \phi(r_k) - \psi(r_k) = 0$ for an infinite sequence (r_k), where $r_k > 1$ for $k \longrightarrow +\infty$. If (r_k) is unbounded, then $(r_k) \longrightarrow +\infty$ as $k \longrightarrow +\infty$. (For simplicity, we do not pass to subsequences.) If the sequence is bounded, i.e., $1 < r_k \leqslant p_0$ for all k, then (r_k) has a limit point in the interval $[1, p_0]$. Now, let us recall that any power series $g(r)$ on the real axis $\mathbb{R}$ vanishing for a sequence (s_k) with a limit point must also be vanishing on the whole axis $\mathbb{R}$ (see, e.g., Maurin [15, Chap. XV, Sect. 2]). Since sums of exponential functions are representable in the form of a power series, letting $g(r) = f(r)$ and $s_k = r_k$, we finally obtain that $f(r) = 0$ on $\mathbb{R}$. Hence, we can assume that $\phi(t_k) = \psi(t_k)$ for some infinite sequence t_k such that $t_k \longrightarrow +\infty$ for $k \longrightarrow +\infty$.

Since sequences (α_i) and (β_i) do not coincide, without loss of generality we can assume that β_n is maximal among all elements and $\alpha_i < \beta_n$ for $1 \leqslant i \leqslant n$. Then for the mentioned above sequence $t_k \longrightarrow +\infty$ we have

$$\sum_{i=0}^{n-1} \left(\frac{\alpha_i}{\beta_n}\right)^{t_k} + \left(\frac{\alpha_n}{\beta_n}\right)^{t_k} = \sum_{i=0}^{n-1} \left(\frac{\beta_i}{\beta_n}\right)^{t_k} + 1, \qquad (19.20)$$

with all quotients strictly less than one. We achieve a contradiction with the statement ending the previous paragraph, since both sides of (19.20) tend to different limits. □

Theorem 19.10. (Gawiejnowicz and Kurc [10]) *If $a \neq \bar{a}$, then solutions to problem (P_p) are symmetric only for a finite number of $p \geqslant 1$. In particular, there exists $p_0 > 1$ such that for all $p > p_0$ the inequality $\|C(a)\|_p \neq \|C(\bar{a})\|_p$ holds, whenever $a \neq \bar{a}$.*

Proof. Let us consider functions $\phi(r)$ and $\psi(r)$ from Lemma 19.9 with $\alpha_i = C_i(a)$ and $\beta_i = C_i(\bar{a})$, where $1 \leqslant i \leqslant n$. Let us notice that sequences (α_i) and (β_i) are strictly increasing for a composed of distinct elements and they are not identical. Indeed, assume that $a \neq \bar{a}$ but $C_i(a) = C_i(\bar{a})$ for $1 \leqslant i \leqslant n$. Then $C(a) = C(\bar{a})$, which implies that $A(a)C(a) = d$ and $A(\bar{a})C(\bar{a}) = d$. Hence, $(A(a) - A(\bar{a}))C(a) = 0$ which implies that $(a_i - a_{n-i+1})C_{i-1}(a) = 0$ for $1 \leqslant i \leqslant n$. Since $C_i(a) > 0$ for all i, this implies in turn that $a = \bar{a}$. A contradiction. To complete the proof, it is sufficient to notice that $\|C(a)\|_p = \phi(p)^{\frac{1}{p}}$ and $\|C(\bar{a})\|_p = \psi(p)^{\frac{1}{p}}$, and to apply Lemma 19.9. □

19.3.5 V-shapeness

In this section, we show that optimal solutions to problem (P_p) for any index $1 \leqslant p \leqslant +\infty$ may have only a few forms of a common shape.

The first form of the common shape was indicated in [16], where it was shown that solutions to problem (P_1) are *V-shaped*.

Definition 19.11. (A V-shaped sequence)
A sequence $a \in \mathcal{P}(a^\circ)$ is said to be V-shaped if a can be divided into two (empty or not) subsequences such that the first of them is non-increasing, while the second one is non-decreasing.

The main result of this section is based on a few auxiliary results.
Let us denote

$$\Delta(a, q, p) := \sum_{i=0}^{q-1} a_{i+1} \dots a_{q-1} - \sum_{i=q+1}^{n} \left(\frac{C_i^\theta(a)}{C_q^\theta(a)} \right)^{p-1} a_{q+2} \dots a_i,$$

where $a \in \mathcal{P}(a^\circ)$, $1 \leqslant q \leqslant n - 1$, $1 \leqslant p < +\infty$ and $C^\theta(a) := \theta C(a) + (1 - \theta)C(b)$ for $\theta \in [0, 1]$. Let $b = a(a_q \longleftrightarrow a_{q+1})$ denote the sequence obtained from a by mutually replacing elements a_q and a_{q+1} for $1 \leqslant q \leqslant n - 1$.

Lemma 19.12. (Gawiejnowicz and Kurc [10]) *If $A(a)C(a) = A(b)C(b) = d$, then for each $1 \leqslant p < +\infty$ and $1 \leqslant q \leqslant n - 1$ there exists $\theta \in [0, 1]$ such that*

$$\|C(b)\|_p - \|C(a)\|_p = \left(\frac{C_q^\theta(a)}{\|C^\theta(a)\|_p} \right)^{p-1} (a_{q+1} - a_q) \cdot \Delta(a, q, p). \quad (19.21)$$

Proof. See [10, Lemma 3].   ◇

The next result is a consequence of Lemma 19.12.

Lemma 19.13. (Gawiejnowicz and Kurc [10]) *If $a^\star$ is an optimal solution to an instance of problem (P_p), then for each $1 \leqslant q \leqslant n-1$ there exists $\theta \in [0, 1]$ such that either (a) $a_{q+1} - a_q \leqslant 0$ and $\Delta(a, q, p) \leqslant 0$, or (b) $a_{q+1} - a_q \geqslant 0$ and $\Delta(a, q, p) \geqslant 0$.*

Proof. See [10, Lemma 4].   ◇

Let us denote

$$\Delta^\star(a, q, p) := \sum_{j=0}^{q-1} a_{j+1} \dots a_{q-1} - \left(1 + \frac{1}{\sigma_n(a^\circ)} \right)^{p-1} \sum_{i=q+1}^{n} a_{q+2} \dots a_i$$

and

$$\Delta_\star(a, q, p) := \sum_{j=0}^{q-1} a_{j+1} \dots a_{q-1} - (\sigma_n(a^\circ))^{p-1} \sum_{i=q+1}^{n} a_{q+2} \dots a_i.$$

The properties of these two functions are summarized in the following lemma.

Lemma 19.14. (Gawiejnowicz and Kurc [10]) (a) *For all* $a \in \mathcal{P}(a^\circ)$, $1 \leqslant q \leqslant n-1$ *and* $1 \leqslant p < +\infty$, *the inequalities* $\Delta_\star(a,q,p) \leqslant \Delta(a,q,p) \leqslant \Delta^\star(a,q,p)$ *hold.* (b) $\Delta^\star(a,q,p)$ *is strictly increasing with respect to* q. (c) $\Delta_\star(a,q,p)$ *is strictly decreasing with respect to* p. (d) $\Delta^\star(a,1,p) < -\frac{1}{\sigma_n(a^\circ)} < 0$. (e) $\Delta_\star(a,n-1,1) > 0$, *whenever* $n \geqslant 2$.

Proof. See [10, Lemma 5].   ◇

Given $a \in \mathcal{P}(a^\circ)$ and $q \in \{1,2,\ldots,n-1\}$, let us consider $p_\star(a,q) = p$ such that $\Delta_\star(a,q,p) = 0$ and $p^\star(a,q) = p$ such that $\Delta^\star(a,q,p) = 0$. Then

$$p_\star(a,q) = 1 + \frac{\ln \sum_{j=0}^{q-1} a_{j+1} \ldots a_{q-1} - \ln \sum_{i=q+1}^{n} a_{q+2} \ldots a_i}{\ln(\sigma_n(a^\circ))} \quad (19.22)$$

and

$$p^\star(a,q) = 1 + \frac{\ln \sum_{j=0}^{q-1} a_{j+1} \ldots a_{q-1} - \ln \sum_{i=q+1}^{n} a_{q+2} \ldots a_i}{\ln\left(1 + \frac{1}{\sigma_n(a^\circ)}\right)}. \quad (19.23)$$

Let us denote

$$p_\star(q) := \min_{a \in \mathcal{P}(a^\circ)} \{p_\star(a,q)\}$$

and

$$p^\star(q) := \max_{a \in \mathcal{P}(a^\circ)} \{p^\star(a,q)\}.$$

Lemma 19.15. (Gawiejnowicz and Kurc [10]) (a) *Functions* $q \mapsto p_\star(a,q)$ *and* $q \mapsto p^\star(a,q)$ *are strictly increasing with respect to* $1 \leqslant q \leqslant n-1$. (b) *For each* $q \in \{1,2,\ldots,n-1\}$, *both minimum* $p_\star(q)$ *and maximum* $p^\star(q)$ *can be determined in* $O(n\log n)$ *time, and are obtained whenever* $a \in \mathcal{P}(a^\circ)$ *is such that*

$$a_{q-1} \leqslant \cdots \leqslant a_1 \leqslant a_q \wedge a_{q+1} \leqslant a_q \vee a_{q+1} \leqslant a_n \leqslant \cdots \leqslant a_{q+2}$$

and

$$a_{q+2} \leqslant \cdots \leqslant a_n \leqslant a_q \wedge a_{q+1} \leqslant a_q \vee a_{q+1} \leqslant a_1 \leqslant \cdots \leqslant a_{q-1},$$

respectively. (c) $p^\star(a,n-1) > p_\star(a,n-1) > 1$, *whenever* $n \geqslant 2$.

Proof. The result is a consequence of [9, Lemma 12.5].   ◇

Now we introduce two new forms of V-shaped sequences.

Definition 19.16. (A weakly V-shaped sequence)
A sequence $a \in \mathcal{P}(a^\circ)$ *is said to be* weakly V-shaped *if* a *can be divided into three (empty or not) subsequences such that the first of them is non-increasing, while the third one is non-decreasing.*

Remark 19.17. Let us notice that each V-shaped sequence is weakly V-shaped as well.

Definition 19.18. (A k-weakly V-shaped sequence)
A sequence $a \in \mathcal{P}(a^\circ)$ is said to be k-weakly V-shaped if a can be divided into three (empty or not) subsequences such that the first subsequence of a is non-increasing for $1 \leqslant i \leqslant n - k$ and its third subsequence is non-decreasing for at most k indices, where $0 \leqslant k \leqslant n - 1$.

Remark 19.19. Let us notice that each non-increasing (non-decreasing) subsequence is V-shaped and 0-weakly (n-weakly) V-shaped as well.

Let us denote, for brevity,

$$p_k := p^\star(n - k).$$

Let us notice that

$$p_k = \max_{a \in \mathcal{P}(a^\circ)} \{p^\star(a, n - k)\}.$$

Lemma 19.20. (Gawiejnowicz and Kurc [10]) *Let $k \in \{1, 2, \ldots, n - 1\}$ and p be such that $1 \leqslant p_k \leqslant p \leqslant p_1$ and let $a^p \in \mathcal{P}(a^\circ)$ be a solution to problem (P_p) with this p. Then (a) if $1 \leqslant q \leqslant n - k$, then $\Delta(a^p, q, p) \leqslant 0$; (b) if $\Delta(a^p, q, p) > 0$, then $p \leqslant p^\star(a^p, q)$. Hence, $n - k \leqslant q \leqslant n - 1$.*

Proof. See [10, Lemma 7].   ◇

Let us denote

$$p_\infty := \min_{a \in \mathcal{P}(a^\circ)} \{p_\star(a, n - 1)\}.$$

Let us notice that $p_\infty > 1$ and $p_k < p_\infty$.

Now, applying Lemma 19.20, we are ready to prove the main result in the section.

Theorem 19.21. (Gawiejnowicz and Kurc [10]) *Let $a^\circ = (a_1, \ldots, a_n)$ be such that $a_i > 1$, $a_i \neq a_j$ whenever $1 \leqslant i \neq j \leqslant n$ and let a^p be an optimal solution to problem (P_p). If for a given $k \in \{1, 2, \ldots, n - 1\}$ we have $p_k = p^\star(n - k) \geqslant 1$, then (a) if $1 \leqslant p \leqslant p_\infty$, then a^p is weakly V-shaped and the time complexity of problem (P_p) cannot be less than that of (P_1); (b) if $p_k \leqslant p \leqslant p_1$, then a^p is k-weakly V-shaped and the time complexity of problem (P_p) is $O(n^k + n \log n)$; (c) if $p_1 < p$, then a^p is 0-weakly V-shaped and the time complexity of problem (P_p) is $O(n \log n)$.*

Proof. (a) Let $1 \leqslant p \leqslant p_\infty$. In this case, $\Delta^\star(a, q, p) \leqslant 0$ for some $1 \leqslant q$ and $0 \leqslant \Delta_\star(a, q, p)$ for some $q \leqslant n - 1$. Hence, in view of Lemma 19.13, sequence a^p decreases at the beginning and increases at the end, i.e., it is weakly V-shaped. Since the number of weakly V-shaped sequences is not lower than the number of V-shaped sequences, problem (P_p) for such p is not easier than problem (P_1).

(b) From previous considerations in this section it follows that for a given p and the corresponding solution a^p of problem (P_p), $\Delta(a^p, q, p)$ changes their sign from negative to positive when $1 \leqslant q \leqslant n - 1$. If $\Delta(a^p, q, p) > 0$ then, according to Lemma 19.20, $n - k \leqslant q \leqslant n - 1$ since $p_k \leqslant p \leqslant p_1$. This is still true for $q \leqslant q' \leqslant n - 1$, since $\Delta(a^p, q, p)$ is non-decreasing with respect to q. On the other hand, Lemma 19.20 implies that $\Delta(a^p, q, p) \leqslant 0$ for $1 \leqslant q \leqslant n - k$. Thus, for $1 \leqslant q \leqslant n - 1$ the optimal sequence a^p is non-increasing and then non-decreasing for at most k last components, i.e., it is k-weakly V-shaped. To complete the proof of this assertion, it is sufficient to notice that such sequences can be determined in $O(n^k + n \log n)$ time, starting with a base sequence a°.

(c) In order to prove this assertion, it is sufficient to notice that if $p_1 < p$, then $\Delta(a^p, q, p) \leqslant 0$ for $1 \leqslant q \leqslant n - 1$. In consequence, sequence a^p is 0-weakly V-shaped, i.e., non-increasing, and thus it can be determined in $O(n \log n)$ time.   ■

Remark 19.22. We have proved assertion (b) of Theorem 19.21 under assumption that k is a constant. If for a particular instance of problem (P_p) we have $k = k(n)$, then solving the instance needs $O(n^{k(n)} + n \log n)$ time.

We complete the section with two numerical examples which illustrate the applications of Theorem 19.21.

Example 19.23. (Gawiejnowicz and Kurc [10]) Let us consider $a^\circ = (1 + \frac{1}{100}, 1 + \frac{2}{100}, \ldots, 1 + \frac{10}{100})$. Then, by Lemma 19.15, $p_1 = 41.40320$. Hence, by Theorem 19.21, for all $p > 41.40320$ the optimal value of criterion $W_{P_p}(a) = \|C(a)\|_p$ is obtained for sequence $a^\star = (1 + \frac{10}{100}, 1 + \frac{9}{100}, \ldots, 1 + \frac{1}{100})$, which one can construct in $O(n \log n)$ time by arranging a° in non-increasing order.

On the other hand, by Lemma 19.15, for a° we have $p_\infty = 1.84914$. Hence, for all $1 \leqslant p < 1.84914$, the optimal value of criterion $W_{P_p}(a)$ can be obtained in $O(2^{n-1})$ time by considering all V-shaped sequences and applying the symmetry property whenever $p = 1$. In the case of a° this means that we have to consider 512 V-shaped sequences.   ◆

Example 19.24. (Gawiejnowicz and Kurc [10]) Let us consider $a^\circ = (1, 2, \ldots, 10)$. Then $p_1 = 1.45727 \times 10^8$, while $p_\infty = 1.66699$. In the cases when $p_\infty < p < p_1$, problem (P_p) can be solved by considering all k-weakly V-shaped sequences which can be done in $O(n^k + n \log n)$ time for some k such that $q = n - k$ is 'close' to $n - 1$.   ◆

There exist various time-dependent scheduling problems which have similar properties as their counterparts with fixed job processing times. Moreover, schedules for corresponding to each other problems of this kind have similar structure as well, differing only in the completion times of scheduled jobs.

These similarities can be explained using the notion of *mutually related scheduling problems*. In subsequent sections, we discuss three classes of such scheduling problems, called *equivalent, conjugate* and *isomorphic* time-dependent scheduling problems.

19.4 Equivalent scheduling problems

Some authors (e.g., Cheng and Ding [22, 21], Cheng et al. [23], Gawiejnowicz et al. [24]) observed that there exist pairs of time-dependent scheduling problems which have similar properties.

Example 19.25. The single machine problem of scheduling jobs with processing times in the form of $p_j = b_j t$ and the $\sum C_j$ criterion is optimally solved by scheduling jobs in non-decreasing order of b_j rates, while the single machine problem of scheduling jobs with processing times in the form of $p_j = 1+b_j t$ and the $C_{\max}$ criterion is optimally solved by scheduling jobs in the non-increasing order of the b_j values, $1 \leqslant j \leqslant n$.   ♦

Similar examples concern other single and multi-machine time-dependent scheduling problems with proportional or linear processing times, and with the total completion time or the maximum completion time criteria.

In this section, we show that the particular cases are, in fact, special cases of a more general principle, which we explain using the notion of *equivalent problems*, proved by Gawiejnowicz et al. [26, 25]. In order to define equivalent problems, we will introduce a general transformation of an arbitrary instance of a time-dependent scheduling problem (called the *initial* problem) with the total weighted starting time criterion into an instance of another time-dependent scheduling problem (called the *transformed* problem) with the criterion of the same type but with other job processing times and job weights. Both problems can concern either a single or parallel machine environment, with or without restrictions on job completion times, and a schedule is optimal for the initial problem if and only if a schedule constructed by this transformation is optimal for the transformed problem.

19.4.1 The initial problem

Throughout this section, we consider a few different cases of the parallel-machine problem of minimizing the *total weighted starting time* of all jobs, $Pm|p_j = a_j + b_j t| \sum w_j S_j$, which will be called an *initial* problem.

Remark 19.26. Since $S_j = C_{j-1}$ for $1 \leqslant j \leqslant n$, the applied criterion $\sum w_j S_j$ can be replaced with a special version of the total weighted completion time criterion in which weight w_{j+1} is assigned to the completion time C_j, i.e. $\sum w_j S_j := \sum_{j=0}^n w_{j+1} C_j$. However, since such a form of the criterion $\sum w_j C_j$ may lead to misunderstanding, we will only use the criterion $\sum w_j S_j$.

Now, following [25], we describe the form of an arbitrary schedule for the initial problem, separately for single and parallel machine cases.

Single machine initial problems

Let $\beta_j := 1 + b_j$ for $1 \leqslant j \leqslant n$ and let sequences $(\beta_1, \ldots, \beta_n)$, $(a_1, \ldots, a_n)$ and $(w_1, \ldots, w_n)$ be given. Any schedule for problem $1|p_j = a_j + b_j t| \sum w_j S_j$ will be identified with a sequence $\sigma = ((a_1, \beta_1, w_1), \ldots, (a_n, \beta_n, w_n))$ of the triples (a_i, β_i, w_i), $1 \leqslant i \leqslant n$. Minimization of $\sum w_j S_j$ will be carried over all $\sigma \in \mathfrak{S}_n(\sigma^\circ)$, where $\mathfrak{S}_n(\sigma^\circ)$ denotes the set of all permutations of the initial sequence σ° of the triples.

An arbitrary schedule σ can be represented in another way by the following table:

$$T(\sigma) := \begin{bmatrix} a_0 & a_1 & a_2 & \ldots & a_n \\ & \beta_1 & \beta_2 & \ldots & \beta_n \\ & w_1 & w_2 & \ldots & w_n & w_{n+1} \end{bmatrix}, \tag{19.24}$$

in which $a_0 := t_0^1$ is the time at which the machine starts the processing of jobs and the weight w_{n+1} has a special meaning, defined in (19.30). Any other schedule $\sigma \in \mathfrak{S}_n(\sigma^\circ)$ can be obtained by a permutation of these columns of table $T(\sigma)$ which correspond to triples (a_i, β_i, w_i), $1 \leqslant i \leqslant n$.

Given a schedule $\sigma \in \mathfrak{S}_n(\sigma^\circ)$, the completion times of jobs in the schedule are given by the recurrence equation $C_j(\sigma) = \beta_j C_{j-1}(\sigma) + a_j$, where $1 \leqslant j \leqslant n$ and $C_0(\sigma) := a_0$. Applying the matrix approach introduced in Section 19.2, we obtain the following form of the initial problem:

$$(P^1) \quad \begin{cases} \text{minimize } W_{P^1}(\sigma) := w^\mathsf{T} C(\sigma) \\ \text{subject to } A(\sigma) C(\sigma) = a, \ \sigma \in \mathfrak{S}_n(\sigma^\circ), \end{cases} \tag{19.25}$$

where $w = (w_1, \ldots, w_{n+1})^\mathsf{T}$, $a = (a_0, \ldots, a_n)^\mathsf{T}$ and $C(\sigma) = (C_0, \ldots, C_n)^\mathsf{T}$. The non-singular matrix $A(\sigma) \in \mathcal{M}_{(n+1) \times (n+1)}(\mathbb{R})$ is defined as in Sect. 19.2.1.

Example 19.27. (Gawiejnowicz et al. [26]) Let us consider the following instance of the single-machine problem $1|p_j = b_j + \alpha_j t| \sum w_j S_j$. We are given $n = 2$ jobs with processing times $p_1 = 1 + 2t$, $p_2 = 2 + 3t$ and weights $w_1 = 5$, $w_2 = 6$. Let us assume that the machine is available from $t_0 = 0$, and let $w_3 = 1$.

Then only two schedules are possible: $\sigma_1 = ((1, 3, 5)|(2, 4, 6))$ and $\sigma_2 = ((2, 4, 6)|(1, 3, 5))$. The tables (19.24) of schedules σ_1 and σ_2 are as follows:

$$T(\sigma_1) = \begin{bmatrix} 0 & 1 & 2 \\ & 3 & 4 \\ & 5 & 6 & 1 \end{bmatrix} \quad \text{and} \quad T(\sigma_2) = \begin{bmatrix} 0 & 2 & 1 \\ & 4 & 3 \\ & 6 & 5 & 1 \end{bmatrix}. \tag{19.26}$$

Given the table $T(\sigma_1)$, we have $C_0(\sigma_1) = 0$, $C_1(\sigma_1) = 3 \times 0 + 1 = 1$, $C_2(\sigma_1) = 4 \times 1 + 2 = 6$. Hence $\sum w_j S_j(\sigma_1) = 5 \times 0 + 6 \times 1 + 1 \times 6 = 12$.

Similarly, for $T(\sigma_2)$ we have $C_0(\sigma_2) = 0$, $C_1(\sigma_2) = 4 \times 0 + 2 = 2$, $C_2(\sigma_2) = 3 \times 2 + 1 = 7$. Hence $\sum w_j S_j(\sigma_2) = 6 \times 0 + 5 \times 2 + 1 \times 7 = 17$. ◆

Parallel machine initial problems

Now we pass to the parallel machine problem $Pm|p_j = a_j + b_j t| \sum w_j S_j$. In this case, the schedule $\sigma = (\sigma^1, \ldots, \sigma^m)$ is composed of subschedules σ^k, $1 \leqslant k \leqslant m$. The subschedule σ^k corresponds to machine M_k and it is in the form of $\sigma^k = ((a_1^k, \beta_1^k, w_1^k), \ldots, (a_{n_k}^k, \beta_{n_k}^k, w_{n_k}^k))$, where $1 \leqslant k \leqslant m$ and $\sum_{k=1}^m n_k = n$.

The subschedule σ^k can be presented by the table

$$T(\sigma^k) = \begin{bmatrix} a_0^k & a_1^k & a_2^k & \ldots & a_{n_k}^k \\ & \beta_1^k & \beta_2^k & \ldots & \beta_{n_k}^k \\ & w_1^k & w_2^k & \ldots & w_{n_k}^k & w_{n_k+1}^k \end{bmatrix}, \qquad (19.27)$$

where $a_0^k := t_0^k \geqslant 0$, $1 \leqslant k \leqslant m$, is the time at which machine M_k starts the processing of jobs and the weight $w_{n_k+1}^k$ has a special meaning, defined in (19.30). Any other schedule for the problem can be obtained by permuting or mutually exchanging the triples $(a_i^k, \beta_i^k, w_i^k)$, $(a_j^l, \beta_j^l, w_j^l)$ for any possible $1 \leqslant k, l \leqslant m$, $1 \leqslant i \leqslant n_k$ and $1 \leqslant j \leqslant n_l$, including $k = l$.

Given a subschedule σ^k, $1 \leqslant k \leqslant m$, the completion times of jobs in the subschedule are given by the recurrence equation $C_j^k(\sigma^k) = \beta_j^k C_{j-1}^k(\sigma^k) + a_j^k$, where $1 \leqslant j \leqslant n_k$ and $C_0^k(\sigma^k) := a_0^k$.

The matrix form of the problem can be written as follows:

$$(P^m) \quad \begin{cases} \text{minimize } W_{Pm}(\sigma) := w^\mathsf{T} C(\sigma) \\ \text{subject to } A(\sigma)C(\sigma) = a, \ \sigma \in \mathfrak{S}_n(\sigma^\circ). \end{cases} \qquad (19.28)$$

Example 19.28. (Gawiejnowicz et al. [25]) Let us consider the following instance of problem $P2|p_j = b_j + \alpha_j t| \sum w_j S_j$. We are given $m = 2$ machines and $n = 5$ jobs with processing times $p_1 = 2 + 3t$, $p_2 = 1 + 2t$, $p_3 = 4t$, $p_4 = 1 + 3t$ and $p_5 = 2 + t$, and weights $w_1 = 4$, $w_2 = 2$, $w_3 = 1$, $w_4 = 0$, $w_5 = 3$.

Let us assume also that the machine M_1 is available from time $t_0^1 = a_0^1 = 1$, the machine M_2 is available from time $t_0^2 = a_0^2 = 3$.

Let us consider two schedules, $\sigma_1 = (\sigma_1^1, \sigma_1^2)$ and $\sigma_2 = (\sigma_2^1, \sigma_2^2)$, where σ_i^j, for $1 \leqslant i, j \leqslant 2$, are described by the tables of the type (19.27). Then

$$\sigma_1^1 = \begin{bmatrix} 1 & 2 & 0 & 1 \\ 4 & 5 & 4 \\ 4 & 1 & 0 & 1 \end{bmatrix}, \sigma_1^2 = \begin{bmatrix} 3 & 1 & 2 \\ 3 & 2 \\ 2 & 3 & 4 \end{bmatrix},$$

$$\sigma_2^1 = \begin{bmatrix} 1 & 0 & 1 \\ 5 & 4 \\ 1 & 0 & 1 \end{bmatrix}, \sigma_2^2 = \begin{bmatrix} 3 & 2 & 1 & 2 \\ 2 & 3 & 4 \\ 3 & 2 & 4 & 4 \end{bmatrix}.$$

♦

Remark 19.29. The definitions of single and parallel machine initial problems, as well as their matrix representations (19.24) and (19.27), can be extended by adding *critical lines* which are vector counterparts of ready times and deadlines (see [25, Sect. 3] for details).

In the next section, we show how to construct from any instance of the initial problem an instance of another problem, called the *transformed* problem, in such a way that both these problems are *equivalent* in the sense described in Sect. 19.4.2.

In order to obtain the transformed problem, we replace the formula $W_{P^1}(\sigma) := w^{\mathsf{T}} C(\sigma)$, where $A(\sigma)C(\sigma) = a$, by a dual formula, separately for every schedule $\sigma \in \mathfrak{S}_n(\sigma^\circ)$.

19.4.2 The transformed problem

In this section, we describe how to transform instances of the initial problem into instances of the transformed problem.

Single machine transformed problems

Let us consider problem $1|p_j = a_j + b_j t| \sum w_j S_j$ in the form of (P^1). Define problem (D^1), corresponding to (P^1), as follows:

$$(D^1) \quad \begin{cases} \text{minimize } W_{D^1}(\overline{\sigma}) := \overline{a}^{\mathsf{T}} C(\overline{\sigma}) \\ \text{subject to } A(\overline{\sigma})C(\overline{\sigma}) = \overline{w}, \ \overline{\sigma} \in \mathfrak{S}_n(\overline{\sigma^\circ}), \end{cases} \quad (19.29)$$

where $\overline{a} = (a_n, a_{n-1}, \ldots, a_0)^{\mathsf{T}}$ and $\overline{w} = (w_{n+1}, w_n, \ldots, w_1)^{\mathsf{T}}$.

For simplicity of further presentation, let us introduce the following definition.

Definition 19.30. (Single machine equivalent problems)
Given a schedule $\sigma = ((a_1, \beta_1, w_1), \ldots, (a_n, \beta_n, w_n)) \in \mathfrak{S}_n(\sigma^\circ)$, let $\overline{\sigma} \in \mathfrak{S}_n(\overline{\sigma^\circ})$ be a schedule such that $\overline{\sigma} = ((w_n, \beta_n, a_n), \ldots, (w_1, \beta_1, a_1))$. Then
(a) the correspondence $\sigma \longleftrightarrow \overline{\sigma}$ will be called a transformation *of the schedule σ for the problem (P^1) into the schedule $\overline{\sigma}$ for the corresponding problem (D^1) and vice versa,*
(b) both corresponding problems, (P^1) and (D^1), will be called equivalent *problems.*

Given a schedule $\sigma \in \mathfrak{S}_n(\sigma^\circ)$ described by the table $T(\sigma)$, the transformed schedule $\overline{\sigma}$ is fully described by the table

$$T(\overline{\sigma}) := \begin{bmatrix} w_{n+1} & w_n & w_{n-1} & \cdots & w_1 \\ & \beta_n & \beta_{n-1} & \cdots & \beta_1 \\ & a_n & a_{n-1} & \cdots & a_1 & a_0 \end{bmatrix}, \quad (19.30)$$

where $C_0(\overline{\sigma}) := w_{n+1}$ is the time at which the machine starts the processing of jobs in the problem (D^1), while $C_0(\sigma) := a_0$ is the time at which the machine starts the processing of jobs in the problem (P^1).

In Definition 19.30 we have defined an equivalence between the problems (P^1) and (D^1), based on the transformation $\sigma \longleftrightarrow \overline{\sigma}$. The equivalence is justified by the following result.

Theorem 19.31. (Gawiejnowicz et al. [26]) *Let* $\sigma \in \mathfrak{S}_n(\sigma^\circ)$ *and* $\overline{\sigma} \in \mathfrak{S}_n(\overline{\sigma}^\circ)$, *where* $\sigma = ((a_1, \beta_1, w_1), \dots, (a_n, \beta_n, w_n))$, $\overline{\sigma} = ((w_n, \beta_n, a_n), \dots, (w_1, \beta_1, a_1))$. *Then*

(a) *if* $\overline{\sigma}$ *has been obtained by the transformation* $\sigma \longleftrightarrow \overline{\sigma}$, *then the equality*

$$W_{P^1}(\sigma) = w^\mathsf{T} C(\sigma) = \overline{a}^\mathsf{T} C(\overline{\sigma}) = W_{D^1}(\overline{\sigma}) \tag{19.31}$$

holds;

(b) $\sigma^\star$ *is an optimal schedule for the problem* (P^1) *if and only if* $\overline{\sigma}^\star$ *is an optimal schedule for the problem* (D^1); *moreover, the equality* $W_{P^1}(\sigma^\star) = W_{D^1}(\overline{\sigma}^\star)$ *holds.*

Proof. (a) The implication follows from Lemma 19.1 for $H \equiv A(\sigma)^{-1}$, $u \equiv w$ and $v \equiv a$ with $p = q = n + 1$.

(b) Let $\sigma^\star$ be an optimal schedule for the problem (P^1) and let there exist a schedule $\overline{p} \in \mathfrak{S}_n(\overline{\sigma}^\circ)$ for the problem (D^1), $\overline{p} \neq \overline{\sigma}^\star$, such that $W_{D^1}(\overline{\sigma}^\star) > W_{D^1}(\overline{p})$. Let us consider a schedule $\rho \in \mathfrak{S}_n(\sigma^\circ)$ for (P^1), equivalent to $\overline{p}$. Then $W_{P^1}(\rho) = W_{D^1}(\overline{p}) < W_{D^1}(\overline{\sigma}^\star) = W_{P^1}(\sigma^\star)$. A contradiction. The converse implication can be proved in an analogous way. The equality $W_{P^1}(\sigma^\star) = W_{D^1}(\overline{\sigma}^\star)$ follows from (a).   ∎

Example 19.32. Let us consider the data from Example 19.27 again. Then in the transformed problem, $1|p_j = w_j + a_j t| \sum b_j S_j$, we have two jobs with processing times $p_1 = 6 + 3t$, $p_2 = 5 + 2t$, with weights $w_1 = 2, w_2 = 1$. The machine is available from $t_0 = 1$, and $w_3 = 0$.

The tables of type (19.30) are as follows:

$$T(\overline{\sigma_1}) = \begin{bmatrix} 1 & 6 & 5 \\ & 4 & 3 \\ & 2 & 1 & 0 \end{bmatrix} \quad \text{and} \quad T(\overline{\sigma_2}) = \begin{bmatrix} 1 & 5 & 6 \\ & 3 & 4 \\ & 1 & 2 & 0 \end{bmatrix}. \tag{19.32}$$

Since $C_0(\overline{\sigma_1}) = 1$, $C_1(\overline{\sigma_1}) = 4 \times 1 + 6 = 10$, $C_2(\overline{\sigma_1}) = 3 \times 10 + 5 = 35$, we have $\sum w_j S_j(\overline{\sigma_1}) = 2 \times 1 + 1 \times 10 + 0 \times 35 = 12 = \sum w_j S_j(\sigma_1)$.

Similarly, since $C_0(\overline{\sigma_2}) = 1$, $C_1(\overline{\sigma_2}) = 3 \times 1 + 5 = 8$, $C_2(\overline{\sigma_2}) = 4 \times 8 + 6 = 38$, we have $\sum w_j S_j(\overline{\sigma_2}) = 1 \times 1 + 2 \times 8 + 0 \times 38 = 17 = \sum w_j S_j(\sigma_2)$.

It is easy to see that the optimal initial schedule, σ_1, corresponds to the transformed schedule, $\overline{\sigma_1}$, and vice versa.   ♦

Parallel machine transformed problems

Let us consider problem $Pm|p_j = a_j + b_j t| \sum w_j S_j$ in the form of (P^m). Let us define problem (D^m), corresponding to (P^m), as follows:

$$(D^m) \quad \begin{cases} \text{minimize } W_{D^m}(\overline{\sigma}) := \overline{a}^{\mathsf{T}} C(\overline{\sigma}) \\ \text{subject to } A(\overline{\sigma}) C(\overline{\sigma}) = \overline{w}, \ \overline{\sigma} \in \mathfrak{S}_n(\overline{\sigma^\circ}), \end{cases} \tag{19.33}$$

where $\overline{a} = (\overline{a^1}, \dots, \overline{a^m})^{\mathsf{T}}$ and $\overline{w} = (\overline{w^1}, \dots, \overline{w^m})^{\mathsf{T}}$, $\overline{a^k} = (a_{n_k}^k, a_{n_k-1}^k, \dots, b_0^k)^{\mathsf{T}}$ and $\overline{w^k} = (w_{n_k+1}^k, \dots, w_1^k)$ for $1 \leqslant k \leqslant m$.

The following definition is an extension of Definition 19.30.

Definition 19.33. (Parallel machine equivalent problems)
Given a schedule $\sigma = (\sigma^1, \dots, \sigma^m) \in \mathfrak{S}_n(\sigma^\circ)$, where $\sigma^k = ((a_1^k, \beta_1^k, w_1^k), \dots, (a_{n_k}^k, \beta_{n_k}^k, w_{n_k}^k))$, let $\overline{\sigma} = (\overline{\sigma^1}, \dots, \overline{\sigma^m}) \in \mathfrak{S}_n(\overline{\sigma^\circ})$ be a schedule such that $\overline{\sigma^k} = ((w_{n_k}^k, \beta_{n_k}^k, a_{n_k}^k), \dots, (w_1^k, \beta_1^k, a_1^k)), 1 \leqslant k \leqslant m$. Then
(a) the correspondence $\sigma \longleftrightarrow \overline{\sigma}$ will be called a transformation of the schedule σ for problem (P^m) into the schedule $\overline{\sigma}$ for the corresponding problem (D^m) and vice versa,
(b) both corresponding problems, (P^m) and (D^m), will be called equivalent problems.

Given a schedule σ for the initial problem (P^m), described by the tables $T(\sigma^k)$, $1 \leqslant k \leqslant m$, the schedule $\overline{\sigma}$ for the transformed problem (D^m) is described by the tables

$$T(\overline{\sigma^k}) := \begin{bmatrix} w_{n_k+1}^k & w_{n_k}^k & w_{n_k-1}^k & \cdots & w_1^k \\ & \beta_{n_k}^k & \beta_{n_k-1}^k & \cdots & \beta_1^k \\ & a_{n_k}^k & a_{n_k-1}^k & \cdots & a_1^k & a_0^k \end{bmatrix}, \tag{19.34}$$

where $1 \leqslant k \leqslant m$, $C_0^k(\overline{\sigma^k}) := w_{n_k+1}^k$ is the time at which the machine M_k starts the processing of jobs in the problem (D^m) and $C_0^k(\sigma^k) := a_0^k$ is the time at which the machine M_k starts the processing of jobs in the original problem (P^m), $1 \leqslant k \leqslant m$.

The next theorem, which concerns the equivalence of problems (P^m) and (D^m), is a counterpart of Theorem 19.31 for the case of m machines.

Theorem 19.34. (Gawiejnowicz et al. [26]) *Let $\sigma = (\sigma^1, \dots, \sigma^m)$, $\sigma^k = ((a_1^k, \beta_1^k, w_1^k), \dots, (a_{n_k}^k, \beta_{n_k}^k, w_{n_k}^k))$, be an arbitrary schedule from $\mathfrak{S}_n(\sigma^\circ)$ and $\overline{\sigma} \in \mathfrak{S}_n(\overline{\sigma^\circ})$ be the transformed schedule of σ in the form of $\overline{\sigma} = (\overline{\sigma^1}, \dots, \overline{\sigma^m})$, $\overline{\sigma^k} = ((w_{n_k}^k, \beta_{n_k}^k, a_{n_k}^k), \dots, (w_1^k, \beta_1^k, a_1^k)), 1 \leqslant k \leqslant m$. Then*
(a) if $\overline{\sigma}$ has been obtained by the transformation $\sigma \longleftrightarrow \overline{\sigma}$, then the equality

$$W_{P^m}(\sigma) = w^{\mathsf{T}} C(\sigma) = \overline{a}^{\mathsf{T}} C(\overline{\sigma}) = W_{D^m}(\overline{\sigma}) \tag{19.35}$$

holds;
(b) $\sigma^\star$ is an optimal schedule for problem (P^m) if and only if $\overline{\sigma^\star}$ is an optimal schedule for problem (D^m); moreover, the equality $W_{P^m}(\sigma^\star) = W_{D^m}(\overline{\sigma^\star})$ holds.

Proof. (a) (b) Similar to the proof of Theorem 19.31. $\square$

Example 19.35. Let us consider the data from Example 19.32 again. Then, two schedules, $\overline{\sigma_1} = (\overline{\sigma_1^1}, \overline{\sigma_1^2})$ and $\overline{\sigma_2} = (\overline{\sigma_2^1}, \overline{\sigma_2^2})$, equivalent to schedules $\sigma_1 = (\sigma_1^1, \sigma_1^2)$ and $\sigma_2 = (\sigma_2^1, \sigma_2^2)$, respectively, where $\overline{\sigma_i^j}$, for $1 \leqslant i, j \leqslant 2$, are described by the tables (19.34), i.e.,

$$T(\overline{\sigma_1^1}) = \begin{bmatrix} 1 & 0 & 1 & 4 \\ 4 & 5 & 4 \\ 1 & 0 & 2 & 1 \end{bmatrix}, T(\overline{\sigma_1^2}) = \begin{bmatrix} 4 & 3 & 2 \\ 2 & 3 \\ 2 & 1 & 3 \end{bmatrix},$$

$$T(\overline{\sigma_2^1}) = \begin{bmatrix} 1 & 0 & 1 \\ 4 & 5 \\ 1 & 0 & 1 \end{bmatrix}, T(\overline{\sigma_2^2}) = \begin{bmatrix} 4 & 4 & 2 & 3 \\ 4 & 3 & 2 \\ 2 & 1 & 2 & 3 \end{bmatrix}.$$

$\blacklozenge$

19.4.3 Properties of equivalent problems

We start with the result saying that in multi-machine problems the total weighted starting time criterion, $\sum w_j S_j$, is equivalent to the total machine load criterion, $\sum C_{\max}^{(k)}$.

Remark 19.36. Since in the considered case $\sum w_j S_j \equiv \sum w_j C_j$, we will write $\sum w_j C_j$ instead of $\sum w_j S_j$.

Theorem 19.37. (Gawiejnowicz et al. [25]) *If $t_0^k > 0$ for $1 \leqslant k \leqslant m$, then problems $Pm|p_j = b_j t| \sum wC_j$ and $Pm|p_j = w + b_j t| \sum a_0^k C_{\max}^{(k)}$ are equivalent.*

Proof. The result is a corollary from Theorem 19.34 for $w_j = w$, $a_j^k = 0$ and $a_0^k := t_0^k > 0$ for $1 \leqslant j \leqslant n_k$ and $1 \leqslant k \leqslant m$. $\square$

As a corollary from Theorem 19.37, we obtain the following result, which gives three main examples of equivalent time-dependent scheduling problems.

Corollary 19.38. (Gawiejnowicz et al. [25]) *The following pairs of time-dependent scheduling problems are equivalent: (a) $1|p_j = b_j t| \sum \beta_j C_j$ and $1|p_j = 1 + b_j(1+t)|C_{\max}$; (b) $1|p_j = a_j + bt| \sum C_j$ and $1|p_j = 1 + bt| \sum a_j C_j$; (c) $1|p_j = bt| \sum a_j C_j$ and $1|p_j = a_j + bt|C_{\max}$, where $\beta_j = 1 + b_j$ for $1 \leqslant j \leqslant n$.*

The next properties show common features of equivalent problems with $m \geqslant 2$ machines. First, such problems have the same lower bound on the optimal value of a criterion.

Property 19.39. (Gawiejnowicz et al. [25]) *If $\omega = 1$, $t_0^k = 1$ for $1 \leqslant k \leqslant m$, $h = \lfloor \frac{n}{m} \rfloor$ and $r = n - hm$, the optimal total machine load for problem $Pm|p_j = 1 + b_j t| \sum C_{\max}^{(k)}$ is not less than $m \sum_{i=1}^{h} \sqrt[m]{\prod_{j=1}^{im+r} \beta_j} + \sum_{j=1}^{r} \beta_j$.*

Proof. The result follows from Theorem 19.37 and the lower bound for problem $Pm|p_j = b_j t| \sum C_j$ (Jeng and Lin [27]). □

Second, equivalent problems have the same time complexity status.

Theorem 19.40. (Gawiejnowicz et al. [25]) *Let* $t_0^k = a_0 > 0$ *and* $\omega > 0$ *be common machine starting times for machines* M_k, $1 \leqslant k \leqslant m$, *in problem* $Pm|p_j = b_j t| \sum \omega C_j$ *and in problem* $Pm|p_j = \omega + b_j t| \sum a_0 C_{\max}^{(k)}$, *respectively. Then the first problem has the same time complexity as the second problem. In particular, if* $m \geqslant 2$, *both these problems are* $\mathcal{NP}$-*hard in the ordinary sense.*

Proof. The result follows from Theorem 19.37 and ordinary $\mathcal{NP}$-hardness of problem $P2|p_j = b_j t| \sum C_j$ (Chen [20], Kononov [28]). □

Third, equivalent problems have the same approximability status.

Theorem 19.41. (Gawiejnowicz et al. [25]) *There is no polynomial-time approximation algorithm with a constant worst-case bound for problem* $P|p_j = \omega + b_j t| \sum a_0 C_{\max}^{(k)}$, *unless* $\mathcal{P} = \mathcal{NP}$.

Proof. The result follows from Theorem 19.37 and the theorem about non-approximability of problem $P|p_j = b_j t| \sum C_j$ (Chen [20]). □

Finally, if for one of the equivalent problems there exists an FPTAS, then an FPTAS also exists for the second of these problems.

Theorem 19.42. (Gawiejnowicz et al. [25]) *For problem* $Pm|p_j = \omega + b_j t| \sum a_0^k C_{\max}^{(k)}$ *there exists an FPTAS.*

Proof. The result follows from Theorem 19.37 and a theorem about existence of an FPTAS for problem $Pm|p_j = b_j t| \sum w_j C_j$ (Woeginger [29]). □

19.5 Conjugate scheduling problems

Besides equivalent problems, in time-dependent scheduling there exist also other pairs of mutually related time-dependent scheduling problems. Cheng and Ding [31] observed a symmetry between single machine problems with linearly decreasing (increasing) processing times and release times (deadlines) and the maximum completion time criterion. This symmetry allows us to define a *symmetry reduction* thanks to which we are able to obtain a schedule for a problem with deadlines using a schedule for its counterpart with ready times, and vice versa. For example, given a schedule for an instance of problem $1|p_j = a_j - b_j t, r_j|C_{\max}$, we can take it as a schedule for a corresponding instance of problem $1|p_j = a_j + b_j t, d_j|C_{\max}$, viewed from the reverse direction.

Lemma 19.43. (Cheng and Ding [31]) *There exists a symmetry reduction between problem* $1|p_j = a_j - b_j t, r_j|C_{\max}$ *and problem* $1|p'_j = a'_j + b'_j t, d_j|C_{\max}$ *such that any schedule for the first problem defines a schedule for the second one and vice versa.*

Proof. Given an instance I of the problem $1|p_j = a_j - b_j t, r_j|C_{\max}$ and threshold $G > \max_{1 \leqslant j \leqslant n}\{r_j\}$, an instance II of the problem $1|p_j = a_j + b_j t, d_j|C_{\max}$ can be constructed in the following way: $a'_j = \frac{a_j - b_j G}{1 - b_j}$, $b'_j = \frac{b_j}{1 - b_j}$ and $d_j = G - r_j$ for $1 \leqslant j \leqslant n$.

Since it is sufficient to consider only schedules without idle times and since the above reduction can be done in polynomial time, the result follows. $\square$

By Lemma 19.43, we obtain the following result.

Lemma 19.44. (Cheng and Ding [31]) *There exists a symmetry reduction between the following pairs of problems:*
(a) $1|p_j = a_j + b_j t, r_j|C_{\max}$ *and* $1|p'_j = a'_j - b'_j t, d_j|C_{\max}$,
(b) $1|p_j = a_j + bt, r_j|C_{\max}$ *and* $1|p'_j = a'_j - b'_j t, d_j|C_{\max}$,
(c) $1|p_j = a_j - bt, r_j|C_{\max}$ *and* $1|p'_j = a'_j + b'_j t, d_j|C_{\max}$,
(d) $1|p_j = a_j + bt, r_j \in \{0, R\}|C_{\max}$ *and* $1|p'_j = a'_j - b't, d_j \in \{D_1, D_2\}|C_{\max}$,
(e) $1|p_j = a_j - bt, r_j \in \{0, R\}|C_{\max}$ *and* $1|p'_j = a'_j + b't, d_j \in \{D_1, D_2\}|C_{\max}$,
such that any schedule for the first problem from a pair defines a schedule for the second one and vice versa.

Results of Lemmas 19.43 and 19.44 suggest that a schedule for the single machine problem of minimizing the maximum completion time for a set of jobs with linearly decreasing processing times and non-zero ready times, $p_j = b_j - \alpha_j t$ and $r_j > 0$, where $1 \leqslant j \leqslant n$, $0 \leqslant \alpha_j < 1$ and $\alpha_i \sum_{j=1}^{n} b_j < b_i$ for all $1 \leqslant i \leqslant n$, can be considered as a schedule for the single machine problem of minimizing the maximum completion time for a set of jobs with linearly increasing processing times and finite deadlines, $p_j = b_j + \alpha_j t$ and $d_j < +\infty$ for $1 \leqslant j \leqslant n$, using the first schedule and reading it backwards.

In this section, we explain the results using the notion of *conjugate* time-dependent scheduling problems introduced by Gawiejnowicz et al. [32]. In order to define the problems, we introduce a *conjugacy formula* which relates an arbitrary instance of a time-dependent scheduling problem (called the *initial* problem) with an instance of another time-dependent scheduling problem (called the *conjugated* problem) in such a way that the values of criteria of both these problems are equal. We also introduce *composite* problems, using another transformation which generates from the initial problem another problem called the *adjoint* problem, and we show their basic properties.

19.5.1 The initial problem

Throughout this section, we consider the following problem. We are given a set of n time-dependent jobs, $J_1, J_2, \ldots, J_n$, which have to be processed on a single machine starting from time $t_0 \geqslant 0$. The jobs are independent, non-preemptable and there are no ready times or deadlines defined, i.e., $r_j = 0$ and $d_j = +\infty$ for $1 \leqslant j \leqslant n$. The actual processing time p_j of job J_j, $1 \leqslant j \leqslant n$, is a linear function of the starting time t of the job, $p_j = b_j + \alpha_j t$, where $b_j \geqslant 0$, $\alpha_j \in (-1, +\infty) \setminus \{0\}$ and $t \geqslant t_0 := b_0$. The objective is to minimize the total weighted completion time of all jobs, $\sum w_j C_j$, where C_j and $w_j \geqslant 0$ denote the completion time of job J_j and a given weight related to J_j, $1 \leqslant j \leqslant n$, respectively. The problem will be denoted in short as $1|p_j = b_j + \alpha_j t| \sum w_j C_j$ and called the *initial* problem. Bachman et al. [30] proved that the problem, under some additional assumptions, is $\mathcal{NP}$-hard in the ordinary sense.

Since each job J_j corresponds to a triple (b_j, a_j, w_j), $1 \leqslant j \leqslant n$, any schedule for the initial problem can be identified with the table

$$T(\sigma) := \begin{bmatrix} b_0 & b_1 & b_2 & \dots & b_n \\ & a_1 & a_2 & \dots & a_n \\ w_0 & w_1 & w_2 & \dots & w_n \end{bmatrix}, \tag{19.36}$$

where $a_j = 1 + \alpha_j$ for $1 \leqslant j \leqslant n$, b_0 is the machine starting time and w_0 is an artificial weight corresponding to b_0.

Given a schedule σ° for the initial problem, any other schedule σ can be obtained from σ° by permutations of columns $j = 1, 2, \dots, n$ of the respective table (19.36), with the first column (for $j = 0$) unchanged. The set of all such schedules will be denoted by $\mathfrak{S}_n(\sigma^\circ)$.

Let $\sigma \in \mathfrak{S}_n(\sigma^\circ) \neq \emptyset$. Since the set $\mathfrak{S}_n(\sigma^\circ)$ is composed of non-delay schedules, the completion time of job J_j in schedule σ is given by the recurrence equation

$$C_j(\sigma) = C_{j-1}(\sigma) + p_j(C_{j-1}(\sigma)) = a_j C_{j-1}(\sigma) + b_j, \tag{19.37}$$

where $1 \leqslant j \leqslant n$ and $C_0(\sigma) := b_0$. Let us define $w(\sigma) := (w_0, w_1, \dots, w_n)^\intercal$, $b(\sigma) := (b_0, b_1, \dots, b_n)^\intercal$ and $C(\sigma) := (C_0, C_1, \dots, C_n)^\intercal$.

If we define $W_P(\sigma) := w^\intercal C(\sigma) = \sum_{j=0}^n w_j C_j(\sigma)$, then problem $1|p_j = b_j + \alpha_j t| \sum w_j C_j$ can be formulated in matrix form as

$$(P) \quad \begin{cases} W_P := \min \{W_P(\sigma) : \sigma \in \mathfrak{S}_n(\sigma^\circ)\}, \\ \text{subject to } A(\sigma)C(\sigma) = b, \end{cases} \tag{19.38}$$

where $A(\sigma), C(\sigma)$ and b are defined as in Sect. 19.2.1, respectively. Therefore, from now on, the term *the initial problem* will be used for both, problem (P) and problem $1|p_j = b_j + \alpha_j t| \sum w_j C_j$.

The following lemma introduces the *conjugacy formula*, which plays a crucial role in our further considerations.

Lemma 19.45. (Gawiejnowicz et al. [32]) *For any $\sigma \in \mathfrak{S}_n(\sigma^\circ)$ the following identity holds:*

$$\left(\sum_{i=0}^n b_i \prod_{j=i+1}^n a_j\right)\left(\sum_{k=0}^n w_k \prod_{j=k+1}^n \frac{1}{a_j}\right) + \sum_{i=0}^n w_i b_i = \tag{19.39}$$

$$\sum_{k=0}^{n} w_k \left(\sum_{i=0}^{k} b_i \prod_{j=i+1}^{k} a_j \right) + \sum_{i=0}^{n} b_i \left(\sum_{k=0}^{i} w_k \prod_{j=k+1}^{i} \frac{1}{a_j} \right).$$

Proof. Let us notice that for any $k = 0, 1, \ldots, n$ we have

$$C_n(\sigma) = \left(\prod_{j=k+1}^{n} a_j \right) \sum_{i=0}^{k} b_i \prod_{j=i+1}^{k} a_j + \sum_{i=k+1}^{n} b_i \prod_{j=i+1}^{n} a_j. \tag{19.40}$$

By multiplying both sides of (19.40) by $\left(\prod_{j=k+1}^{n} a_j \right)^{-1}$ and $w_k \geqslant 0$, and next by summing over $k = 0, 1, \ldots, n$, we obtain

$$C_n(\sigma) \sum_{k=0}^{n} w_k \prod_{j=k+1}^{n} \frac{1}{a_j} = \sum_{k=0}^{n} w_k \left(\sum_{i=0}^{k} b_i \prod_{j=i+1}^{k} a_j \right) + \tag{19.41}$$

$$\sum_{k=0}^{n-1} w_k \left(\sum_{i=k+1}^{n} b_i \prod_{j=k+1}^{i} \frac{1}{a_j} \right).$$

Interchanging the order of summation in the second term of (19.41) we obtain, finally,

$$C_n(\sigma) \sum_{k=0}^{n} w_k \prod_{j=k+1}^{n} \frac{1}{a_j} = \sum_{k=0}^{n} w_k \left(\sum_{i=0}^{k} b_i \prod_{j=i+1}^{k} a_j \right) + \tag{19.42}$$

$$\sum_{i=0}^{n} b_i \left(\sum_{k=0}^{i} w_k \prod_{j=k+1}^{i} \frac{1}{a_j} \right) - \sum_{i=0}^{n} w_i b_i. \qquad \blacksquare$$

19.5.2 The composite problem

Let f denote a transformation of schedule $\sigma \in \mathfrak{S}_n(\sigma^\circ)$, represented by the table of the form of (19.36), i.e.,

$$T(f(\sigma)) := \begin{bmatrix} h_0 & h_1 & h_2 & \ldots & h_n \\ g_1 & g_2 & \ldots & g_n \\ u_0 & u_1 & u_2 & \ldots & u_n \end{bmatrix}. \tag{19.43}$$

We will assume that the transformation f is *idempotent*, i.e., $\sigma = f(f(\sigma))$ or $f \circ f = I$, where I is the *identity transformation* and $\circ$ denotes a superposition of two functions.

Table (19.43) will be called *adjoint* to table (19.36), and the schedule corresponding to $f(\sigma)$ will be called the *adjoint schedule* to the schedule corresponding to σ.

Given an adjoint to $\sigma \in \mathfrak{S}_n(\sigma°)$, we can consider the *adjoint problem* to (P), i.e.,

$$(P_f) \quad \begin{cases} W_{P_f} := \min \{W_P(f(\sigma)) : \sigma \in \mathfrak{S}_n(\sigma°)\}, \\ \text{subject to } A(f(\sigma))C(f(\sigma)) = h. \end{cases} \tag{19.44}$$

Matrix $A(f(\sigma))$ differs from matrix $A(\sigma)$ in the subdiagonal, where now we have $-g_j$ instead of $-a_j$, $1 \leqslant j \leqslant n$. The criterion function $W_P(f(\sigma))$ in problem (P_f) is equal to $\sum_{j=0}^{n} u_j C_j(f(\sigma))$, where $u := (u_0, u_1, \ldots, u_n)^\mathsf{T}$.

Remark 19.46. The criterion function $W_P(f(\sigma))$ will be denoted in short as $W_P(f(\sigma)) = u^\mathsf{T} C(f(\sigma))$. Let us notice that by definition $W_{P_f}(\sigma) = W_P(f(\sigma))$.

We want to consider problems (P) and (P_f) simultaneously, i.e., we are interested in properties of the following *composite problem*

$$(P, P_f) \quad \begin{cases} W_{P,P_f} := \min \{W_{P,P_f}(\sigma) : \sigma \in (\sigma°)\}, \\ \text{subject to } W_{P,P_f}(\sigma) := W_P(\sigma) + W_P(f(\sigma)). \end{cases} \tag{19.45}$$

Below we summarize the main properties of composite problem (P, P_f), omitting immediate proofs of the properties.

Property 19.47. (Gawiejnowicz et al. [32])
(a) For composite problem (P, P_f) there always exists an optimal solution $\sigma^\star \in (\sigma°)$. (b) The symmetry property for composite problem (P, P_f) holds, i.e., $W_{P,P_f}(\sigma) = W_{P,P_f}(f(\sigma))$ for any $\sigma \in \mathfrak{S}_n(\sigma°)$. (c) If $\sigma^\star$ is an optimal solution to composite problem (P, P_f), then $f(\sigma^\star)$ is also an optimal solution to composite problem (P, P_f). (d) If problem (P) has the symmetry property, i.e., $W_P(\sigma) = W_P(f(\sigma))$, then problems (P), (P_f) and (P, P_f) have the same solutions. (e) For all $\sigma \in \mathfrak{S}_n(\sigma°)$ the equality $W_{P,P_{f_1 \circ f_2}}(\sigma) = W_{P_{f_1} P_{f_2}}(\varrho)$ holds, where $W_{P_{f_1} P_{f_2}}(\varrho) = W_P(f_1(\varrho)) + W_P(f_2(\varrho))$, $\varrho = f_2(\sigma)$ and $f_1 \circ f_2$ is the superposition of transformations f_1 and f_2.

In the next sections, we consider two types of transformation f. The first transformation is $f(\sigma) := \sigma^{-1}$ and the second one is $f(\sigma) := \overline{\sigma}$ where, for a given $\sigma \in \mathfrak{S}_n(\sigma°)$,

$$T(\sigma^{-1}) := \begin{bmatrix} w_0 & w_1 & w_2 & \ldots & w_n \\ & a_1^{-1} & a_2^{-1} & \ldots & a_n^{-1} \\ b_0 & b_1 & b_2 & \ldots & b_n \end{bmatrix} \tag{19.46}$$

and

$$T(\overline{\sigma}) := \begin{bmatrix} w_0 & w_n & w_{n-1} & \ldots & w_1 \\ & a_n & a_{n-1} & \ldots & a_1 \\ b_0 & b_n & b_{n-1} & \ldots & b_1 \end{bmatrix} \tag{19.47}$$

Remark 19.48. Let us notice that σ and $\overline{\sigma}$ (as well as σ and σ^{-1}) are adjoint to each other. The following results mainly concern problem (P, P_f) in the case when $f(\sigma) := \sigma^{-1}$.

19.5.3 The conjugate problem

In this section, we consider the case when $f(\sigma) := \sigma^{-1}$. We will show that transformation of this form leads to a specific subclass of composite problems. The composite problems with $f(\sigma) := \sigma^{-1}$ will be called *conjugated problems* and are defined as follows.

Definition 19.49. (Conjugate problems, Gawiejnowicz et al. [32])
Problem $1|p_j = h_j + \gamma_j t| \sum u_j C_j$ *is said to be* conjugated *to problem* $1|p_j = b_j + \alpha_j t| \sum w_j C_j$ *if there exists transformation* f *such that*

$$C_n(\sigma)C_n(\varrho) + \sum_{j=0}^{n} w_j(\sigma)u_j(\varrho) = \sum_{j=0}^{n} w_j(\sigma)C_j(\sigma) + \sum_{j=0}^{n} u_j(\varrho)C_j(\varrho), \quad (19.48)$$

where $\varrho = f(\sigma)$ *for* $\sigma \in \mathfrak{S}_n(\sigma^{\circ})$, $C_j(\varrho) = g_j C_{j-1}(\varrho) + h_j$ *for* $1 \leqslant j \leqslant n$, $C_0(\varrho) := h_0$ *and* $g_j = 1 + \gamma_j$.

In other words, an initial time-dependent scheduling problem is conjugated to another time-dependent scheduling problem if equality (19.48) holds for any pair of corresponding schedules to the problems.

For transformation f defined by $f(\sigma) = \sigma^{-1}$, where $\sigma \in \mathfrak{S}_n(\sigma^{\circ})$, let us consider adjoint problem (P_f). The problem will be called *problem* (C) and is defined as

$$(C) \quad \begin{cases} W_C := \min\{W_P(\sigma^{-1}) : \sigma \in \mathfrak{S}_n(\sigma^{\circ})\}, \\ \text{subject to } A(\sigma^{-1})C(\sigma^{-1}) = w. \end{cases} \quad (19.49)$$

Criterion $W_P(\sigma^{-1}) = b^{\mathsf{T}} C(\sigma^{-1}) = \sum_{j=0}^{n} b_j C_j(\sigma^{-1})$, with $C_0(\sigma^{-1}) := w_0$. The matrix $A^{-1}(\sigma^{-1})$ exists and differs from $A^{-1}(\sigma)$ in this that products $a_i \cdots a_j$ are now replaced by inverses $(a_i \cdots a_j)^{-1}$.

Remark 19.50. Let us notice that in view of the form of $A^{-1}(\sigma^{-1})$, we have $C_j(\sigma^{-1}) := \sum_{i=0}^{j} w_i \prod_{k=i+1}^{j} a_k^{-1}$ for $j = 0, 1, \ldots, n$.

The next result, showing that problem (C) is conjugated to problem (P), gives the main pair of conjugated problems.

Theorem 19.51. (Gawiejnowicz et al. [32]) *Problem* $1|p_j = w_j - \beta_j t| \sum b_j C_j$, *where* $\beta_j = \frac{\alpha_j}{1+\alpha_j}$ *and* $\sum b_j C_j = W_P(\sigma^{-1})$, *is conjugated to problem* $1|p_j = b_j + \alpha_j t| \sum w_j C_j$, *where* $\alpha_j > -1$ *and* $\sum w_j C_j = W_P(\sigma)$, *i.e., the equality*

$$C_n(\sigma)C_n(\sigma^{-1}) + w^{\mathsf{T}} b = W_P(\sigma) + W_P(\sigma^{-1}) \quad (19.50)$$

is satisfied for all schedules $\sigma \in \mathfrak{S}_n(\sigma^{\circ})$, *where* $W_P(\sigma) = w^{\mathsf{T}} C(\sigma)$ *and* $W_P(\sigma^{-1}) = b^{\mathsf{T}} C(\sigma^{-1})$.

Proof. We apply identity (19.39) from Lemma 19.45. Indeed, given schedule σ^{-1} with $a_j = 1 + \alpha_j$ and $a_j^{-1} = 1 - \beta_j$, where $\beta_j = \frac{\alpha_j}{1+\alpha_j}$, conjugacy formula (19.39) takes the form

$$C_n(\sigma)C_n(\sigma^{-1}) + \sum_{j=0}^{n} w_j b_j = \sum_{j=0}^{n} w_j C_j(\sigma) + \sum_{j=0}^{n} b_j C_j(\sigma^{-1}), \qquad (19.51)$$

where $C_0(\sigma^{-1}) := w_0$, $C_j(\sigma^{-1}) = a_j^{-1} C_{j-1}(\sigma^{-1}) + w_j$ for $1 \leqslant j \leqslant n$. The second term is nothing else than the value of the criterion function in problem (C). ∎

Remark 19.52. Let us notice that the explicit form of job completion times is $C_j(\sigma^{-1}) = \sum_{k=0}^{j} w_k \prod_{i=k+1}^{j} a_i^{-1}$ for $0 \leqslant j \leqslant n$.

The next result shows a relation between the $C_{\max}$ and $\sum w_j C_j$ criteria.

Property 19.53. (Gawiejnowicz et al. [32]) If for problems (P) and (C) the respective completion times $C_j(\sigma)$ and $C_j(\sigma^{-1})$ are non-decreasing for $j = 0, 1, \dots, n$, then the formula

$$C_{\max}(\sigma)C_{\max}(\sigma^{-1}) + w^{\mathsf{T}} b = \sum_{j=0}^{n} w_j C_j(\sigma) + \sum_{j=0}^{n} b_j C_j(\sigma^{-1}) \qquad (19.52)$$

holds.

Proof. Formula (19.52) follows from formula (19.50), since in the case when the completion times $C_j(\sigma)$ and $C_j(\sigma^{-1})$ are non-decreasing for $j = 0, 1, \dots, n$, we have $C_{\max}(\sigma) = C_n(\sigma)$ and $C_{\max}(\sigma^{-1}) = C_n(\sigma^{-1})$. To complete the proof, it is sufficient to notice that $W_P(\sigma) = \sum_{j=0}^{n} w_j C_j(\sigma)$ and $W_P(\sigma^{-1}) = \sum_{j=0}^{n} b_j C_j(\sigma^{-1})$. ∎

In the case when $f(\sigma) := \sigma^{-1}$ for $\sigma \in \mathfrak{S}_n(\sigma^\circ)$, the composite problem (P, P_f) is in the form of

$$(P, C) \quad \begin{cases} W_{P,C} := \min\{W_{P,C}(\sigma) : \sigma \in \mathfrak{S}_n(\sigma^\circ)\}, \\ \text{subject to } W_{P,C}(\sigma) := W_P(\sigma) + W_P(\sigma^{-1}). \end{cases} \qquad (19.53)$$

Therefore, in this case, we will denote the problem (P, P_f) as (P, C).

Property 19.54. (Gawiejnowicz et al. [32]) The criterion function $W_{P,C}(\sigma)$ in composite problem (P, C) can be expressed in the form of

$$W_{P,C}(\sigma) = |a|^{-1} C_n(\sigma) C_n(\overline{\sigma}) + w^{\mathsf{T}} b, \qquad (19.54)$$

where $|a| = a_1 a_2 \cdots a_n$, $|a|^{-1} C_n(\overline{\sigma}) = C_n(\sigma^{-1})$ and $\sigma \in \mathfrak{S}_n(\sigma^\circ)$.

Proof. Notice that for transformation $f(\sigma) = \overline{\sigma}$ defined by table (19.47) the following identity holds:

$$|a| \cdot C_n(\sigma^{-1}) = \sum_{k=0}^{n} w_k \prod_{j=k+1}^{n} \frac{|a|}{a_j} = \sum_{k=0}^{n} w_k \prod_{j=1}^{k} a_j = C_n(\overline{\sigma}). \qquad (19.55)$$

Applying the equality (19.50) from Lemma 19.51, we have $W_{P,C}(\sigma) = C_n(\sigma)C_n(\sigma^{-1})+w^\top b$. Hence, in view of (19.55), $W_{P,C}(\sigma) = |a|^{-1}C_n(\sigma)C_n(\overline{\sigma})+ w^\top b$. ∎

19.5.4 Properties of conjugate problems

In this section, we present several properties (cf. Gawiejnowicz et al. [32]) concerning the relations between initial problem (P) and adjoint problem (C) in conjugated problem (P, C). In order to simplify the presentation, we will use explicit forms of criterion functions $W_P(\sigma)$ and $W_P(\sigma^{-1})$.

Property 19.55. The conjugated problem to problem (C) is initial problem (P). Moreover, the conjugate schedule to schedule σ^{-1} is the original schedule σ.

Proof. The property follows immediately from the idempotency of the transformation $f(\sigma) := \sigma^{-1}$. ∎

The next properties concern the relations between optimal solutions to the initial problem and its counterparts to the conjugate problem. We will use the following notation. Let

$$C_n(\underline{\sigma})C_n(\underline{\sigma}^{-1}) = \min\{C_n(\sigma)C_n(\sigma^{-1}) : \sigma \in \mathfrak{S}_n(\sigma^\circ)\},$$

$$w^\top C(\overset{\bullet}{\sigma}) = \min\{w^\top C(\sigma) : \sigma \in \mathfrak{S}_n(\sigma^\circ)\}, w^\top C(\overset{\circ}{\sigma}) = \max\{w^\top C(\sigma) : \sigma \in \mathfrak{S}_n(\sigma^\circ)\},$$

$$b^\top C(\overset{\wedge}{\sigma}^{-1}) = \min\{b^\top C(\sigma^{-1}) : \sigma \in \mathfrak{S}_n(\sigma^\circ)\}$$

and

$$b^\top C(\overset{\vee}{\sigma}^{-1}) = \max\{b^\top C(\sigma^{-1}) : \sigma \in \mathfrak{S}_n(\sigma^\circ)\}.$$

The first of these properties summarizes the relations between extremal values of the respective optimality criteria.

Property 19.56. The following inequalities hold:

$$w^\top C(\overset{\bullet}{\sigma}) \leqslant w^\top C(\underline{\sigma}) \leqslant w^\top C(\overset{\wedge}{\sigma}), \qquad (19.56)$$

$$b^\top C(\overset{\wedge}{\sigma}^{-1}) \leqslant b^\top C(\underline{\sigma}^{-1}) \leqslant b^\top C(\overset{\bullet}{\sigma}^{-1}) \qquad (19.57)$$

and

$$b^\top C(\overset{\wedge}{\sigma}^{-1}) - w^\top b \leqslant C_n(\underline{\sigma})C_n(\underline{\sigma}^{-1}) - w^\top C(\overset{\bullet}{\sigma}) \leqslant b^\top C(\overset{\bullet}{\sigma}^{-1}) - w^\top b. \qquad (19.58)$$

Proof. In order to prove (19.56), notice that in view of definition of schedules $\overset{\bullet}{\sigma}, \underline{\sigma}, \overset{\wedge}{\sigma}$ and in view of definition of the conjugated schedule σ^{-1}, we have

$$w^\mathsf{T}C(\overset{\wedge}{\sigma}) + b^\mathsf{T}C(\overset{\wedge}{\sigma}{}^{-1}) = C_n(\overset{\wedge}{\sigma})C_n(\overset{\wedge}{\sigma}{}^{-1}) + w^\mathsf{T}b \geqslant$$

$$C_n(\underline{\sigma})C_n(\underline{\sigma}^{-1}) + w^\mathsf{T}b \geqslant w^\mathsf{T}C(\underline{\sigma}) + b^\mathsf{T}C(\overset{\wedge}{\sigma}{}^{-1}).$$

Hence $w^\mathsf{T}C(\overset{\wedge}{\sigma}) \geqslant w^\mathsf{T}C(\underline{\sigma})$, which together with inequality $w^\mathsf{T}C(\underline{\sigma}) \geqslant w^\mathsf{T}C(\overset{\bullet}{\sigma})$ completes the proof. The proof of (19.57) can be done in a similar way.

In order to prove (19.58), it is sufficient to apply the conjugacy formula (19.39) and inequalities (19.56) and (19.57).   ∎

The next property shows relations which hold between $C_n(\underline{\sigma})C_n(\underline{\sigma}^{-1})$, $w^\mathsf{T}C(\overset{\bullet}{\sigma})$ and $b^\mathsf{T}C(\overset{\vee}{\sigma}{}^{-1})$.

Property 19.57. The following inequality holds:

$$C_n(\underline{\sigma})C_n(\underline{\sigma}^{-1}) \leqslant \min_{\sigma \in \mathfrak{S}_n(\sigma^\circ)} w^\mathsf{T}C(\sigma) + \max_{\sigma \in \mathfrak{S}_n(\sigma^\circ)} b^\mathsf{T}C(\sigma^{-1}) - b^\mathsf{T}w \leqslant C_n(\widetilde{\sigma})C_n(\widetilde{\sigma}^{-1}),$$

$$(19.59)$$

where $C_n(\widetilde{\sigma})C_n(\widetilde{\sigma}^{-1}) = \max\{C_n(\sigma)C_n(\sigma^{-1}) : \sigma \in \mathfrak{S}_n(\sigma^\circ)\}$.

Proof. Indeed, we have $C_n(\underline{\sigma})C_n(\underline{\sigma}^{-1}) + w^\mathsf{T}b = w^\mathsf{T}C(\underline{\sigma}) + b^\mathsf{T}C(\underline{\sigma}^{-1})$. But, for schedule $\overset{\bullet}{\sigma}$, the inequality $C_n(\underline{\sigma})C_n(\underline{\sigma}^{-1}) \leqslant w^\mathsf{T}C(\overset{\bullet}{\sigma}) + b^\mathsf{T}C(\overset{\bullet}{\sigma}{}^{-1}) - b^\mathsf{T}w$ with $b^\mathsf{T}C(\overset{\bullet}{\sigma}{}^{-1}) \leqslant b^\mathsf{T}C(\overset{\vee}{\sigma}{}^{-1})$ holds.

The second inequality can be proved in a similar way.   ∎

The next example gives a negative answer to the question whether sum $b^\mathsf{T}C(\overset{\bullet}{\sigma}{}^{-1})$ is maximal for conjugate problem (C), provided an optimal schedule $\overset{\bullet}{\sigma}$ for problem (P) is given.

Example 19.58. (Gawiejnowicz et al. [32]) Let $n = 5$, $b = w = (1, 1, 1, 1, 1)^\mathsf{T}$ and $a = (2, 3, 4, 5, 6)$. The optimal schedule to problem (P) is V-shaped schedule $\overset{\bullet}{a} = (5, 4, 2, 3, 6)$, with $w^\mathsf{T}C(\overset{\bullet}{a}) = 1162$. On the other hand, the maximum for problem (C) is realized for schedule $\overset{\vee}{a}{}^{-1} = (\frac{1}{5}, \frac{1}{3}, \frac{1}{2}, \frac{1}{4}, \frac{1}{6})$ which gives as the maximum $b^\mathsf{T}C(\overset{\vee}{a}{}^{-1}) = 7.9625$. Thus $\overset{\vee}{a}{}^{-1}$ is Λ-shaped, but $\overset{\vee}{a}{}^{-1} \neq \overset{\bullet}{\sigma}{}^{-1}$, since $\overset{\bullet}{\sigma}{}^{-1} = (\frac{1}{5}, \frac{1}{4}, \frac{1}{2}, \frac{1}{3}, \frac{1}{6})$.   ♦

Analysis of sum $b^\mathsf{T}C(\sigma^{-1}) = \sum_{k=0}^n b_k C_k(\sigma^{-1})$, where factor $C_k(\sigma^{-1}) := \sum_{i=0}^k w_i \prod_{j=i+1}^k a_j^{-1}$, shows that the values of $b^\mathsf{T}C(\sigma^{-1})$ can be small compared to $w^\mathsf{T}C(\sigma)$, whenever $a_j \gg 1$. In this case, we can approximate $W_P(\overset{\bullet}{\sigma}) = w^\mathsf{T}C(\overset{\bullet}{\sigma})$ by $W_{P,C}(\underline{\sigma}) = C_n(\underline{\sigma})C_n(\underline{\sigma}^{-1}) + b^\mathsf{T}w$. Conversely, if $0 < a_j \ll 1$ we can approximate $W_P(\overset{\wedge}{\sigma}{}^{-1}) = b^\mathsf{T}C(\overset{\wedge}{\sigma}{}^{-1})$ by $W_{P,C}(\underline{\sigma}) = C_n(\underline{\sigma})C_n(\underline{\sigma}^{-1}) + b^\mathsf{T}w$, since in this case $w^\mathsf{T}C(\sigma)$ is small compared to $b^\mathsf{T}C(\sigma^{-1})$.

The following result gives necessary estimations of the quality of the approximation in the case when $w = b = (1, 1, \ldots, 1)^\mathsf{T}$.

Remark 19.59. Since in this case σ can be fully identified with a, we will write a and a^{-1} instead of σ and σ^{-1}, respectively. Similarly, we will write $a \in \mathfrak{S}_n(a^\circ)$ instead of $\sigma \in \mathfrak{S}_n(\sigma^\circ)$.

Let us notice that in this case the conjugacy formula (19.50) takes the form

$$C_n(a)C_n(a^{-1}) + (n+1) = \sum_{j=0}^{n} C_j(a) + \sum_{j=0}^{n} C_j(a^{-1}). \tag{19.60}$$

Lemma 19.60. (Gawiejnowicz et al. [32]) *Given $a \in \mathfrak{S}_n(a^\circ)$, let*

$$a_\triangle := \min\{a_i : 1 \leqslant i \leqslant n\} \quad and \quad a_\triangledown := \max\{a_i : 1 \leqslant i \leqslant n\}.$$

Let us define $\lambda_n(x) := \frac{x}{x-1}(\frac{x}{x-1}(x^n - 1) - n)$ for $x > 1$. Then the following inequalities hold:

$$\lambda_n(a_\triangledown^{-1}) + (n+1) \leqslant W_P(a^{-1}) = \sum_{j=0}^{n} C_j(a^{-1}) \leqslant \lambda_n(a_\triangle^{-1}) + (n+1), \tag{19.61}$$

$$\lambda_n(a_\triangle) + (n+1) \leqslant W_P(a) = \sum_{j=0}^{n} C_j(a) \leqslant \lambda_n(a_\triangledown) + (n+1). \tag{19.62}$$

Moreover, $\lambda_n(x^{-1}) = \frac{1}{x-1}(n - \frac{1}{x-1}(1 - \frac{1}{x^n}))$ and for each $n \geqslant 1$ the inequalities $\lambda_n(x^{-1}) = O(n)$ if $x \geqslant \frac{1+\sqrt{5}}{2}$, and $\lambda_n(x) = O(x^{n+2})$ if $x \geqslant 2$ hold.

Proof. Let us denote $s = a_\triangle$, $S = a_\triangledown$. In order to prove inequality (19.61) let us notice that

$$\sum_{j=0}^{n} C_j(a^{-1}) = \frac{1}{|a|} \sum_{k=1}^{n} \sum_{i=1}^{k} (a_1 \cdots a_{i-1})(a_{k+1} \cdots a_n) + (n+1)$$

$$= \sum_{k=1}^{n} \frac{1}{a_k}\left(1 + \frac{1}{a_{k-1}} + \cdots + \frac{1}{a_1 \cdots a_{k-1}}\right) + (n+1).$$

Hence

$$\frac{1}{S} \sum_{k=1}^{n} \frac{1 - \frac{1}{S^k}}{1 - \frac{1}{S}} + (n+1) \leqslant \sum_{j=0}^{n} C_j(a^{-1}) \leqslant \frac{1}{s} \sum_{k=1}^{n} \frac{1 - \frac{1}{s^k}}{1 - \frac{1}{s}} + (n+1). \tag{19.63}$$

After further summations of geometric sequences, applying the formula for $\lambda_n(x^{-1})$, we obtain the desired inequalities. In order to prove the second pair of inequalities, we proceed in a similar way.

In order to prove inequality (19.62), we proceed in the following way. Let $x \geqslant \frac{1+\sqrt{5}}{2}$. Then the inequality $\lambda_n(\frac{1}{x}) \leqslant n(1 + \frac{1}{x})$ holds or, equivalently, $-\frac{x}{x-1}(1 - \frac{1}{x^n}) \leqslant n(x^2 - x - 1)$. Hence, $\lambda_n(x^{-1}) = O(n)$, where $n \geqslant 1$.

In order to prove that $\lambda_n(x) = O(x^{n+2})$, where $x \geqslant 2$, it is sufficient to observe that in this case $\lambda_n(x) \leqslant x^2(x^n - 1) - nx \leqslant x^{n+2}$.

The next property illustrates the impact of particular terms of criterion function $W_{P,C}(a)$ of composite problem (P, C) on the total value of the criterion. We omit the technical proof.

Property 19.61. (Gawiejnowicz et al. [32]) If elements of sequence a are large, i.e., $a_\triangle \gg 1$, then the impact of the term $W_P(a^{-1})$ in $W_{P,C}(a)$ becomes smaller while the term $W_P(a)$ becomes larger, and vice versa.

Now, we will apply Lemma 19.60 to the estimation of the relative error of approximation of the optimal value $W_P(\overset{\bullet}{a})$ of criterion $W_P(a) = \sum_{j=0}^{n} C_j(a)$ for initial problem (P) by means of the optimal value $W_{P,C}(\underline{a})$ decreased by $n+1$ for composite problem (P, C). We will assume that $w = b = (1, 1, \dots, 1)^{\mathsf{T}}$.

Let us consider the relative error of the approximation, defined as

$$\delta_a := \frac{(W_{P,C}(\underline{a}) - n - 1) - W_P(\overset{\bullet}{a})}{W_P(\overset{\bullet}{a})} = \frac{C_n(\underline{a})C_n(\underline{a}^{-1}) - W_P(\overset{\bullet}{a})}{W_P(\overset{\bullet}{a})}.$$

Let also

$$\delta(a) := \frac{W_{P,C}(a) - W_P(a)}{W_P(a)} \quad \text{and} \quad \Lambda_n(x) := \frac{\lambda_n(x^{-1})}{\lambda_n(x) + n + 1}, \quad \text{where } x > 1.$$

Let us notice that $\Lambda_n(x) \searrow$ for $x \in (1, +\infty)$. Moreover,

$$\lim_{x \to 1+} \Lambda_n(x) = \frac{n}{n+2}$$

and

$$\lim_{x \to +\infty} \Lambda_n(x) = 0.$$

To prove that, it is sufficient to note that for $x > 1$ we have

$$\Lambda_n(x) = \frac{x^{-n} - (n + 1 - nx)}{(n + 1 - nx) - 2x + x^{n+2}}.$$

Hence we obtain

Property 19.62. (Gawiejnowicz et al. [32]) For any $a \in \mathfrak{S}_n(\sigma^\circ)$ the inequalities $\delta(\underline{a}) \leqslant \delta_a \leqslant \delta(\overset{\bullet}{a})$ hold. Moreover,

$$0 < \Lambda_n(a_\triangledown) \leqslant \delta(a) \leqslant \Lambda_n(a_\triangle) \leqslant \frac{n}{n+2}. \tag{19.64}$$

We complete the section by three examples (cf. Gawiejnowicz et al. [32]), which illustrate the range of the relative error δ_a for different types of a.

Example 19.63. Let $a_j = 10^j$ for $1 \leqslant j \leqslant 6$. Then $a_\triangle = 10$ and $a_\triangledown = 10^6$. Estimating $\lambda_n(a_\triangle^{-1})$ and $\lambda_n(a_\triangledown^{-1})$ we have $\sum_{j=0}^{n} C_j(a^{-1}) \in [7.000006, 7.654321]$. Estimating $\lambda_n(a_\triangle)$ and $\lambda_n(a_\triangledown)$ we have $\sum_{j=0}^{n} C_j(a) \in [1234567, 1000002 \times 10^{30}]$. Applying the inequality (19.64) we have $5.99 \times 10^{-49} \leqslant \delta_a \leqslant 5.3 \times 10^{-7}$.

◆

Example 19.64. Let $a_j = j + 1$ for $1 \leqslant j \leqslant 6$. In this case, $a_\triangle = 2$ and $a_\triangledown = 7$. Estimation of $\lambda_n(a_\triangle^{-1})$ and $\lambda_n(a_\triangledown^{-1})$ gives $\sum_{j=0}^{n} C_j(a^{-1}) \in [7.9722224, 12.015625]$, while $\sum_{j=0}^{n} C_j(a) \in [247, 160132]$. Hence $0.0713 \times 10^{-6} \leqslant \delta_a \leqslant 0.02031$.   ◆

Example 19.65. If $a_\triangle > 1$ approaches 1, the approximation of $C_n(\underline{a})C_n(\underline{a}^{-1})$ by $\sum_{j=0}^{n} C_j(\overset{\bullet}{a})$ can be less satisfactory. In the boundary case, when $a_j = 1$, for $1 \leqslant j \leqslant n$, we have $\sum_{j=0}^{n} C_j(\sigma) = \sum_{j=0}^{n} C_j(\sigma^{-1}) = \frac{1}{2}(n+1)(n+2)$. As a result, we have $\delta_a = \frac{n}{n+2} < 1$.   ◆

19.6 Isomorphic scheduling problems

Gawiejnowicz and Kononov [35] observed that time-dependent scheduling problems with proportional-linearly deteriorating jobs and their counterparts with fixed job processing times have similar properties. Moreover, schedules for the time-dependent scheduling problems have similar structure as those for scheduling problems with fixed job processing time. A similar behaviour can be observed for the worst-case performance measures of approximation algorithms for time-dependent scheduling problems with proportional-linearly deteriorating jobs and their counterparts with fixed job processing times.

In this section, following [35], we consider *isomorphic* scheduling problems, the third class of pairs of mutually related scheduling problems studied in the chapter, which allows us to explain the thought-provoking similarities. Any pair of isomorphic problems is composed of a scheduling problem with fixed job processing times and its time-dependent counterpart with processing times that are proportional-linear functions of the job starting times. The isomorphic problems are defined by a transformation between scheduling problems with fixed job processing times and their time-dependent counterparts with proportionally-linear job processing times. Besides giving an explanation of the similarities mentioned above, the results on isomorphic problems show us how to convert polynomial algorithms for scheduling problems with fixed job processing times into polynomial algorithms for proportional-linear counterparts of the original problems and indicate how approximation algorithms for isomorphic problems are related.

19.6.1 Generic problem

In this section, we adopt some definitions proposed by Baptiste et al. [33], to introduce a generic problem with fixed job processing times. The problem can be defined as follows and it will be denoted as $GP||C_{\max}$.

Let $\mathcal{O} = \{o_1, \ldots, o_n\}$ denote a finite set of *operations* that compose *jobs*. The set of all jobs will be denoted by $\mathcal{J} = \{J_1, \ldots, J_j\}$, where $j \leqslant n$. All operations have to be processed on a set of *machines* and assumptions concerning these machines will be specified separately for each considered case.

Each operation $o_i \in \mathcal{O}$ has a specified *processing time* $p_i = b_i(a + bt)$, $1 \leqslant i \leqslant n$. The finite set of all possible *modes* of the execution of operation $o_i \in \mathcal{O}$ will be denoted by $\mathcal{M}(o_i)$. Each particular mode $\mu_i \in \mathcal{M}(o_i)$ is specific to an operation and establishes a machine (or a set of machines) on which the operation can be executed, i.e., sets $\mathcal{M}(o_i)$ are disjoint. The set of all modes will be denoted by $\mathcal{M} = \bigcup\limits_{o_i \in \mathcal{O}} \mathcal{M}(o_i)$.

Let $0 \leqslant \tau_1 < \ldots < \tau_D$ be a finite sequence of fixed *dates* τ_k, $1 \leqslant k \leqslant D$. The processing of all given operations must occur during time interval $[\tau_1, \tau_D]$.

There are defined *precedence constraints* between operations and *simultaneity constraints* between modes of the execution of the operations.

Simultaneity constraints are defined for each time interval $[\tau_k, \tau_{k+1}]$, $1 \leqslant k \leqslant D - 1$, by associated families $\mathcal{P}_k \subseteq 2^{\mathcal{M}}$ of subsets of $\mathcal{M}$. Each $P \in \mathcal{P}_k$ is a subset of modes which may be simultaneously used at any time $t \in [\tau_k, \tau_{k+1}]$, $1 \leqslant k \leqslant D - 1$, and is called a *feasible pattern*. The set of all families will be denoted by $\mathcal{P} = \bigcup\limits_{1 \leqslant k \leqslant D} \mathcal{P}_k$.

Simultaneity constraints may prevent simultaneous processing of

1) two different modes, μ_1 and μ_2, for the same operation, if $\{\mu_1, \mu_2\} \not\subseteq P$ for all $P \in \mathcal{P}$;
2) two operations, o_1 and o_2, such that if $\mathcal{M}(o_1) \cap P \neq \emptyset$, then $\mathcal{M}(o_2) \cap P = \emptyset$;
3) two modes on the same *machine* or *processor* with unit capacity;
4) *mode date constraints*, where mode $\mu \in \mathcal{M}$ cannot be used before the given release date of an operation or after its deadline.

Throughout the section, a *schedule* σ is an assignment of each operation $o_i \in \mathcal{O}$ to a time interval $[S_i(\sigma), C_i(\sigma)]$ in one possible mode. For a given schedule σ, the values $S_i(\sigma)$ and $C_i(\sigma)$ will be called *the starting time* and *the completion time* of operation $o_i \in \mathcal{O}$ in schedule σ, respectively. The corresponding *scheduling problem* will be understood as the problem of finding a schedule satisfying precedence or simultaneity constraints such that $C_i(\sigma) = S_i(\sigma) + p_i$ for all $o_i \in \mathcal{O}$.

Remark 19.66. If there is no ambiguity, the symbol σ is omitted and we will write S_i and C_i instead of $S_i(\sigma)$ and $C_i(\sigma)$, respectively.

19.6.2 The (γ, θ)-reducibility

In this section, we introduce the notion of isomorphic problems and we prove some properties of these problems.

We start with the following modified version of the definition proposed by Kononov [41].

Definition 19.67. ((γ, θ)-reducibility, Gawiejnowicz and Kononov [35])
Let I_Π, $\sigma = (s_1, \ldots, s_k, C_1, \ldots, C_k, \mu_1, \ldots, \mu_k)$ and $f_\Pi(C_1, \ldots, C_k)$ denote an instance of an optimization problem Π, a feasible solution to the instance

and the value of its criterion function, respectively. Problem Π_1 is said to be (γ, θ)–reducible to problem Π_2 if there exist two strictly increasing continuous functions, $\gamma : \mathbb{R}_+ \to \mathbb{R}_+$ and $\theta : \mathbb{R}_+ \to \mathbb{R}_+$, such that the following two conditions hold:

1) for any instance I_{Π_1} of problem Π_1 there exists an instance I_{Π_2} of problem Π_2 such that function γ transforms any feasible solution σ of instance I_{Π_1} into feasible solution $\sigma_d = (\gamma(s_1), \ldots, \gamma(s_k), \gamma(C_1), \ldots, \gamma(C_k), \mu_1, \ldots, \mu_k)$ of instance I_{Π_2}, and for any feasible solution $\tau_d = (s'_1, \ldots, s'_k, C'_1, \ldots, C'_k, \mu'_1, \mu'_2, \ldots, \mu'_k)$ of instance I_{Π_2} solution $\tau = (\gamma^{-1}(s'_1), \ldots, \gamma^{-1}(s'_k), \gamma^{-1}(C'_1), \ldots, \gamma^{-1}(C'_k), \mu'_1, \ldots, \mu'_k)$ is a feasible solution of instance I_{Π_1};

2) for any feasible solution σ of instance I_{Π_1} criterion functions f_{Π_1} and f_{Π_2} satisfy equality $f_{\Pi_2}(\gamma(C_1), \ldots, \gamma(C_k)) = \theta(f_{\Pi_1}(C_1, \ldots, C_k))$.

Property 19.68. If problem Π_1 is (γ, θ)–reducible to problem Π_2, then problem Π_2 is $(\gamma^{-1}, \theta^{-1})$–reducible to problem Π_1.

Proof. The property follows from the fact that each strictly increasing continuous function possesses a strictly increasing continuous inverse function.   □

Though Definition 19.67 is general enough, it concerns only optimization problems. The counterpart of this definition for decision problems is as follows.

Definition 19.69. (γ-reducibility, Gawiejnowicz and Kononov [35])
Let I_Π and σ denote an instance of a decision problem Π and a feasible solution to I_Π, respectively. Problem Π_1 is said to be γ–reducible to problem Π_2 if there exists a strictly increasing continuous function $\gamma : \mathbb{R}_+ \to \mathbb{R}_+$ such that for any instance I_{Π_1} of problem Π_1 there exists an instance I_{Π_2} of problem Π_2 such that function γ transforms any feasible solution σ of instance I_{Π_1} into feasible solution σ_d of instance I_{Π_2}, and for any feasible solution τ_d of instance I_{Π_2} solution τ is a feasible solution of instance I_{Π_1}.

Property 19.70. If problem Π_1 is γ–reducible to problem Π_2, then problem Π_2 is γ^{-1}–reducible to problem Π_1.

Proof. Similar to the proof of Property 19.68.   □

Definition 19.71. (Isomorphic scheduling problems)
Scheduling problems satisfying Definition 19.67 or Definition 19.69 are called isomorphic problems.

From Definition 19.71 it follows that two scheduling problems are isomorphic if one of them is a scheduling problem with fixed operation processing times and the second one is a time-dependent counterpart of the problem, obtained from the first one in such a way that conditions specified in Definition 19.67 or Definition 19.69 hold for arbitrary, corresponding instances of these two scheduling problems.

The following general result is the main tool in proofs of results concerning isomorphic scheduling problems.

Lemma 19.72. (Gawiejnowicz and Kononov [35]) *Let scheduling problem Π_2 be (γ, θ)–reducible to scheduling problem Π_1. Then if schedule $\sigma^* = (s_1^*, \ldots, s_k^*, C_1^*, \ldots, C_k^*, \mu_1^*, \ldots, \mu_k^*)$ is optimal for instance I_{Π_1} of problem Π_1, then schedule $\sigma_d^* = (\gamma(s_1^*), \ldots, \gamma(s_k^*), \gamma(C_1^*), \ldots, \gamma(C_k^*), \mu_1^*, \ldots, \mu_k^*)$ is optimal for instance I_{Π_2} of problem Π_2 and vice versa.*

Proof. Let σ^* be an optimal schedule for instance I_{Π_1} of problem Π_1 and let Π_1 be (γ, θ)–reducible to Π_2. Then σ_d^*, by condition 1) of Definition 19.67, is a feasible schedule for instance I_{Π_2} of problem Π_2. Since the equality $f_{\Pi_2}(\sigma_d^*) = \theta(f_{\Pi_1}(\sigma^*))$ holds, and since θ is a strictly increasing function, σ_d^* is an optimal schedule by condition 2) of Definition 19.67 as well. The second part of Lemma 19.72 follows from Property 19.68. ∎

Now, we prove some properties of the generic problem from Sect. 19.6.1.

Theorem 19.73. (Gawiejnowicz and Kononov [35]) *Problem $GP||C_{\max}$ is (γ, θ)–reducible to problem $GP|p_j = b_j(a + bt)|C_{\max}$ with $\gamma = \theta = 2^x - \frac{a}{b}$.*

Proof. Let Π_1 and Π_2 denote, respectively, the generic problem with fixed processing times of operations, $GP||C_{\max}$, and its counterpart with time-dependent processing times of operations, $GP|p_j = b_j(a + bt)|C_{\max}$. Let us recall that any instance I_{Π_1} of problem Π_1 is characterized by a finite sequence of fixed dates $\tau_k(\Pi_1)$, $1 \leqslant k \leqslant D$, a set of precedence constraints between operations, simultaneity constraints between modes for each interval $[\tau_k(\Pi_1), \tau_{k+1}(\Pi_1)]$, $1 \leqslant k \leqslant D - 1$, and processing times p_i of each operation o_i, $1 \leqslant i \leqslant n$.

We construct instance I_{Π_2}, isomorphic to instance I_{Π_1}, as follows. We set $\tau_k(\Pi_2) = 2^{\tau_k(\Pi_1)} - \frac{a}{b}$ for $1 \leqslant k \leqslant D$ and $v_i = \frac{2^{p_i} - 1}{b}$ for $1 \leqslant i \leqslant n$. We keep for instance I_{Π_2} the same set of constraints as in instance I_{Π_1}.

Let σ be a feasible schedule for I_{Π_1}. Let each operation $o_i \in \mathcal{O}$ be executed in a time interval $[S_i(\sigma), C_i(\sigma)]$ in a mode $\mu_i \in \mathcal{M}(o_i)$. By condition 1) of Definition 19.67 function γ transforms interval $[s_i(\sigma), C_i(\sigma)]$ into interval $[2^{s_i(\sigma)} - \frac{a}{b}, 2^{C_i(\sigma)} - \frac{a}{b}]$. Let $\tau_1 = t_1 < \ldots < t_n = \tau_k$ be an ordered sequence of all fixed dates, starting times and completion times of all operations in schedule σ. Let P_i be the only pattern used in time interval $[t_i, t_{i+1}]$, where $1 \leqslant i \leqslant n - 1$. Let us notice that the feasibility of schedule σ implies the feasibility of all patterns P_i, $1 \leqslant i \leqslant n - 1$.

Since γ is a strictly increasing function, this transformation preserves all constraints between patterns. To end the proof we must show that the transformation also preserves the relation between the starting time and the completion time of each operation.

Let C_i' denote the completion time of operation o_i in a schedule for instance I_{Π_2}. From condition 1) of Definition 19.67 we obtain

$$C_i' = s_i' + b_i(a + bs_i') = 2^{s_i}(1 + b_i b) - \frac{a}{b} = 2^{s_i + p_i} - \frac{a}{b} = 2^{C_i} - \frac{a}{b}.$$

Equivalence of criterion functions follows from the equality

$$\max_{1 \leqslant i \leqslant n} \left\{ 2^{C_i} \right\} - \frac{a}{b} = 2^{\max_{1 \leqslant i \leqslant n} \{C_i\}} - \frac{a}{b}.$$

Thus, condition 2) of Definition 19.67 is satisfied as well. ∎

Applying similar reasoning as above, we can prove a similar result for the case of the $\sum w_j U_j$ criterion.

Theorem 19.74. (Gawiejnowicz and Kononov [35]) *Problem $GP || \sum w_j U_j$ is (γ, θ)–reducible to problem $GP | p_j = b_j(a + bt) | \sum w_j U_j$ with $\gamma = 2^x - \frac{a}{b}$ and $\theta = x$.*

Unfortunately, the reasoning from the proof of Theorem 19.73 cannot be applied to other maximum criteria. For example, we cannot prove a counterpart of Theorem 19.74 for the L_{max} criterion, since the criterion value is equal to the maximal difference of two values. We can, however, apply the following transformation of the L_{max} criterion into the decision version of the C_{max} criterion. The transformation is based on assuming a value of L_{max}, adding it to the values of d_i and looking for a schedule that achieves exactly the same value of C_{max}. If such a schedule exists, it means that $d_i + C_{max}$ is an upper bound on the value of L_{max}. Thus, the following result holds.

Theorem 19.75. (Gawiejnowicz and Kononov [35]) *The decision version of problem $GP || L_{max}$ is γ–reducible to the decision version of problem $GP | p_j = b_j(a + bt) | L_{max}$ with $\gamma = 2^x - \frac{a}{b}$.*

A similar reasoning leads to a similar result for the T_{max} criterion.

19.6.3 Algorithms for isomorphic problems

In this section, we show that if we have a polynomial algorithm for a scheduling problem with fixed processing times of operations, we can convert it into a polynomial algorithm for the counterpart of the problem with proportional–linear processing times of operations, provided that both scheduling problems are isomorphic. We also show how approximation algorithms for isomorphic scheduling problems are related.

Preliminary definitions

Let us separate the set of parameters and variables of a scheduling problem into the following two subsets. The first subset is composed of all parameters which concern time: processing times, release and due dates, and the starting times and the completion times of operations. Notice that in our definition of the generic problem all time-related parameters are given by a finite sequence of fixed dates. We will call these parameters and variables *time parameters* and *time variables*, respectively. An arithmetic operation on the values of time parameters or variables will be called a *time arithmetic operation*.

The second subset is composed of all other parameters and variables such as weights of operations, resource requirements, variables of the assignment of operations to machines, etc. We will call these parameters and variables *general parameters* and *general variables*, respectively.

Definition 19.76. (Time vs. algorithms, Gawiejnowicz and Kononov [35])
(a) *An algorithm* mixes up *parameters of a scheduling problem if this algorithm performs arithmetic operations on two numbers, one of which is a value of a time parameter (or a time variable) and the other one is a value of a general parameter or an assignment of a value of a general variable (parameter) to a time variable, or vice versa.*
(b) *An algorithm for a problem is called* time-independent *if the running time of this algorithm does not depend on the values of time parameters of the problem.*

We complete this section by introducing two auxiliary operations.

Definition 19.77. (Artificial operations, Gawiejnowicz and Kononov [35])
Operations called artificial addition $\hat{+}$ *and* artificial subtraction $\hat{-}$ *are defined as follows:*

$$x \hat{+} y := \left(x + \frac{a}{b} \right) \left(y + \frac{a}{b} \right) - \frac{a}{b}$$

and

$$x \hat{-} y := \frac{x + \frac{a}{b}}{y + \frac{a}{b}} - \frac{a}{b}$$

for any feasible values of x and y.

Remark 19.78. Let us notice that both artificial operations from Definition 19.77 consist of five elementary operations.

Polynomial algorithms for isomorphic problems

Let I_{Π_1} be an instance of problem $GP||C_{\max}$ and let I_{Π_2} be an instance of problem $GP|p_j = b_j(a + bt)|C_{\max}$. Let us suppose that I_{Π_1} is (θ, γ)–reducible to I_{Π_2} with $\theta = \gamma = 2^x - \frac{a}{b}$. Let algorithm A produce a feasible solution for instance I_{Π_1}, and let us assume that algorithm A does not mix up parameters of the scheduling problem and does not use multiplication and division operations with the values of time parameters or variables. Under these assumptions, we can use A to design a new algorithm $\bar{A}$ by a transformation of time arithmetic operations as follows:

- if these time arithmetic operations are additions, we replace each addition with an artificial addition;
- if these time arithmetic operations are subtractions, we replace each subtraction with an artificial subtraction.

As a result of these changes, in all places where algorithm A uses the value of processing time p_j, algorithm $\bar{A}$ uses the value of *artificial processing time* $l_j = b_j b + 1 - \frac{a}{b}$ for $1 \leqslant j \leqslant n$.

Lemma 19.79. (Gawiejnowicz and Kononov [35]) *Let I_{Π_1} and I_{Π_2} be instances defined as in the proof of Lemma 19.73 and let algorithm A produce a feasible solution σ for instance I_{Π_1}. Then algorithm $\bar{A}$ produces a feasible solution σ_d for instance I_{Π_2} and the equality $\gamma(\sigma) = \sigma_d$ holds.*

Proof. Let I_{Π_1} and I_{Π_2} denote instances of scheduling problems Π_1 and Π_2, respectively. We remember that all time parameters of problem Π_1, except the processing times of operations, are given by a finite sequence of fixed dates τ_k, $1 \leqslant k \leqslant D$. From this observation it follows that $\tau_k(\Pi_2) = 2^{\tau_k(\Pi_1)} - \frac{a}{b}$ for all $1 \leqslant k \leqslant D$. We will show that all operations of algorithm $\bar{A}$ retain this property, i.e., if x is the value of a time variable after an elementary step of algorithm A, then $X = 2^x - \frac{a}{b}$ is the value of an appropriate time variable after the corresponding elementary step of algorithm $\bar{A}$.

First, let us suppose that the new value of a time variable is obtained by addition $x + y$, where x and y are two values of time parameters or variables. We have by mathematical induction, $X = 2^x - \frac{a}{b}$, $Y = 2^y - \frac{a}{b}$. Then

$$X \hat{+} Y = \left(X + \frac{a}{b}\right)\left(Y + \frac{a}{b}\right) - \frac{a}{b} = 2^{x+y} - \frac{a}{b}.$$

Second, let us suppose that $y = p_i$. Then, by definition of algorithm $\bar{A}$, we have $Y = l_i$. From definition of I_{Π_1} and I_{Π_2} we also have that $b_i = \frac{2^{p_i}}{b} - 1$. In this case,

$$X \hat{+} Y = \left(X + \frac{a}{b}\right)\left(Y + \frac{a}{b}\right) - \frac{a}{b} = 2^x\left(l_i - \frac{a}{b}\right) - \frac{a}{b}$$
$$= 2^x(b_i b + 1) - \frac{a}{b} = 2^x 2^{p_i} - \frac{a}{b}$$
$$= 2^{x+p_i} - \frac{a}{b} = 2^{x+y} - \frac{a}{b}.$$

Finally, let us suppose that $x = p_i$ and $y = p_j$. Then,

$$X \hat{+} Y = \left(l_i - \frac{a}{b}\right)\left(l_j - \frac{a}{b}\right) = 2^{p_i + p_j} - \frac{a}{b} = 2^{x+y} - \frac{a}{b}.$$

Let us notice that we obtain a similar result if we consider a subtraction instead of an addition. Since $2^x - \frac{a}{b}$ is a strictly increasing function, then the comparison of any two numbers gives the same result for both algorithms.

Let algorithm A produce a schedule $\sigma = (u_1, \ldots, u_k, v_1, \ldots, v_l)$, where $u_1, \ldots, u_k$ and $v_1, \ldots, v_l$ denote the values of time variables and the values of general variables, respectively. Since we do not change any general parameters and variables, and each elementary step preserves the relation between time parameters and variables, algorithm $\bar{A}$ produces at its output the schedule $\sigma_d = (\gamma(u_1), \ldots, \gamma(u_k), v_1, \ldots, v_l)$. ∎

By Lemma 19.79 we can prove the following main result of this section.

Theorem 19.80. (Gawiejnowicz and Kononov [35]) *Let algorithm A produce an optimal schedule for any instance of problem $GP||C_{\max}$. Then algorithm $\bar{A}$ produces an optimal schedule for any instance of problem $GP|p_j = b_j(a + bt)|C_{\max}$. Moreover, if algorithm A is time-independent and $O(f(n))$ is its running time, then the running time of algorithm $\bar{A}$ is $O(f(n) + n)$, where n is the number of operations.*

Proof. The optimality of algorithm $\bar{A}$ follows directly from Lemma 19.45 and Lemma 19.79. Since the running time of algorithm A does not depend on the values of time parameters of problem $GP||C_{\max}$ and algorithm $\bar{A}$ repeats all elementary steps of algorithm A, the number of elementary steps (including artificial operations) of algorithm $\bar{A}$ is equal to the number of elementary steps of algorithm A. Since by Remark 19.78 each artificial operation consists of five elementary operations, artificial operations increase the running time of algorithm $\bar{A}$ at most five times. Finally, let us notice that we need $O(n)$ time to evaluate the values of l_i. ∎

Approximation algorithms for isomorphic problems

The following result shows a relation which holds between approximation algorithms for two isomorphic scheduling problems.

Theorem 19.81. (Gawiejnowicz and Kononov [35]) *If A is an approximation algorithm for problem $GP||C_{\max}$ such that*

$$\frac{C_{\max}^A}{C_{\max}^\star} \leqslant r_A < +\infty,$$

then for its time-dependent counterpart $\bar{A}$ for problem $GP|p_i = v_i(a+bt)|C_{\max}$ the equality

$$\frac{\log(C_{\max}^{\bar{A}} + \frac{a}{b})}{\log(C_{\max}^{\bar{\star}} + \frac{a}{b})} = \frac{C_{\max}^A}{C_{\max}^\star}$$

holds.

Proof. Let σ^A and $\sigma^\star$ denote a schedule constructed by algorithm A and an optimal schedule for problem $GP||C_{\max}$, respectively. Let

$$C_{\max}^A := C_{\max}(\sigma^A) = \max_{o_i \in \mathcal{O}}\{C_i(\sigma)\}$$

and

$$C_{\max}^\star := C_{\max}(\sigma^\star) = \max_{o_i \in \mathcal{O}}\{C_i(\sigma^\star)\}.$$

Then, by assumption, $\frac{C_{\max}^A}{C_{\max}^\star} \leqslant r_A$.

Since by Theorem 19.73 problem $GP||C_{\max}$ is (γ, θ)–reducible to problem $GP|p_j = b_j(a + bt)|C_{\max}$ with $\gamma = \theta = 2^x - \frac{a}{b}$, there exists a schedule σ_d^A

constructed by algorithm $\bar{A}$ and an optimal schedule $\sigma_d^{\star}$ such that $C_i(\sigma_d^A) = 2^{C_i(\sigma)} - \frac{a}{b}$ and $C_i(\sigma_d^{\star}) = 2^{C_i^{\star}} - \frac{a}{b}$ for all $o_i \in \mathcal{O}$. Hence,

$$C_{\max}^{\bar{A}} := C_{\max}(\sigma_d^A) = \max_{o_i \in \mathcal{O}}\left\{2^{C_i(\sigma)}\right\} - \frac{a}{b} = 2^{\max_{o_i \in \mathcal{O}}\{C_i(\sigma)\}} - \frac{a}{b} = 2^{C_{\max}^A} - \frac{a}{b}.$$

Similarly,

$$C_{\max}^{\bar{\star}} := C_{\max}(\sigma_d^{\star}) = \max_{o_i \in \mathcal{O}}\left\{2^{C_i(\sigma^{\star})}\right\} - \frac{a}{b} = 2^{\max_{o_i \in \mathcal{O}}\{C_i(\sigma^{\star})\}} - \frac{a}{b} = 2^{C_{\max}^{\star}} - \frac{a}{b}.$$

Thus,

$$\frac{\log(C_{\max}^{\bar{A}} + \frac{a}{b})}{\log(C_{\max}^{\bar{\star}} + \frac{a}{b})} = \frac{C_{\max}^A}{C_{\max}^{\star}} \leqslant r_A.$$

∎

Remark 19.82. Let us notice that Theorem 19.81 concerns the *logarithmic worst-case ratio*, not the ordinary one. Thus, by Theorem 19.81 we cannot state that the latter ratio is bounded (see Example 14.9).

By Theorem 19.81 we immediately obtain a number of results concerning time-dependent adaptations of heuristic algorithms for scheduling problems with fixed processing times of operations. We illustrate this claim with the following examples.

Example 19.83. (Gawiejnowicz and Kononov [35]) Let us consider the list scheduling algorithm LS, proposed by Graham [37] for problem $Pm|prec|C_{\max}$ (cf. Sect. 14.1). As long as there are jobs to be scheduled, the algorithm LS assigns the first available job to the first available machine.

Let us apply the time-dependent counterpart of this algorithm, $\overline{LS}$, to problem $Pm|p_j = b_j t|C_{\max}$. By Theorem 19.81, assuming $a = 0$ and $b = 1$, we immediately obtain the following result.

Corollary 19.84. *For the algorithm $\overline{LS}$ applied to problem $Pm|p_j = b_j t|C_{\max}$ the inequality $\frac{\log C_{\max}^{\overline{LS}}}{\log C_{\max}^{\star}} \leqslant 2 - \frac{1}{m}$ holds.* ♦

Example 19.85. Let us consider the algorithm $\overline{LPT}$ which is a time-dependent adaptation of the algorithm LPT for problem $Pm||C_{\max}$ (Graham [38], cf. Sect. 14.1). In the algorithm $\overline{LPT}$, we first order all jobs in non-increasing order of job deterioration rates and next we apply the algorithm $\overline{LS}$.

If we apply the algorithm $\overline{LPT}$ to problem $Pm|p_j = b_j t|C_{\max}$, then assuming $a = 0$ and $b = 1$, by Theorem 19.81 we obtain the following result.

Corollary 19.86. *For the algorithm $\overline{LPT}$ applied to problem $Pm|p_j = b_j t|C_{\max}$ the inequality $\frac{\log C_{\max}^{\overline{LPT}}}{\log C_{\max}^{\star}} \leqslant \frac{4}{3} - \frac{1}{3m}$ holds.* ♦

Remark 19.87. Cheng et al. [34] obtained the results of Corollaries 19.84 and 19.86 by direct computation (see Sect. 14.2.1).   ◇

Remark 19.88. Let us notice that similar results as those of Corollaries 19.84 and 19.86 can be obtained for any subproblem of the generic problem $GP||C_{\max}$ for which there exists an approximation algorithm with a known worst-case ratio.

Below we show examples of scheduling problems with proportional-linear processing times of operations which may be solved in polynomial time by using modified algorithms for the problems' counterparts with fixed processing times of operations.

Examples of single machine isomorphic scheduling problems

Throughout this section, we assume that we are given a set of jobs $\mathcal{J} = \{J_1, \ldots, J_n\}$ to be processed on a single machine. Each job $J_j \in \mathcal{J}$ has processing time p_j and for the job there may be defined weight w_j, release date r_j and due date d_j, $1 \leqslant j \leqslant n$. In the case when release dates (due dates) are defined, jobs are numbered in such a way that $r_1 \leqslant \cdots \leqslant r_n$ ($d_1 \leqslant \cdots \leqslant d_n$). The first two examples concern the criterion $C_{\max}$.

Theorem 19.89. *Problem* $1|r_j, p_j = b_j(a + bt)|C_{\max}$ *is solvable in* $O(n \log n)$ *time.*

Proof. The result follows immediately from Theorem 19.73.   □

Theorem 19.90. (Kononov [42]) *Problem* $1|p_j = b_j(a+bt), prec|C_{\max}$ *is solvable in* $O(n^2)$ *time.*

Proof. The result follows from a result by Lawler [43] for problem $1|prec|C_{\max}$.   □

The next two examples concern the criterion $\sum w_j U_j$. Let us assume that all $w_j = 1$.

Theorem 19.91. (Gawiejnowicz and Kononov [35]) *Problem* $1|p_j = b_j(a + bt)|\sum U_i$ *is solvable in* $O(n \log n)$ *time.*

Proof. Let us notice that in order to realize Moore's algorithm [45], it is sufficient to evaluate the completion time of added jobs and to compare this time with the due date of the last added job. Since this algorithm uses only 'comparisons' and 'additions', by Theorem 19.80, the result follows.   □

Now, let us assume that w_j are arbitrary for $1 \leqslant j \leqslant n$.

Theorem 19.92. (Gawiejnowicz and Kononov [35]) *Problem* $1|p_j = b_j(a + bt)|\sum w_j U_j$ *is solvable in* $O(n \sum_{j=1}^{n} w_j)$ *time.*

Proof. Problem $1||\sum w_j U_j$ was proved to be ordinarily $\mathcal{NP}$-hard by Karp [40]. Two different pseudo-polynomial algorithms for this problem, both based on dynamic programming, were proposed by Lawler and Moore [44] and Sahni [47].

The first of these algorithms recursively evaluates function $F_k(s)$ of the minimum weighted number of late jobs for the subproblem involving jobs $J_1, \ldots, J_k$, where the last on-time job completes at time s. Since this algorithm is not time-independent, it cannot be used for problem $1|p_j = b_i(a + bt)|\sum w_j U_j$.

The second of these algorithms recursively evaluates function $F_j(w)$ of the minimum total processing time over all feasible subsets of $\{J_1, \ldots, J_n\}$ whose the total weight is equal to w.

Using initialization $F_0(0) = r_0$ and assuming that $F_j(w) = +\infty$ for $1 \leqslant j \leqslant n$, where r_0 is the common release date for all jobs and $w \in \{-w_{\max}, -w_{\max} + 1, \ldots, \sum_{j=1}^{n} w_j\}$, for $1 \leqslant j \leqslant n - 1$ we have the following recursion:

$$F_{j+1}(w) = \min\{F_j(w), F_j(w - w_{j+1}) + p_{j+1}\}. \tag{19.65}$$

We can calculate correct values of state variables $F_j(w)$ in constant time. Hence this dynamic programming algorithm can be implemented to run in $O(n\sum_{j=1}^{n} w_j)$ time and space. The value of the optimum solution is then equal to $\sum_{j=1}^{n} w_j - w^*$, where w^* corresponds to the maximal value of w such that $F_n(w) < +\infty$. The schedule corresponding to this value is determined by backtracking.

Let us notice that the second algorithm is time-independent and it uses only operations of 'comparison' and 'addition' for time variables. Therefore, we can solve problem $1|p_j = b_j(a + bt)|\sum w_j U_j$ with a time-dependent counterpart of the algorithm. The new form of recursion (19.65) is

$$\begin{aligned} F_{j+1}(w) &= \min\left\{F_j(w), F_j(w - w_{j+1}) \hat{+} l_{j+1}\right\} \\ &= \min\left\{F_j(w), F_j(w - w_{j+1})(w_{j+1}b + 1) + w_{j+1}a\right\} \end{aligned} \tag{19.66}$$

for $1 \leqslant j \leqslant n - 1$. Using Eq. (19.66) we can calculate the values of state variables $F_i(w)$, where $1 \leqslant j \leqslant n$ and $w \in \{-w_{max}, -w_{max}+1, \ldots, \sum_{j=1}^{n} w_j\}$, in constant time. Hence, by Theorem 19.80, the result follows. $\qquad\square$

Examples of dedicated machine isomorphic scheduling problems

In the following results we assume that each job $J_j \in \mathcal{J}$ consists of two operations o_{ij} with processing times p_{ij}, operation o_{ij} must be processed on machine M_i, each machine can process at most one operation at a time and any two operations of a job cannot run simultaneously, $1 \leqslant i \leqslant 2, 1 \leqslant j \leqslant n$. There may be defined precedence relations between operations of all jobs, specific for each considered problem. The optimality criterion is the maximum completion time, $C_{\max}$.

Theorem 19.93. (Kononov [41], Mosheiov [46]) *Problem* $F2|p_{ij} = b_{ij}(a + bt)|C_{\max}$ *is solvable in* $O(n \log n)$ *time.*

Proof. Johnson [39] proved that problem $F2||C_{\max}$ can be solved in $O(n \log n)$ time by the following Johnson's algorithm: first, schedule jobs with $p_{1j} \leqslant p_{2j}$ in non-decreasing order of p_{1j}'s and then schedule the remaining jobs in non-decreasing order of p_{2j}'s, see Algorithm 9.1 (cf. Theorem 9.2). Since the algorithm uses 'comparisons' for the construction of an optimal sequence of jobs and 'additions' to evaluate the starting time of each job in the optimal schedule, by Theorem 19.80 the time-dependent counterpart of this algorithm generates an optimal schedule for problem $F2|p_{ij} = b_{ij}(a + bt)|C_{\max}$. □

Theorem 19.94. (Kononov [41], Mosheiov [46]) *Problem* $O2|p_{ij} = b_{ij}(a + bt)|C_{\max}$ *is solvable in* $O(n)$ *time.*

Proof. Gonzalez and Sahni [36] proved that problem $O2||C_{\max}$ can be solved in $O(n)$ time as follows. Let us define two sets of jobs, $\mathcal{J}_1 = \{J_j \in \mathcal{J}|p_{1j} \leqslant p_{2j}\}$ and $\mathcal{J}_2 = \{J_j \in \mathcal{J}|p_{1j} > p_{2j}\}$. Without loss of generality we assume that

$$p_{1r} = \max\{\max\{p_{1j}|J_j \in \mathcal{J}_1\}, \max\{p_{2j}|J_j \in \mathcal{J}_2\}\},$$

since if the maximum is achieved on set $\mathcal{J}_2$, we can interchange machines M_1 and M_2 and sets $\mathcal{J}_1$ and $\mathcal{J}_2$. An optimal schedule is constructed by scheduling 1) on machine M_1 first all jobs in $\mathcal{J}_1 \setminus \{J_r\}$ in an arbitrary order, then all jobs in $\mathcal{J}_2$ in an arbitrary order, followed by job J_r; 2) on machine M_2 first job J_r, then all jobs in $\mathcal{J}_1 \setminus \{J_r\}$ and $\mathcal{J}_2$ in the same order as on machine M_1; 3) first job J_r on machine M_2 and then on machine M_1, while all remaining jobs are first scheduled on machine M_1.

Again, the algorithm uses 'comparisons' for the construction of an optimal sequence of operations on each machine and 'additions' to evaluate the starting time of each operation in the optimal schedule. By Theorem 19.80, the time-dependent counterpart of this algorithm generates an optimal schedule for problem $O2|p_{ij} = b_{ij}(a + bt)|C_{\max}$. □

Theorem 19.95. (Gawiejnowicz and Kononov [35]) *Problem* $Dag2|n_j \leqslant 2, p_{ij} = b_{ij}(a + bt)|C_{\max}$ *is solvable in* $O(n \log n)$ *time.*

Proof. Problem $Dag2|n_j \leqslant 2|C_{\max}$ is a shop scheduling problem with two machines, in which there are arbitrary relations between operations and each job has at most two operations. Strusevich [48] has shown that the problem can be solved in $O(n \log n)$ time by an algorithm which uses only 'comparisons' and 'additions'. Hence, by Theorem 19.80, the result follows. □

References

The matrix approach

1. S. Gawiejnowicz, *Time-Dependent Scheduling*. Berlin-Heidelberg: Springer 2008.

2. S. Gawiejnowicz, W. Kurc and L. Pankowska, A greedy approach for a time-dependent scheduling problem. *Lecture Notes in Computer Science* **2328** (2002), 79–86.

3. S. Gawiejnowicz, W. Kurc and L. Pankowska, Analysis of a time-dependent scheduling problem by signatures of deterioration rate sequences. *Discrete Applied Mathematics* **154** (2006), no. 15, 2150–2166.

4. S. Gawiejnowicz, W. Kurc and L. Pankowska, The equivalence of linear time-dependent scheduling problems. Report 127/2005, Adam Mickiewicz University, Faculty of Mathematics and Computer Science, Poznań, December 2005.

Minimizing the l_p norm

5. Y. Azar, L. Epstein, Y. Richter and G. J. Woeginger, All-norm approximation algorithms. *Journal of Algorithms* **52** (2004), no. 2, 120–133.

6. N. Bansal and K. Pruhs, Server scheduling in the L_p norm: a rising tide lifts all boats. *Proceedings of the STOC 2003*, 242–250.

7. I. Caragiannis, Better bounds for online load balancing on unrelated machines. *Proceedings of the 19th Annual ACM-SIAM Symposium on Discrete Algorithms*, 2008, 972–981.

8. Y. H. Chung, H.C. Liu, C. C. Wu and W. C. Lee, A deteriorating jobs problem with quadratic function of job lateness. *Computers and Industrial Engineering* **57** (2009), no. 4, 1182–1186.

9. S. Gawiejnowicz, *Time-Dependent Scheduling*. Berlin-Heidelberg: Springer 2008.

10. S. Gawiejnowicz and W. Kurc, Structural properties of time-dependent scheduling problems with the l_p norm objective. *Omega* **57** (2015), 196–202.

11. A. K. Gupta. and A. I. Sivakumar, Multi-objective scheduling of two-job families on a single machine. *Omega* **33** (2005), no. 5, 399–405.

12. W-H. Kuo and D-L. Yang, Parallel-machine scheduling with time dependent processing times. *Theoretical Computer Science* **393** (2008), no. 1–3, 204–210.

13. M. Liu, C-B. Chu, Y-F. Xu and J-Z. Huo, An optimal online algorithm for single machine scheduling to minimize total general completion time. *Journal of Combinatorial Optimization* **23** (2012), no. 2, 189–195.

14. K. Maurin, *Analysis. Part I: Elements*. Dordrecht: D. Reidel/Polish Scientific Publishers 1976.

15. K. Maurin, *Analysis. Part II: Integration, distributions, holomorphic functions, tensor and harmonic analysis*. Dordrecht: D. Reidel/Polish Scientific Publishers 1980.

16. G. Mosheiov, V-shaped policies in scheduling deteriorating jobs. *Operations Research* **39** (1991), no. 6, 979–991.

17. J-B. Wang, Single-machine scheduling problems with the effects of learning and deterioration. *Omega* **35** (2007), no. 4, 397–402.

18. C-M. Wei and J-B. Wang, Single machine quadratic penalty function scheduling with deteriorating jobs and group technology. *Applied Mathematical Modelling* **34** (2010), no. 11, 3642–3647.

19. S. Yu and P. W. H. Wong, Online scheduling of simple linear deteriorating jobs to minimize the total general completion time. *Theoretical Computer Science* **487** (2013), 95–102.

Equivalent scheduling problems

20. Z-L. Chen, Parallel machine scheduling with time dependent processing times. *Discrete Applied Mathematics* **70** (1996), no. 1, 81–93. (Erratum: *Discrete Applied Mathematics* **75** (1997), no. 1, 103.)

21. T-C. E. Cheng and Q. Ding, Single machine scheduling with deadlines and increasing rates of processing times. *Acta Informatica* **36** (2000), no. 9–10, 673–692.

22. T-C. E. Cheng and Q. Ding, The complexity of scheduling starting time dependent tasks with release times. *Information Processing Letters* **65** (1998), no. 2, 75–79.

23. T-C. E. Cheng, Q. Ding and B. M-T. Lin, A concise survey of scheduling with time-dependent processing times. *European Journal of Operational Research* **152** (2004), no. 1, 1–13.

24. S. Gawiejnowicz, W. Kurc and L. Pankowska, Analysis of a time-dependent scheduling problem by signatures of deterioration rate sequences. *Discrete Applied Mathematics* **154**, no. 15, 2150–2166.

25. S. Gawiejnowicz, W. Kurc and L. Pankowska, Equivalent time-dependent scheduling problems. *European Journal of Operational Research* **196** (2009), no. 3, 919–929.

26. S. Gawiejnowicz, W. Kurc and L. Pankowska, The equivalence of linear time-dependent scheduling problems, report 127/2005, Adam Mickiewicz University, Faculty of Mathematics and Computer Science, Poznań, December 2005.

27. A. A-K. Jeng and B. M-T. Lin, Makespan minimization in single-machine scheduling with step-deterioration of processing times. *Journal of the Operational Research Society* **55** (2004), no. 3, 247–256.

28. A. Kononov, Scheduling problems with linear increasing processing times. In: U. Zimmermann et al. (eds.), *Operations Research 1996*. Berlin-Heidelberg: Springer 1997, pp. 208–212.

29. G. J. Woeginger, When does a dynamic programming formulation guarantee the existence of a fully polynomial time approximation scheme (FPTAS)? *INFORMS Journal on Computing* **12** (2000), no. 1, 57–74.

Conjugate scheduling problems

30. A. Bachman, A. Janiak and M. Y. Kovalyov, Minimizing the total weighted completion time of deteriorating jobs, *Information Processing Letters* **81** (2002), no. 2, 81–84.

31. T-C. E. Cheng and Q. Ding, The complexity of scheduling starting time dependent tasks with release times. *Information Processing Letters* **65** (1998), no. 2, 75–79.

32. S. Gawiejnowicz, W. Kurc and L. Pankowska, Conjugate problems in time-dependent scheduling. *Journal of Scheduling* **12** (2009), no. 5, 543–553.

Isomorphic scheduling problems

33. P. Baptiste, J. Carlier, A. Kononov, M. Queyranne, S. Sevastianov and M. Sviridenko, Structural properties of optimal preemptive schedules. *Discrete Analysis and Operations Research* **16** (2009), 3–36 (in Russian).

34. M-B. Cheng, G-Q. Wang and L-M. He, Parallel machine scheduling problems with proportionally deteriorating jobs, *International Journal of Systems Science* **40** (2009), no. 1, 53–57.

35. S. Gawiejnowicz and A. Kononov, Isomorphic scheduling problems. *Annals of Operations Research* **213** (2014), no. 1, 131–145.

36. T. Gonzalez and S. Sahni, Open shop scheduling to minimize finish time. *Journal of the ACM* **23** (1976), no. 4, 665–679.

37. R. L. Graham, Bounds for certain multiprocessing anomalies. *Bell System Technology Journal* **45** (1966), no. 9, 156–1581.

38. R. L. Graham, Bounds on multiprocessing timing anomalies. *SIAM Journal on Applied Mathematics* **17** (1969), no. 2, 416–429.

39. S. M. Johnson, Optimal two- and three-stage production schedules with setup times included. *Naval Research Logistics Quarterly* **1** (1954), no. 1, 61–68.

40. R. M. Karp, Reducibility among combinatorial problems. In: R. E. Miller and J. W. Thatcher (eds.), *Complexity of Computer Computations.* New York: Plenum Press, pp. 85–103.

41. A. Kononov, Combinatorial complexity of scheduling jobs with simple linear deterioration. *Discrete Analysis and Operations Research* **3** (1996), no. 2, 15–32 (in Russian).

42. A. Kononov, Single machine scheduling problems with processing times proportional to an arbitrary function. *Discrete Analysis and Operations Research* **5** (1998), 17–37 (in Russian).

43. E. L. Lawler, Optimal sequencing of a single machine subject to precedence constraints. *Management Science* **19** (1973), no. 5, 544–546.

44. E. L. Lawler and J. M. Moore, A functional equation and its applications to resource allocation and sequencing problems. *Management Science* **16** (1969), no. 1, 77–84.

45. J. M. Moore, An n job, one machine sequencing algorithm for minimizing the number of late jobs. *Management Science* **15** (1968), no. 1, 102–109.

46. G. Mosheiov, Complexity analysis of job-shop scheduling with deteriorating jobs. *Discrete Applied Mathematics* **117** (2002), no. 1–3, 195–209.

47. S. Sahni, Algorithms for scheduling independent tasks. *Journal of the ACM* **23** (1976), 116–127.

48. V. Strusevich, Two-machine super-shop scheduling problem. *Journal of the Operational Research Society* **42** (1991), no. 6, 479–492.

20

Bi-criteria time-dependent scheduling

In previous chapters, we considered time-dependent scheduling problems with a *single* optimality criterion. However, advanced scheduling problems require a multi-criteria approach. In this chapter, we consider *bi-criteria* time-dependent scheduling problems, in which a schedule is evaluated with *two* criteria, minimized either in the scalar or in the Pareto sense.

Chapter 20 is composed of five sections. In Sect. 20.1, we introduce two single machine bi-criteria time-dependent scheduling problems considered in the chapter and give some preliminary results. In Sect. 20.2, we address the results concerning bi-criteria Pareto optimality. In Sect. 20.3, we discuss time-dependent scheduling problems with bi-criteria scalar optimality. In Sect. 20.4, we summarize the results of computational experiments related to bi-criteria Pareto optimality. In Sect. 20.5, we present the results of time-dependent flow shop scheduling with hierarchical minimization of two criteria. The chapter is completed with a list of references.

20.1 Preliminaries

In this section, we formulate two problems discussed throughout the chapter. First, we state the assumptions which hold for both problems.

20.1.1 Problems formulation

We are given a single machine and a set of $n + 1$ linearly deteriorating jobs to be processed on the machine. The processing times of jobs are in the form of $p_j = 1 + b_j t$, where $b_j > 0$ for $j = 0, 1, \ldots, n$. All jobs are available for processing at time $t_0 = 0$.

Input data for the problems are described by the sequence $(b_0, b_1, \ldots, b_n)$ of job deterioration rates. For simplicity of further presentation, however, instead of the sequence $(b_0, b_1, \ldots, b_n)$ we use the sequence $\hat{\beta} = (\beta_0, \beta_1, \ldots, \beta_n)$, where

© Springer-Verlag GmbH Germany, part of Springer Nature 2020
S. Gawiejnowicz, *Models and Algorithms of Time-Dependent Scheduling*,
Monographs in Theoretical Computer Science. An EATCS Series,
https://doi.org/10.1007/978-3-662-59362-2_20

$\beta_j = b_j + 1$ for $j = 0, 1, \ldots, n$. (The elements β_j will be called *deterioration coefficients* in order to distinguish them from deterioration rates b_j.)

The first problem is to find such a schedule $\beta^\star$ that the pair

$$\left(\sum_j C_j(\beta^\star), \max_j \{ C_j(\beta^\star) \} \right)$$

of values of the total completion time and the maximum completion time criteria for this schedule is Pareto optimal, i.e.,

$$\beta^\star \equiv \hat{\beta}_{\pi^\star} = \arg \min_{\hat{\beta}_\pi} \left\{ \left(\sum_j C_j(\hat{\beta}_\pi), \max_j \{ C_j(\hat{\beta}_\pi) \} \right) : \pi \in \mathfrak{S}_n \right\},$$

where the minimum, with respect to the order relation $\prec$, is taken in the sense of Pareto optimum.

Remark 20.1. The order relation $\prec$ and the Pareto optimality were introduced in Definition 1.5 and Definition 1.26, respectively.

Let $\| \cdot \|_{(\lambda)}$ denote a convex combination of the $\sum C_j$ and $C_{\max}$ criteria, i.e.,

$$\| C(\hat{\beta}_\pi) \|_{(\lambda)} := \lambda \sum_{j=0}^n C_j(\hat{\beta}_\pi) + (1 - \lambda) \max_{0 \leqslant j \leqslant n} \{ C_j(\hat{\beta}_\pi) \},$$

where $C(\hat{\beta}_\pi) = [C_0(\hat{\beta}_\pi), C_1(\hat{\beta}_\pi), \ldots, C_n(\hat{\beta}_\pi)]$ is the vector of job completion times for a given sequence $\hat{\beta}_\pi$ and $\lambda \in \langle 0, 1 \rangle$ is an arbitrary but fixed number.

The second problem is to find a schedule $\beta^\star$ for which the value of criterion $\| \cdot \|_{(\lambda)}$ is minimal, i.e.,

$$\beta^\star \equiv \hat{\beta}_{\pi^\star} = \arg \min_{\hat{\beta}_\pi} \left\{ \| C(\hat{\beta}_\pi) \|_{(\lambda)} : \pi \in \mathfrak{S}_n \right\},$$

where the minimum is taken with respect to the ordinary relation $\leqslant$.

We refer to the first and second problem as the *TDPS* (Time-Dependent Pareto-optimal Scheduling) and the *TDBS* (Time-Dependent Bi-Criterion Scheduling) problem, respectively. Optimal schedules for these problems will be called *TDPS-optimal* and *TDBS-optimal* schedules, respectively.

20.1.2 Preliminary results

In this section, we prove some preliminary results that will be used in the next two sections. The results refer to both problems under consideration. Introductory results concerning only one of the problems, either TDPS or TDBS, are presented in Sect. 20.2 and Sect. 20.3, respectively.

The following result is a generalization of Property A.2.

Lemma 20.2. (Gawiejnowicz et al. [2]) *Let* $\beta_{\max} = \max\{\beta_j : j = 0, 1, \ldots, n\}$ *for a given sequence* $\hat{\beta}$. *Then in any TDPS-optimal (TDBS-optimal) schedule for* $\hat{\beta}$, *the job corresponding to deterioration coefficient* $\beta_{\max}$ *is scheduled as the first one.*

Proof. First, we consider the TDPS problem. Let us assume that there exists a Pareto optimal schedule β^* for the TDPS problem in which the job corresponding to the greatest deterioration coefficient, $\beta_{\max}$, is not scheduled as the first one, $\beta_{[0]} \neq \beta_{\max}$. Let the first job in β^* have the coefficient β_m, $\beta_{[0]} = \beta_m$, where $\beta_m < \beta_{\max} = \beta_{[k]}$. Let us consider schedule β' obtained by switching in β^* the first job and the kth job. Since the deterioration coefficient of the first scheduled job does not influence the values of the $\sum C_j$ and $C_{\max}$ criteria and, furthermore, by Remark 7.15 both these criteria are monotonically non-decreasing with respect to $\beta_{[j]}$ for $j = 1, 2, \ldots, n$, the above switching will decrease the values of $\sum C_j$ and $C_{\max}$. Thus, in view of Definition 1.26, schedule β^* cannot be optimal. A contradiction.

Let us consider now the TDBS problem. Since the value of criterion $\| \cdot \|_{(\lambda)}$ decreases with the decreasing values of the $\sum C_j$ and $C_{\max}$ criteria, then by applying similar reasoning as above we complete the proof. ∎

Remark 20.3. From now on, we will assume that $\beta_{[0]}$ has been established according to Lemma 20.2 and we will denote sequence $\hat{\beta}$ without the maximal element $\beta_{[0]}$ by $\beta = (\beta_1, \beta_2, \ldots, \beta_n)$.

Remark 20.4. We will assume that $n > 2$, i.e., sequence $\beta = (\beta_1, \beta_2, \ldots, \beta_n)$ contains at least three elements such that $\beta_j > 1$ for $j = 1, \ldots, n$.

Let $\beta(\beta_q \leftrightarrow \beta_r)$ denote sequence β with elements β_q and β_r mutually interchanged. The next preliminary result is the following lemma.

Lemma 20.5. (Gawiejnowicz et al. [2]) *Let* $\beta' = \beta(\beta_q \leftrightarrow \beta_r)$ *and* $1 \leqslant q < r \leqslant n$. *Then for* $0 \leqslant j \leqslant n$ *the following equality holds:*

$$
C_j(\beta') - C_j(\beta) = \begin{cases} \left(\frac{\beta_q - \beta_r}{\beta_r}\right) \sum\limits_{i=q}^{r-1} \prod\limits_{k=i+1}^{j} \beta_k \, , & 1 \leqslant q < r \leqslant j \leqslant n, \\[2ex] \left(\frac{\beta_r - \beta_q}{\beta_q}\right) \sum\limits_{i=0}^{q-1} \prod\limits_{k=i+1}^{j} \beta_k \, , & 1 \leqslant q \leqslant j < r \leqslant n, \\[2ex] 0 & , \ 0 \leqslant j < q < r \leqslant n. \end{cases}
$$

Proof. The case $j = 0$ is clear, since $C_0(\beta) = C_0(\beta') = 1$. In the case when $0 < j < q < r \leqslant n$ we have $C_j(\beta) = C_j(\beta') = \sum\limits_{i=0}^{j} \prod\limits_{k=i+1}^{j} \beta_k$, since $j < q$.

Let $r \leqslant j \leqslant n$ and $1 \leqslant q < r \leqslant n$. From formula (7.9) for $a_{\sigma_i} = 1$ for $1 \leqslant i \leqslant n$ it follows that

$$
C_j(\beta) = \sum_{i=0}^{q-1} \beta_{i+1} \ldots \beta_q \ldots \beta_r \ldots \beta_j + \sum_{i=q}^{r-1} \beta_{i+1} \ldots \beta_r \ldots \beta_j + \sum_{i=r}^{j} \beta_{i+1} \ldots \beta_j.
$$

Next, in the corresponding formula for $C_j(\beta')$ we have β_q and β_r mutually interchanged in the first sum. Clearly, in this case the first sum remains unchanged. In the second sum in $C_j(\beta')$ factor β_r must be replaced by factor β_q. Finally, in the third sum we have no changes related to the transition from β to β'. Therefore, for $1 \leqslant q < r \leqslant j \leqslant n$, the equality

$$C_j(\beta') - C_j(\beta) = \left(\frac{\beta_q - \beta_r}{\beta_r}\right) \sum_{i=q}^{r-1} \prod_{k=i+1}^{j} \beta_k$$

holds. The case when $1 \leqslant q \leqslant j < r \leqslant n$ can be proved in a similar way. ∎

Lemma 20.6. (Gawiejnowicz et al. [1, 2]) *Let* $\beta' = \beta(\beta_q \leftrightarrow \beta_{q+1})$*, where* $q = 1, 2, \ldots, n-1$*. Then for* $j = 0, 1, \ldots, n$ *the following equality holds:*

$$C_j(\beta') - C_j(\beta) = \begin{cases} (\beta_q - \beta_{q+1}) \displaystyle\prod_{k=q+2}^{j} \beta_k & , 1 \leqslant q < j \leqslant n, \\ (\beta_{q+1} - \beta_q) \displaystyle\sum_{i=0}^{q-1} \prod_{k=i+1}^{q-1} \beta_k & , j = q, \\ 0 & , 0 \leqslant j < q. \end{cases}$$

Proof. The result follows from Lemma 20.5 by letting $r = q + 1$. ∎

Lemma 20.7. (Gawiejnowicz et al. [1, 2]) *Let* $\beta' = \beta(\beta_q \leftrightarrow \beta_{q+1})$*, where* $q = 1, 2, \ldots, n-1$*. Then the following equality holds:*

$$\sum_{j=0}^{n} C_j(\beta') - \sum_{j=0}^{n} C_j(\beta) = (\beta_{q+1} - \beta_q) \left(\sum_{j=0}^{q-1} \prod_{k=j+1}^{q-1} \beta_k - \sum_{i=q+1}^{n} \prod_{k=q+2}^{i} \beta_k \right).$$

Proof. By Lemma 20.6, summing the differences $C_j(\beta') - C_j(\beta)$ for $0 \leqslant j \leqslant n$, we obtain the result. ∎

Lemma 20.8. (Gawiejnowicz et al. [1, 2]) *Let* $\beta' = \beta(\beta_q \leftrightarrow \beta_{q+1})$*, where* $q = 1, 2, \ldots, n-1$*. Then the following equality holds:*

$$\max_{0 \leqslant j \leqslant n} \{C_j(\beta')\} - \max_{0 \leqslant j \leqslant n} \{C_j(\beta)\} = (\beta_q - \beta_{q+1}) \prod_{k=q+2}^{n} \beta_k.$$

Proof. Since $\max_{0 \leqslant j \leqslant n} \{C_j(\beta)\} = C_n(\beta)$, by letting $j = n$ in Lemma 20.6, the result follows. ∎

20.2 Pareto optimality

In this section, we consider the TDPS problem, i.e. the problem of finding a schedule which is Pareto optimal with respect to the $\sum C_j$ and $C_{\max}$ criteria.

20.2.1 Basic definitions

Let X denote a set of all solutions of a bi-criterion scheduling problem. In our case, $X = \{\hat{\beta}_\pi : \pi \in \mathfrak{S}_n\}$ is discrete and consists of all permutations of the original sequence $\hat{\beta}$. Let us also recall (see Definition 1.26) that for a given bi-criterion optimization problem, X_{Par} ($X_{\text{w-Par}}$) denotes the set of all Pareto (weak Pareto) optimal solutions.

Let us notice that in view of Lemmas 20.7 and 20.8, if $\beta' = \beta(\beta_q \leftrightarrow \beta_{q+1})$ is a pairwise transposition of sequence β, where $q = 1, 2, \ldots, n-1$, we have:

$$\|C(\beta')\|_1 - \|C(\beta)\|_1 = (\beta_{q+1} - \beta_q) \left(\sum_{j=0}^{q-1} \prod_{k=j+1}^{q-1} \beta_k - \sum_{i=q+1}^{n} \prod_{k=q+2}^{i} \beta_k \right) \quad (20.1)$$

and

$$\|C(\beta')\|_\infty - \|C(\beta)\|_\infty = -(\beta_{q+1} - \beta_q) \prod_{k=q+2}^{n} \beta_k. \quad (20.2)$$

20.2.2 Main results

First, we prove a sufficient condition for $\beta^\star \in X$ to be a (weakly) Pareto optimal solution to the TDPS problem. (A necessary condition for $\beta \in X$ to be a (weakly) Pareto optimal solution to the TDPS problem will be given later.)

Theorem 20.9. (Gawiejnowicz et al. [2]) (a) *A sufficient condition for sequence* $\beta^\star \in X$ *to be weakly Pareto optimal is that* $\beta^\star$ *is optimal with respect to the scalar criterion* $\|\cdot\|_{(\lambda)}$, *where* $0 \leq \lambda \leq 1$. (b) *A sufficient condition for sequence* $\beta^\star \in X$ *to be Pareto optimal is that* $\beta^\star$ *is optimal with respect to the scalar criterion* $\|\cdot\|_{(\lambda)}$, *where* $0 \leq \lambda < 1$. *In particular, the sequence obtained by non-increasing ordering of* $\beta^\star$ *is Pareto optimal for the TDPS problem.*

Proof. (a) The statement immediately follows from inclusion $X_{(\lambda)} \subset X_{\text{w-Par}}$, where $0 \leq \lambda \leq 1$.

(b) The statement follows from inclusion $X_{(\lambda)} \subset X_{\text{Par}}$, whenever $0 < \lambda < 1$. To end the proof it is sufficient to consider the case $\lambda = 0$, i.e., the case of criterion $\|\cdot\|_\infty$. Let $\beta^\star \in X$ be non-increasing. The sequence is Pareto optimal when relation

$$(\|C(\beta')\|_1, \|C(\beta')\|_\infty) \prec (\|C(\beta^\star)\|_1, \|C(\beta^\star)\|_\infty)$$

does not hold for any $\beta' \in X, \beta' \neq \beta^\star$. By Lemma 1.6, this means that either the disjunction

$$\|C(\beta')\|_1 - \|C(\beta^\star)\|_1 > 0 \quad \text{or} \quad \|C(\beta')\|_\infty - \|C(\beta^\star)\|_\infty > 0 \quad (20.3)$$

or the conjuction

$$\|C(\beta')\|_1 = \|C(\beta^\star)\|_1 \text{ and } \|C(\beta')\|_\infty = \|C(\beta^\star)\|_\infty \qquad (20.4)$$

holds. If $\beta^\star$ contains only distinct elements, then $\|C(\beta')\|_\infty - \|C(\beta^\star)\|_\infty > 0$ and the disjunction (20.3) holds. In the opposite case, either $\|C(\beta')\|_\infty - \|C(\beta^\star)\|_\infty > 0$, and then the disjunction (20.3) holds, or $\|C(\beta')\|_\infty - \|C(\beta^\star)\|_\infty = 0$, and then the conjuction (20.4) holds, since in this case we also have $\|C(\beta')\|_1 - \|C(\beta^\star)\|_1 = 0$ by the uniqueness of $\beta^\star$ up to the order of equal elements.

Concluding, a non-increasing $\beta^\star \in X$ must be Pareto optimal for the TDPS problem. ∎

Example 20.10. Let us consider sequence $\hat{\beta} = (6, 3, 4, 5, 2)$. Then $\sum C_j(\hat{\beta}) = 281$ and $C_{\max}(\hat{\beta}) = 173$. By Theorem 20.9 we know that each non-increasing sequence is Pareto optimal for the TDPS problem. Thus $\beta' = (6, 5, 4, 3, 2)$ is Pareto optimal for this problem, with $\sum C_j(\beta') = 261$ and $C_{\max}(\beta') = 153$. ♦

Now we prove the necessary condition for $\beta^\star \in X$ to be a (weakly) Pareto optimal solution to the TDPS problem. We start with the following result.

Lemma 20.11. (Gawiejnowicz et al. [2]) *Let*

$$\Delta_q(\beta) := \sum_{j=0}^{q-1} \prod_{k=j+1}^{q-1} \beta_k - \sum_{i=q+1}^{n} \prod_{k=q+2}^{i} \beta_k$$

for a given sequence β. Then for any permutation of β there exists a unique number q_0, $1 \leqslant q_0 \leqslant n - 1$, such that q_0 is the greatest number for which $\Delta_q(\beta)$ $q = q_0$ is negative, i.e., $\Delta_q(\beta) < 0$ for $1 \leqslant q \leqslant q_0$ and $\Delta_q(\beta) \geqslant 0$ for $q_0 < q \leqslant n - 1$.

Proof. First, note that

$$\Delta_1(\beta) = -\sum_{i=3}^{n} \prod_{k=3}^{i} \beta_k < 0$$

and

$$\Delta_{n-1}(\beta) = \sum_{j=0}^{n-3} \prod_{k=j+1}^{n-2} \beta_k > 0.$$

Moreover, sequence $\Delta_q(\beta)$ is strictly increasing for $q = 1, 2, \ldots, n - 1$, since

$$\Delta_{q+1}(\beta) - \Delta_q(\beta) = \sum_{j=0}^{q} \prod_{k=j+1}^{q} \beta_k - \sum_{i=q+2}^{n} \prod_{k=q+3}^{i} \beta_k - \sum_{j=0}^{q-1} \prod_{k=j+1}^{q-1} \beta_k +$$

$$+ \sum_{i=q+1}^{n} \prod_{k=q+2}^{i} \beta_k$$

$$= (\beta_q - 1) \sum_{j=0}^{q-1} \prod_{k=j+1}^{q-1} \beta_k + (\beta_{q+2} - 1) \sum_{i=q+2}^{n} \prod_{k=q+3}^{i} \beta_k + 2 > 0.$$

Thus there must exist a maximal integer q_0 such that $\Delta_q(\beta) < 0$ for $1 \leqslant q \leqslant q_0$ and $\Delta_q(\beta) \geqslant 0$ for $q_0 < q \leqslant n - 1$. ∎

Now, assume that $\beta^\star$ is a Pareto optimal solution of the TDPS problem, $\beta^\star \in X_{\mathrm{Par}}$. Then, by Lemma 20.11, there exists a $q_0^\star$ such that

$$\Delta_q(\beta^\star) < 0 \quad \text{for} \quad 1 \leqslant q \leqslant q_0^\star \tag{20.5}$$

and

$$\Delta_q(\beta^\star) \geqslant 0 \quad \text{for} \quad q_0^\star < q \leqslant n - 1. \tag{20.6}$$

Knowing that there exists a $q_0^\star$ which is the point of change of sign of $\Delta_q(\beta^\star)$, we can prove the following result.

Theorem 20.12. (Gawiejnowicz et al. [2]) *Let* $\beta^\star = (\beta_1^\star, \beta_2^\star, \ldots, \beta_n^\star) \in X_{\mathrm{Par}}$ *and* $q_0^\star$ *be specified by conditions* (20.5) *and* (20.6). *Then for* $q = 1, 2, \ldots, n-1$ *the inequality* $\beta_q^\star \geqslant \beta_{q+1}^\star$ *or, if* $q_0^\star < q \leqslant n - 1$, *the inequality* $\beta_q^\star \leqslant \beta_{q+1}^\star$ *holds.*

Proof. Let $\beta^\star = (\beta_1^\star, \beta_2^\star, \ldots, \beta_n^\star)$ be Pareto optimal solution of the TDPS problem. Then there does not exist $\beta' \in X$ such that

$$(\|C(\beta')\|_1, \|C(\beta')\|_\infty) \prec (\|C(\beta^\star)\|_1, \|C(\beta^\star)\|_\infty)$$

or, equivalently, for each $\beta' \in X$ there does not hold the relation

$$(\|C(\beta')\|_1 - \|C(\beta^\star)\|_1, \|C(\beta')\|_\infty - \|C(\beta^\star)\|_\infty) \prec 0.$$

In particular, this relation does not hold for transpositions $\beta' = \beta^\star(\beta_q^\star \leftrightarrow \beta_{q+1}^\star)$ of the optimal sequence for $q = 1, 2, \ldots, n - 1$. Now, applying Lemma 1.6, (20.1) and (20.2), we see that for $q = 1, 2, \ldots, n - 1$ either the disjunction

$$\|C(\beta')\|_1 - \|C(\beta^\star)\|_1 = (\beta_{q+1}^\star - \beta_q^\star)\, \Delta_q(\beta^\star) > 0 \tag{20.7}$$

or

$$\|C(\beta')\|_\infty - \|C(\beta^\star)\|_\infty = -(\beta_{q+1}^\star - \beta_q^\star) \prod_{k=q+2}^{n} \beta_k^\star > 0 \tag{20.8}$$

or the conjuction

$$\|C(\beta')\|_1 - \|C(\beta^\star)\|_1 = 0 \text{ and } \|C(\beta')\|_\infty - \|C(\beta^\star)\|_\infty = 0 \tag{20.9}$$

hold. If $\beta^\star$ contains distinct elements only, then the conjuction (20.9) cannot be satisfied. Thus there must hold the disjunction (20.7) or (20.8). Hence, for $q = 1, 2, \ldots, n - 1$, the inequality $\beta_q^\star > \beta_{q+1}^\star$, if $q_0^\star < q \leqslant n - 1$, the inequality $\beta_q^\star < \beta_{q+1}^\star$ holds.

If not all elements of $\beta^\star$ are distinct, then apart from the disjunction (20.7) or (20.8) also the conjuction (20.9) can be satisfied. Hence for $q = 1, 2, \ldots, n-1$ there holds the inequality $\beta_q^\star \geqslant \beta_{q+1}^\star$ or, if $q_0^\star < q \leqslant n - 1$, the inequality $\beta_q^\star \leqslant \beta_{q+1}^\star$. $\blacksquare$

We now introduce a definition (cf. Gawiejnowicz et al. [2]) that allows us to formulate the previous result in a more concise way.

Definition 20.13. (A weakly V-shaped sequence with respect to $\Delta_q(\beta)$)
A sequence $\beta = (\beta_1, \ldots, \beta_n)$ *is said* to have a weak V-shape (to be weakly
V-shaped) with respect to $\Delta_q(\beta)$ *if* β *is non-increasing for indices* q *for which
inequality* $\Delta_q(\beta) < 0$ *holds.*

Let us notice that from Definition 20.13 and from properties of function $\Delta_q(\beta)$ it follows that, in general, weakly V-shaped sequences are non-
increasing for $1 \leqslant q \leqslant q_0$ and can vary in an arbitrary way for $q_0 < q \leqslant n-1$,
with appropriate $1 \leqslant q_0 \leqslant n-1$.

Applying Definition 20.13, we can now reformulate Theorem 20.12 as follows.

Theorem 20.12′. (Gawiejnowicz et al. [2]) *A necessary condition for sequence*
$\beta^\star \in X$ *to be a Pareto optimal solution to the TDPS problem is that* $\beta^\star$ *must
be weakly V-shaped with respect to* $\Delta_q(\beta^\star)$.

We illustrate applications of Theorem 20.12′ by two examples (cf. [2]).

Example 20.14. Let sequence $\hat{\beta} = (\beta_0, \beta_1, \ldots, \beta_n)$ be such that $\beta_0 = \max_i\{\beta_i\}$
and $\beta_1 = \min_i\{\beta_i\}$. Since $\Delta_1(\hat{\beta}) < 0$ (by Lemma 20.11), we have $q_0 \geqslant 1$ and by
Theorem 20.12′ no sequence which is in the form of $(\beta_0, \beta_1, \beta_{\pi_2}, \beta_{\pi_3}, \ldots, \beta_{\pi_n})$,
where $\beta_{\pi_i} \in \{\beta_2, \beta_3, \ldots, \beta_n\}$ for $2 \leqslant i \leqslant n$ and $\pi_i \neq \pi_j$ for $i \neq j$, can be
Pareto optimal. $\blacklozenge$

Example 20.15. Let $\hat{\beta} = (7, 3, 2, 4, 5, 6)$. In order to check if $\hat{\beta}$ can be a
solution to the TDPS problem, we must determine the value of q_0. Af-
ter calculations we have: $\Delta_1(\hat{\beta}) = -144$, $\Delta_2(\hat{\beta}) = -32$, $\Delta_3(\hat{\beta}) = 2$.
Hence $q_0 = 2$ and sequence $\hat{\beta}$, according to Theorem 20.12′, cannot be
a Pareto optimal solution to the TDPS problem. Moreover, all schedules
which are in the form of $(7, 3, 2, \beta_{\pi_3}, \beta_{\pi_4}, \beta_{\pi_5})$, where $\beta_{\pi_i} \in \{4, 5, 6\}$ for
$3 \leqslant i \leqslant 5$ and $\pi_3 \neq \pi_4 \neq \pi_5$, cannot be Pareto optimal, either. There are six
such schedules: $(7, 3, 2, 4, 5, 6)$, $(7, 3, 2, 4, 6, 5)$, $(7, 3, 2, 5, 4, 6)$, $(7, 3, 2, 5, 6, 4)$,
$(7, 3, 2, 6, 4, 5)$ and $(7, 3, 2, 6, 5, 4)$. $\blacklozenge$

20.3 Scalar optimality

In this section, we consider the TDBS problem of finding a schedule which is
optimal for criterion $\| \cdot \|_{(\lambda)}$.

20.3.1 Basic definitions

Let us recall that criterion $\| \cdot \|_{(\lambda)}$ is a convex combination of the total com-
pletion time $\sum C_j$ and the maximum completion time $C_{\max}$, i.e.,

$$\|C(\beta)\|_{(\lambda)} := \lambda \sum_{j=0}^{n} C_j(\beta) + (1 - \lambda) \max_{0 \leqslant j \leqslant n} \{C_j(\beta)\}, \tag{20.10}$$

where $\lambda \in \langle 0, 1 \rangle$ is arbitrary but fixed. Our aim is to find such a sequence $\beta^\star$ for which the value of $\|C(\beta^\star)\|_{(\lambda)}$ is minimal.

Remark 20.16. Let us notice that if we are interested in minimizing the combination of the $\sum C_j$ and $C_{\max}$ criteria with arbitrary weights, criterion $\| \cdot \|_{(\lambda)}$ is general enough, since for any real numbers $\alpha > 0, \beta > 0$ we have

$$\alpha \sum C_j + \beta C_{\max} = (\alpha + \beta) \left(\tfrac{\alpha}{\alpha+\beta} \sum C_j + \tfrac{\beta}{\alpha+\beta} C_{\max} \right) = (\alpha + \beta)\| \cdot \|_{(\lambda)} \text{ with}$$

$\lambda = \frac{\alpha}{\alpha+\beta}$.

Remark 20.17. Notice also that since criterion $\| \cdot \|_{(\lambda)}$ is a convex combination of the $\sum C_j$ and $C_{\max}$ criteria, which are particular cases of the l_p norm (cf. Definition 1.19), $\| \cdot \|_{(\lambda)}$ is also a norm.

In other words, the sequence $\beta^\star = (\beta_1^\star, \beta_2^\star, \ldots, \beta_n^\star)$ is optimal for the problem TDBS if $\|C(\beta^\star)\|_{(\lambda)} = \min\{\|C(\beta_\pi\|_{(\lambda)} : \pi \in \mathfrak{S}_n\}$, i.e., for all $\pi \in \mathfrak{S}_n$ the inequality

$$0 \leqslant \|C(\beta_\pi)\|_{(\lambda)} - \|C(\beta^\star)\|_{(\lambda)} \tag{20.11}$$

holds.

Remark 20.18. Let us notice that we can modify (20.10); since $C(\beta)$ is non-decreasing, we have $\max\limits_{1 \leqslant j \leqslant n} \{C_j(\beta)\} = C_n(\beta)$. Thus, to define the norm $\| \cdot \|_{(\lambda)}$ we can also use the formula $\|C(\beta)\|_{(\lambda)} = \lambda \sum_{j=0}^{n-1} C_j(\beta) + C_n(\beta)$.

20.3.2 Main results

In view of the form of $\| \cdot \|_{(\lambda)}$, some relations between this criterion and the $\sum C_j$ and $C_{\max}$ criteria hold.

Lemma 20.19. (Gawiejnowicz et al. [5]) *The following inequalities hold:*
(a) $\|C(\beta)\|_{(\lambda)} - \|C(\beta)\|_\infty \leqslant \lambda(n-1)\|C(\beta)\|_\infty$,
(b) $0 \leqslant \frac{\|C(\beta)\|_{(\lambda)} - \|C(\beta)\|_\infty}{\|C(\beta)\|_\infty} \leqslant \lambda(n-1)$,
(c) $0 \leqslant \frac{\|C(\beta)\|_1 - \|C(\beta)\|_{(\lambda)}}{\|C(\beta)\|_1} \leqslant (1 - \lambda)\frac{n-1}{n}$.

Proof. (a) By (20.10) and since $\|C\|_\infty \leqslant \|C\|_1 \leqslant n\|C\|_\infty$, the inequality follows.

(b) The inequalities follow from (20.10) and from (a).

(c) Similar to the proof of (a). □

Theorem 20.20. (Gawiejnowicz et al. [5]) *If $\lambda \in \langle 0, 1 \rangle$ and n is any natural number, then the inequalities*

$$\|C(\beta)\|_{(\lambda)} \leqslant \|C(\beta)\|_1 \leqslant \frac{n}{1 + \lambda(n-1)}\|C(\beta)\|_{(\lambda)}$$

hold.

Proof. The result follows from Lemma 20.19. □

From (20.10), Lemma 20.7 and Lemma 20.8 we get a formula describing the behaviour of $\|C(\beta)\|_{(\lambda)}$ under transpositions $\beta' = \beta(\beta_q \leftrightarrow \beta_{q+1})$.

Theorem 20.21. (Gawiejnowicz et al. [6]) *Let* $\beta' = \beta(\beta_q \leftrightarrow \beta_{q+1})$. *Then for* $q = 1, 2, \ldots, n-1$ *the following equality holds:*

$$\|C(\beta')\|_{(\lambda)} - \|C(\beta)\|_{(\lambda)} =$$

$$(\beta_{q+1} - \beta_q) \left(\lambda \left(\sum_{j=0}^{q-1} \prod_{k=j+1}^{q-1} \beta_k - \sum_{j=q+1}^{n-1} \prod_{k=q+2}^{j} \beta_k \right) - \prod_{k=q+2}^{n} \beta_k \right).$$

Proof. From (20.10) it follows that

$$\|C(\beta')\|_{(\lambda)} - \|C(\beta)\|_{(\lambda)} = \lambda \left(\sum_{j=0}^{n} C_j(\beta') - \sum_{j=0}^{n} C_j(\beta) \right) +$$

$$(1 - \lambda) \left(\max_{0 \leqslant j \leqslant n} \{C_j(\beta')\} - \max_{0 \leqslant j \leqslant n} \{C_j(\beta)\} \right).$$

Consequently, in view of Lemma 20.7 and Lemma 20.8, we obtain

$$\|C(\beta')\|_{(\lambda)} - \|C(\beta)\|_{(\lambda)} = \lambda(\beta_{q+1} - \beta_q) \left(\sum_{j=0}^{q-1} \prod_{k=j+1}^{q-1} \beta_k - \sum_{j=q+1}^{n-1} \prod_{k=q+2}^{j} \beta_k \right) +$$

$$(1 - \lambda)(\beta_q - \beta_{q+1}) \prod_{k=q+2}^{n} \beta_k =$$

$$= (\beta_{q+1} - \beta_q) \left(\lambda \left(\sum_{j=0}^{q-1} \prod_{k=j+1}^{q-1} \beta_k - \sum_{j=q+1}^{n-1} \prod_{k=q+2}^{j} \beta_k \right) - \prod_{k=q+2}^{n} \beta_k \right). \qquad \blacksquare$$

Now, we will prove that for infinitely many values of the parameter $\lambda \in \langle 0, \lambda_0 \rangle$, for some $0 < \lambda_0 < 1$, the TDBS problem can be solved in $O(n \log n)$ time. We will also show that there exist infinitely many values of $\lambda \in \langle \lambda_1, 1 \rangle$, for some λ_1, where $\lambda_0 < \lambda_1 < 1$, such that the optimal schedule for this problem has a V-shape.

Let $q = 1, 2, \ldots, n-1$ and let $\lambda \in \langle 0, 1 \rangle$ be arbitrary but fixed. For a given $\beta = (\beta_1, \beta_2, \ldots, \beta_n)$, define function $\Lambda_q(\lambda)$ as follows:

$$\Lambda_q(\lambda) := \lambda \left(\sum_{j=0}^{q-1} \prod_{k=j+1}^{q-1} \beta_k - \sum_{j=q+1}^{n-1} \prod_{k=q+2}^{j} \beta_k \right) - \prod_{k=q+2}^{n} \beta_k. \qquad (20.12)$$

The behaviour of function $\Lambda_q(\lambda)$ is crucial for further considerations. We begin with a necessary condition for the sequence $\beta = (\beta_1, \beta_2, \ldots, \beta_n)$ to be optimal with respect to criterion $\| \cdot \|_{(\lambda)}$.

Lemma 20.22. (Gawiejnowicz et al. [6]) *Let sequence* $\beta^\star = (\beta_1^\star, \beta_2^\star, \ldots, \beta_n^\star)$ *be optimal with respect to criterion* $\| \cdot \|_{(\lambda)}$ *and let* $\beta' = \beta^\star (\beta_q^\star \leftrightarrow \beta_{q+1}^\star)$. *Then for* $q = 1, 2, \ldots, n-1$ *the following inequality holds:*

$$0 \leqslant \|C(\beta')\|_{(\lambda)} - \|C(\beta^\star)\|_{(\lambda)} = (\beta_{q+1}^\star - \beta_q^\star)\,\Lambda_q(\lambda). \qquad (20.13)$$

Proof. In view of (20.11), (20.12) and Theorem 20.21, the result follows. $\square$

In view of Lemma 20.22, it is important to know the behaviour of the sign of function $\Lambda_q(\lambda)$ for $q = 1, 2, \ldots, n-1$ and $\lambda \in \langle 0, 1 \rangle$, since then, by (20.13), we can control the sign of difference $\|C(\beta')\|_{(\lambda)} - \|C(\beta)\|_{(\lambda)}$.

Let us notice that $\Lambda_1(\lambda)$ is always strictly less than 0, while the sign of $\Lambda_q(\lambda)$, for $q = 2, \ldots, n-1$, depends on λ. In fact, from definition of $\Lambda_q(\lambda)$ we obtain the following lemma.

Lemma 20.23. (Gawiejnowicz et al. [6]) *Let* $\lambda \in \langle 0, 1 \rangle$ *be arbitrary but fixed. Then* $\Lambda_1(\lambda) < 0$ *and the following inequalities hold:*

$$\Lambda_1(\lambda) \leqslant \Lambda_2(\lambda) \leqslant \ldots \leqslant \Lambda_{n-1}(\lambda).$$

Proof. Indeed, these inequalities hold since for $q = 1, 2, \ldots, n-1$ we have

$$\Lambda_{q+1}(\lambda) - \Lambda_q(\lambda) = \lambda \left((\beta_q - 1) \sum_{j=0}^{q-1} \prod_{k=j+1}^{q-1} \beta_k + 1 \right) +$$

$$\lambda \left((\beta_{q+2} - 1) \sum_{j=q+2}^{n-1} \prod_{k=q+3}^{j} \beta_k + 1 \right) +$$

$$\prod_{k=q+3}^{n} \beta_k\,(\beta_{q+2} - 1) \geqslant 0.$$

To end the proof, it is sufficient to note that for $q = 1$ the equality

$$\Lambda_1(\lambda) = -\lambda \sum_{j=3}^{n-1} \prod_{k=3}^{j} \beta_k - \prod_{k=3}^{n} \beta_k < 0$$

holds. $\blacksquare$

In view of these results, given the sequence $\beta = (\beta_1, \beta_2, \ldots, \beta_n)$, the fundamental problem is to determine λ_0 and λ_1, $0 < \lambda_0 < \lambda_1 < 1$, such that $\Lambda_q(\lambda) \leqslant 0$ for all $\lambda \in \langle 0, \lambda_0 \rangle$ and $q = 1, 2, \ldots, n-1$, and $\Lambda_{n-1}(\lambda) \geqslant 0$ for all $\lambda \in \langle \lambda_1, 1 \rangle$. In the first case, sequence $\Lambda_q(\lambda)$ has only non-positive elements and the non-increasing ordering of sequence β is, by (20.13), a necessary condition for optimality of β. In the second case, there is the change of sign in sequence $\Lambda_q(\lambda)$, and the sequence β must have a V-shape.

We now prove the following result, in which strongly restrictive formulae for λ_0 and λ_1 are used. We start with a definition (cf. [6]).

Definition 20.24. (Numbers λ_0 and λ_1)
Let $\bar{\beta} := \max\{\beta_1, \beta_2, \ldots, \beta_n\}$ and $\underline{\beta} := \min\{\beta_1, \beta_2, \ldots, \beta_n\}$. Define λ_0 and λ_1 as follows:

$$\lambda_0 := \frac{\bar{\beta} - 1}{\bar{\beta}^{n-1} - 1} \tag{20.14}$$

and

$$\lambda_1 := \frac{\underline{\beta} - 1}{\underline{\beta}^{n-1} - 1}. \tag{20.15}$$

Lemma 20.25. (Gawiejnowicz et al. [6]) *Let there be given a sequence $\beta = (\beta_1, \beta_2, \ldots, \beta_n)$. Then for $q = 1, 2, \ldots, n-1$ the inequality $\Lambda_q(\lambda) \leqslant 0$ holds, where $\lambda \in \langle 0, \lambda_0 \rangle$ and $\lambda_0 > 0$ is defined by (20.14).*

Proof. Let us notice that function $\Lambda_q(\lambda)$ is non-decreasing for each $\lambda \in \langle 0, 1 \rangle$ and $q = 1, 2, \ldots, n-1$. Therefore, $\Lambda_q(\lambda) \leqslant 0$ for $q = 1, 2, \ldots, n-1$ if and only if $\Lambda_{n-1}(\lambda) \leqslant 0$. This, in turn, is equivalent to

$$0 \leqslant \lambda \leqslant \frac{1}{\sum_{j=0}^{n-2} \prod_{k=j+1}^{n-2} \beta_k}.$$

Since

$$\sum_{j=0}^{n-2} \prod_{k=j+1}^{n-2} \beta_k \leqslant \sum_{i=0}^{n-2} \bar{\beta}^i = \frac{\bar{\beta}^{n-1} - 1}{\bar{\beta} - 1} \equiv \frac{1}{\lambda_0},$$

it is sufficient for $\Lambda_q(\lambda) \leqslant 0$ that $\lambda \in \langle 0, \lambda_0 \rangle$, where $q = 1, 2, \ldots, n-1$. ∎

Lemma 20.26. (Gawiejnowicz et al. [6]) *Let there be given a sequence $\beta = (\beta_1, \beta_2, \ldots, \beta_n)$. Then for each $\lambda \in \langle \lambda_1, 1 \rangle$ the inequality $\Lambda_{n-1}(\lambda) \geqslant 0$ holds, where $\lambda \in \langle \lambda_1, 1 \rangle$ and $\lambda_1 < 1$ is defined by formula (20.15).*

Proof. $\Lambda_{n-1}(\lambda) \geqslant 0$ if and only if

$$\lambda \geqslant \frac{1}{\sum_{j=0}^{n-2} \prod_{k=j+1}^{n-2} \beta_k}.$$

Since the equality

$$\sum_{j=0}^{n-2} \prod_{k=j+1}^{n-2} \beta_k \geqslant \sum_{i=0}^{n-2} \underline{\beta}^i = \frac{\underline{\beta}^{n-1} - 1}{\underline{\beta} - 1} \equiv \frac{1}{\lambda_1}$$

holds, it is sufficient for $\Lambda_{n-1}(\lambda) \geqslant 0$ that $\lambda \in \langle \lambda_1, 1 \rangle$. ∎

The following result is a corollary from Lemmas 20.22, 20.25 and 20.26.

Theorem 20.27. (Gawiejnowicz et al. [6]) *Let sequence $\beta^\star = (\beta_1^\star, \beta_2^\star, \ldots, \beta_n^\star)$ be optimal with respect to criterion $\|\cdot\|_{(\lambda)}$ and let λ_0 and λ_1 be defined by formulae (20.14) and (20.15), respectively. Then $0 < \lambda_0 \leqslant \lambda_1 < 1$ and the following implications hold:*
(a) *if $\lambda \in \langle 0, \lambda_0 \rangle$, then $\beta^\star$ is non-increasing;*
(b) *if $\lambda \in \langle \lambda_1, 1 \rangle$, then $\beta^\star$ is V-shaped.*
Moreover, if sequence $\beta^\star$ contains distinct elements, then $\lambda_0 < \lambda_1$.

Proof. (a) Let sequence $\beta^\star = (\beta_1^\star, \beta_2^\star, \ldots, \beta_n^\star)$ be optimal with respect to criterion $\|\cdot\|_{(\lambda)}$, let λ_0 be defined by formula (20.14) and $\lambda \in \langle 0, \lambda_0 \rangle$ be arbitrary but fixed. Then, by Lemma 20.25, for $q = 1, 2, \ldots, n-1$ the inequality $\Lambda_q(\lambda) \leqslant 0$ holds. But this means, by Lemma 20.22, that for $q = 1, 2, \ldots, n-1$ we have $\beta_{q+1}^\star - \beta_q^\star \leqslant 0$. Hence the sequence $\beta^\star$ is non-increasing.
(b) Let sequence $\beta^\star = (\beta_1^\star, a_2^\star, \ldots, \beta_n^\star)$ again be optimal with respect to criterion $\|\cdot\|_{(\lambda)}$, let λ_1 be defined by (20.15) and $\lambda \in \langle \lambda_1, 1 \rangle$ be arbitrary but fixed. Then, by Lemma 20.26, the inequality $\Lambda_{n-1}(\lambda) \geqslant 0$ holds. But we know, by Lemma 20.23, that $\Lambda_1(\lambda) < 0$ and the sequence $\Lambda_q(\lambda)$, for $q = 1, 2, \ldots, n-1$, is non-decreasing. Hence there must exist $1 < r < n-1$ such that $\Lambda_{r-1}(\lambda) \leqslant 0$ but $\Lambda_r(\lambda) \geqslant 0$. But this implies, by Lemma 20.22, that for $q = 1, 2, \ldots, r-1$ the inequality $\beta_{q+1}^\star - \beta_q^\star \leqslant 0$ holds and for $q = r, r+1, \ldots, n-1$ the inequality $\beta_{q+1}^\star - \beta_q^\star \geqslant 0$ holds. Thus the sequence $\beta^\star$ must be V-shaped.
To end the proof, it is sufficient to notice that if $\beta^\star$ contains distinct elements, then $\underline{\beta}^\star \neq \bar{\beta}^\star$ and $\lambda_0 < \lambda_1$. ∎

We can formulate a stronger version of Theorem 20.27, which gives more precise conditions for the monotonicity and the V-shapeness of the optimal sequence $\beta^\star$. The version requires, however, $O(n \log n)$ additional operations to determine the respective values of λ_0 and λ_1.
Before we prove the main result, we introduce a definition (cf. [6]).

Definition 20.28. (Numbers $\lambda(\beta)$, $\lambda_\bullet$ and $\lambda^\bullet$)
Given a sequence $\beta = (\beta_1, \beta_2, \ldots, \beta_n)$, define $\lambda(\beta)$ as follows:

$$\lambda(\beta) := \frac{1}{\displaystyle\sum_{j=0}^{n-2} \prod_{k=j+1}^{n-2} \beta_k}. \tag{20.16}$$

Moreover, let

$$\lambda_\bullet := \min_{\pi \in \mathfrak{S}_n} \{\lambda(\beta_\pi)\}$$

and

$$\lambda^\bullet := \max_{\pi \in \mathfrak{S}_n} \{\lambda(\beta_\pi)\}.$$

Theorem 20.29. (Gawiejnowicz et al. [6]) *Let sequence $\beta^\star = (\beta_1^\star, \beta_2^\star, \ldots, \beta_n^\star)$ be optimal with respect to criterion $\|\cdot\|_{(\lambda)}$, and let λ_0 and λ_1 be defined by formulae (20.14) and (20.15), respectively. Then the following implications hold:*

(a) *if* $\lambda \in \langle 0, \lambda_\bullet \rangle$, *then* $\beta^\star$ *is non-increasing;*
(b) *if* $\lambda \in \langle \lambda^\bullet, 1 \rangle$, *then* $\beta^\star$ *is V-shaped.*
   Moreover, $0 < \lambda_0 \leqslant \lambda_\bullet$ *and* $\lambda^\bullet \leqslant \lambda_1 < 1$, *and these inequalities are strict, whenever sequence* $\beta^\star$ *contains only distinct elements.*

Proof. Similar to the proof of Theorem 20.27.   □

Remark 20.30. Let us notice that if $n = 2$, then we have $\lambda_0 = \lambda_1 = 1$. Moreover, for $n = 2, 3$ the equalities $\lambda_0 = \lambda_\bullet$ and $\lambda_1 = \lambda^\bullet$ hold. Let us also notice that calculating the minimum (the maximum) in the definition of $\lambda_\bullet$ ($\lambda^\bullet$) needs only $O(n \log n)$ time, since by Lemma 1.2 (a) the denominator in formula (20.16) is maximized (minimized) by ordering β non-decreasingly (non-increasingly). Finally, there exists only one sequence (up to the order of equal elements) that maximizes (minimizes) this denominator.

   Now we present a few examples (cf. Gawiejnowicz et al. [6]) that illustrate some consequences and applications of Theorem 20.27.

Example 20.31. Let us consider the sequence $\hat{\beta} = (5, 3, 2, 4)$. Then $\underline{\beta} = 2$, $\bar{\beta} = 4$ and $\lambda_0 = \lambda_\bullet = \frac{1}{5}$, $\lambda_1 = \lambda^\bullet = \frac{1}{3}$. By Theorem 20.27, for any $\lambda \in \langle 0, \frac{1}{5} \rangle$ the optimal schedule for the TDBS problem is non-increasing, $\beta^\star = (5, 4, 3, 2)$, while for any $\lambda \in \langle \frac{1}{3}, 1 \rangle$ the optimal schedule for the problem has a V-shape.
◆

   Theorem 20.27 is also useful in the case when the form of criterion $\| \cdot \|_{(\lambda)}$ is known in advance and we want to check if a given sequence is optimal with respect to this particular criterion.

Example 20.32. Let $\|C(\hat{\beta})\|_{(\lambda)} = \frac{1}{7} \sum C_j(\hat{\beta}) + \frac{6}{7} C_{\max}(\hat{\beta})$ and $\hat{\beta} = (2, 3, 4, 5)$. Then the sequence $(5, 4, 3, 2)$, by Theorem 20.27, is optimal for criterion $\| \cdot \|_{(\lambda)}$, since $\lambda = \frac{1}{7} < \lambda_0 = \frac{1}{5}$.
◆

Example 20.33. Let $\|C(\hat{\beta})\|_{(\lambda)} = \frac{6}{7} \sum C_j(\hat{\beta}) + \frac{1}{7} C_{\max}(\hat{\beta})$ and $\hat{\beta} = (2, 3, 4, 5)$. Then, since $\lambda = \frac{6}{7} > \lambda_1 = \frac{1}{3}$, any optimal solution is a V-shaped sequence. There are three such V-shaped sequences: $(5, 4, 2, 3)$, $(5, 2, 3, 4)$, and $(5, 3, 2, 4)$. The first sequence is the optimal solution.
◆

   The next example shows the main difference between the values of λ_0 and λ_1 and the values of $\lambda_\bullet$ and $\lambda^\bullet$: in order to calculate $\lambda_\bullet$ and $\lambda^\bullet$ we must know all elements of sequence $\hat{\beta}$, while in order to calculate λ_0 and λ_1 we need only the values of $\underline{\beta}$ and $\bar{\beta}$.

Example 20.34. Let $\hat{\beta} = (1.5, 1.3, 1.1, 1.2, 1.4)$. Then we have $\underline{\beta} = 1.1$, $\bar{\beta} = 1.4$ and $\lambda_0 = 0.23 < \lambda_\bullet = 0.24$, $\lambda^\bullet = 0.28 < \lambda_1 = 0.30$. (Note that we have the same values of λ_0 and λ_1 for all sequences with $n = 4$ elements, in which $\bar{\beta} = 1.4$ and $\underline{\beta} = 1.1$.) If we knew only the values of $\underline{\beta}$ and $\bar{\beta}$, we still could calculate λ_0 and λ_1 but we would not be able to calculate $\lambda_\bullet$ and $\lambda^\bullet$.   ◆

The results presented in Theorem 20.27 and Theorem 20.29 were *necessary* conditions, i.e., we assumed that β is an optimal sequence, and we showed its properties. Now we give a *sufficient* condition for sequence $\beta = (\beta_1, \beta_2, \ldots, \beta_n)$ to be the optimal solution to the TDBS problem.

Theorem 20.35. (Gawiejnowicz et al. [6]) *A sufficient condition for a sequence* $\beta = (\beta_1, \ldots, \beta_n)$ *to be optimal with respect to criterion* $\| \cdot \|_{(\lambda)}$, $\lambda \in \langle 0, 1 \rangle$, *is that* β *is non-increasing and* $0 \leqslant \lambda \leqslant \lambda_\bullet$.

Proof. Let $0 \leqslant \lambda \leqslant \lambda_\bullet$. Then, by Theorem 20.29, any sequence β which is optimal with respect to criterion $\| \cdot \|_{(\lambda)}$ must be non-increasing. Moreover, there exists only one (up to the order of equal elements) such optimal sequence. Thus, since this monotonic sequence is unique (again up to the order of equal elements), it must coincide with the optimal sequence. ∎

20.4 Computational experiments

In this section, we present selected results of computational experiments related to the TDPS problem.

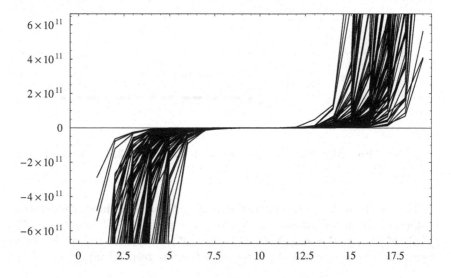

Fig. 20.1: Behaviour of q_0 for 100 random sequences

In the first computational experiment the average behaviour of q_0 (cf. Lemma 20.11) was investigated. In the experiment 100 random sequences β have been generated, each with $n = 20$ elements taken from the interval $\langle 4, 30 \rangle$.

The results of this experiment (see Fig. 20.1, cf. [6]) suggest that for random sequences composed of elements generated from uniform distribution, the values of q_0 concentrate in the middle part of the domain of $\Delta_q(\beta)$.

In order to obtain some insight into the structure of the set $Y = f(X)$ of all solutions of the TDPS problem, where $f = (\Sigma C_j, C_{\max})$ and $X = \{\hat{\beta}_\pi : \pi \in \mathfrak{S}_n\}$ for a given $\hat{\beta}$, several other computational experiments have been conducted.

Examples of such a structure are given in Fig. 20.2 and Fig. 20.3 (cf. [6]). The box '□' denotes a V-shaped solution, the diamond '♦' denotes a weakly V-shaped solution, the circle '○' denotes a Pareto optimal solution and the symbol '×' denotes a solution which is neither V-shaped nor Pareto optimal.

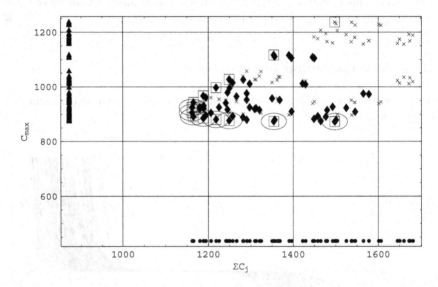

Fig. 20.2: The structure of set Y for $\hat{a} = (2, 3, 4, 5, 6)$

The results of these experiments suggest that Pareto optimal schedules can be found only in the triangle with vertices in points $(\sum C_j(\beta^\star), C_{\max}(\beta^\star))$, $(\sum C_j(\beta^\bullet), C_{\max}(\beta^\bullet))$ and $(\sum C_j(\beta^\star), C_{\max}(\beta^\bullet))$, where $\beta^\star$ and $\beta^\bullet$ denote an optimal schedule for the $\sum C_j$ and $C_{\max}$ criterion, respectively.

20.5 Other results

Some authors continued the research of bi-criteria time-dependent scheduling problems, initiated by [6]. A problem of *hierarchical scheduling* in a two-machine flow shop with proportional jobs and the objective to minimize the

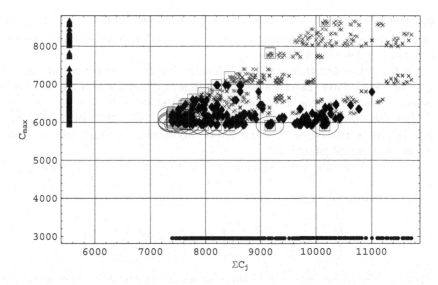

Fig. 20.3: The structure of set Y for $\hat{a} = (2, 3, 4, 5, 6, 7)$

$\sum C_j$ criterion subject to minimizing the $C_{\max}$ criterion was addressed by Cheng et al. [3]. Let us recall that hierarchical scheduling with respect to criteria f_1 and f_2 consists in finding an optimal schedule with respect to criterion f_2, provided that the schedule is searched for only among those schedules which are optimal with respect to criterion f_1.

Remark 20.36. We will denote hierarchical minimization of criteria f_1 and f_2 by the symbol $f_1 \to f_2$ in the third field of the $\alpha|\beta|\gamma$ notation.

Theorem 20.37. (Cheng et al. [3]) *Let $b_{2j} = b$ for $1 \leqslant j \leqslant n$. Then problem $F2|p_{ij} = b_{ij}t|C_{\max} \to \sum C_j$ is solvable in $O(n \log n)$ time by scheduling jobs in non-decreasing order of b_{1j} values.*

Proof. Because the processing times of operations on the second machine are equal, the order of operations on the first machine decides about the optimality of a schedule. Let us denote $b_{1,\min} := \min_{1 \leqslant j \leqslant n}\{b_{1j}\}$ and $b_{1,\max} := \max_{1 \leqslant j \leqslant n}\{b_{1j}\}$. To complete the proof it is sufficient to consider three cases: $b_{1,\min} \geqslant b$ (Case 1), $b \geqslant b_{1,\max}$ (Case 2) and $b_{1,\min} \leqslant b \leqslant b_{1,\max}$ (Case 3).

In Case 1, $C_{2n} = (1 + b) \prod_{j=1}^{n}(1 + b_{1j})$ and the value of $C_{\max}$ does not depend on the order of jobs (cf. Theorem 7.1). Hence, the value of the total completion time $\sum_{j=1}^{n} C_j(\sigma) = (1 + b) \sum_{j=1}^{n} \prod_{i=1}^{j}(1 + b_{1i})$ is minimized by scheduling jobs in non-decreasing order of the deterioration rates b_{1j} (cf. Theorem 7.61).

In Case 2, $C_{2n} = (1 + b_{1,[1]})(1 + b)^n$ and the value of $C_{\max}$ is minimized by scheduling jobs in non-decreasing order of the deterioration rates

b_{1j}. Similarly to Case 1, the value of the total completion time $\sum_{j=1}^{n} C_j(\sigma) = (1+b_{1,[1]})\frac{1+b}{b}((1+b)^n - 1)$ is minimized by scheduling jobs in non-decreasing order of the deterioration rates b_{1j}.

In Case 3, either there exists a job such that its completion time on the first machine is later than the completion time of its predecessor on the second machine (Case 3 (a)) or such a job does not exist (Case 3 (b)).

In Case 3 (a), the value of $C_{\max}$ is the same as in Case 2. Hence, scheduling jobs in non-decreasing order of the deterioration rates b_{1j} minimizes the value of the total completion time.

In Case 3 (b), the value of $C_{\max}$ is as in Case 1. However, since $b_{1,\min} \leqslant b \leqslant b_{1,\max}$, there exists a job J_r such that $C_{1r} \geqslant C_{2,r-1}$. Hence, the total completion time of all jobs equals

$$(1+b_{1,[1]})(1+b)\left(\frac{(1+b)^{r-1}-1}{b} + \prod_{i=2}^{r}(1+b_{1,[i]})(1 + \sum_{j=r+1}^{n}\prod_{i=r+1}^{j}(1+b_{1,[i]})\right).$$

Applying the pairwise job interchange technique, we can show that scheduling jobs in non-decreasing order of the deterioration rates b_{1j} minimizes this value. ∎

Theorem 20.38. (Cheng et al. [3]) *Let $b_{1j} = b$ for $1 \leqslant j \leqslant n$. Then problem $F2|p_{ij} = b_{ij}t|C_{\max} \rightarrow \sum C_j$ is solvable in $O(n\log n)$ time by scheduling jobs (a) in non-decreasing order of the deterioration rates b_{2j} if $b \leqslant b_{1,\min}$ and (b) in non-increasing order of the deterioration rates b_{2j} if $b \geqslant b_{1,\min}$.*

Proof. Similar to the proof of Theorem 20.37. □

Remark 20.39. Cheng et al. [3] also give two results on polynomial cases of similar flow shop problems with dominant machines (cf. Ho and Gupta [7]); see [3, Theorems 3–4]. ◇

Remark 20.40. A similar approach was applied by Cheng et al. [4] to a two-machine time-dependent flow shop problem with the weighted sum of the $C_{\max}$ and $\sum C_j$ criteria. The authors proved counterparts of Theorems 20.37–20.38 for criterion $\lambda C_{\max} + (1 - \lambda)\sum C_j$, where $\lambda \in \langle 0, 1\rangle$ (see [4, Theorems 1–2]). ◇

References

Pareto optimality

1. S. Gawiejnowicz, W. Kurc and L. Pankowska, Bicriterion approach to a single machine time-dependent scheduling problem. In: *Operations Research Proceedings 2001*, P. Chamoni et al. (eds.), Berlin: Springer 2002, pp. 199–206.
2. S. Gawiejnowicz, W. Kurc and L. Pankowska, Pareto and scalar bicriterion scheduling of deteriorating jobs. *Computers and Operations Research* 33 (2006), no. 3, 746–767.

Scalar optimality

3. M-B. Cheng, P. R. Tadikamalla, J. Shang and S-Q. Zhang, Bicriteria hierarchical optimization of two-machine flow shop scheduling problem with time-dependent deteriorating jobs. *European Journal of Operational Research* **234** (2014), no. 3, 650–657.

4. M-B. Cheng, P. R. Tadikamalla, J. Shang and B. Zhang, Two-machine flow shop scheduling with deteriorating jobs: minimizing the weighted sum of makespan and total completion time. *Journal of the Operational Research Society* **66** (2015), no. 5, 709–719.

5. S. Gawiejnowicz, W. Kurc and L. Pankowska, Polynomial-time solutions in a bi-criterion time-dependent scheduling problem. Report 115/2002, Adam Mickiewicz University, Faculty of Mathematics and Computer Science, Poznań, October 2002.

6. S. Gawiejnowicz, W. Kurc and L. Pankowska, Pareto and scalar bicriterion scheduling of deteriorating jobs. *Computers and Operations Research* **33** (2006), no. 3, 746–767.

7. J. Ho and J. N. D. Gupta, Flowshop scheduling with dominant machines. *Computers and Operations Research* **22** (1995), no. 2, 237–246.

New topics in time-dependent scheduling

N ew topics continue to emerge in time-dependent scheduling. In this chapter, closing the book, we present selected new topics which have appeared in time-dependent scheduling in the last fifteen years.

Chapter 21 is composed of five sections. In Sect. 21.1, we consider time-dependent scheduling on machines which are available for processing only in some periods of time. In Sect. 21.2, we study two-agent time-dependent scheduling problems, where jobs are scheduled by independent agents competing for access to the available machines. In Sect. 21.3, we review time-dependent scheduling problems with job rejection, where some jobs may be not scheduled at all. In Sect. 21.4, we focus on time-dependent scheduling with mixed deterioration, in which jobs from the same set may have different forms of processing times. Finally, in Sect. 21.5, we briefly describe time-dependent scheduling games, in which time-dependent scheduling problems are solved using game-theoretic tools. The chapter is completed with a list of references.

21.1 Time-dependent scheduling on machines with limited availability

Considering time-dependent scheduling problems, we so far have assumed that machines are continuously available. However, this assumption does not allow us to model situations in which machines need maintenance or repair. In order to cope with scheduling problems in which machines are not continuously available, in 1984 Schmidt [25] introduced the concept of non-availability periods with respect to machines and established the main properties of several scheduling problems with limited machine availability. The literature on classical scheduling on machines with limited availability is reviewed by Lee [13, 14], Lee et al. [15], Schmidt [26] and Błażewicz et al. [1, Chap. 11].

Remark 21.1. Another concept related to limited machine availability is the concept of *rate-modifying activity*, where job processing times vary depending on whether they are scheduled before or after such an activity. The

© Springer-Verlag GmbH Germany, part of Springer Nature 2020   451
S. Gawiejnowicz, *Models and Algorithms of Time-Dependent Scheduling*,
Monographs in Theoretical Computer Science. An EATCS Series,
https://doi.org/10.1007/978-3-662-59362-2_21

idea of rate-modifying activity of a fixed length was introduced in 2001 by Lee and Leon [16]. Next, Lee and Lin [17] modified the idea to include the case of *maintenance activities* of a fixed length, where a machine is allowed to break down if a maintenance activity is not scheduled before a certain time. Finally, Mosheiov and Sidney [22] introduced rate-modifying activities of variable length. We refer the reader to the reviews by Potts and Strusevich [23] and Rustogi and Strusevich [24], and the monograph by Strusevich and Rustogi [27] for detailed discussions of different variants of rate-modifying and maintenance activities.

In this section, we present the main results on *time-dependent scheduling with limited machine availability* initiated in 2003 by Wu and Lee [29].

21.1.1 Preliminaries

Throughout the section, we will assume that machines are not continuously available and for each of them there are given $k \geqslant 1$ disjoint *periods of machine non-availability*. These periods will be described by time intervals $\langle W_{i1}, W_{i2} \rangle$, where $W_{11} > t_0$ and $W_{i1} < W_{i2}$ for $1 \leqslant i \leqslant k < n$.

Remark 21.2. Symbol h_{ik} in the α field of the $\alpha|\beta|\gamma$ notation, where $1 \leqslant i \leqslant m$ is the number of machines and $k \geqslant 1$ is the number of non-availability periods, denotes that on machine M_i there are k non-availability periods.

Example 21.3. (a) Symbol $1, h_{11}|p_j = b_j t|C_{\max}$ denotes a single machine time-dependent scheduling problem with proportional jobs, criterion $C_{\max}$ and a single non-availability period.
(b) Symbol $1, h_{1k}|p_j = b_j t|C_{\max}$ denotes a single machine time-dependent scheduling problem with proportional jobs, criterion $C_{\max}$ and $k \geqslant 1$ non-availability periods.   ◆

Since any non-availability period can interrupt the processing of a job, we have to specify how to proceed when the job is interrupted by the start time of a non-availability period (cf. Lee [13]).

Definition 21.4. (Non-resumable and resumable jobs)
(a) *A job is said to be* non-resumable *if in the case when the job has been interrupted by the start time of a non-availability period, this job has to be restarted after the machine becomes available again.*
(b) *A job is said to be* resumable *if in the case when the job has been interrupted by the start time of a non-availability period, this job does not need to be restarted and can be completed after the machine becomes available again.*

Remark 21.5. The fact that jobs are non-resumable (resumable) will be denoted in field β of the $\alpha|\beta|\gamma$ notation by the symbol *nres* (*res*).

Example 21.6. (a) Symbol $1, h_{11}|p_j = b_j t, nres|C_{\max}$ denotes a single machine problem from Example 21.3 (a) with non-resumable jobs.
(b) Symbol $1, h_{1k}|p_j = b_j t, res|C_{\max}$ denotes a single machine problem from Example 21.3 (b) with resumable jobs.   ◆

21.1.2 Proportional deterioration

In this section, we present the main results concerning time-dependent scheduling on machines with limited availability and proportional jobs.

Among time-dependent scheduling problems with limited machine availability the most studied is problem $1, h_{1k}|p_j = b_j t|f$ of single machine scheduling of proportional jobs, $k \geqslant 1$ non-availability periods and criterion f.

The problem of scheduling proportional resumable jobs on a machine with a single non-availability period and criterion $f = C_{\max}$ was introduced in 2003 by Wu and Lee [29]. The authors considered two cases of the problem: Case 1, when the completion time of the last job processed before the beginning of a non-availability period is strictly earlier than W_{11}, and Case 2, when the completion time is equal to W_{11}.

Property 21.7. (Wu and Lee [29]) Let $[i]$ be the index of the last job processed before the beginning of a non-availability period in a schedule σ. Then
(a) the maximum completion time of schedule σ in Case 1 equals

$$C_{\max}^1(\sigma) = t_0 \prod_{j=1}^n (1 + b_{[j]}) + (W_{12} - W_{11}) \prod_{j=i+2}^n (1 + b_{[j]}); \qquad (21.1)$$

(b) the maximum completion time of schedule σ in Case 2 equals

$$C_{\max}^2(\sigma) = t_0 \prod_{j=1}^n (1 + b_{[j]}) + (W_{12} - W_{11})b_{[i+1]} \prod_{j=i+2}^n (1 + b_{[j]}). \qquad (21.2)$$

Proof. By direct computation; see [29, Sect. 3].   ◇

Wu and Lee [29] also formulated the problem $1, h_{11}|p_j = b_j t, nres|C_{\max}$ in terms of 0-1 integer programming. Let us denote by x_j a binary variable such that $x_j = 1$ if job J_j is scheduled before a non-availability period and $x_j = 0$ otherwise, $1 \leqslant j \leqslant n$. Let $J_{\max}$ and $\mathcal{J}' := \mathcal{J} \setminus \{J_{\max}\}$ denote a job with deterioration rate $b_{\max}$ defined as in Sect. 6.2.2 and the set of all jobs except $J_{\max}$, respectively. Let $B_j := \ln(1+b_j)$ and $\theta := \ln \frac{W_{11}}{t_0}$. Then problem $1, h_{11}|p_j = b_j t, res|C_{\max}$ can be formulated as follows:

$$\max \textstyle\sum_{J_j \in \mathcal{J}'} B_j x_j$$
$$\text{subject to } B_j x_j < \theta, j \in \mathcal{J}'.$$

The time complexity of the problem, however, was left by the authors as an open problem.

Gawiejnowicz [4, 5] and Ji et al. [11] proved that scheduling non-resumable jobs with proportional processing times and one period of machine non-availability is a computationally intractable problem.

Theorem 21.8. (Gawiejnowicz [4, 5], Ji et al. [11]) *The decision version of problem* $1, h_{11}|p_j = b_j t, nres|C_{\max}$ *is $\mathcal{NP}$-complete in the ordinary sense.*

Proof. Gawiejnowicz [4, 5] uses the following reduction from the SP problem: $n = p$, $t_0 = 1$, $b_j = y_j - 1$ for $1 \leqslant j \leqslant n$, $k = 1$, $W_{11} = B, W_{12} = 2B$ and the threshold value is $G = 2Y$, where $Y = \prod_{j=1}^{p} y_j$.

Let us notice that by Theorem 7.1 we can check in polynomial time whether $C_{\max}(\sigma) \leqslant G$ for a given schedule σ for the above instance of problem $1|p_j = b_j t|C_{\max}$ with a single non-availability period. Therefore, the decision version of this problem belongs to the $\mathcal{NP}$ class.

Since the above reduction is polynomial, in order to complete the proof it is sufficient to show that the SP problem has a solution if and only if there exists a feasible schedule σ for the above instance such that $C_{\max}(\sigma) \leqslant G$.

Ji et al. [11] use the following reduction from the SP problem: $n = p$, t_0 arbitrary, $W_{11} = t_0 B$, $W_{12} > W_{11}$ arbitrary, $b_j = y_j - 1$ for $1 \leqslant j \leqslant n$ and the threshold value is $G = W_{12}\frac{Y}{B}$, where $Y = \prod_{j=1}^{p} y_j$. The remaining steps of the proof are as above. □

Luo and Ji [21] considered a single machine time-dependent scheduling problem with a variable *mandatory maintenance*, which is a generalization of problem $1, h_{11}|p_j = b_j t$, $nres|C_{\max}$. The maintenance must start before a given deadline s_d and it has a variable length $f(s)$, where $f(s)$ is a non-decreasing function of the starting time s of this maintenance.

Remark 21.9. Scheduling problems with variable mandatory maintenance will be denoted by symbol VM in the α field of the three-field notation.

Theorem 21.10. (Luo and Chen [20]) *Problem* $1, VM|p_j = b_j t, nres|C_{\max}$ *is* $\mathcal{NP}$-*hard in the ordinary sense.*

Proof. Luo and Chen [20] use the following reduction from the SP problem: $n = p$, $t_0 = 1$, $b_j = y_j - 1$ for $1 \leqslant j \leqslant n$, $s_d = B$, $f(s) = B$ and the threshold is $G = 2Y$, where $Y = \prod_{j=1}^{p} y_j$.

In order to complete the proof it is sufficient to show that the SP problem has a solution if and only if there exists a feasible schedule σ for the above instance, such that $C_{\max}(\sigma) \leqslant G$. □

Luo and Chen [20] also proposed for problem $1, VM|p_j = b_j t$, $nres|C_{\max}$ an FPTAS, based on an approximation scheme for a classical scheduling problem. The main idea of this FPTAS is to transform a given instance of problem $1, VM|p_j = b_j t$, $nres|C_{\max}$ into an instance of problem $1|d_j = s| \sum w_j U_j$, and then to apply to the transformed instance the FPTAS proposed by Gens and Levner [8]. Hence, the former FPTAS is composed of two steps. In the first step, a given instance of problem $1, VM|p_j = b_j t$, $nres|C_{\max}$ is transformed into an instance of problem $1|d_j = s| \sum w_j U_j$ by assuming a common deadline s and by mapping job deterioration rates b_j into job processing times p_j and job weights w_j, where

$$s := \ln \frac{s_d}{t_0}$$

and

$$p_j = w_j := \ln(1 + b_j)$$

for $1 \leqslant j \leqslant n$. In the second step, the FPTAS by Gens and Levner [8] for problem $1|d_j = s| \sum w_j U_j$ is applied. The pseudo-code of the FPTAS by Luo and Chen [20] can be formulated as follows.

Algorithm 21.1. for problem $1, VM|p_j = b_j t, nres|C_{\max}$

---

1 **Input** : sequence $(b_1, b_2, \ldots, b_n)$, numbers s_d, s, ϵ
2 **Output:** a suboptimal schedule
  ▷ Step 1
3 $B \leftarrow \prod_{j=1}^{n}(1 + b_j)$;
4 **if** $(t_0 B \leqslant s_d)$ **then**
5 | Schedule all jobs in an arbitrary order starting from t_0;
6 | Schedule the maintenance;
7 | **return**;
  end
  ▷ Step 2
8 Construct a corresponding instance I of problem $1|d_j = s| \sum w_j U_j$;
9 $\delta \leftarrow \log_B(1 + \epsilon)$;
10 Apply to I the FPTAS by Gens and Levner [8];
11 $X_1 \leftarrow \{J_j : U_j = 0\}$;
12 $X_2 \leftarrow \{J_j : U_j = 1\}$;
  ▷ Step 3
13 Schedule jobs from set X_1 in an arbitrary order;
14 Schedule maintenance;
15 Schedule jobs from set X_2 in an arbitrary order;
16 **return**.

Theorem 21.11. (Luo and Chen [20]) *Algorithm 21.1 is an FPTAS for problem* $1, VM|p_j = b_j t, nres|C_{\max}$ *and runs in* $O(n^3 \epsilon^{-1} L)$ *time, where*

$$L := \log \max \left\{ s, d, \max_{1 \leqslant j \leqslant n} \{1 + b_j\} \right\}.$$

Proof. See [20, Theorem 2.3].   ◊

Ji et al. [11] analyzed online and offline approximation algorithms for problem $1, h_{11}|p_j = b_j t, nres|C_{\max}$, which are variations of approximation algorithms for jobs with fixed processing times. In most cases, the variations consist in the replacement of job processing times with job deterioration rates. This simple replacement, however, leads to other results than those for scheduling problems with fixed job processing times.

First we present a result concerning the competitive ratio of the time-dependent counterpart of algorithm LS (Graham [9]), discussed in Chaps. 14 and 19 as algorithm $\overline{LS}$ (see Algorithm 21.2). Let us recall that both algorithms, LS and $\overline{LS}$, assign the first available job to the first available machine as long as there are jobs to be scheduled.

Algorithm 21.2. for problem $1, h_{11}|p_j = b_j t, nres|C_{\max}$

---

1 **Input** : sequence $(b_1, b_2, \ldots, b_n)$
2 **Output:** a suboptimal schedule
  ▷ **Step 1**
3 **while** (*there are jobs to be scheduled*) **do**
4 | Assign the first available job to the first available machine;
  end
  ▷ **Step 2**
5 **return.**

Theorem 21.12. (Ji et al. [11]) *If* $t_0 \leqslant W_{11}$, *then Algorithm 21.2 is* $\frac{W_{11}}{t_0}$-*competitive for problem* $1, h_{11}|p_j = b_j t, nres|C_{\max}$.

Proof. Let $N_{\mathcal{J}_1}$ ($N_{\mathcal{J}_2}$) denote the set of indices of jobs scheduled before (after) a non-availability period in an optimal schedule. Let $C^\star_{\max}$ and $C^{A21.2}_{\max}$ denote the length of the optimal schedule and the schedule constructed by Algorithm 21.2, respectively.

If $N_{\mathcal{J}_2} = \emptyset$, then by (7.1) we have $C^\star_{\max} = C^{A21.2}_{\max} = t_0 \prod_{j=1}^n (1 + b_j)$ and $\frac{C^{A21.2}_{\max}}{C^\star_{\max}} = 1 \leqslant \frac{W_{11}}{t_0}$.

Let us assume that $N_{\mathcal{J}_2} \neq \emptyset$. Then

$$t_0 \prod_{j \in N_{\mathcal{J}_1}} (1 + b_j) \leqslant W_{11} \tag{21.3}$$

and

$$C^\star_{\max} = W_{12} \prod_{j \in N_{\mathcal{J}_2}} (1 + b_j), \tag{21.4}$$

where $W_{12} > W_{11}$ denotes the end time of the non-availability period.

From (21.4) it follows that

$$W_{12} \prod_{j=1}^n (1 + b_j) = C^\star_{\max} \prod_{j \in N_{\mathcal{J}_1}} (1 + b_j). \tag{21.5}$$

Hence, by (21.4) and (21.5), we have

$$C^{A21.2}_{\max} \leqslant W_{12} \prod_{j=1}^n (1 + b_j). \tag{21.6}$$

From (21.6), by (21.3) and (21.5), it follows that $C^{A21.2}_{\max} \leqslant \frac{W_{11}}{t_0} C^\star_{\max}$. ∎

For problem $1, h_{11}|p_j = b_j t, nres|C_{\max}$, Ji et al. [11] also analyzed an off-line Algorithm 21.3 (see Remark 14.11) which is an adaptation of algorithm

LPT (Graham [10]), discussed in Chaps. 14 and 19. The pseudo-code of the new algorithm is as follows.

Algorithm 21.3. for problem $1, h_{11}|p_j = b_j t, nres|C_{\max}$

1 **Input** : sequence $(b_1, b_2, \ldots, b_n)$
2 **Output:** a suboptimal schedule
  ▷ Step 1
3 Arrange jobs in non-increasing order of the deterioration rates b_j;
  ▷ Step 2
4 Apply Algorithm 21.2 to the list of jobs obtained in Step 1;
  ▷ Step 3
5 **return.**

Theorem 21.13. (Ji et al. [11]) *If* $1 + b_{\min} \leqslant \frac{W_{11}}{t_0}$, *where* $b_{\min}$ *is defined as in inequality (6.39), then the worst-case ratio of Algorithm 21.3 for problem* $1, h_{11}|p_j = b_j t, nres|C_{\max}$ *equals* $1 + b_{\min}$; *otherwise it equals* 1.

Proof. If $1 + b_{\min} > \frac{W_{11}}{t_0}$, then all jobs must be processed after the non-availability period. Thus, the schedule generated by Algorithm 21.3 is optimal.

Let us now assume that $1 + b_{\min} \leqslant \frac{W_{11}}{t_0}$ and let us suppose that there exists a schedule which violates the ratio $1 + b_{\min}$. Let $I = (\mathcal{J}, t_0, b_1, b_2)$ denote the instance of problem $1, h_{11}|p_j = b_j t, nres|C_{\max}$, corresponding to the schedule with the fewest possible number of jobs.

First, let us notice that if a job J_j is scheduled before (after) the non-availability period in the schedule generated by Algorithm 21.3, then the job must also be scheduled after (before) the non-availability period in an optimal schedule. Indeed, if a job J_j is scheduled before the non-availability period in both the schedule generated by Algorithm 21.3 and an optimal schedule, then we can construct from instance I a new instance $I' = (\mathcal{J} \setminus \{J_j\}, t_0, \frac{b_1}{1+b_j}, \frac{b_2}{1+b_j})$. Then $C_{\max}^{A21.3}(I') = \frac{C_{\max}^{A21.3}(I)}{1+b_j}$ and $C_{\max}^{\star}(I') = \frac{C_{\max}^{\star}(I)}{1+b_j}$. Hence,

$$\frac{C_{\max}^{A21.3}(I')}{C_{\max}^{\star}(I')} \geqslant \frac{C_{\max}^{A21.3}(I)}{C_{\max}^{\star}(I)}$$

which implies that I' is a smaller instance than I. A contradiction.

Second, in a similar way as above we can prove that job J_i with $b_i = b_{\min}$ from instance I defined as above must be scheduled in a schedule generated by Algorithm 21.3 after the non-availability period.

Based on the two claims, we can show by contradiction that the result holds; see [11, Theorem 4] for details. □

Yu et al. [37] generalized Theorem 14.6 to case of a single non-availability period on one of two parallel identical machines. Let $b_{\max}$ be defined as in Sect. 6.2.2.

Theorem 21.14. (Yu et al. [37]) *For problem* $P2, h_{1,1}|p_j = b_j t, nres|C_{\max}$ *no online algorithm is better than* $(\max\{\frac{W_{1,1}}{t_0}, 1 + b_{\max}\})$*-competitive.*

Proof. Similar to the proof of Theorem 14.8; see [37, Sect. 3] for details.   ◇

An FPTAS for problem $1, h_{11}|p_j = b_j t, nres|C_{\max}$ was proposed by Ji et al. [11]. The FPTAS is based on the observation that in order to determine which jobs are scheduled after the non-availability period we can apply any FPTAS for the KP problem, e.g., the FPTAS by Gens and Levner [7]. In order to determine the jobs which should be scheduled after time W_{12}, from any instance of problem $1, h_{11}|p_j = b_j t, nres|C_{\max}$ we can construct an instance of the KP problem as follows: given $\epsilon > 0$, we set $D := \prod_{j=1}^{n}(1 + b_j)$ and $\delta := \log_D(1 + \epsilon)$. Next, for each job J_j we define an item of the KP problem with profit u_j and weight w_j, where

$$u_j = w_j := \ln(1 + b_j) \tag{21.7}$$

for $1 \leqslant j \leqslant n$. Finally, we define the capacity of the knapsack as

$$U := \ln \frac{W_{11}}{t_0}. \tag{21.8}$$

The pseudo-code of this FPTAS can be formulated as follows.

Algorithm 21.4. for problem $1, h_{11}|p_j = b_j t, res|C_{\max}$

---

1 **Input** : sequence $(b_1, b_2, \ldots, b_n)$, numbers W_{11}, W_{12}, ϵ
2 **Output:** a suboptimal schedule
  ▷ Step 1
3 **if** $(t_0 \prod_{j=1}^{n}(1 + b_j) \leqslant W_{11})$ **then**
4 | $C_{\max} \leftarrow t_0 \prod_{j=1}^{n}(1 + b_j)$;
  end
  ▷ Step 2
5 **for** $j \leftarrow 1$ **to** n **do**
6 | Define u_j and w_j according to (21.7);
  end
7 Define U according to (21.8);
8 Determine set $\mathcal{J}_{KP}$ of jobs scheduled after the non-avalability period;
9 $C_{\max} \leftarrow W_{12} \prod_{J_j \in \mathcal{J}_{KP}}(1 + b_j)$;
  ▷ Step 3
10 **return** $C_{\max}$.

Theorem 21.15. (Ji et al. [11]) *Algorithm 21.4 is an FPTAS for problem* $1, h_{11}|p_j = b_j t, res|C_{\max}$ *and runs in* $O(n^2 \epsilon^{-1})$ *time.*

Proof. See [11, Theorem 7].   ◇

Gawiejnowicz and Kononov [6] proved that the problem with a single non-availability period remains intractable also for resumable jobs.

Theorem 21.16. (Gawiejnowicz and Kononov [6]) *The decision version of problem* $1, h_{11}|p_j = b_j t, res|C_{\max}$ *is* $\mathcal{NP}$-*complete in the ordinary sense.*

Proof. The reduction from the SP problem is as follows: $n = p + 1$, $t_0 = 1$, $\alpha_j = y_j - 1$ for $1 \leqslant j \leqslant p$, $\alpha_{p+1} = B - 1$, $k = 1$, $W_{11} = B + 1, W_{12} = 2B + 1$ and the threshold value is $G = (B + 1)Y$, where $Y = \prod_{j=1}^{p} y_j$.

In order to complete the proof it is sufficient to show that the SP problem has a solution if and only if there exists a feasible schedule σ for the above instance of problem $1, h_{11}|p_j = b_j t, res|C_{\max}$ with the non-availability period $\langle W_{11}, W_{12} \rangle$ such that $C_{\max}(\sigma) \leqslant G$. □

Gawiejnowicz and Kononov [6] also proposed a dynamic programming algorithm for the problem. Let us denote $\Delta_1 := W_{12} - W_{11}$. Then the algorithm can be formulated as follows.

Algorithm 21.5. for problem $1, h_{11}|p_j = b_j t, res|C_{\max}$

---

1   $X_0 \leftarrow \{[t_0, \Delta_1]\}$;
2   $\tau \leftarrow t_0 \prod_{j=1}^{n}(1 + b_j)$;
3   **for** $k \leftarrow 1$ **to** n **do**
4      $X_k \leftarrow \emptyset$;
5   **for each** $x \in X_{k-1}$ **do**
6          $F_1(b_k, t_1, t_2) \leftarrow [t_1, t_2(1 + b_k)]$;
7   **if** $(t_1(1 + b_k) < W_{11})$ **then**
8             $F_2(b_k, t_1, t_2) \leftarrow [t_1(1 + b_k), t_2]$;
         end
9          $X_k \leftarrow X_k \cup \{F_1(b_k, t_1, t_2)\} \cup \{F_2(b_k, t_1, t_2)\}$;
     end
  end
10 **for each** $x \in X_n$ **do**
11      $C_{\max}(x) \leftarrow \tau + t_2$;
  end
12 **return** $\min \{C_{\max}(x) : x \in X_n\}$.

The main idea of Algorithm 21.5 is as follows. The algorithm goes through n phases. The kth phase, $1 \leqslant k \leqslant n$, produces a set X_k of states. Any state in X_k is a vector x which encodes a partial schedule for the first k jobs. The sets $X_1, X_2, \ldots, X_k$ are constructed iteratively. The initial set X_0 contains an initial state. The set X_k, where $1 \leqslant k \leqslant n$, is obtained from the set X_{k-1} via some functions which translate the states of set X_{k-1} into the states of set X_k. The number and the form of the functions depend on the problem under consideration. The optimal schedule, i.e. the one with the minimal value of criterion $C_{\max}$, is an element of set X_n.

Woeginger [28] showed how to transform a dynamic programming algorithm for a problem into an FPTAS for this problem, provided this algorithm satisfies some conditions (cf. [28, Lemma 6.1, Theorem 2.5]). Because Algorithm 21.5 satisfies conditions mentioned above, it can be transformed into an FPTAS for problem $1, h_{11}|p_j = b_j t, res|C_{\max}$.

The problem with $k \geqslant 2$ non-availability periods is hard to approximate.

Theorem 21.17. (Gawiejnowicz and Kononov [6]   *There does not exist a polynomial approximation algorithm with a constant worst-case ratio for problem $1, h_{1k}|p_j = b_j t, res|C_{\max}, k \geqslant 2$, unless $\mathcal{P} = \mathcal{NP}$.*

Proof. The proof of Theorem 21.17 is obtained by contradiction: the existence of an approximation algorithm with a constant worst-case ratio for problem $1, h_{1k}|p_j = b_j t, res|C_{\max}, k \geqslant 2$ would allow us to solve the SP problem in polynomial time, which is impossible unless $\mathcal{P} = \mathcal{NP}$ (see [3, Chap. 5]). □

Problem $1|p_j = b_j t, nres|f$ remains intractable also when $f = \sum C_j$.

Theorem 21.18. (Ji et al. [11]) *The decision version of problem $1, h_{11}|p_j = b_j t, nres|\sum C_j$ is $\mathcal{NP}$-complete in the ordinary sense.*

Proof. Ji et al. [11] use the following reduction from the SP problem: $n = p + 4$, arbitrary t_0, $W_{11} = t_0 B^5$, arbitrary $W_{12} > W_{11}$, $b_j = y_j - 1$ for $1 \leqslant j \leqslant p$, $b_{p+1} = YB - 1$, $b_{p+2} = \frac{Y^2}{B} - 1$, $b_{p+3} = b_{p+4} = Y^3 - 1$ and threshold value is $G = (p+2)W_{12}B^2 + (t_0 + W_{12})B^5$, where $Y = \prod_{j=1}^{p} y_j$.

In order to complete the proof it is sufficient to show that the SP problem has a solution if and only if there exists a feasible schedule σ for the above instance of problem $1|p_j = b_j t, nres|\sum C_j$ with a non-availability period $\langle W_{11}, W_{12}\rangle$ such that $\sum C_j(\sigma) \leqslant G$. □

Luo and Ji [21] considered problem $1, VM|p_j = b_j t, nres|\sum C_j$ with a variable mandatory maintenance.

Theorem 21.19. Luo and Chen [20]) *Problem $1, VM|p_j = b_j t, nres|\sum C_j$ is $\mathcal{NP}$-hard in the ordinary sense.*

Proof. Luo and Chen [20] use the following reduction from the SP problem: $n = p + 4$, $t_0 = 1$, $b_j = y_j - 1$ for $1 \leqslant j \leqslant n$, $b_{p+1} = YA - 1$, $b_{p+2} = YB - 1$, $b_{p+3} = b_{p+4} = Y^4 - 1$, $s_d = Y^5$ and $f(s) = Y^5$. The threshold value is $G = Y^2 + Y^5 + 2(r+1)Y^7 + Y^{10}$, where $Y = AB = \prod_{j=1}^{p} y_j$.

In order to complete the proof it is sufficient to show that the SP problem has a solution if and only if there exists a feasible schedule σ for the above instance of problem $1, VM|p_j = b_j t, nres|\sum C_j$ with a non-availability period $\langle s, s + f(s)\rangle$ such that $\sum C_j(\sigma) \leqslant G$. □

Luo and Chen [20], applying a similar approach as the one for problem $1, VM|p_j = b_j t, nres|C_{\max}$ (cf. Theorem 21.19), proposed an FPTAS for problem $1, VM|p_j = b_jt,\ nres|\sum C_j$. In this FPTAS, the interval $\langle t_0, s_d \rangle$ where the starting time s of the mandatory maintenance is located, is divided into a number of subintervals, and for each such a subinterval the approach by Woeginger [28] is applied to perform an FPTAS for problem $1, h_{11}|p_j = b_jt,\ nres|\sum C_j$. Then, the subinterval which corresponds to an optimal schedule is chosen as the final solution. The FPTAS by Luo and Chen [20] runs in $O(n^6\epsilon^{-4}L^4)$ time, where

$$L := \log \max \left\{ s_d, d, \max_{1 \leqslant j \leqslant n} \{1 + b_j\} \right\}. \tag{21.9}$$

Kacem and Levner [12] improved the FPTAS by Luo and Chen [20], using a simpler dynamic programming algorithm for problem $1, VM|p_j = b_jt$, $nres|\sum C_j$ and converting it into an FPTAS without the necessity of solving a set of auxiliary problems. The FPTAS by Kacem and Levner [12] runs in $O(n^6\epsilon^{-3}L^3)$ time, where L is defined by (21.9).

Fan et al. [2] proved a counterpart of Theorem 21.18 for resumable jobs.

Theorem 21.20. (Fan et al. [2], Luo and Chen [19]) *The decision version of problem* $1, h_{11}|p_j = b_jt, res|\sum C_j$ *is* $\mathcal{NP}$-*complete in the ordinary sense.*

Proof. Luo and Chen [19] use the following reduction from the SP problem: $n = p + 5$, $t_0 = 1$, $W_{11} = X^5 + 1$, $W_{12} = X^{11} + X^5 + 1$, $b_j = y_j - 1$ for $1 \leqslant j \leqslant p$, $b_{p+1} = YA - 1$, $b_{p+2} = YB - 1$, $b_{p+3} = b_{p+4} = Y^3 - 1$ and $b_{p+5} = Y^6 - 1$, where $Y = AB = \prod_{j=1}^{p} y_j$. The threshold value is $G = Y^2 + Y^5 + Y^{11} + 2(p+1)Y^{13} + 2Y^{16}$.

Fan et al. [2] use the following reduction from the SP problem: $n = p + 3$, $t_0 = 1$, $W_{11} = Y^2 + 1$, $W_{12} = Y^7 + Y^2 + 1$, $b_j = y_j - 1$ for $1 \leqslant j \leqslant p$, $b_{p+1} = YA - 1$, $b_{p+2} = YB - 1$, $b_{p+3} = Y^5 - 1$. The threshold value is $G = 2Y^9 + Y^8$, where $Y = AB = \prod_{j=1}^{n} y_j$.

In order to complete the proof it is sufficient to show that the SP problem has a solution if and only if there exists a feasible schedule σ for the above instance of problem $1|p_j = b_jt, res|\sum C_j$ with the non-availability period $\langle W_{11}, W_{12} \rangle$ such that $\sum C_j(\sigma) \leqslant G$. $\qquad \square$

For problem $1, h_{11}|p_j = b_jt, res|\sum C_j$, Fan et al. [2] proposed the following dynamic programming Algorithm 21.6, which can be transformed into an FPTAS using the approach by Woeginger [28]. The algorithm is similar to Algorithm 21.5, i.e., the final schedule is constructed in n phases, where in the kth phase, $1 \leqslant k \leqslant n$, a set X_k of states is produced. The initial set X_0 contains an initial state, and the remaining sets $X_1, X_2, \ldots, X_n$ are constructed iteratively. The set X_k, $1 \leqslant k \leqslant n$, is obtained from the set X_{k-1} via some functions which translate the states of set X_{k-1} into the states of set X_k.

There are also some differences between Algorithms 21.5 and 21.6. The two main differences concern the dimension of the vector sets $X_0, X_1, \ldots,$ X_n and the meaning of components which compose the vectors. In the case of Algorithm 21.5, the vectors are composed of two components, t_1 and t_2.

Component t_1 is the completion time of the last of the jobs which have been completed before time W_{11}. Component t_2 is the total processing time of the jobs which start after time W_{12}. In the case of Algorithm 21.6, the vectors are composed of four components, ℓ_1, z_1, ℓ_2 and z_2. Component ℓ_1 is the completion time of the last job among all jobs completed before time W_{11}, while component z_1 is the initial total completion time of these jobs. Component ℓ_2 is the completion time of the job which started before time W_{11} but completed after time W_{12}, while component z_2 is the initial total completion time of all jobs which started not earlier than W_{12}. The pseudo-code of Algorithm 21.6 is as follows.

Algorithm 21.6. for problem $1, h_{11}|p_j = b_j t, res| \sum C_j$

1 $X_0 \leftarrow \{[t_0, 0, W_{11}, 0]\};$
2 **for** $k \leftarrow 1$ **to** n **do**
3 $\quad$ **if** $(k \neq j)$ **then**
4 $\quad\quad$ $A_k \leftarrow t_0 \prod_{i=1, i \neq j}^{k} (1 + b_j);$
$\quad$ **end**
end
5 **for** $k \leftarrow 1$ **to** n **do**
6 $\quad$ **if** $(k \neq j)$ **then**
7 $\quad\quad$ $X_k \leftarrow \emptyset;$
8 $\quad\quad$ **for each** $x \in X_{k-1}$ **do**
9 $\quad\quad\quad$ $X_k \leftarrow X_k \cup \left\{[\ell_1, z_1, \ell_2, z_2 + \frac{A_k}{\ell_1}]\right\};$
$\quad\quad\quad$ **if** $(\ell_1(b_k + 1)(b_j + 1) \leqslant W_{11})$ **then**
10 $\quad\quad\quad\quad$ $X_k \leftarrow X_k \cup \{[\ell_1(b_k + 1), z_1 + \ell_1(b_k + 1), \ell_2, z_2]\};$
$\quad\quad\quad$ **end**
$\quad\quad\quad$ **if** $((\ell_1(b_k + 1) < W_{11})$ **and** $(\ell_1(b_k + 1)(b_j + 1) > W_{11}))$
$\quad\quad\quad$ **then**
$\quad\quad\quad\quad$ $X_k \leftarrow X_k \cup$
$\quad\quad\quad\quad$ $\{[\ell_1(b_k + 1), z_1 + \ell_1(b_k + 1), \ell_1(b_k + 1)(b_j + 1) + W_{11}, z_2]\};$
$\quad\quad\quad$ **end**
$\quad\quad$ **end**
$\quad$ **end**
end
11 **for all** $x \in X_k$ **do**
12 $\quad$ $G(\ell_1, z_1, \ell_2, z_2) \leftarrow t_0 z_1 + \max\{\ell_2, W_{12}\} z_2 + \ell_2;$
end
13 **return** $\min \{G(\ell_1, z_1, \ell_2, z_2) : [\ell_1, z_1, \ell_2, z_2] \in X_n\}.$

Theorem 21.21. (Fan et al. [2]) *Algorithm 21.6 is an FPTAS for problem* $1, h_{11}|p_j = b_j t, res| \sum C_j$.

Proof. See [2, Theorem 5.1]. $\diamond$

Luo and Chen [19], based on the approach by Woeginger [28], proposed another dynamic programming algorithm for problem $1, h_{11}|p_j = b_jt, res|\sum C_j$, which can be transformed into an FPTAS for this problem. This FPTAS runs in $O(n^9 L^4 \epsilon^{-4})$ time, where

$$L := \log \max \left\{ W_{11}, \max_{1 \leqslant j \leqslant n} \{1 + b_j\} \right\}.$$

Some results are also known for time-dependent scheduling problems with limited availability of parallel machines. Lee and Wu [18] considered problem $Pm, h_{m1}|p_j = b_jt|C_{\max}$, where for each machine a single non-availablity period is defined. For this problem, the authors proposed the following two lower bounds on the optimal value of the $C_{\max}$ criterion.

Lemma 21.22. (Lee and Wu [18]) *Let W_{1i} and W_{2i} denote the start time and the end time of a non-availability period on machine $M_i, 1 \leqslant i \leqslant m$, respectively. Let x_i be a binary variable such that $x_i := 1$ if*

$$W_{2i} \leqslant \frac{t_0 \sqrt[m]{\prod_{j=1}^n (1 + b_jt)}}{\prod_{j=1}^{m-1}(1 + b_j)}$$

and $x_i := 0$ otherwise. Then the optimal value of the maximum completion time for problem $Pm, h_{m1}|p_j = b_jt|C_{\max}$ is not less than

(a) $t_0 \left(\sqrt[m]{\prod_{j=1}^n (1 + b_j)} \right) \left(1 + \sqrt[m]{\prod_{i=1}^m \frac{x_i(W_{2i} - W_{1i})}{W_{1i}(1 + b_i)}} \right)$, *if jobs are resumable;*

(b) $t_0 \left(\sqrt[m]{\prod_{j=1}^n (1 + b_j)} \right) \left(1 + \sqrt[m]{\prod_{i=1}^m \frac{x_i(W_{2i} - W_{1i})}{W_{1i}}} \right)$, *if jobs are non-resumable.*

Proof. (a) By direct computation; see [18, Proposition 1].
    (b) By direct computation; see [18, Proposition 2].   ◇

21.1.3 Linear deterioration

In this section, we present the main results concerning time-dependent scheduling on machines with limited availability and linear jobs.

Luo and Ji [21] considered problem $1, VM|p_j = a_j + b_jt, nres|C_{\max}$ with linearly deteriorating jobs and a mandatory maintenance.

Theorem 21.23. (Luo and Ji [21]) *Problem $1, VM|p_j = a_j + b_jt, nres|C_{\max}$ is $\mathcal{NP}$-hard in the ordinary sense.*

Proof. Luo and Ji [21] use the following reduction from the SP problem: $n = p$, $t_0 = 0$, $a_j = b_j = y_j - 1$ for $1 \leqslant j \leqslant n$, $s_d = B - 1$, $f(s) = B$ and the threshold value is $G = 2Y - 1$, where $Y = \prod_{j=1}^p y_j$.
    Since the above reduction is polynomial, in order to complete the proof it is sufficient to show that the SP problem has a solution if and only if there exists a feasible schedule σ for the above instance such that $C_{\max}(\sigma) \leqslant G$.   □

Luo and Ji [21], applying the approach by Woeginger [28], proposed an FPTAS for problem $1, VM|p_j = a_j + b_jt|C_{\max}$, running in $O(n^4 L^2 \epsilon^{-2})$ time, where

$$L := \log \max \{s, d, a_j, \max \{1 + b_j : 1 \leqslant j \leqslant n\}\}.$$

The authors also considered the case of the $\sum C_j$ criterion.

Theorem 21.24. (Luo and Ji [21]) *Problem* $1, VM|p_j = a_j + b_jt, nres| \sum C_j$ *is $\mathcal{NP}$-hard in the ordinary sense.*

Proof. Luo and Ji [21] use the following reduction from the SP problem: $n = p + 3$, $t_0 = 0$, $a_j = b_j = y_j - 1$ for $1 \leqslant j \leqslant n - 3$, $a_{n-2} = b_{n-2} = Y^2 A - 1$, $a_{n-1} = b_{n-1} = Y^2 B - 1$, $a_n = b_n = Y^4 - 1$, $s_d = Y^3 - 1$, $f(s) = Y^3$ and the threshold value is $G = Y^3 - 1 + 2(p+1)(Y^6 - \frac{1}{2}) + 2Y^{10} - 1$, where $Y = AB = \prod_{j=1}^{p} y_j$.

Since the above reduction is polynomial, in order to complete the proof it is sufficient to show that the SP problem has a solution if and only if there exists a feasible schedule σ for the above instance such that $\sum C_j(\sigma) \leqslant G$. $\square$

Remark 21.25. Some authors considered single machine time-dependent scheduling problems with *maintenance activities*, another form of machine non-availability, close to rate-modifying activities (see Remark 21.1). For example, Rustogi and Strusevich [24] considered several variations of the single machine problem $1|p_j = a_j + bt|C_{\max}$, where performance of the machine deteriorates during operation and a maintenance activity of a fixed length is necessary. The authors proved that each of these problems can be solved in polynomial time by a reduction to a linear assignment problem with a product matrix.

21.2 Time-dependent two-agent scheduling

In time-dependent scheduling problems considered so far, there was no competition among the processed jobs. In this section, we will consider *two-agent scheduling problems*, where a set of jobs is divided between two agents, denoted as *agent A* and *agent B*. Each agent possesses their own optimality criterion and competes for access to available machines. The aim is to find a schedule which, in a predefined sense, satisfies both agents.

Two-agent scheduling problems were introduced by Baker and Smith [32] and Agnetis et al. [31]. The authors considered several scheduling problems with fixed job processing times and two competing agents, presenting $\mathcal{NP}$-completeness results and polynomial algorithms for these problems. We refer the reader to the review by Perez-Gonzalez and Framinan [44] and the monograph by Agnetis et al. [30] for detailed discussions of two-agent scheduling.

The fact that two-agent scheduling problems model competing manufacturing systems caused a rapid development of this domain. This, in turn, resulted in the appearance in 2007–2008 of *time-dependent two-agent scheduling problems*, introduced by Ding et al. [34] and Liu and Tang [41]. In this section, we review the main results from this domain.

21.2.1 Preliminaries

Throughout this section, we will consider variations of the following problem. Two agents, A and B, have to schedule n jobs from a set $\mathcal{J}$ on a common machine (common machines). The sets of jobs of agents A and B will be denoted as $\mathcal{J}_A$ and $\mathcal{J}_B$, respectively. We will assume that $\mathcal{J}_A \cap \mathcal{J}_B = \emptyset$, i.e., the sets do not intersect. Jobs of agent A and B will be denoted as $J_1^A, J_2^A, \ldots, J_{n_A}^A$ and $J_1^B, J_2^B, \ldots, J_{n_B}^B$, respectively, where $n_A + n_B = n$. The processing time of job J_k^X equals p_k^X, where $X \in \{A, B\}$. The objective of agent $X \in \{A, B\}$ is to minimize criterion f^X, with or without restrictions.

We denote two-agent time-dependent scheduling problems using the three-field notation with some extensions. For brevity, if jobs of both agents are of the same form and share the same additional requirements, we use the same notation as for single-agent problems. Otherwise, we describe the job characteristics separately.

Example 21.26. (a) The symbol $1|p_j = b_j t| \sum w_j^A C_j^A + L_{\max}^B$ denotes a single-machine, two-agent time-dependent scheduling problem in which *both* agents have proportional job processing times and the objective is to minimize the *sum* of the total weighted completion time of jobs of agent A and the maximum lateness of jobs of agent B, $\sum w_j^A C_j^A + L_{\max}^A$.
(b) The symbol $1|p_j^A = b_j^A t, p_j^B = a_j^B + b_j^B t| \sum w_j^A C_j^A + L_{\max}^B$ denotes the same problem as the one described in Example 21.26 (a), with the difference that now deteriorating jobs of agent B have linear processing times.
(c) The symbol $1|p_j = b_j t, \ C_{\max}^B \leqslant Q|L_{\max}^A$ denotes a single-machine, two-agent time-dependent scheduling problem in which both agents have proportional jobs and we have to find a schedule such that the maximum lateness $L_{\max}^A$ for jobs of agent A is minimal, provided that the maximum completion time $C_{\max}^B$ for jobs of agent B does not exceed a given upper bound $Q \geqslant 0$. ◆

Remark 21.27. In this section, we consider only two-agent time-dependent scheduling problems in which the sets of jobs of the agents do not intersect. Hence, we apply an abbreviated form of the three-field notation symbols, without symbol 'CO' in the β field of the notation, which indicates that $\mathcal{J}_A \cap \mathcal{J}_B = \emptyset$. For example, symbol $1|p_j = b_j t, \ C_{\max}^B \leqslant Q|L_{\max}^A$ is an abbreviated form of symbol $1|CO, p_j = b_j t, \ C_{\max}^B \leqslant Q|L_{\max}^A$. We refer the reader to Agnetis et al. [30, Sect. 1.4] for more details on the three-field notation for two-agent scheduling problems.   ◇

21.2.2 Proportional deterioration

In this section, we consider two-agent time-dependent scheduling problems with proportionally deteriorating job processing times.

Liu and Tang [41] considered problem $1|p_j = b_j t, \ C_{\max}^B \leqslant Q|L_{\max}^A$, where we have to find a schedule such that the maximum lateness $L_{\max}^A$ for jobs of agent A is minimal, provided that the maximum completion time $C_{\max}^B$ for jobs of agent B does not exceed a given upper bound $Q \geqslant 0$.

An optimal algorithm for problem $1|p_j = b_j t,\ C_{\max}^B \leqslant Q|L_{\max}^A$ is based on the following counterparts of properties for problem $1|C_{\max}^B \leqslant Q|L_{\max}^A$.

Property 21.28. (Liu and Tang [41]) An optimal schedule for problem $1|p_j = b_j t,\ C_{\max}^B \leqslant Q|L_{\max}^A$ is a non-idle schedule.

Proof. The property holds, since both criteria $L_{\max}^A$ and $C_{\max}^B$ are regular.   □

Property 21.29. (Liu and Tang [41]) Given an instance of problem $1|p_j = b_j t,\ C_{\max}^B \leqslant Q|L_{\max}^A$, the maximum completion time for agent B is given by formula (7.2) and the value does not depend on the schedule.

Proof. The property follows from Theorem 7.1.   □

Property 21.30. (Liu and Tang [41]) For problem $1|p_j = b_j t,\ C_{\max}^B \leqslant Q|L_{\max}^A$, there exists an optimal schedule in which the jobs of agent A are scheduled in non-decreasing order of due dates d_j^A.

Proof. The property follows from Theorem 7.84.   □

Property 21.31. (Liu and Tang [41]) For problem $1|p_j = b_j t,\ C_{\max}^B \leqslant Q|L_{\max}^A$, there exists an optimal schedule in which jobs of agent B are scheduled consecutively in a single block.

Proof. By the pairwise job interchange argument.   □

Let $C_{J_B}(\sigma)$ denote the completion time of the artificial job J_B in a schedule σ. Based on Properties 21.28–21.31, we can formulate the following optimal algorithm for problem $1|p_j = b_j t,\ C_{\max}^B \leqslant Q|L_{\max}^A$.

Algorithm 21.7. for problem $1|p_j = b_j t,\ C_{\max}^B \leqslant Q|L_{\max}^A$

---

1 **Input** : numbers n_A, n_B, sequences $(b_1^A, b_2^A, \ldots, b_{n_A}^A)$,
$(d_1^A, d_2^A, \ldots, d_{n_A}^A)$, $(b_1^B, b_2^B, \ldots, b_{n_B}^B)$
2 **Output:** an optimal schedule
  ▷ Step 1
3 $u \leftarrow t_0 \prod_{j=1}^{n_A}(1 + b_{[j]}^A) \prod_{j=1}^{n_B}(1 + b_{[j]}^B)$;
4 Arrange all jobs of agent A in non-decreasing order of due dates d_j^A;
5 Create artificial job J_B composed of all jobs of agent B;
6 **for** $i \leftarrow n_A + 1$ **downto** 1 **do**
7 | Create schedule σ of all jobs of agent A in which job J_B is in
| position i;
8 | **if** $(C_{J_B}(\sigma) \leqslant Q)$ **then**
9 | | **return** σ;
| **end**
  end
  ▷ Step 2
10 **write** 'Input instance is not feasible';
11 **return.**

The main idea of Algorithm 21.7 is as follows. We arrange the jobs of $\mathcal{J}^A$ in non-decreasing order of their due dates and, by Property 21.30, we replace all jobs from $\mathcal{J}^B$ with a single *artificial job* J_B. Next, at each iteration, we schedule job J_B in a given position, starting from position $n_A + 1$, and check whether the completion time of job J_B does not exceed the upper bound Q. If so, we return the schedule; otherwise, we decrease the position of J_B by 1 and pass to the next iteration.

Theorem 21.32. (Liu and Tang [41]) *Problem* $1|p_j = b_j t, C_{\max}^B \leqslant Q|L_{\max}^A$ *is solvable in* $O(n_A \log n_A + n_A n_B)$ *time by Algorithm 21.7.*

Proof. The correctness of Algorithm 21.7 follows from Properties 21.28–21.30. Lines 3 and 4 need $O(n_A + n_B)$ and $O(n_B \log n_B)$ time, respectively. The creation of the artificial job J_B in line 5 needs $O(n_B)$ time. Loop **for** in lines 6–9 is performed $O(n_A)$ times, while creating schedule σ in line 7 and checking the condition in sentence **if** in line 8 can be done in a constant time and $O(n_B)$ time, respectively. Hence, the overall running time of Algorithm 21.7 is equal to $n_A + n_B + n_A \log n_A + n_A n_B = O(n_A \log n_A + n_A n_B)$. ■

We illustrate the application of Algorithm 21.7 by an example.

Example 21.33. (Agnetis et al. [30]) Let us consider the following instance of problem $1|p_j = b_j t, C_{\max}^B \leqslant Q|L_{\max}^A$, in which $t_0 = 1$, $n_A = 2$, $n_B = 1$, $p_1^A = t$, $d_1^A = 6$, $p_2^A = 2t$, $d_2^A = 12$, $p_1^B = 3t$.

All possible job sequences for this instance, their job completion times and the values of criteria $L_{\max}^A$ and $C_{\max}^B$ are given in Table 21.1, where $L_{\max}^A(\sigma)$ and $C_{\max}^B(\sigma)$ denote the value of $L_{\max}^A$ and $C_{\max}^B$ for schedule σ, respectively.

Table 21.1: Sequences for an instance of problem $1|p_j = b_j t, C_{\max}^B \leqslant Q|L_{\max}^A$

Sequence σ	$p_{[1]}$	$C_{[1]}$	$p_{[2]}$	$C_{[2]}$	$p_{[3]}$	$C_{[3]}$	$L_{\max}^A(\sigma)$	$C_{\max}^B(\sigma)$
(J_1^A, J_2^A, J_1^B)	1	2	4	6	18	24	12	6
(J_1^A, J_1^B, J_2^A)	1	2	6	8	16	24	-4	24
(J_2^A, J_1^A, J_1^B)	2	3	3	6	18	24	12	3
(J_2^A, J_1^B, J_1^A)	2	3	9	12	12	24	18	3
(J_1^B, J_1^A, J_2^A)	3	4	4	8	16	24	2	24
(J_1^B, J_2^A, J_1^A)	3	4	8	12	12	24	18	12

Let us notice that in all schedules corresponding to sequences from Table 21.1 we have $C_{[3]} = 24$, since by Theorem 7.1 the value of $C_{\max}$ does not depend on schedule of time-dependent proportional jobs. Let us also notice that different values of Q may or may not be restrictive. For example, if $Q < 3$ then no schedule is feasible; if $3 \leqslant Q \leqslant 6$ then three schedules are feasible and two of them, (J_1^A, J_2^A, J_1^B) and (J_2^A, J_1^A, J_1^B), are optimal; if $Q \geqslant 24$ then all schedules are feasible but only one, (J_1^A, J_1^B, J_2^A), is optimal.   ◆

Liu and Tang [41] also considered problem $1|p_j = b_j t, f_{\max}^B \leqslant Q| \sum C_j^A$. This problem, in turn, is a time-dependent counterpart of the two-agent scheduling problem with fixed job processing times, $1|f_{\max}^B \leqslant Q| \sum C_j^A$, considered earlier by Agnetis et al. [31].

An optimal algorithm for problem $1|p_j = b_j t, f_{\max}^B \leqslant Q| \sum C_j^A$ is based on Properties 21.28–21.29, which still hold for the problem, and two new properties. Hereafter, we assume that the cost functions f_i^B of jobs of $\mathcal{J}^B$ are regular, and their values can be computed in a constant time. Let u be defined as in Algorithm 21.7.

Property 21.34. (Liu and Tang [41]) If in a feasible instance of problem $1|p_j = b_j t, f_{\max}^B \leqslant Q| \sum C_j^A$ there is a job $J_k^B \in \mathcal{J}^B$ such that $f_k^B(u) \leqslant Q$, then there exists an optimal schedule for the instance in which job J_k^B is scheduled in the last position, and there is no optimal schedule in which a job of $\mathcal{J}^A$ is scheduled in the last position.

Proof. By contradiction.   □

Property 21.35. (Liu and Tang [41]) If in a feasible instance of problem $1|p_j = b_j t, f_{\max}^B \leqslant Q| \sum C_j^A$ for all jobs $J_k^B \in \mathcal{J}^B$ it holds $f_k^B(u) > Q$, then in any optimal schedule job $J_l^A \in \mathcal{J}^A$ with the largest deterioration rate is scheduled in the last position.

Proof. By the pairwise job interchange argument.   □

Properties 21.34–21.35 imply that in any optimal schedule for problem $1|p_j = b_j t, f_{\max}^B \leqslant Q| \sum C_j^A$, jobs of $\mathcal{J}^A$ are arranged in non-decreasing order of their deterioration rates. Given a value of Q, for each job J_j^B one can define a '*deadline*' D_i^B such that $f_i^B(C_i^B) \leqslant Q$ if $C_i^B \leqslant D_i^B$ and $f_i^B(C_i^B) > Q$ otherwise. Each D_i^B can be computed in constant time if the inverse functions f_i^{B-1} are available, otherwise it requires $O(\log n_B)$ time.

Example 21.36. Let us consider an instance of problem $1|p_j = b_j t, f_{\max}^B \leqslant Q| \sum C_j^A$, where $n_A = 2$, $n_B = 2$, $t_0 = 1$, $b_1^B = 3$, $b_2^B = 2$, $b_1^A = 1$, $b_2^A = 4$, $f_1^B = C_1^B + 2$, $f_2^B = 3C_2^B + 1$ and $Q = 10$. Then $D_1^B = 8$ and $D_2^B = 3$, and $D_1^B > D_2^B$. However, if we assume that $f_1^B = \frac{1}{2}(C_1^B)^2$ and $f_2^B = \sqrt{(C_2^B)^2 + 36}$, then for the same value of Q we have $D_1^B = 2\sqrt{5}$ and $D_2^B = 8$, and $D_1^B < D_2^B$.   ◆

Based on the above properties, we can formulate an algorithm for problem $1|p_j = b_j t, f_{\max}^B \leqslant Q| \sum C_j^A$. This algorithm (see Algorithm 21.8) is a time-dependent variation of algorithm for problem $1|f_{\max}^B \leqslant Q| \sum C_j^A$ (cf. [31, Sect. 5.2]) and it can be formulated as follows.

Algorithm 21.8. for problem $1|p_j = b_j t, f_{\max}^B \leqslant Q| \sum C_j^A$

---

1 **Input** : numbers n_A, n_B, sequences $(b_1^A, b_2^A, \ldots, b_{n_A}^A)$,
$\qquad\qquad (b_1^B, b_2^B, \ldots, b_{n_B}^B)$, $(f_1^B, f_2^B, \ldots, f_{n_B}^B)$

2 **Output:** an optimal schedule

▷ **Step 1**

3 $J \leftarrow \mathcal{J}^A \cup \mathcal{J}^B$;

4 $u \leftarrow t_0 \prod_{j=1}^{n_A}(1 + b_{[j]}^A) \prod_{j=1}^{n_B}(1 + b_{[j]}^B)$;

5 Arrange all jobs of agent A in non-decreasing order of the
  deterioration rates b_j^A;

6 **for** $j \leftarrow 1$ **to** n_B **do**

7 $\quad$| Define for job J_j^B 'deadline' D_j^B;

$\quad$**end**

8 Arrange all jobs of agent B in non-decreasing order of the 'deadlines' D_j^B;

▷ **Step 2**

9 $\sigma \leftarrow \emptyset$;

10 **while** (*there exist in J unscheduled jobs*) **do**

11 $\quad$**if** (*there exists job $J_k^B \in \mathcal{J}^B$ such that $f_k^B(u) \leqslant Q$*) **then**

12 $\quad\quad$| $J_{sel} \leftarrow J_k^B$;

$\quad$**else**

13 $\quad\quad$**if** (*all jobs of agent B have been scheduled*) **then**

14 $\quad\quad\quad$**write** 'Input instance is infeasible';

15 $\quad\quad\quad$**return**;

$\quad\quad$**else**

16 $\quad\quad\quad$| $J_{sel} \leftarrow$ a job of agent A with the largest deterioration rate;

$\quad\quad$**end**

$\quad$**end**

17 $\quad$Schedule job J_{sel} in the last position in σ;

18 $\quad$$J \leftarrow J \setminus \{J_{sel}\}$;

19 $\quad$$u \leftarrow \frac{u}{1 + b_{sel}}$;

end

▷ **Step 3**

20 **return** σ.

Algorithm 21.8 is composed of the following three steps. In the first step, we calculate the length u of a schedule of all jobs of both agents, next we arrange all jobs of agent A in non-decreasing order of their deterioration rates, calculate the values of 'deadlines' of all jobs of agent B and arrange the jobs in non-decreasing order of these 'deadlines'.

In the second step, at each iteration we select an unscheduled job to be scheduled in the last position. If it is possible, we select a job of agent B, otherwise we select a job of agent A with the largest deterioration rate. We proceed as long as there are unscheduled jobs in $\mathcal{J}$. If all jobs of agent A have already been scheduled, and no job of agent B can be feasibly scheduled in the current last position, the instance is infeasible.

In the third step, we return the schedule generated in the second step. The pseudo-code of this algorithm is as follows.

Theorem 21.37. (Liu and Tang [41]) *Problem* $1|p_j = b_j t, f_{\max}^B \leqslant Q| \sum C_j^A$ *is solvable in* $O(n_A \log n_A + n_B \log n_B)$ *time by Algorithm 21.8.*

Proof. The correctness of Algorithm 21.8 follows from Properties 21.28–21.29 and 21.34–21.35. Lines 3 and 4 both need $O(n_A + n_B)$ time. Lines 5 and 8 need $O(n_A \log n_A)$ and $O(n_B \log n_B)$ time, respectively. Loop **for** in lines 6–7 need $O(n_B)$ time. Loop **while** in lines 10–18 is performed $O(n_A + n_B)$ times, and each such iteration needs constant time, since a single job is selected. Hence, the overall running time of Algorithm 21.8 is $O(n_A \log n_A + n_B \log n_B)$. $\square$

Liu et al. [42] proposed a polynomial algorithm for an extension of problem $1|p_j = b_j t, f_{\max}^B \leqslant Q| \sum C_j$ to a batch scheduling problem with time-dependent setup times.

Remark 21.38. Time-dependent setup times in time-dependent batch scheduling problems will be denoted similarly to time-dependent job processing times. For example, proportional setup times will be denoted by symbol '$s_i = \theta_i t$' in the β field of the three-field notation.

This new problem, $1|p_j = b_j t, s_i = \theta_i t, GT, f_{\max}^B \leqslant Q| \sum C_j$, is solvable by a similar algorithm as the one for problem $1|p_j = b_j t, f_{\max}^B \leqslant Q| \sum C_j$. Therefore, we present only the main result.

Theorem 21.39. (Liu et al. [42]) *Problem* $1|p_j = b_j t, s_i = \theta_i t, f_{\max}^B \leqslant Q| \sum C_j$ *is solvable in* $O(n_A \log n_A)$ *time.*

Proof. See [42, Theorem 1].   ◇

Gawiejnowicz et al. [36] considered the problem of minimizing the total tardiness $\sum T_j^A$ of jobs of agent A, given that no job of agent B is tardy. For this problem, the authors proposed a branch-and-bound algorithm based on Properties 21.40–21.43.

Property 21.40 gives conditions under which a schedule σ dominates σ', i.e., when $\sum_j T_j^A(\sigma) \leqslant \sum_j T_j^A(\sigma')$.

Property 21.40. (Gawiejnowicz et al. [36]) Let $B_i := 1 + b_i$, $B_j := 1 + b_j$ and $B_{ij} = B_{ji} := (1 + b_i)(1 + b_j)$. Schedule σ dominates schedule σ' if *any* of the following holds:

(a) $J_i \in \mathcal{J}^A$ and $J_j \in \mathcal{J}^A$ are such that $B_{ij} t < d_j^A$;

(b) $J_i \in \mathcal{J}^A$ and $J_j \in \mathcal{J}^A$ are such that $B_i t \geqslant d_i^A$, $B_j t \geqslant d_j^A$ and $b_i < b_j$;

(c) $J_i \in \mathcal{J}^B$ and $J_j \in \mathcal{J}^A$ are such that $B_{ij} t \geqslant d_j^A$;

(d) $J_i \in \mathcal{J}^B$ and $J_j \in \mathcal{J}^A$ are such that $B_i t \geqslant d_i^A$ and $B_{ji} t \geqslant d_i^A$ or

(e) $J_i \in \mathcal{J}^B$ and $J_j \in \mathcal{J}^B$ are such that $B_i t \geqslant d_i^A$, $B_{ij} t \geqslant d_j^A$.

Proof. By the pairwise job interchange argument. □

The next two results, Property 21.41 and 21.42, allow us to determine a sequence of unscheduled jobs and the feasibility of a given schedule, and are used to speed up the search of the tree of all possible schedules.

Let π denote a sequence of unscheduled jobs. Also, let $(\pi, [k+1], \ldots, [n])$ be the schedule in which the order of the last $n - k$ jobs has been determined backwards and let $(\pi^{\nearrow}, [k+1], \ldots, [n])$ be the schedule in which unscheduled jobs are arranged in non-decreasing order of due dates.

Property 21.41. (Gawiejnowicz et al. [36]) If in sequence $\pi^{\nearrow}$ there are no tardy jobs, then schedule $(\pi^{\nearrow}, [k+1], \ldots, [n])$ dominates any schedule in the form of $(\pi, [k+1], \ldots, [n])$.

Property 21.42. (Gawiejnowicz et al. [36]) If job $J_{[i]}$ of agent B is such that

$$t_0 \prod_{j \in \pi}(1 + b_j) \prod_{j=k+1}^{i}(1 + b_{[j]}) \geqslant d_{[i]}^B$$

for $k + 1 \leqslant i \leqslant n$, then $(\pi, [k+1], \ldots, [n])$ is not a feasible schedule.

Proofs of Properties 21.41–21.42 follow from definitions of schedule dominance and feasibility.

The next result, Property 21.43, gives a lower bound on the total tardiness for jobs in $\mathcal{J}^A$. Let $\sigma = (\pi, [k+1], \ldots, [n])$ denote a schedule in which the order of the last $n - k$ jobs has been determined backwards, and assume that among the unscheduled jobs there are k_A jobs from $\mathcal{J}^A$ and k_B jobs from $\mathcal{J}^B$, where $k_A + k_A = k$.

Property 21.43. (Gawiejnowicz et al. [36]) Given $\sigma = (\pi, [k+1], \ldots, [n])$, we have

$$\sum T_j^A(\pi) \geqslant \sum_{i=1}^{k_A} \max\left\{C_{(i)}^A(\pi) - d_{(i)}^A, 0\right\}$$

and

$$\sum T_j^A(\sigma) \geqslant \sum_{i=1}^{k_A} \max\left\{C_{(i)}^A(\sigma) - d_{(i)}^A, 0\right\} + \sum_{k_A+1 \leqslant i \leqslant k, J_{[i]} \in \mathcal{J}^A} T_{[i]}^A(\sigma).$$

Proof. Since $C_{[j]}(\sigma) = t_0 \prod_{i=1}^{j}(1 + b_{[i]}) \geqslant t_0 \prod_{i=1}^{j}(1 + b_{(i)})$ for $1 \leqslant j \leqslant k$, $t_0 \prod_{i=1}^{j}(1 + b_{(i)})$ is a lower bound on the completion time of job $J_{[j]}$ in σ. Moreover, since the jobs of agent B cannot be tardy, we should complete the jobs in $\mathcal{J}^B$ as late as possible but before their due dates. We can do this by checking whether the completion time of a given job does not exceed its deadline and if so, putting the job in the right place in a schedule. Repeating this procedure a number of times, we obtain the given two bounds. ■

For this problem, the same authors proposed a genetic algorithm (see Algorithm 21.9), using the TEAC library developed by Gawiejnowicz et al. [37]. In this algorithm, implemented in C#, the operator of mutation was called with probability 0.02, the operator of crossing was called with probability 0.80, the size of the offspring population was equal to 50 individuals, and tournament preselection (with tournament size equal to 5) and random postselection were used. The stop condition was passing 100 generations. The pseudo-code of the algorithm is as follows.

Algorithm 21.9. for problem $1|p_j = b_j t, \sum U_j^B = 0| \sum T_j^A$

---

1 **Input** : numbers n_A, n_B, sequences $(b_1^A, b_2^A, \ldots, b_{n_A}^A)$,
        $(d_1^A, d_2^A, \ldots, d_{n_A}^A)$, $(b_1^B, b_2^B, \ldots, b_{n_B}^B)$

2 **Output:** a suboptimal schedule

  ▷ Step 1

3 Create base population P_0;

4 Evaluate P_0;

5 **while** (*does not hold stop condition*) **do**

6 | Create temporary population T_i by using a preselection operator to P_i;

7 | Create offspring population O_i by using crossing and mutation operators to T_i;

8 | Evaluate O_i;

9 | Create a new population P_{i+1} by using a postselection operator to P_i and O_i;

  end

  ▷ Step 2

10 **return** *the best solution in* P_i.

Liu et al. [42] considered a batch scheduling problem $1|p_j = b_j t, s_i = \theta_i t, GT, f_{\max}^B \leqslant Q| \sum C_j^A$, which is an extension of problem $1|p_j = b_j t, f_{\max}^B \leqslant Q| \sum C_j^A$. As opposed to the latter case, where all jobs belong to the same group, now jobs are divided into a number of distinct groups, each group corresponds to a batch, and a sequence-dependent setup time is needed between jobs from different groups.

Problem $1|p_j = b_j t, s_i = \theta_i t, GT, f_{\max}^B \leqslant Q| \sum C_j^A$ is not more difficult than problem $1|p_j = b_j t, f_{\max}^B \leqslant Q| \sum C_j^A$, since it can still be solved in $O(n_A \log n_A + n_A n_B)$ time by a slightly modified Algorithm 21.7.

21.2.3 Proportional-linear deterioration

In this section, we consider two-agent time-dependent scheduling problems with proportional-linearly deteriorating job processing times.

Liu et al. see [43] considered problems $1|p_j = b_j(a + bt), C_{\max}^B \leqslant Q| L_{\max}^A$ and $1| p_j = b_j(a + bt), f_{\max}^B \leqslant Q| \sum C_j^A$ which are extensions of problems

$1|p_j = b_jt, C_{\max}^B \leqslant Q|L_{\max}^A$ and $1|p_j = b_jt, f_{\max}^B \leqslant Q|\sum C_j$, presented in Sect. 21.2.2, to the case of proportional-linear job processing times.

The replacement of proportional job processing times with proportional-linear job processing times does not increase the time complexity of these problems. Moreover, algorithms for these two problems are based on the same properties and have almost the same form as their counterparts for proportional job processing times. For example, problem $1|p_j = b_j(a + bt), C_{\max}^B \leqslant Q|L_{\max}^A$ can be solved in $O(n_A \log n_A + n_A n_B)$ time by an appropriately modified Algorithm 21.7 for problem $1|p_j = b_jt, C_{\max}^B \leqslant Q|L_{\max}^A$ (cf. Sect. 21.2.2). A similar claim concerns problem $1|\ p_j = b_j(a + bt), f_{\max}^B \leqslant Q|\sum C_j^A$ which can be solved in $O(n_A \log n_A + n_B \log n_B)$ time by an appropriately modified Algorithm 21.8 for problem $1|p_j = b_jt, f_{\max}^B \leqslant Q|\sum C_j^A$ (cf. Sect. 21.2.2).

For problem $1|p_j = b_j(a+bt)|T_{\max}^A + \sum T_j^B$, where $T_{\max}^X(\sigma) := \max\{T_j(\sigma)\}$, $T_j^X(\sigma) := \max\{0, C_j(\sigma) - d_j\}$ and $\sum T_j^X(\sigma) := \sum_{J_j \in \mathcal{J}^X} T_j^X(\sigma)$ for a given schedule σ and $X \in \{A, B\}$, Wu et al. [50] proposed a branch-and-bound algorithm, an ACO algorithm and a SA algorithm.

Yin et al. [52] considered problem $1|p_j = b_j(a + bt), f_{\max}^B \leqslant Q|f_{\max}^A$, a time-dependent counterpart of problem $1|f_{\max}^B \leqslant Q|f_{\max}^A$ considered earlier by Agnetis et al. [31]. Yin et al. [52] proved a few counterparts of properties proved in [31, Sect. 4] and on the basis of these results they showed that the problem can be solved in $O(n_A^2 + n_B \log n_B)$ time.

The same authors also proved the following $\mathcal{NP}$-hardness result.

Theorem 21.44. (Yin et al. [52]) *Problem* $1|p_j = b_j(a + bt), C_{\max}^B \leqslant Q|\sum w_j^A(1 - e^{-rC_j^A})$ *is* $\mathcal{NP}$-*hard in the ordinary sense.*

Proof. Yin et al. [52] proved the result by showing the $\mathcal{NP}$-hardness of problem $1||\sum w_j^A(1 - e^{-rC_j^A})$ via the following reduction from the KP problem: $n_A = k$, $p_j^A = u_j$ and $w_j^A = w_j$ for $1 \leqslant j \leqslant n_A$, $n_B = 1$ and $p_1^B = uw$, where $u := \sum_{j=1}^k u_j$ and $w := \sum_{j=1}^k w_j$. Threshold values are

$$U_A = (1 - e^{-r})(uw + (w - W)p_1^B) = (1 - r^{-r})(1 + w - W)p_1^B$$

and

$$U_B = b + p_1^B,$$

where r is a constant such that $0 < r < 1$.

In order to complete the proof it is sufficient to show that an instance of the KP problem has a solution if and only if the above instance of the decision version of problem $1||\sum w_j^A(1 - e^{-rC_j^A})$ has a solution, and to notice that the latter problem is a special case of problem $1|p_j = b_j(a + bt), C_{\max}^B \leqslant Q|\sum w_j^A(1 - e^{-rC_j^A})$; we refer the reader to [52, Theorem 3.9] for details.   □

Wang et al. [48] proved several results which are time-dependent counterparts of two-agent scheduling problems with fixed job processing times, considered earlier by Baker and Smith [32]. Proofs of the results are based on time-dependent counterparts of properties proved in [32], similar to those mentioned in the following property.

Property 21.45. (Wang et al. [48]) (*a*) If agent $X \in \{A, B\}$ possesses the criterion $C_{\max}^X$, then there exists an optimal schedule in which jobs of the agent are processed consecutively. (*b*) If agent $X \in \{A, B\}$ possesses the criterion $L_{\max}^X$, then there exists an optimal schedule in which jobs of the agent are processed in the EDD order.

Proof. (*a*)(*b*) By the adjacent job interchange technique.   □

Based on Property 21.45, Wang et al. [48] proved the following result.

Theorem 21.46. (Wang et al. [48])
(*a*) *Problem* $1|p_j = b_j(a + bt)|C_{\max}^A + \theta C_{\max}^B$ *is solvable in* $O(n_A + n_B)$ *time.*
(*b*) *Problem* $1|p_j = b_j(a + bt)|L_{\max}^A + \theta L_{\max}^B$ *is solvable in* $O(n_A^2 n_B + n_A n_B^2)$ *time.*
(*c*) *Problem* $1|p_j = b_j(a + bt)|C_{\max}^A + \theta L_{\max}^B$ *is solvable in* $O(n_B \log n_B)$.
(*d*) *Problem* $1|p_j = b_j(a + bt)|\sum w_j^A C_j^A + \theta L_{\max}^B$ *is* $\mathcal{NP}$-*hard in the strong sense.*

Proof. (*a*) Since by Property 21.45 (a) jobs of agents A and B compose two blocks of jobs, in order to find an optimal schedule we should only check which order of the blocks is better; see [48, Sect. 3.1].
(*b*) Since by Property 21.45(b) jobs of agents A and B are in the EDD order, we have only to decide how the jobs should be interleaved; see [48, Sect. 3.2].
(*c*) Since by Property 21.45(a)(b) jobs of agent A compose a single block, while jobs of agent B are in the EDD order, in order to find an optimal schedule we have only to choose where the block of jobs of agent A should be located; see [48, Sect. 3.4].
(*d*) A special case of this problem is a strongly $\mathcal{NP}$-hard problem $1||\sum w_j^A C_j^A + \theta L_{\max}^B$ [32, Theorem 1].   □

He and Leung [39] considered the problem of finding all Pareto optimal schedules (cf. Definition 1.26) for several two-agent time-dependent scheduling problems, applying a general method of finding all Pareto-optimal points, called *ϵ-constrained approach*. This approach is as follows. To obtain the first Pareto-optimal point (x^1, y^1) we compute the minimum value y^1 of criterion f_2, and then we compute the minimum value x^1 of criterion f_1 under the constraint that $f_2 = y^1$. Given the generated Pareto-optimal point (x^i, y^i), we generate the next Pareto-optimal point (x^{i+1}, y^{i+1}), where y^{i+1} is obtained by minimizing f_2 subject to the constraint that $f_1 < x^i$, while x^{i+1} is obtained by minimizing f_1 subject to the constraint that $f_2 \leqslant y^{i+1}$. We proceed in this way as long as we can generate new points, otherwise we stop.

Remark 21.47. A more detailed description of the ϵ-constrained approach may be found in the literature; see, e.g., Agnetis et al. [30, Chap. 1], T'kindt and Billaut [46, Sect. 3.6].

All Pareto optimal schedules for a time-dependent two-agent scheduling problem can be generated using the ϵ-constrained approach, in two steps. In the first step, we generate the first Pareto optimal schedule. Next, starting from the previous Pareto optimal schedule, we generate the next Pareto optimal schedule, etc.

He and Leung [39] applied ϵ-constrained approach to solve several two-agent time-dependent scheduling problems, using algorithms composed of two steps. In the first step, we find a Pareto optimal schedule for problem $1p_j = b_j(a + bt), f_{\max}^B \leqslant Q | \sum C_j^A$, assuming that jobs in set $\mathcal{J}^A$ are arranged in non-decreasing order of deterioration rates b_j^A. This step can be done with the following Algorithm 21.10.

Algorithm 21.10. for problem $1p_j = b_j(a + bt), f_{\max}^B \leqslant Q | \sum C_j^A$

1 **Input** : numbers Q, n_A, n_B, a, b, t_0, sequences $(b_1^A, b_2^A, \ldots, b_{n_A}^A)$,
$\quad\quad\quad (b_1^B, b_2^B, \ldots, b_{n_B}^B)$,
2 **Output:** a Pareto optimal schedule
$\quad \triangleright$ Step 1
3 $\sigma \leftarrow (\phi); t \leftarrow (t_0 + \frac{a}{b}) \prod_{j=1}^{n_A}(1 + bb_j^A) \prod_{j=1}^{n_B}(1 + bb_j^B) - \frac{a}{b};$
4 $\mathcal{J}^{(A)} \leftarrow \mathcal{J}^A; \mathcal{J}^{(B)} \leftarrow \mathcal{J}^B;$
$\quad \triangleright$ Step 2
5 **while** $(\mathcal{J}^{(B)} \neq \emptyset)$ **do**
6 $\quad$ **if** $(n_B \neq 0)$ **then**
7 $\quad\quad$ Find job $J_k^B \in \mathcal{J}^{(B)}$ such that
$\quad\quad\quad f_k^B(t) = \min\{f_j^B(t) \leqslant Q : J_j \in \mathcal{J}^{(B)}\};$
8 $\quad\quad \sigma \leftarrow (J_k | \sigma);$
9 $\quad\quad t \leftarrow \frac{t - bb_j^B}{1 + bb_j^B};$
10 $\quad\quad \mathcal{J}^{(B)} \leftarrow \mathcal{J}^{(B)} \setminus \{J_k^B\};$
$\quad$ **else**
$\quad\quad$ **if** $(|\mathcal{J}^{(A)}| = 0)$ **then**
11 $\quad\quad\quad$ write 'Infeasible instance';
$\quad\quad$ **else**
12 $\quad\quad\quad \sigma \leftarrow (J_{|n_A|}^A | \sigma);$
13 $\quad\quad\quad t \leftarrow \frac{t - bb_j^A}{1 + bb_j^A};$
14 $\quad\quad\quad \mathcal{J}^{(A)} \leftarrow \mathcal{J}^{(A)} \setminus \{J_{|n_A|}^A\};$
15 $\quad\quad\quad n_A \leftarrow n_A - 1;$
$\quad\quad$ **end**
$\quad$ **end**
end
$\quad \triangleright$ Step 3
16 **return** σ.

Theorem 21.48. (He and Leung [39]) *Algorithm 21.10 generates a Pareto-optimal schedule for problem* $1|p_j = b_j(a+bt), f_{max}^B \leqslant Q| \sum C_j^A$ *in* $O(n_A + n_B^2)$ *time.*

Proof. See [39, Theorem 2.5]. ◇

In the second step of solving problem $1p_j = b_j(a + bt)|(\sum C_j^A, f_{max}^B)$ with the ϵ-constrained approach, we generate all Pareto optimal schedules for this problem using the following Algorithm 21.11.

Algorithm 21.11. for problem $1p_j = b_j(a + bt)|(\sum C_j^A, f_{max}^B)$

---

1 **Input** : numbers n_A, n_B, a, b, t_0, ϵ, sequences $(b_1^A, b_2^A, \ldots, b_{n_A}^A)$,
      $(b_1^B, b_2^B, \ldots, b_{n_B}^B)$,
2 **Output:** all Pareto optimal schedules
  ▷ Step 1
3 $C \leftarrow (t_0 + \frac{a}{b}) \prod_{j=1}^{n_A}(1 + bb_j^A) \prod_{j=1}^{n_B}(1 + bb_j^B) - \frac{a}{b}$;
4 $Q \leftarrow \max\{f_j^B(C) : 1 \leqslant j \leqslant n_B\}$;
5 $i \leftarrow 0$;
  ▷ Step 2
6 Call Algorithm 21.10;
7 **if** $(\sigma = (\phi))$ **then**
8 | write 'Infeasible instance';
  else
9 | $\sigma^{i+1} \leftarrow \sigma$;
10 | $i \leftarrow i + 1$;
11 | $Q \leftarrow \max\{f_j^B(\sigma^i) - \epsilon\}$;
  end
  ▷ Step 3
12 **return** $\sigma^1, \sigma^2, \ldots, \sigma^i$.

Theorem 21.49. (He and Leung [39]) *All Pareto-optimal schedules for problem* $1p_j = b_j(a + bt)|(\sum C_j^A, f_{max}^B)$ *can be generated in* $O(n_A n_B(n_B^2 + n_B))$ *time by Algorithm 21.11.*

Proof. See [39, Theorems 2.5–2.6]. ◇

He and Leung [39] applied the same approach to the case when both agents apply the maximum cost criteria.

Theorem 21.50. (He and Leung [39]) *All Pareto-optimal schedules for problem* $1p_j = b_j(a+bt)|(f_{max}^A, f_{max}^B)$ *can be generated in* $O(n_A n_B(n_A^2 + n_B^2))$ *time by the modified Algorithm 21.11.*

Proof. See [39, Theorems 2.11–2.12]. ◇

Tang et al. [45] considered several two-agent single-machine batch scheduling problems with proportional-linearly deteriorating jobs. The authors assumed that capacity c of the machine may be either bounded, $c < n$, or unbounded, $c = +\infty$. Two types of agents were analyzed: *incompatible agents*, when each batch contains only jobs of the same agent, and *compatible agents*, when batches may be composed of jobs of both agents. Tang et al. [45] assumed that the machine is a *parallel-batching machine*, i.e., the size of a batch equals the maximal processing time among all jobs assigned to the batch.

Remark 21.51. (a) Compatible and incompatible agents will be denoted in the β field of the three-field notation by symbols '*IF*' and '*CF*', respectively. (b) Parallel batching will be denoted in the β field by symbol '*p-batch*'.

Let C and Δ be defined as follows:

$$C := \left(t_0 + \frac{a}{b}\right) \prod_{j=1}^{n_A} \left(1 + bb_j^A\right) \prod_{j=1}^{n_B} \left(1 + bb_j^B\right) - \frac{a}{b} \qquad (21.10)$$

and

$$\Delta := \max \{f_j(C) : 1 \leqslant j \leqslant n\}. \qquad (21.11)$$

Tang et al. [45] noticed that there is an optimal schedule for problem $1|p\text{-}batch, p_j = b_j(a + bt), c = +\infty, IF, f_1^B \leqslant Q|f_2^A$, where jobs (batches) of each agent are scheduled in non-decreasing order of deterioration rates b_j, and that a similar property holds for problem $1|p\text{-}batch, p_j = b_j(a + bt), c = +\infty, CF, f_1^B \leqslant Q|f_2^A$ (see [45, Lemmas 3.1–3.2]). Based on the properties, they proved the following result for the unbounded case.

Theorem 21.52. (Tang et al. [45]) *Let C and Δ be defined by Eqs. (21.10) and (21.11), respectively. Then*
(a) *problem $1|p\text{-}batch, p_j = b_j(a + bt), c = +\infty, IF, f_{\max}^B \leqslant Q|\sum f_j^A$ is solvable in $O(n_A^3 n_B n^2 \Delta^2)$ time by a dynamic programming algorithm;*
(b) *problem $1|p\text{-}batch, p_j = b_j(a + bt), c = +\infty, CF, f_{\max}^B \leqslant Q|\sum f_j^A$ is solvable in $O(n^5 \Delta^2)$ time by a dynamic programming algorithm;*
(c) *problem $1|p\text{-}batch, p_j = b_j(a + bt), c = +\infty, IF, \sum f_j^B \leqslant Q|\sum f_j^A$ is solvable in $O(n_A n_B n \Delta Q(n_A^2 n \Delta + n_B^2 Q))$ time by a dynamic programming algorithm;*
(c) *problem $1|p\text{-}batch, p_j = b_j(a + bt), c = +\infty, CF, \sum f_j^B \leqslant Q|\sum f_j^A$ is solvable in $O(n^5 \Delta^2 Q^2)$ time by a dynamic programming algorithm.*

Proof. (a) See [45, Theorem 3.1.1]. (b) See [45, Theorem 3.1.2]. (c) See [45, Theorem 3.1.3]. (d) See [45, Theorem 3.1.4].   ◇

Tang et al. [45] proved that problem $1|p\text{-}batch, p_j = b_j(a + bt), r_j, c = +\infty, IF, C_{\max}^B \leqslant Q|C_{\max}^A$ can be solved in $O(n_A n_B n \log(Q_U - Q_L))$ time by a weakly polynomial (cf. Sect. 2.2.4) dynamic programming algorithm, where Q_U and Q_L are the upper and lower bounds on the $C_{\max}$ criterion value, respectively. For the same problem, Gao et al. [35] proposed a strongly polynomial algorithm (cf. Sect. 2.2.4), running in $O(n^3)$ time.

Tang et al. [45] claimed that problem $1|p\text{-}batch, p_j = b_j(a+bt), r_j, c = +\infty,$ $CF, C_{\max}^B \leqslant Q|C_{\max}^A$ can be solved in $O(\max\{n^2, n^2 n_A, n^2 n_B\}) = O(n^3)$ time. This result was improved by Gao et al. [35], who proposed solving the latter problem with an $O(n^2 \log n)$ time algorithm.

Tang et al. [45] also proved $\mathcal{NP}$-completeness of decision versions of some two-agent time-dependent scheduling problems.

Theorem 21.53. (Tang et al. [45]) *(a) The decision version of problem* $1|p\text{-}batch, p_j = b_j(a + bt), c < n, CF, C_{\max}^A \leqslant Q_A|C_{\max}^B$ *is $\mathcal{NP}$-complete in the ordinary sense even if $c = 2$. (b) The decision version of problem* $1|p\text{-}batch, p_j = b_j(a + bt), c < n, \sum U_j^A \leqslant Q_A|\sum U_j^B$ *is $\mathcal{NP}$-complete in the ordinary sense for both incompatible and compatible agents, even if $c = 2$.*

Proof. Tang et al. [45] applied the following reductions from the EPP problem. (a) $n_A = 3q$, $n_B = q$, $a = 0$, $b = 1$, $t_0 = 1$, $c = 2$, $b_j^A = E^{4j} z_j - 1$ for $1 \leqslant j \leqslant q$, $b_{q+j}^A = b_{2q+j} = \frac{E^{4j}}{z_j} - 1$ for $1 \leqslant j \leqslant q$ and $b_j^B = E^{4j} - 1$ for $1 \leqslant j \leqslant q$, where $E = \sqrt{\prod_{j=1}^q z_j}$. Threshold values are $Q_A = E^{4(q^2+q)+1}$ and $Q_B = E^{2(q^2+q)+1}$.

In order to complete the proof it is sufficient to show that an instance of the PP problem has a solution if and only if there exists a feasible schedule σ for the above instance such that $C_{\max}^A(\sigma) \leqslant Q^A$ and $C_{\max}^B(\sigma) \leqslant Q^B$.

(b) The result is proved by showing that a single-agent problem $1|p\text{-}batch,$ $p_j = b_j(a+bt), c < n|\sum U_j$ is $\mathcal{NP}$-complete in the ordinary sense. The latter is proved by the following reduction from the EPP problem: $n = 4q$, $a = 0$, $b = 1$, $t_0 = 1$, $c = 2$, $b_j = E^{4j} z_j - 1$ for $1 \leqslant j \leqslant q$, $b_{q+j} = b_{2q+j} = \frac{E^{4j}}{z_j} - 1$ and $b_{3q+j} = E^{4j} - 1$ for $1 \leqslant j \leqslant q$, $d_j = d_{q+j} = d_{2q+j} = E^{4(q^2+q)+1}$ and $d_{3q+j} = E^{2(q^2+q)+1}$ for $1 \leqslant j \leqslant q$, where $E = \sqrt{\prod_{j=1}^q z_j}$. The threshold value is $G = 0$. The remaining part of the proof is similar to that of Theorem 21.53 (a).

Since problem $1|p\text{-}batch, p_j = b_j(a + bt), c < n|\sum U_j$ is a special case of problem $1|p\text{-}batch, p_j = b_j(a+bt), c < n, \sum U_j^B \leqslant Q|\sum U_j^A$, the latter is $\mathcal{NP}$-hard in the ordinary sense. □

Theorem 21.54. *(a) (Tang et al. [45]) If job ready times and job due dates are agreeable, then the decision version of problem* $1|p\text{-}batch, p_j = b_j(a+bt), r_j,$ $c < n, \sum U_j^A \leqslant Q_A|\sum U_j^B$ *is $\mathcal{NP}$-complete in the ordinary sense, even if $c = 2$.*
(b) (Gao et al. [35]) The decision version of problem $1|p\text{-}batch, p_j = b_j(a + bt), r_j, c < n, C_{\max}^B \leqslant Q_B|C_{\max}^A$ *is $\mathcal{NP}$-complete in the ordinary sense even if $c = 1$ and $n_A = 1$.*

Proof. (a) Tang et al. [45] proved the result by showing that the decision version of single-agent problem $1|p\text{-}batch, p_j = b_j(a + bt), r_j, c < n|\sum U_j$ is $\mathcal{NP}$-complete in the ordinary sense if job ready times and job deadlines are agreeable. The latter is proved by the following reduction from the 4-P problem: $n = 10p$, $a = 0$, $b = 1$, $t_0 = 1$, $c = 2$, job deterioration rates $b_j^A = D - 1$ for $1 \leqslant j \leqslant p$, $b_{p+j}^A = u_{\lceil \frac{1}{2}(j-p+1) \rceil} - 1$ for $1 \leqslant j \leqslant 9p$, $b_{9p+j}^B = D - 1$ for $1 \leqslant j \leqslant p$, job ready times $r_j = t_0$ for $1 \leqslant j \leqslant 9p$, $r_{9p+j} = D^{2j-1} - 1$ for $1 \leqslant j \leqslant p$ and job deadlines $d_j = D^{2j}$ for $1 \leqslant j \leqslant p$ and $d_{p+j} = D^{2p}$ for $1 \leqslant j + 9p$. The threshold value is $G = 0$.

In order to complete the proof it is sufficient to show that the 4-P problem has a solution if and only if there exists a feasible schedule σ for the above instance, such that $\sum U_j(\sigma) \leqslant G$.

In view of the above, we can conclude that when the threshold value Q^B of agent B is sufficiently large, the decision version of two-agent problem $1|p\text{-}batch, p_j = b_j(a + bt), r_j, c < n, \sum U_j^A \leqslant Q_A | \sum U_j^B$ is equivalent to single-agent problem $1|p\text{-}batch, p_j = b_j(a + bt), r_j, c < n | \sum U_j$ and hence the former is $\mathcal{NP}$-complete in the ordinary sense as well.

(b) Gao et al. [35] applied the following reduction from the SP problem: $n_A = 1$, $n_B = p$, $a = 0$, $b = 1$, $c = 1$, job J_1^A has ready time $r_1^A = B$, deterioration rate $b_1^A = h - 1$, where $2 \leqslant h \leqslant B$, jobs J_j^B have ready times $r_j^B = 1$ and deterioration rates $b_j^B = y_j - 1$ for $1 \leqslant j \leqslant p$. Threshold values are $Q_A = hB$ and $Q_B = hB^2$.

In order to complete the proof it is sufficient to show that the SP problem has a solution if and only if there exists a feasible schedule σ for the above instance such that $C_{\max}^A(\sigma) \leqslant Q^A$ and $C_{\max}(\sigma) \leqslant Q^B$. $\square$

21.2.4 Linear deterioration

In this section, we consider two-agent time-dependent scheduling problems with linearly deteriorating job processing times.

Ding et al. [34] described several results concerning single machine two-agent time-dependent scheduling problems, where all jobs have the same linear job processing times in the form of $p_j = 1 + bt$. The results, given without proofs, are time-dependent counterparts of the results on two-agent scheduling by Baker and Smith [32].

Lee et al. [40] considered problem $1|p_j = a_j + bt, \sum U_j^B = 0| \sum w_j^A C_j^A$, where jobs from set $\mathcal{J}^B$ must be completed before their due dates, while jobs of agent A want to minimize their total weighted completion time. This problem is strongly $\mathcal{NP}$-hard, because it includes a strongly $\mathcal{NP}$-hard problem $1| \sum U_j^B = 0| \sum w_j^A C_j^A$ as a special case (see [30, Chap. 3]).

For this problem, based on the following result, Lee et al. [40] proposed a branch-and-bound algorithm.

Property 21.55. (Lee et al. [40]) Let $P_i := a_i + bt$, $P_j := a_j + bt$ and $P_{ij} := a_i(1 + b) + (1 + b)^2 t$. Schedule σ dominates schedule σ' if *any* of the following holds:

(a) $J_i \in \mathcal{J}^A$ and $J_j \in \mathcal{J}^A$ are such that $w_j P_i < w_i P_j$ and $a_i \leqslant a_j$;

(b) $J_i \in \mathcal{J}^A$ and $J_j \in \mathcal{J}^B$ are such that $P_{ij} + a_j < d_j$ and $a_i < a_j$;

(c) $J_i \in \mathcal{J}^A$ and $J_j \in \mathcal{J}^A$ are such that $(1+b)t + a_i \leqslant d_i$, $P_{ij} + a_j \leqslant d_j$ and $a_i < a_j$.

Proof. By the pairwise job interchange argument.   □

Let us notice that, given a schedule σ, the feasibility of σ can be checked by verifying whether there exists a job J_j^B such that $S_j^B(\sigma)(1+b) + a_j^B > d_j^B$. If such a job exists, then σ is not feasible.

Lee et al. [40] applied the following approach to find a lower bound on the optimal value of the $\sum w_j^A C_j^A$ criterion in problem $1|p_j = a_j + bt$, $\sum U_j^B = 0|\sum w_j^A C_j^A$. Let us assume that a lower bound of the total weighted completion time of a schedule σ in which the order of the first k jobs was determined is known. Let in the set of unscheduled jobs there be k_A jobs of agent A and k_B jobs of agent B, $k_A + k_B = n - k$. Then the lower bound is equal to the value

$$\sum_{1 \leqslant i \leqslant k, J_{[i]} \in \mathcal{J}^A} w_{[i]} C_{[i]}(\sigma) + LB_{US}, \tag{21.12}$$

where LB_{US} is calculated as follows (see Algorithm 21.12).

Algorithm 21.12. for calculation of lower bound LB_{US}
for problem $1|\, p_j = a_j + bt,\, \sum U_j^B = 0|\sum w_j^A C_j^A$

---

1 **Input** : numbers n_A, n_B, b, sequences $(a_1^A, a_2^A, \ldots, a_{n_A}^A)$,
$(w_1^A, w_2^A, \ldots, w_{n_A}^A)$, $(a_1^B, a_2^B, \ldots, a_{n_B}^B)$, $(d_1^B, d_2^B, \ldots, d_{n_B}^B)$
2 **Output:** lower bound LB_{US}
▷ Step 1
3 **for** $j \leftarrow 1$ **to** $n - k$ **do**
4 $\quad \hat{C}_{[k+j]} \leftarrow C_{[k]}(1+b)^j + \sum_{i=1}^{j} a_{(k+i)}(1+b)^{j-i}$;
end
5 Arrange all jobs of agent A in non-decreasing order of the weights w_i^A;
6 Arrange all jobs of agent B in non-decreasing order of the deadlines d_i^B;
7 $ic \leftarrow 0;\ ia \leftarrow n_A;\ ib \leftarrow n_B$;
8 **while** $(ic \geqslant n - k)$ **do**
9 $\quad LB_{US} \leftarrow \sum_{j=1}^{n_A} w_{(n_A-j+1)}^A C_{(j)}^A$;
10 $\quad$ **if** $(\hat{C}_{[n-k]} \leqslant d_{(ib)}^B)$ **then**
11 $\quad\quad C_{(ib)}^B \leftarrow \hat{C}_{[n-ic]};\ ib \leftarrow ib - 1$;
$\quad$ **else**
12 $\quad\quad C_{(ia)}^A \leftarrow \hat{C}_{[n-k]};\ ia \leftarrow ia - 1$;
$\quad$ end
13 $\quad ic \leftarrow ic + 1$;
end
▷ Step 2
14 **return** LB_{US}.

Wu et al. [49] considered problem $1|p_j = a_j + bt, L_{\max}^B \leqslant Q|\sum w_j^A U_j^A$. For this problem, the authors proposed a branch-and-bound algorithm, based on the following result.

Property 21.56. (Wu et al. [49]) Let $P_i := a_i + (1 + b)t$, $P_j := a_j + (1 + b)t$, $P_{ij} := a_j + a_i(1 + b) + (1 + b)^2 t$ and $P_{ji} := a_i + a_j(1 + b) + (1 + b)^2 t$. Schedule σ dominates schedule σ' if *any* of the following holds:

(a) $J_i \in \mathcal{J}^A$ and $J_j \in \mathcal{J}^A$ are such that $P_{ji} > d_i \geqslant P_i$, $a_i \leqslant a_j$ and $P_{ij} \leqslant d_j$;

(b) $J_i \in \mathcal{J}^A$ and $J_j \in \mathcal{J}^B$ are such that $P_j > d_j$, $P_{ji} > d_i \geqslant P_i$ and $a_i \leqslant a_j$;

(c) $J_i \in \mathcal{J}^A$ and $J_j \in \mathcal{J}^A$ are such that $P_{ij} > d_j \geqslant P_j$, $P_{ji} > d_i \geqslant P_i$, $w_i > w_j$ and $a_i < a_j$;

(d) $J_i \in \mathcal{J}^A$ and $J_j \in \mathcal{J}^A$ are such that $P_j > d_j$, $P_i < d_i$ and $a_i < a_j$;

(e) $J_i \in \mathcal{J}^A$ and $J_j \in \mathcal{J}^A$ are such that $P_{ij} \leqslant d_j$, $P_{ji} \leqslant d_i$ and $a_i < a_j$;

(f) $J_i \in \mathcal{J}^B$ and $J_j \in \mathcal{J}^B$ are such that $P_i - d_i \leqslant Q < P_{ji} - d_i$ and $P_{ij} - d_j \leqslant Q$;

(g) $J_i \in \mathcal{J}^B$ and $J_j \in \mathcal{J}^B$ are such that $P_{ji} - d_i \leqslant Q$, $P_{ij} - d_j \leqslant Q$ and $a_i < a_j$;

(h) $J_i \in \mathcal{J}^A$ and $J_j \in \mathcal{J}^B$ are such that $P_{ji} > d_i$ and $P_{ij} - d_j \leqslant Q$.

Proof. By the pairwise job interchange argument.   □

As in Sect. 21.2.3, the feasibility of a schedule σ can be checked by verifying whether there exists a job J_j^B such that $S_j^B(\sigma)(1 + b) + a_j - d_j^B > Q$. If such a job exists, schedule σ is not feasible.

To establish a lower bound of the total weighted number of tardy jobs of a schedule σ in which k jobs have been fixed, Wu et al. [49] applied the value

$$\sum_{1 \leqslant i \leqslant k, J_i \in \mathcal{J}^A} w_{[i]} U_{[i]}(\sigma) + LB_{US},$$

where LB_{US} is computed with an algorithm (see [49, Algorithm 1]) similar to Algorithm 21.12. For the considered problem, the authors also proposed a TS and a SA algorithms (cf. Chap. 17); see [49, Sect. 4] for details.

Applying a similar approach as Lee et al. [49], Cheng [33] proved that problems $1|p_j = a_j + bt, d_j^A = d_j^B = D, E_{\max}^B \leqslant Q|\sum E_j^A$ and $1|p_j = a_j + bt, d_j^A = d_j^B = D, E_{\max}^B \leqslant Q|\sum w_j^A E_j^A$, where $E_j^X(\sigma) := \max\{0, d_j^X - C_j(\sigma)\}$ and $\sum E_j^X(\sigma) := \sum_{J_j \in \mathcal{J}^X} E_j(\sigma)$ for $X \in \{A, B\}$ and a given schedule σ, are solvable in $O(n_A \log n_A + n_B \log n_B)$ time.

Gawiejnowicz and Suwalski [38] considered a single machine problem $1|p_j = a_j + b_j t|\sum w_j^A C_j^A + \theta L_{\max}^B$ with linear jobs. This problem is a time-dependent counterpart of the $\mathcal{NP}$-hard problem $1||\sum_{A_1} w_j C_j + \theta L_{\max}^A$ considered by Baker and Smith [32], and it is related to $\mathcal{NP}$-hard time-dependent scheduling problems $1|p_j = a_j + b_j t|\sum_{A_1} w_j C_j$ and $1|p_j = a_j + b_j t|L_{\max}^{A_2}$ (see Chap. 10) and to polynomially solvable time-dependent scheduling problems $1|p_j = b_j t|\sum_{A_1} w_j C_j$ and $1|p_j = b_j t|L_{\max}^{A_2}$ (see Chap. 7).

Gawiejnowicz and Suwalski [38] established the $\mathcal{NP}$-hardness of the problem, using the following auxiliary result.

Lemma 21.57. (Gawiejnowicz and Suwalski [38]) *Let* $w_j = \frac{b_j}{b_j+1}$ *and* $B_j = 1 + b_j$ *for* $1 \leqslant j \leqslant n$. *Then* $\sum_{J_j \in \mathcal{J}} w_j C_j(\sigma) = t_0(B_1 B_2 \dots B_n - 1)$.

Proof. By direct computation; see [38, Lemma 1].   ◇

Based on Lemma 21.57, we can prove the following $\mathcal{NP}$-hardness result.

Theorem 21.58. (Gawiejnowicz and Suwalski [38]) *Problem* $1|p_j = a_j + b_j t| \sum w_j^A C_j^A + \theta L_{\max}^B$ *is* $\mathcal{NP}$-hard in the ordinary sense, even if $\theta = 1$ and agent B has only two jobs.

Proof. Gawiejnowicz and Suwalski [38] used the following reduction from the EPP problem: $t_0 = \frac{1}{2}$, $n_A = q$, $b_j^A = z_j - 1$ and $w_j^A = \frac{b_j^A}{b_j^A+1}$ for $1 \leqslant j \leqslant n_A$, $n_B = 2$, $b_1^B = b_2^B = 1$, $d_1^B = 1$, $d_2^B = 2E$ and $\theta = 1$, where $E = \sqrt{\prod_{j=1}^{q} z_j}$. The threshold value is $G = 2E^2 - E - 1$.

To complete the proof it is sufficient to show that an instance of the EPP problem has a solution if and only if for the constructed instance of the decision version of problem $1|p_j = a_j + b_j t| \sum w_j^A C_j^A + \theta L_{\max}^B$ there exists a schedule σ such that $\sum w_j^A C_j^A(\sigma) + \theta L_{\max}^B(\sigma) \leqslant G$; see [38, Theorem 1] for details.   □

Gawiejnowicz and Suwalski [38] also proved the following property.

Property 21.59. (Gawiejnowicz and Suwalski [38])
(a) If some jobs of agent A have equal weights, then there exists an optimal schedule for problem $1|p_j = a_j + b_j t| \sum w_j^A C_j^A + \theta L_{\max}^B$ in which the jobs are scheduled in non-decreasing order of their deterioration rates b_j.
(b) If some jobs of agent A have equal deterioration rates, then there exists an optimal schedule for problem $1|p_j = a_j + b_j t| \sum w_j^A C_j^A + \theta L_{\max}^B$ in which the jobs are scheduled in non-decreasing order of their weights w_j.
(c) If deterioration rates and weights of adjacent jobs J_i and J_j of agent A are such that $\frac{b_i}{(1+b_i)w_i} \leqslant \frac{b_j}{(1+b_j)w_j}$, then there exists an optimal schedule for problem $1|p_j = a_j + b_j t| \sum w_j^A C_j^A + \theta L_{\max}^B$ in which job J_i precedes job J_j.
(d) If deterioration rates and weights of jobs of agent A are such that $b_i \leqslant b_j$ whenever $\frac{b_i}{(1+b_i)w_i} \leqslant \frac{b_j}{(1+b_j)w_j}$ for $1 \leqslant i \neq j \leqslant n$, then there exists an optimal schedule for problem $1|p_j = a_j + b_j t| \sum w_j^A C_j^A + \theta L_{\max}^B$ in which the jobs are scheduled in non-decreasing order of the ratios $\frac{b_j}{(1+b_j)w_j}$.
(e) If some jobs of agent B have equal due-dates, then in an optimal schedule for problem $1|p_j = a_j + b_j t| \sum w_j^A C_j^A + \theta L_{\max}^B$ the order of the jobs is immaterial.
(f) In an optimal schedule for problem $1|p_j = a_j + b_j t| \sum w_j^A C_j^A + \theta L_{\max}^B$ jobs of agent B are scheduled in non-decreasing order of their due-dates d_j.

Proof. See [38, Properties 1–6].   ◇

Remark 21.60. Based on Property 21.59, Gawiejnowicz and Suwalski proposed for problem $1|p_j = a_j + b_j t| \sum w_j^A C_j^A + \theta L_{\max}^B$ a branch-and-bound algorithm and a meta-heuristic; we refer the reader to [38, Sects. 4–6] for details.

21.2.5 Proportional-linear shortening

In this section, we consider two-agent time-dependent scheduling problems with proportional-linearly shortening job processing times.

Yin et al. [51] considered problem $1|p_j = b_j(1 - bt)|f_{max}^A, f_{max}^B$, a time-dependent counterpart of problem $1|f_{max}^B \leqslant Q|f_{max}^A$ considered earlier by Agnetis et al. [31]. The introduction of variable job processing times does not make problem $1|p_j = b_j(1 - bt), f_{max}^B \leqslant Q|f_{max}^A$ more difficult than $1|f_{max}^B \leqslant Q|f_{max}^A$, since the former problem can be solved by an appropriately modified algorithm for the latter one. The modified algorithm uses a counterpart of Property 21.28 for shortening job processing times and the following property similar to that concerning problem $1||f_{max}^A, f_{max}^B$.

Property 21.61. (Yin et al. [51])
(a) If in a given schedule the last job is completed at time t and if there is a job J_j^B such that $f_{max}^B(\max\{0, d_j^B - (t + b_j^B(1 - bt))\}) \leqslant Q$, then there exists an optimal schedule in which a job $J_j \in \mathcal{J}^B$ completes at time t and there exists no optimal schedule in which a job $J_i \in \mathcal{J}^A$ completes at time t.
(b) If in a given schedule the last job completes at time t, and for any job J_j^B one has $f_j^B(\max\{0, d_j^B - (t + b_j^B(1 - bt))\}) > Q$, then there exists an optimal schedule in which job J_i^A with the minimum cost is scheduled at time t.
(c) In an optimal schedule, jobs in $\mathcal{J}^B$ are scheduled in non-decreasing order of the artificial 'deadlines' D_j^B.

Based on these and a few other but similar properties, we can prove the following result.

Theorem 21.62. (Yin et al. [51]) (a) *Problem* $1|p_j = b_j(1 - bt), f_{max}^B \leqslant Q|f_{max}^A$ *is solvable in* $O(n_A^2 + n_B \log n_B)$ *time.* (b) *Problem* $1|p_j = b_j(1 - bt), d_j^A = d^A, f_{max}^B \leqslant Q| \sum w_j^A E_j^A$ *is solvable in* $O(n_A \log n_A + n_B^2)$ *time.*

Proof. (a) See [51, Theorem 3.6]. (b) See [51, Theorem 4.2]. (c) See [51, Theorem 4.5]. ◇

He and Leung [39] and Wan et al. [47] applied the ε-constrained approach to solve some time-dependent two-agent scheduling problems with shortening jobs and the $\sum C_j$, $\sum E_j$ and f_{max} criteria.

Theorem 21.63. (a) (He and Leung [39]) *All Pareto-optimal schedules for problem* $1|p_j = b_j(1 - bt)|(\sum C_j^A, f_{max}^B)$ *can be generated in* $O(n_A n_B(n_B^2 + n_A))$ *time by the modified Algorithm 21.10.*
(b) (He and Leung [39], Wan et al. [47]) *All Pareto-optimal schedules for problem* $1|p_j = b_j(1 - bt)|(f_{max}^A, f_{max}^B)$ *can be generated in* $O(n_A n_B(n_A^2 + n_B^2))$ *time by the modified Algorithm 21.11.*
(c) (Wan et al. [47]) *All Pareto-optimal schedules for problem* $1|p_j = b_j(1 - bt)|(\sum E_j^A, f_{max}^B)$ *can be generated in* $O(n_A n_B(n_A \log n_A + n_B^2))$ *time.*

Proof. (*a*) See [39, Theorem 3.6].
(*b*) See [39, Theorem 3.12], [47, Theorem 3.1].
(*c*) See [47, Theorem 4.1].   ◇

21.3 Time-dependent scheduling with job rejection

In the classical scheduling we assume that all jobs have to be scheduled. In the case of some scheduling problems, however, this assumption is too strong and should be weakened. For example, when the processing of certain jobs on our machines is too costly, we process the jobs on external machines, paying a fee for that. Hence, in such problems a job may be scheduled at no cost or *rejected* at the cost of a *rejection penalty*. The aim is to find a schedule which minimizes both the applied optimality criterion and the total rejection cost.

Scheduling problems of this type, called *scheduling problems with job rejection*, were introduced in 1996 by Bartal and Leonardi [53], and today are an independent domain in the classical scheduling theory (see Shabtay et al. [59] for a review). Growing interest in such scheduling problems has led to the appearance of *time-dependent scheduling with job rejection* introduced in 2009 by Cheng et al. [54]. In this section, we briefly review the main results related to this domain of time-dependent scheduling.

21.3.1 Preliminaries

Throughout the section, we consider time-dependent scheduling problems of the following form. We are given n jobs which have to be scheduled on $m \geqslant 1$ parallel identical machines. Each job is defined by a time-dependent processing time p_j, a ready time r_j, a deadline d_j and a weight w_j, $1 \leqslant j \leqslant n$. Job J_k may be *accepted*, and then it is scheduled on one of the machines, or *rejected*, and then it is not scheduled at all but we have to pay a *rejection penalty* e_k for that job. The aim is to find a schedule minimizing the sum of the value of criterion f for the set S of accepted jobs and the total rejection cost for the set $\bar{S}$ of rejected jobs.

Remark 21.64. Time-dependent scheduling problems with job rejection are denoted by symbol rej in the β field of three-field notation.

Example 21.65. (*a*) The symbol $1|p_j = a_j + b_j t, rej|C_{\max} + \sum e_j$ denotes a single machine time-dependent scheduling problem with linear jobs, job rejection and criterion $f = C_{\max} + \sum w_j$.
(*b*) The symbol $Pm|p_j = a_j + bt, rej|\sum C_j + \sum e_j$ denotes the problem of scheduling linear jobs with the same deterioration rates on parallel identical machines with criterion $f = \sum C_j + \sum e_j$.   ♦

21.3.2 Main results

Time-dependent variations of scheduling problems with job rejection are usually intractable. This is caused by the fact that the problems are similar to the KP problem which is $\mathcal{NP}$-hard (Karp [55]).

Cheng and Sun [54] demonstrated, by reductions from the SP problem, the ordinary $\mathcal{NP}$-hardness of single machine problems $1|rej, p_j = b_jt|f$, where $f \in \{C_{\max} + \sum e_j, T_{\max} + \sum e_j, \sum w_jC_j + \sum e_j\}$. The authors also proposed FPTASes for several single machine time-dependent scheduling problems with proportional jobs and job rejection, based on the approach introduced by Woeginger [60].

Theorem 21.66. (Cheng and Sun [54]) *There exists an FPTAS for problem* $1|rej, p_j = b_jt|f$, *where (a)* $f := C_{\max} + \sum e_j$, *(b)* $f := T_{\max} + \sum e_j$ *and (c)* $f := \sum w_jC_j + \sum e_j$.

Li and Zhao [58] proposed an FPTAS for problem $1, h_{11}|rej, r_j, p_j = b_j(a + bt)|C_{\max} + \sum e_j$. This scheme is a generalization of the FPTAS for problem $1, h_{11}|rej, p_j = b_jt|C_{\max} + \sum e_j$ (cf. Theorem 21.66 (a)). It runs in $O(n^6L^5\epsilon^{-4})$ time, where

$$L = \log \max \left\{ n, \frac{1}{\epsilon}, 1 + bb_{\max}, r_{\max} + \frac{a}{b}, T_2 + \frac{a}{b}, W \right\},$$

T_1 and T_2 denote, respectively, the starting time and the completion time of the machine non-availability interval, $b_{\max}$ is defined as in Sect. 6.2.2 and $r_{\max} := \max\{r_j : 1 \leqslant j \leqslant n\}$.

There are also known some results on parallel machine time-dependent scheduling with job rejection. Li and Yuan [57], applying the approach introduced by Kovalyov and Kubiak [56], proposed an FPTAS for problem $Pm|rej, p_j = b_jt| \sum C_j + \sum e_j$, running in $O(n^{2m+2}L^{m+2}\epsilon^{-(m+1)})$ time, where

$$L = \log \max \left\{ n, \frac{1}{\epsilon}, a_{\max}, \max_{1 \leqslant j \leqslant n} \{1 + b_j\}, e_{\max} \right\},$$

$a_{\max}$ is defined as in Sect. 11.1.3 and $e_{\max} := \max\{e_j : 1 \leqslant j \leqslant n\}$. The authors also proposed an $O(n^2)$ algorithm for the latter problem but with the $\sum C_{\max}^{(k)} + \sum e_j$ criterion (see [57, Sect. 3.2]).

21.4 Time-dependent scheduling with mixed job processing times

Job processing times in time-dependent scheduling problems are usually of the same form, e.g., either all jobs are proportional or all jobs are linear. In 2010, Gawiejnowicz and Lin [64] introduced a new model of time-dependent scheduling, called *mixed deterioration*, where several distinct forms of job processing times may exist in the same job set. In this section, we present the main results concerning this model of job deterioration.

21.4.1 Preliminaries

Throughout the section, we will consider time-dependent scheduling problems of the following form. A set $\mathcal{J}$ of n deteriorating jobs $J_1, J_2, \ldots, J_n$ has to be processed on a single machine, available from time $t_0 > 0$. All jobs are independent, non-preemptable and available for processing at time t_0. Each job $J_j \in \mathcal{J}$ has a due-date $d_j \geqslant 0$ and a weight $w_j > 0$. The processing time p_j of job $J_j \in \mathcal{J}$ can be a constant, proportional or linear function of the job starting time, i.e., $p_j = a_j$, $p_j = b_j t$ or $p_j = A_j + B_j t$, where $a_j > 0, b_j > 0, A_j > 0$ and $B_j > 0$ for $1 \leqslant j \leqslant n$.

Remark 21.67. Various forms of time-dependent scheduling problems with mixed deterioration are denoted in the β field of the $\alpha|\beta|\gamma$ notation by specification of all possible forms of such jobs and numbers indicating how many jobs of a particular type exist.

Example 21.68. (Gawiejnowicz and Lin [64]) The problem of minimizing the $C_{\max}$ criterion for a set of $n = n_1 + n_2$ jobs, among which n_1 jobs have constant processing times and the other n_2 jobs have proportional processing times will be denoted by the symbol $1|p_j \in \{a_j, n_1; b_j t, n_2\}|C_{\max}$.   ◆

21.4.2 Main results

In this section, we present the main results for single machine time-dependent scheduling with mixed job deterioration.

Minimizing the maximum completion time

The case of the $C_{\max}$ criterion is solvable in polynomial time.

Algorithm 21.13. for problem $1|p_j \in \{a_j, n_1; b_j t, n_2; A_j + B_j t, n_3\}|C_{\max}$

---

1 **Input**   : numbers n_1, n_2, n_3, sequences $(a_1, a_2, \ldots, a_{n_1})$,
  $(b_{n_1+1}, b_{n_1+2}, \ldots, b_{n_1+n_2})$, $(A_{n_1+n_2+1}, A_{n_1+n_2+2}, \ldots, A_{n_1+n_2+n_3})$,
  $(B_{n_1+n_2+1}, B_{n_1+n_2+2}, \ldots, B_{n_1+n_2+n_3})$

2 **Output:** optimal schedule $\sigma^\star$
  ▷ Step 1
3 Arrange the proportional jobs in arbitrary order;
4 Denote the obtained sequence by σ^1;
  ▷ Step 2
5 Arrange the linear jobs in non-increasing order of the ratios $\frac{B_j}{A_j}$;
6 Denote the obtained sequence by σ^2;
  ▷ Step 3
7 Arrange the fixed jobs in arbitrary order;
8 Denote the obtained sequence by σ^3;
  ▷ Step 4
9 $\sigma^\star \leftarrow (\sigma^1|\sigma^2|\sigma^3)$;
10 **return** $\sigma^\star$.

Theorem 21.69. (Gawiejnowicz and Lin [64]) *Problem* $1|p_j \in \{a_j, n_1; b_jt, n_2; A_j + B_jt, n_3\}|C_{\max}$ *is solvable by Algorithm 21.13 in* $O(n \log n)$ *time.*

Proof. Let us assume that we are given n_1 fixed jobs, n_2 proportional jobs and n_3 linear jobs. By Lemma 6.4, we know that in an optimal schedule all jobs are scheduled without idle times.

Let σ be some schedule, and let J_i and J_j be two consecutive jobs not abiding by the order specified by Algorithm 21.13. Let us assume that J_i precedes J_j. There are four cases to consider: job J_i and job J_j are linear and $\frac{B_i}{A_i} < \frac{B_j}{A_j}$ (Case 1), job J_i is fixed and job J_j is linear (Case 2), job J_i is fixed and job J_j is proportional (Case 3), job J_i is linear and job J_j is proportional (Case 4). We will show that in each of the four cases we can construct a better schedule, σ', from schedule σ by swapping jobs J_i and J_j.

In Case 1, by Theorem 7.22, we can reduce the length of schedule σ by swapping jobs J_i and J_j.

In Case 2, we have $C_i(\sigma) = t + a_i$ and $C_j(\sigma) = A_j + (B_j + 1)(t + a_i) = A_j + (B_j+1)t + B_ja_i + a_i$. Let σ' be the schedule obtained by swapping jobs J_i and J_j. Then we have $C_j(\sigma') = A_j + (B_j+1)t$ and $C_i(\sigma') = A_j + (B_j+1)t + a_i$. Thus, since $C_i(\sigma') - C_j(\sigma) = -B_ja_i < 0$, schedule σ' is better than schedule σ.

In Case 3, assuming that $A_j = 0$ and $B_j \equiv b_j$ for $1 \leqslant j \leqslant n_2$, and by applying reasoning from Case 2, we also conclude that schedule σ' is better than schedule σ.

In Case 4, we have $C_i(\sigma) = A_i + (B_i + 1)t$ and $C_j(\sigma) = (b_j + 1) \times (A_i + (B_i + 1)t) = b_jA_i + A_i + (B_i + 1)(b_j + 1)t$. By swapping jobs J_i and J_j, we have $C_j(\sigma') = (b_j + 1)t$ and $C_i(\sigma') = A_i + (B_i + 1)(b_j + 1)t$. Thus, since $C_i(\sigma') - C_j(\sigma) = -b_jA_i < 0$, schedule σ' is better than schedule σ.

By swapping other jobs pairwise as above, we obtain a schedule composed of three blocks, each of which contains jobs of the same type. Moreover, in this schedule the block of proportional jobs precedes the block of linear jobs, which is followed by the block of fixed jobs. The order of jobs in the block of linear jobs in the schedule is determined by Theorem 7.22. The order of proportional jobs and the order of fixed jobs are immaterial. The time complexity of Algorithm 21.13 is determined by Step 2, which takes $O(n \log n)$ time. ∎

Minimizing the total completion time

The $\sum C_j$ criterion is more difficult compared to the $C_{\max}$ criterion. We start with an example which shows that an optimal schedule for problem $1|p_j \in \{a_j, n_1; b_jt, n_2\}|\sum C_j$ does not necessarily consist of two disjoint blocks, each of which contains jobs of the same type.

Example 21.70. (Gawiejnowicz and Lin [64]) We are given jobs J_1, J_2, J_3 and J_4 with job processing times $p_1 = 1$, $p_2 = 2$, $p_3 = 2t$ and $p_4 = 3t$, respectively. The initial starting time $t_0 = 1$. There are 24 possible sequences, as shown in

Table 21.2. The sequences corresponding to optimal schedules for the instance
are marked in bold.   ♦

Table 21.2: The list of schedules for data from Example 21.70

$(\sigma_1, \sigma_2, \sigma_3, \sigma_4)$	$(C_{\sigma_1}, C_{\sigma_2}, C_{\sigma_3}, C_{\sigma_4})$	$\sum C_j$
$(1, 2, 3, 4)$	$(2, 4, 12, 48)$	66
$(1, 2, 4, 3)$	$(2, 4, 16, 48)$	70
$(1, 3, 2, 4)$	$(2, 6, 8, 32)$	48
$(1, 3, 4, 2)$	$(2, 6, 24, 26)$	58
$(1, 4, 2, 3)$	$(2, 8, 10, 30)$	50
$(1, 4, 3, 2)$	$(2, 8, 24, 26)$	60
$(2, 1, 3, 4)$	$(3, 4, 12, 48)$	67
$(2, 1, 4, 3)$	$(3, 4, 16, 48)$	71
$(2, 3, 1, 4)$	$(3, 9, 10, 40)$	62
$(2, 3, 4, 1)$	$(3, 9, 36, 37)$	84
$(2, 4, 1, 3)$	$(3, 12, 13, 39)$	67
$(2, 4, 3, 1)$	$(3, 12, 36, 37)$	88
$\mathbf{(3, 1, 2, 4)}$	$\mathbf{(3, 4, 6, 24)}$	$\mathbf{37}$
$(3, 1, 4, 2)$	$(3, 4, 16, 18)$	41
$(3, 2, 1, 4)$	$(3, 5, 6, 24)$	38
$(3, 2, 4, 1)$	$(3, 5, 20, 21)$	49
$(3, 4, 1, 2)$	$(3, 12, 13, 15)$	43
$(3, 4, 2, 1)$	$(3, 12, 14, 15)$	44
$\mathbf{(4, 1, 2, 3)}$	$\mathbf{(4, 5, 7, 21)}$	$\mathbf{37}$
$(4, 1, 3, 2)$	$(4, 5, 15, 17)$	41
$(4, 2, 1, 3)$	$(4, 6, 7, 21)$	38
$(4, 2, 3, 1)$	$(4, 6, 18, 19)$	47
$(4, 3, 1, 2)$	$(4, 12, 13, 15)$	44
$(4, 3, 2, 1)$	$(4, 12, 14, 15)$	45

Example 21.70 shows that the structure of optimal schedules for problem
$1|p_j \in \{a_j, n_1; b_j t, n_2\}| \sum C_j$ is different compared to that for problem $1|p_j \in \{a_j, n_1; b_j t, n_2\}|C_{\max}$. We can show, however, some properties of an optimal
schedule for the former problem.

Property 21.71. (Gawiejnowicz and Lin [64]) In an optimal schedule for prob-
lem $1|p_j \in \{a_j, n_1; b_j t, n_2\}| \sum C_j$, the fixed jobs are scheduled in non-
decreasing order of the job processing times a_j.

Proof. Let us assume that in an optimal schedule σ there exist fixed jobs J_i
and J_j such that $a_i < a_j$ but J_j precedes J_i. Let us denote by $\mathcal{J}^1, \mathcal{J}^2$ and $\mathcal{J}^3$
the sets of jobs scheduled before job J_j, between jobs J_j and J_i, and after job

J_i, respectively. Let σ^l denote the sequence of jobs of set $\mathcal{J}^l$, where $l = 1, 2, 3$. Thus, the considered optimal schedule is in the form of $\sigma = (\sigma^1, j, \sigma^2, i, \sigma^3)$. Let $\sigma' = (\sigma^1, i, \sigma^2, j, \sigma^3)$ be the schedule obtained by swapping jobs J_j and J_i. Then, since $p_i = a_i < a_j = p_j$, we have $C_i(\sigma') < C_j(\sigma)$. Because proportional functions are increasing with respect to job starting times, the actual processing times of the jobs in $\mathcal{J}^2$ either remain the same (fixed jobs) or decrease (proportional jobs). Therefore, the starting time of job J_j in σ' is not later than that of job J_i in σ, i.e., $C_j(\sigma') \leqslant C_i(\sigma)$. But this implies that $\sum C_j(\sigma') < \sum C_j(\sigma)$. A contradiction. ∎

A similar property also holds for proportional jobs.

Property 21.72. (Gawiejnowicz and Lin [64]) In an optimal schedule for problem $1|p_j \in \{a_j, n_1; b_j t, n_2\}|\sum C_j$, proportional jobs are scheduled in nondecreasing order of the deterioration rates b_j.

Proof. Similar to the proof of Property 21.71. □

Properties 21.71–21.72 imply that an optimal schedule for the considered problem is composed of separate blocks of fixed and proportional jobs, and that in each block the jobs of the same type are arranged accordingly. However, the number of jobs in each block and the mutual relations between these blocks are unknown.

Example 21.70 suggests how to construct an optimal schedule for problem $1|p_j \in \{a_j, n_1; b_j t, n_2\}|\sum C_j$, in which proportional jobs are scattered in the schedule among the blocks of fixed jobs. Namely, the optimal schedule can be constructed by optimally interleaving the two sorted sequences of fixed jobs and proportional jobs.

Let us notice that given a sorted sequence of n_1 fixed jobs and a sorted sequence of n_2 proportional jobs, there are $\binom{n_1+n_2}{n_1}$ different interleaving sequences. Hence, if n_1 or n_2 is constant, problem $1|p_j \in \{a_j, n_1; b_j t, n_2\}|\sum C_j$ can be solved in polynomial time. We illustrate the idea with the example of problem $1|p_j \in \{a_j, n-1; b_j t, 1\}|\sum C_j$. For simplicity of further presentation, we will denote fixed jobs by $J_1, J_2, \ldots, J_{n-1}$ and the proportional job by J_n.

Property 21.73. (Gawiejnowicz and Lin [64])
Let $\sigma = (\sigma_1, \sigma_2, \ldots, \sigma_k, \sigma_{k+1}, \sigma_{k+2}, \ldots, \sigma_n)$ be a schedule for problem $1|p_j \in \{a_j, n-1; b_j t, 1\}|\sum C_j$ in which the proportional job J_n is scheduled in the $(k+1)$th position, i.e., $\sigma_{k+1} = n$, $0 \leqslant k \leqslant n-1$. Then the total completion time of schedule σ is equal to

$$\sum C_j(\sigma) = kt_0 + \sum_{j=1}^{n-1}(n-j)a_{\sigma_j} + (n-k)(b_n+1)(t_0 + \sum_{j=1}^{k} a_{\sigma_j}). \quad (21.13)$$

Proof. Let $\mathcal{J}^1$ ($\mathcal{J}^2$) denote the set of jobs scheduled before (after) the proportional job J_n in a given schedule σ, where $|\mathcal{J}^1| = k$ and $|\mathcal{J}^2| = n - k - 1$.

Then,

$$\sum C_j(\sigma) = \sum_{J_j \in \mathcal{J}^1} C_j(\sigma) + C_n(\sigma) + \sum_{J_j \in \mathcal{J}^2} C_j(\sigma).$$

Let us calculate the subsequent terms of the sum $\sum C_j(\sigma)$. First, note that

$$\sum_{J_j \in \mathcal{J}^1} C_j(\sigma) = \sum_{j=1}^{k} C_j(\sigma) = kt_0 + \sum_{j=1}^{k}(k - j + 1)a_{\sigma_j}.$$

Job J_n is completed at the time

$$C_n(\sigma) = C_k(\sigma) + p_n = (1 + b_n)C_k(\sigma), \text{ where } C_k(\sigma) = t_0 + \sum_{j=1}^{k} a_{\sigma_j}.$$

In order to calculate $\sum_{J_j \in \mathcal{J}^2} C_j(\sigma)$, it is sufficient to note that since job J_n is scheduled in the $(k + 1)$th position in schedule σ, the completion times of jobs $J_{\sigma_{k+1}}, J_{\sigma_{k+2}}, \ldots, J_{\sigma_{n-1}}$ increase by p_n units of time. Thus

$$\sum_{J_j \in \mathcal{J}^2} C_j(\sigma) = \sum_{j=k+2}^{n-1} C_j(\sigma) = \sum_{j=k+2}^{n-1} (p_n + (n - j + 1)a_{\sigma_j})$$

$$= (n - k - 1)(1 + b_n)(t_0 + \sum_{j=1}^{k} a_{\sigma_j}) + \sum_{j=k+2}^{n-1} (n - j + 1)a_{\sigma_j}.$$

Collecting all the terms together, we obtain formula (21.13).   ∎

Property 21.73 does not allow us to specify the optimal position of job J_n. However, we can choose the best schedule from among $n + 1$ schedules in which the proportional job is scheduled before (position 0), inside (positions $1, 2, \ldots, n - 1$) or after (position n) the sequence of fixed jobs.

Algorithm 21.14. for problem $1|p_j \in \{a_j, n - 1; b_jt, 1\}|\sum C_j$

▷ Step 1
1 Arrange the fixed jobs in non-decreasing order of the processing times a_j;
2 Denote the obtained sequence by $\sigma = (\sigma_1, \sigma_2, \ldots, \sigma_{n-1})$;
▷ Step 2
3 **for** $k \leftarrow 0$ **to** n **do**
4 | Construct $n + 1$ schedules by scheduling job J_n in the kth position in σ;
end
▷ Step 3
5 Choose the best schedule σ from the schedules generated in Step 2;
6 **return** σ .

Theorem 21.74. (Gawiejnowicz and Lin [64]) *Problem* $1|p_j \in \{a_j, n - 1; b_jt, 1\}|\sum C_j$ *is solvable in* $O(n \log n)$ *time by Algorithm 21.14.*

Proof. The optimality of Algorithm 21.14 follows from Property 21.71. Since Step 1 needs $O(n \log n)$ time, while both Step 2 and Step 3 require $O(n)$ time, the running time of the algorithm is $O(n \log n)$. □

Algorithm 21.14 can be generalized for any fixed number m of proportional jobs, $1 \leqslant m \leqslant n$.

Theorem 21.75. (Gawiejnowicz and Lin [64]) *If $1 \leqslant m \leqslant n$, then problem $1|p_j \in \{a_j, n - m; b_j t, m\}| \sum C_j$ is solvable in $O(\max\{n \log n, n^m\})$ time.*

Proof. In order to solve problem $1|p_j \in \{a_j, n - m; b_j t, m\}| \sum C_j$, we have to construct $(n + 1) \times n \times \cdots \times (n - m + 1) = O(n^m)$ schedules and to choose the best from them. □

Minimizing the total weighted completion time

Similarly to the $\sum C_j$ criterion, $\sum w_j C_j$ is a difficult criterion. We illustrate this claim with the following example.

Example 21.76. (Gawiejnowicz and Lin [64]) We are given jobs J_1, J_2, J_3 with job processing times $p_1 = 1$, $p_2 = 2$, $p_3 = t$ and weights $w_1 = 8$, $w_2 = 1$, $w_3 = 3$, respectively. The initial starting time $t_0 = 1$. There are six possible schedules, as shown in Table 21.3. The sequence corresponding to an optimal schedule is marked in bold. ◆

Table 21.3: The list of all possible schedules for data from Example 21.76

$(\sigma_1, \sigma_2, \sigma_3)$	$(C_{\sigma_1}, C_{\sigma_2}, C_{\sigma_3})$	$\sum w_j C_j$
$(1, 2, 3)$	$(2, 4, 8)$	44
$(\mathbf{1, 3, 2})$	$(\mathbf{2, 4, 6})$	**34**
$(2, 1, 3)$	$(3, 4, 8)$	59
$(2, 3, 1)$	$(3, 6, 7)$	77
$(3, 1, 2)$	$(2, 3, 5)$	35
$(3, 2, 1)$	$(2, 4, 5)$	50

Example 21.76 shows that an optimal schedule for problem $1|p_j \in \{a_j, n_1; b_j t, n_2\}| \sum w_j C_j$ is not necessarily composed of two job blocks, with jobs of the same type in each of them. We can prove, however, some optimality properties concerning the jobs contained in the same block.

Property 21.77. (Gawiejnowicz and Lin [64]) In an optimal schedule for problem $1|p_j \in \{a_j, n_1; b_j t, n_2\}| \sum w_j C_j$, the jobs in a block of fixed jobs are ordered in non-decreasing order of the ratios $\frac{a_j}{w_j}$.

Proof. Let $\sigma = (\sigma^1, j, i, \sigma^2)$ be an optimal schedule in which J_i and J_j are fixed jobs such that $\frac{a_i}{w_i} < \frac{a_j}{w_j}$ and J_j immediately precedes J_i. Let us denote by $\mathcal{J}^1$ ($\mathcal{J}^2$) the set of jobs scheduled before job J_i (after job J_j), and let sequence σ^1 (σ^2) contain the jobs of set $\mathcal{J}^1$ ($\mathcal{J}^2$).

Let us calculate the weighted completion times of jobs J_j and J_i. We have $w_j C_j(\sigma) = w_j(t + a_j)$ and $w_i C_i(\sigma) = w_i(t + a_j + a_i)$. Let us consider now the schedule $\sigma' = (\sigma^1, i, j, \sigma^2)$, obtained by swapping jobs J_j and J_i. Then, $w_i C_i(\sigma') = w_i(t + a_i)$ and $w_j C_j(\sigma') = w_j(t + a_i + a_j)$. Since the swap does not affect the completion of any job of $\mathcal{J}^1 \cup \mathcal{J}^2$, and since $C_j(\sigma) + C_i(\sigma) - C_i(\sigma') - C_j(\sigma') = w_i a_j - w_j a_i = w_i w_j(\frac{a_j}{w_j} - \frac{a_i}{w_i}) > 0$, schedule σ' has a smaller value of the $\sum w_j C_j$ criterion than σ. A contradiction. □

A similar property holds for proportional jobs.

Property 21.78. (Gawiejnowicz and Lin [64]) In an optimal schedule for problem $1|p_j \in \{a_j, n_1; b_j t, n_2\}|\sum w_j C_j$, the jobs in the block of proportional jobs are scheduled in non-decreasing order of the ratios $\frac{b_j}{w_j}$.

Proof. The result follows from the same reasoning as the reasoning in the proof of Property 21.77. □

Similar time-dependent scheduling problems with mixed deterioration and linear jobs are intractable.

Theorem 21.79. (Gawiejnowicz and Lin [64]) *Problems* $1|p_j \in \{a_j, n_1; A_j + B_j t, n_2\}|\sum w_j C_j$, $1|p_j \in \{b_j t, n_1; A_j + B_j t, n_2\}|\sum w_j C_j$ *and* $1|p_j \in \{a_j, n_1; b_j t, n_2; A_j + B_j t, n_3\}|\sum w_j C_j$ *are* $\mathcal{NP}$-hard in the ordinary sense.

Proof. The result follows from the fact that problem $1|p_j = A_j + B_j t|\sum w_j C_j$ is $\mathcal{NP}$-hard in the ordinary sense (Bachman et al. [61]). □

Minimizing the maximum lateness

Single machine time-dependent scheduling with fixed and proportional jobs, mixed deterioration and the $L_{\max}$ criterion is intractable, even if there is only a single fixed job.

Theorem 21.80. (Gawiejnowicz and Lin [64]) *The decision version of problem* $1|p_j \in \{a_j, 1; b_j t, n - 1\}|L_{\max}$ *is* $\mathcal{NP}$-complete in the ordinary sense.

Proof. First, let us formulate the decision version *TDL* of the considered problem: given numbers $t_0 > 0$ and $y \geq 0$, the set $\{1, 2, \ldots, n\}$ for some natural n, a number $b_j > 0$ for each $1 \leq j \leq n-1$, and a number $a_n > 0$, does there exist a single-machine schedule σ for jobs with time-dependent processing times in the form of $p_j = b_j t, 1 \leq j \leq n - 1$, and $p_n = a_n$, such that these jobs are scheduled starting from time t_0 and $L_{\max}(\sigma) \leq y$?

Now, we will show that the SP problem can be reduced to the TDL problem in polynomial time.

Let I_{SP} denote an instance of the SP problem, composed of $B \in \mathbb{Z}^+$, set $P = \{1, 2, \ldots, p\}$ and size $y_j \in \mathbb{Z}^+ \setminus \{1\}$ for each $j \in P$. The instance I_{TDL} of the TDL problem can be constructed as follows: $n = p+1$, $t_0 = 1$, $b_j = y_j - 1$ and $d_j = 2Y$ for $1 \leqslant j \leqslant p$, $a_{p+1} = B$, $d_{m+1} = 2B$, where $Y = \prod_{j=1}^{p} y_j$. The threshold value is $G = 0$.

Given instance I_{TDL} and a schedule σ for it, we can check in polynomial time whether $L_{\max} \leqslant G$. Therefore, the TDL problem is in $\mathcal{NP}$.

There remains to show that the answer to instance I_{SP} is affirmative if and only if the answer to instance I_{TDL} is affirmative.

Let us assume that instance I_{SP} has a desired solution. Then there exists a set $P' \subseteq P$ such that $\prod_{j \in P'} y_j = B$. Let us denote by $\mathcal{J}_{P'}$ ($\mathcal{J}_{P \setminus P'}$) the set of jobs corresponding to the elements of set P' ($P \setminus P'$). Starting from $t_0 = 1$, we schedule the jobs without idle times as follows. First, in an arbitrary order, we schedule the jobs of set $\mathcal{J}_{P'}$. Next, we schedule job J_{p+1}. Finally, in an arbitrary order, we schedule the jobs of set $\mathcal{J}_{P \setminus P'}$. Let us denote the schedule by σ. Let $C(S)$ be the completion time of the last job from a particular job set S.

By Theorem 7.1, we have

$$C(\mathcal{J}_{P'}) = t_0 \prod_{j \in P'} (1 + b_j) = \prod_{j \in P'} y_j = B.$$

Then $C_{p+1} = C(\mathcal{J}_{P'}) + B = 2B$. Since $\prod_{j \in P \setminus P'} (1 + b_j) = \frac{Y}{B}$, we have

$$C(\mathcal{J}_{P \setminus P'}) = C_{p+1} \prod_{j \in P \setminus P'} (1 + b_j) = 2B \times \frac{Y}{B} = 2Y.$$

Therefore, since $d_{m+1} = 2B$ and $d_j = 2Y$ for $j \in P$, we have

$$L_{\max}(\sigma) = \max \{C_j - d_j : j \in P \cup \{p+1\}\} = \max \{2Y - 2Y, 2B - 2B\} = 0.$$

Hence, the instance I_{TDL} of problem TDL has a solution.

Let us now assume that for the instance I_{TDL} there exists a schedule σ such that $L_{\max}(\sigma) \leqslant 0$. Let $\mathcal{J}^1$ ($\mathcal{J}^2$) denote the set of jobs scheduled in σ before (after) job J_{p+1}. We will show that instance I_{SP} of problem SP has a solution in this case.

First, we will show that job J_{p+1} can start only at time B. Indeed, job J_{p+1} cannot start at time t_0, since then

$$\mathcal{J}^1 = \emptyset, \mathcal{J}^2 = \mathcal{J}, C_{\max}(\sigma) = C_{p+1} \prod_{J_j \in \mathcal{J}} (1 + b_j) = BY > 2Y,$$

and

$$L_{\max}(\sigma) = \max\{B - 2B, BY - 2Y\} = \max\{-B, (B-2)Y\} = (B-2)Y > 0.$$

Moreover, job J_{p+1} cannot start later than B, since then $L_{\max}(\sigma) \geqslant C_{p+1} - d_{p+1} > 0$. Finally, this job cannot start earlier than B. Indeed, let us assume that the job starts at time $B - b$, where $0 < b < B$. It means that $\mathcal{J}^1 \neq \emptyset$, $C(\mathcal{J}^1) = B - b$ and $C_{p+1} = B - b + B = 2B - b$. This, in turn, implies that

$$C(\mathcal{J}^2) = (2B - b) \times \frac{Y}{B - b} = 2Y + \frac{bY}{B - b} > 2Y,$$

and $L_{\max}(\sigma) > 0$. Therefore, job J_{p+1} must start at time B and hence $C_{p+1} = B + B = 2B$. By Lemma 6.4, no idle interval is allowed in schedule σ. Therefore, the completion time of the last job in $\mathcal{J}^1$ must be coincident with the start time of job J_{p+1}, implying that $\prod_{j \in P'} y_j = B$. ∎

Remark 21.81. Applying the reduction used in the proof of Theorem 21.80, one can also prove that the decision version of problem $1|p_j \in \{a_j, 1; b_j t, n_2\}| \sum U_j$ is $\mathcal{NP}$-complete in the ordinary sense (see [64, Theorem 7]). ◇

The case of linear jobs is not easier than the case of proportional jobs.

Theorem 21.82. (Gawiejnowicz and Lin [64]) *The following problems are $\mathcal{NP}$-hard in the ordinary sense:* $1|p_j \in \{a_j, n_1; A_j + B_j t, n_2\}|L_{\max}$, $1|p_j \in \{b_j t, n_1; A_j + B_j t, n_2\}|L_{\max}$ *and* $1|p_j \in \{a_j, n_1; b_j t, n_2; A_j + B_j t, n_3\}|L_{\max}$.

Proof. Because all the problems include problem $1|p_j = A_j + B_j t|L_{\max}$ which is known to be $\mathcal{NP}$-hard in the ordinary sense (Kononov [65]), the result follows. □

More general problems, with arbitrary precedence constraints and the criterion $f_{\max}$, were considered by Dębczyński and Gawiejnowicz [63], where it was shown that some of the problems can be solved in polynomial time by using the modified Lawler's algorithm (see [63, Sects. 3–4]).

A polynomial algorithm for a single machine time-dependent scheduling problem with the $f_{\max}$ criterion, mixed processing times and precedence constraints defined by a *k-partite graph*, i.e. a graph whose vertices can be partitioned into $k \geqslant 2$ disjoint sets such that no two vertices within the same set are adjacent, was presented by Dębczyński [62, Sect. 3].

21.5 Time-dependent scheduling games

We complete the chapter with a review of results on time-dependent scheduling problems considered using the concepts of non-cooperative game theory.

Game theory is a domain focused on the construction and analysis of models of interaction between parties, called *players*, in a conflict situation called a *game*. Originated by Zermelo [78], Borel [66], Steinhaus [77] and von Neumann [75], it gained a formal shape in the book by von Neumann and Morgenstern [76] and the papers by Nash [70, 71]. Today game theory is applied in economics, social science and theoretical computer science; we refer the reader to monographs by Owen [73] and Nisan et al. [72] for more details.

Game theory concepts were first applied to scheduling problems with fixed job processing times, which resulted in *scheduling games* (Papadimitriou [74], Koutsopias and Papadimitriou [68]). Scheduling games with time-dependent jobs originated in 2014 by Li et al. [69].

21.5.1 Preliminaries

In this chapter, we consider *time-dependent scheduling games*, where time-dependent scheduling problems with the $C_{\max}$ criterion are expressed in terms of *players*, *utilities of players* and *strategies*. The players are job owners, all jobs have different owners, the strategies are machines and the utility of a player is inversely proportional to the completion time of the last job assigned to a machine selected by the player. Time-dependent scheduling games are *non-cooperative games* (Nash [71]), i.e., players are selfish, wish to optimize only their own utilities and cannot form *coalitions* which are groups of players who act jointly.

There are two main notions in time-dependent scheduling games. The first is the notion of *Nash equilibrium*.

Definition 21.83. (Nash equlibrium, Nash [70])
Nash equilibrium *in a non-cooperative game is a strategy vector such that a player cannot further increase his/her utility by choosing a different strategy when the strategies of other players are fixed.*

In terms of the scheduling theory, a Nash equilibrium is such a schedule that no job in the schedule can reduce its cost by moving to another machine (see, e.g., Koutsoupias and Papadimitriou [68] for details). Notice that for a given scheduling game problem Nash equilibrium may not exist or there may be many Nash equilibria.

The second main notion is the *price of anarchy* of an algorithm, which is a counterpart of the competitive (worst-case) ratio for an online (offline) approximation algorithm (cf. Definitions 2.22 and 2.25).

Definition 21.84. (Price of anarchy, Koutsoupias and Papadimitriou [68])
The price of anarchy *is the ratio of the value of the worst possible Nash equilibrium solution and the value of the optimal solution.*

In terms of the scheduling theory, the price of anarchy is the ratio of the maximum completion time of the worst possible Nash equilibrium and the optimal maximum completion time.

21.5.2 Main results

Li et al. [69] formulated problem $Pm|p_j = b_j t|C_{\max}$ as a non-cooperative game, and proposed a $\Theta(n \log n)$ Algorithm 21.15 for this game, constructing a schedule which converges to a Nash equilibrium in a linear number of rounds. Let $s^i = (s_1, s_2, \ldots, s_n)$, s_i and s_{-i}, where $i \geqslant 0$, denote the strategy vector of all jobs in the ith iteration, the strategy of job J_i and the strategy s^i without s_i, respectively. The pseudo-code of the algorithm is as follows.

Algorithm 21.15. for problem $Pm|p_j = b_j t|C_{\max}$

1 **Input** : numbers n, m, sequence $(b_1, b_2, \ldots, b_n)$
2 **Output:** a strategy vector s^k
  ▷ Step 1
3 Arrange all jobs in non-increasing order of the deterioration rates b_j;
4 **for** $j \leftarrow 1$ **to** n **do**
5 | Randomly assign job J_j to a machine;
  end
6 $k \leftarrow 0$;
  ▷ Step 2
7 $s^0 \leftarrow$ the initial selection vector;
8 **for** $j \leftarrow 1$ **to** n **do**
9 | $C_j(s^0) \leftarrow$ the completion time of the last job scheduled on M_j;
10 | Store the value $C_j(s^0)$ in a min heap;
  end
  ▷ Step 3
11 Choose machine M_j with the smallest current load from the min heap;
12 **repeat**
13 | Let the $(i = k \bmod n + 1)$th player make a decision;
14 | **if** $(C_{s_i}(s) > C_j(s)(1 + b_i))$ **then**
15 | | $s_i \leftarrow j$;
  | **end**
16 | $k \leftarrow k + 1$;
17 | Update the values in the min heap;
  until $(s^k = s^{k-n})$;
18 **return** s^k.

Theorem 21.85. (Li et al. [69]) *The price of anarchy of Algorithm 21.15 equals* $(1 + b_{\max})^{\frac{m-1}{m}}$, *where* $b_{\max}$ *is defined as in Sect. 6.2.2.*

Proof. See [69, Theorems 4.1, 4.3–4.5 and 4.8]   ◇

Using other game theory concepts, Chen et al. [67] considered several parallel machine time-dependent scheduling problems with proportionally deteriorating jobs and the $C_{\max}$, $\sum C_j$ and $\sum C_{\max}^{(k)}$ criteria. These problems are formulated as scheduling games with selfish proportional jobs that wish to minimize their completion times when choosing a machine on which they will be processed. Machines are equipped with *coordination mechanisms*, which help to avoid the chaos caused by competition among jobs. Each machine examines deterioration rates of jobs assigned to it and determines the job processing order, according to a common *scheduling policy*.

Chen et al. [67] considered three scheduling policies: SDR (see Remark 7.62), LDR (see Remark 14.11) and MS (MakeSpan), in which the first available job is assigned to the machine with the smallest makespan.

Before we present the main results on these policies, we define the following two functions.

Definition 21.86. (Functions $K_m(\alpha)$ and $R_m(\alpha)$, Chen et al. [67])
Let for all $\alpha > 1$,

$$K_m(\alpha) := \frac{\sum_{j=0}^{m-1} j\alpha^{\frac{1}{m}}}{\sum_{j=0}^{m-1} \alpha^{\frac{1}{m}}} \qquad (21.14)$$

and

$$R_m(\alpha) := \frac{\lceil K_m(\alpha)\rceil}{m}\alpha^{\frac{\lceil K_m(\alpha)\rceil}{m}-1} + \left(1 - \frac{\lceil K_m(\alpha)\rceil}{m}\right)\alpha^{\frac{\lceil K_m(\alpha)\rceil}{m}}. \qquad (21.15)$$

It can be shown that $\frac{m-1}{2} < K_m(\alpha) < m-1$ (see [67, Sect. 3]).

Definitions (21.14) and (21.15) were used in [67] to obtain the following parametric bounds on the price of anarchy of SDR, LDR and MS policies. Let $b_{\max}$ be defined as in Sect. 6.2.2.

Theorem 21.87. (Chen et al. [67])
(a) The price of anarchy of a time-dependent scheduling game with proportional deterioration, the SDR policy and social costs of minimizing the $C_{\max}$ criterion is $(1 + b_{\max})^{1-\frac{1}{m}}$. The bound is parametrically tight for all $m \geqslant 2$.
(b) The price of anarchy of a time-dependent scheduling game with proportional deterioration, the SDR policy and social costs of minimizing the $\sum C_j^{(k)}$ criterion is $R_m(1 + b_{\max})$. The bound is parametrically tight for all $m \geqslant 2$.
(c) The price of anarchy of a time-dependent scheduling game with proportional deterioration, the LDR policy and social cost of minimizing the $C_{\max}$ criterion is at most $\max_{1\leqslant i\leqslant m}\{(1 + b_{[m+i+1]})^{1-\frac{1}{i}}\}$, and is at least $(1 + b_{[m+i+1]})^{1-\frac{1}{i}}$ for any $1 \leqslant i \leqslant m$. The bound is parametrically tight for $m = 2$.
(d) The price of anarchy of a time-dependent scheduling game with proportional deterioration, the LDR policy and social cost of minimizing the $\sum C_j^{(k)}$ criterion is at least $\dfrac{m+b_{[m+i+1]}}{m-i+i(1+b_{[m+i+1]})^{\frac{1}{i}}}$ for any $1 \leqslant i \leqslant m$. The bound is parametrically tight for $m = 2$.
(e) The price of anarchy of a time-dependent scheduling game with proportional deterioration, the MS policy and social cost of minimizing the $\sum C_j^{(k)}$ is $R_m(1 + b_{\max})$, and the bound is parametrically tight.

Proof. Consecutive parts of Theorem 21.87 are based on some auxiliary results, which estimate the ratios of the cost of schedule generated by a given policy and the optimal cost; see [67, Sects. 3–6] for details.   ◇

This result ends the last part of the book, devoted to the presentation of selected advanced topics in time-dependent scheduling. In the appendix following this part and completing the book, we discuss the most important open problems in time-dependent scheduling.

References

Time-dependent scheduling on machines with limited availability

1. J. Błażewicz, K. Ecker, E. Pesch, G. Schmidt and J. Węglarz, *Handbook on Scheduling*. Berlin-Heidelberg: Springer 2007.
2. B. Fan, S. Li, L. Zhou and L. Zhang, Scheduling resumable deteriorating jobs on a single machine with non-availability constraints. *Theoretical Computer Science* **412** (2011), 275–280.
3. M. R. Garey and D. S. Johnson, *Computers and Intractability: A Guide to the Theory of NP-Completeness*. San Francisco: Freeman 1979.
4. S. Gawiejnowicz, Complexity of scheduling deteriorating jobs with machine or job availability constraints. *The 7th Workshop on Models and Algorithms for Planning and Scheduling Problems*, 2005, pp. 140–141.
5. S. Gawiejnowicz, Scheduling deteriorating jobs subject to job or machine availability constraints. *European Journal of Operational Research* **180** (2007), no. 1, 472–478.
6. S. Gawiejnowicz and A. Kononov, Complexity and approximability of scheduling resumable proportionally deteriorating jobs. *European Journal of Operational Research* **200** (2010), no. 1, 305–308.
7. G. V. Gens and E. V. Levner, Approximate algorithms for certain universal problems in scheduling theory. *Soviet Journal of Computer and Systems Science* **6** (1978), 38–43 (in Russian).
8. G. V. Gens and E. V. Levner, Fast approximate algorithm for job sequencing with deadlines. *Discrete Applied Mathematics* **3** (1981), 313–318.
9. R. L. Graham, Bounds for certain multiprocessing anomalies. *Bell System Technology Journal* **45** (1966), no. 2, 1563–1581.
10. R. L. Graham, Bounds on multiprocessing timing anomalies. *SIAM Journal on Applied Mathematics* **17** (1969), no. 2, 416–429.
11. M. Ji, Y. He and T-C. E. Cheng, Scheduling linear deteriorating jobs with an availability constraint on a single machine. *Theoretical Computer Science* **362** (2006), no. 1–3, 115–126.
12. I. Kacem and E. Levner, An improved approximation scheme for scheduling a maintenance and proportional deteriorating jobs. *Journal of Industrial Management and Optimization* **12** (2016), no. 3, 811–817.
13. C-Y. Lee, Machine scheduling with an availability constraint. *Journal of Global Optimization* **9** (1996), no. 3–4, 363–382.

14. C-Y. Lee, Machine scheduling with availability constraints. In: J. Y-T. Leung (ed.), *Handbook of Scheduling*. Boca Raton: Chapman and Hall/CRC 2004.
15. C-Y. Lee, L. Lei and M. Pinedo, Current trends in deterministic scheduling. *Annals of Operations Research* **70** (1997), no. 1, 1–41.
16. C-Y. Lee and V. J. Leon, Machine scheduling with a rate-modifying activity. *European Journal of Operational Research* **128** (2001), no. 1, 119–128.
17. C-Y. Lee and C-S. Lin, Single-machine scheduling with maintenance and repair rate-modifying activities. *European Journal of Operational Research* **135** (2001), no. 3, 493–513.
18. W-C. Lee and C-C. Wu, Multi-machine scheduling with deteriorating jobs and scheduled maintenance. *Applied Mathematical Modelling* **32** (2008), no. 3, 362–373.
19. W-C. Luo and L. Chen, Approximation scheme for scheduling resumable proportionally deteriorating jobs. *Lecture Notes in Computer Science* **6681** (2011), pp. 36–45.
20. W-C. Luo and L. Chen, Approximation schemes for scheduling a maintenance and linear deteriorating jobs. *Journal of Industrial Management and Optimization* **8** (2012), no. 2, 271–283.
21. W-C. Luo and M. Ji, Scheduling a variable maintenance and linear deteriorating jobs on a single machine. *Information Processing Letters* **115** (2015), no. 1, 33–39.
22. G. Mosheiov and J. B. Sidney, Scheduling a deteriorating maintenance activity on a single machine. *Journal of the Operational Research Society* **61** (2010), no. 5, 882–887.
23. C. N. Potts and V. A. Strusevich, Fifty years of scheduling: a survey of milestones. *Journal of the Operational Research Society* **60** (2009), S41–S68.
24. K. Rustogi and V. A. Strusevich, Single machine scheduling with time-dependent linear deterioration and rate-modifying maintenance. *Journal of the Operational Research Society* **66** (2016), no. 3, 505–515.
25. G. Schmidt, Scheduling on semi-identical processors. *Zeitschrift für Operations Research* **28** (1984), no. 5, 153–162.
26. G. Schmidt, Scheduling with limited machine availability. *European Journal of Operational Research* **121** (2000), no. 1, 1–15.
27. V. S. Strusevich and K. Rustogi, *Scheduling with Times-Changing Effects and Rate-Modifying Activities*. Berlin-Heidelberg: Springer 2017.
28. G. J. Woeginger, When does a dynamic programming formulation guarantee the existence of a fully polynomial time approximation scheme (FPTAS)? *INFORMS Journal on Computing* **12** (2000), no. 1, 57–74.
29. C-C. Wu and W-C. Lee, Scheduling linear deteriorating jobs to minimize makespan with an availability constraint on a single machine. *Information Processing Letters* **87** (2003), no. 2, 89–93.

Two-agent time-dependent scheduling

30. A. Agnetis, J-C. Billaut, S. Gawiejnowicz, D. Pacciarelli and A. Soukhal, *Multi-agent Scheduling: Models and Algorithms*. Berlin-Heidelberg: Springer 2014.
31. A. Agnetis, P. Mirchandani, D. Pacciarelli and A. Pacifici, Scheduling problems with two competing agents. *Operations Research* **52** (2004), no. 2, 229–242.

32. K. Baker and J. C. Smith, A multiple criterion model for machine scheduling. *Journal of Scheduling* **6** (2003), no. 1, 7–16.

33. S-R. Cheng, Scheduling two-agents with a time-dependent deterioration to minimize the minsum earliness measures. *Asia-Pacific Journal of Operational Research* **31** (2014), no. 5, 1450040.

34. G-S. Ding, S-J. Sun and M-B. Cheng, A multiple-criterion model for machine scheduling with constant deteriorating jobs. *Journal of Shanghai University (English Edition)* **11** (2007), no. 6, 541–544.

35. Y. Gao, J-J. Yuan, C-T. Ng and T-C. E. Cheng, A further study on two-agent parallel-batch scheduling with release dates and deteriorating jobs to minimize the makespan. *European Journal of Operational Research* **273** (2019), no. 1, 74–81.

36. S. Gawiejnowicz, W-C. Lee, C-L. Lin and C-C. Wu, Single-machine scheduling of proportionally deteriorating jobs by two agents. *Journal of the Operational Research Society* **62** (2011), no. 11, 1983–1991.

37. S. Gawiejnowicz, T. Onak and C. Suwalski, A new library for evolutionary algorithms. *Lecture Notes in Computer Science* **3911** (2006), 414–421.

38. S. Gawiejnowicz and C. Suwalski, Scheduling linearly deteriorating jobs by two agents to minimize the weighted sum of two criteria. *Computers and Operations Research* **52** (2014), part A, 135–146.

39. C. He and J. Y-T. Leung, Two-agent scheduling of time-dependent jobs. *Journal of Combinatorial Optimization* **34** (2017), no. 2, 362–367.

40. W-C. Lee, W-J. Wang, Y-R. Shiau and C-C. Wu, A single-machine scheduling problem with two-agent and deteriorating jobs. *Applied Mathematical Modelling* **34** (2010), no. 10, 3098–3107.

41. P. Liu and L. Tang, Two-agent scheduling with linear deteriorating jobs on a single machine. *Lecture Notes in Computer Science* **5092** (2008), 642–650.

42. P. Liu, L. Tang and X. Zhou, Two-agent group scheduling with deteriorating jobs on a single machine. *International Journal of Advanced Manufacturing Technology* **47** (2010), no. 5–8, 657–664.

43. P. Liu, N. Yi and X. Zhou, Two-agent single-machine scheduling problems under increasing linear deterioration. *Applied Mathematical Modelling* **35** (2011), no. 5, 2290–2296.

44. P. Perez-Gonzalez and J. M. Framinan, A common framework and taxonomy for multicriteria scheduling problem with interfering and competing jobs: multi-agent scheduling problems. *European Journal of Operational Research* **235** (2014), no. 1, 1–16.

45. L. Tang, X. Zhao, J-Y. Liu and J. Y-T. Leung, Competitive two-agent scheduling with deteriorating jobs on a single and parallel-batching machine. *European Journal of Operational Research* **263** (2017), no. 2, 401–411.

46. V. T'kindt and J-C. Billaut, *Multicriteria Scheduling: Theory, Models and Algorithms*, 2nd ed. Berlin-Heidelberg: Springer 2006.

47. L. Wan, L-J. Wei, N-X. Xiong, J-J. Yuan and J-C. Xiong, Pareto optimization for the two-agent scheduling problems with linear non-increasing deterioration based on Internet of Things. *Future Generation Computer Systems* **76** (2017), 293–300.

48. Z. Wang, C-M. Wei and Y-B. Wu, Single machine two-agent scheduling with deteriorating jobs. *Asia-Pacific Journal of Operational Research* **33** (2016), no. 5, 1650034.

49. W-H. Wu, J-Y. Xu, W-H. Wu, Y-Q. Yin, I-F. Cheng and C-C. Wu, A tabu method for a two-agent single-machine scheduling with deterioration jobs. *Computers & Operations Research* **40** (2013), no. 8, 2116–2127.

50. W-H. Wu, Y-Q. Yin, W-H. Wu, C-C. Wu and P-H. Hsu, A time-dependent scheduling problem to minimize the sum of the total weighted tardiness among two agents. *Journal of Industrial and Management Optimization* **10** (2014), no. 2, 591–611.

51. Y-Q. Yin, T-C. E. Cheng and C.C. Wu, Scheduling problems with two agents and a linear non-increasing deterioration to minimize earliness penalties. *Information Sciences* **189** (2012), 282–292.

52. Y-Q. Yin, T-C. E. Cheng, L. Wan, C-C. Wu and J. Liu, Two-agent single-machine scheduling with deteriorating jobs. *Computers and Industrial Engineering* **81** (2015), 177–185.

Time-dependent scheduling with job rejection

53. Y. Bartal, S. Leonardi, A. Marchetti-Spaccamela, J. Sgal and L. Stougie, Multiprocessor scheduling with rejection. *The 7th Symposium on Discrete Algorithms* 1996, 95–103.

54. Y-S. Cheng and S-J. Sun, Scheduling linear deteriorating jobs with rejection on a single machine. *European Journal of Operational Research* **194** (2009), no. 1, 18–27.

55. R. M. Karp, Reducibility among combinatorial problems. In: R. E. Miller and J. W. Thatcher (eds.), *Complexity of Computer Computations*. New York: Plenum Press 1972, pp. 85–103.

56. M. Y. Kovalyov and W. Kubiak, A fully polynomial approximation scheme for minimizing makespan of deteriorating jobs. *Journal of Heuristics* **3** (1998), no. 4, 287–297.

57. S-S. Li and J-J. Yuan, Parallel-machine scheduling with deteriorating jobs and rejection. *Theoretical Computer Science* **411** (2010), 3642–3650.

58. W-X. Li and C-L. Zhao, Deteriorating jobs scheduling on a single machine with release dates, rejection and a fixed non-availability interval. *Journal of Applied Mathematics and Computation* **48** (2015), no. 1–2, 585–605.

59. D. Shabtay, N. Gaspar and M. Kaspi, A survey of offline scheduling with rejection. *Journal of Scheduling* **16** (2013), no. 1, 3–28. (Erratum: *Journal of Scheduling* **18** (2015), no. 3, 329).

60. G. J. Woeginger, When does a dynamic programming formulation guarantee the existence of a fully polynomial time approximation scheme (FPTAS)? *INFORMS Journal on Computing* **12** (2000), no. 1, 57–74.

Time-dependent scheduling with mixed job processing times

61. A. Bachman, A. Janiak and M. Y. Kovalyov, Minimizing the total weighted completion time of deteriorating jobs, *Information Processing Letters* **81** (2002), 81–84.

62. M. Dębczyński, Maximum cost scheduling of jobs with mixed variable processing times and k-partite precedence constraints. *Optimization Letters* **8** (2014), no. 1, 395–400.

63. M. Dębczyński and S. Gawiejnowicz, Scheduling jobs with mixed processing times, arbitrary precedence constraints and maximum cost criterion. *Computers and Industrial Engineering* **64** (2013), no. 1, 273–279.
64. S. Gawiejnowicz and B. M-T. Lin, Scheduling time-dependent jobs under mixed deterioration. *Applied Mathematics and Computation* **216** (2010), no. 2, 438–447.
65. A. Kononov, Scheduling problems with linear increasing processing times. In: U. Zimmermann et al. (eds.), *Operations Research 1996. Selected Papers of the Symposium on Operations Research*, Berlin-Heidelberg: Springer 1997, pp. 208–212.

Time-dependent scheduling games

66. É. Borel, On games that involve chance and the skill of players, *Théorie des Probabilités*, 1924, 204–24 (in French).
    English translation: *Econometrica* **21** (1953), no. 1, 101–115.
67. Q. Chen, L. Lin, Z. Tan and Y. Yan, Coordination mechanisms for scheduling games with proportional deterioration. *European Journal of Operational Research* **263** (2017), no. 2, 380–389.
68. E. Koutsoupias and C. Papadimitriou, Worst-case equilibria. *Computer Science Review* **3** (2009), no. 2, 65–69.
69. K. Li, C. Liu and K. Li, An approximation algorithm based on game theory for scheduling simple linear deteriorating jobs. *Theoretical Computer Science* **543** (2014), 46–51.
70. J. F. Nash, Equilibrium points in n-person games. *Proceedings of the National Academy of Sciences of the United States of America* **36** (1950), no. 1, 48–49.
71. J. F. Nash, Non-cooperative games. *Annals of Mathematics* **54** (1951), no. 2, 286–295.
72. N. Nisan, T. Roughgarden, E. Tardos and V. V. Vazirani, *Algorithmic Game Theory*. Cambridge: Cambridge University Press 2007.
73. G. Owen, *Game Theory*, 3rd ed. Cambridge: Academic Press 1995.
74. C. Papadimitriou, Algorithms, games and the Internet. *Proceedings of the 33rd Symposium on Theory of Computing*, 2001, 749–753.
75. J. von Neumann, On the theory of parlor games. *Mathematische Annalen* **100** (1928), 295–320 (in German).
76. J. von Neumann and O. Morgenstern, *Theory of Games and Economic Behavior*. Princeton: Princeton University Press 1944.
77. H. Steinhaus, Definitions for a theory of games and pursuit. *Myśl Akademicka* **1** (1925), 13–14 (in Polish).
    English translation: *Naval Research Logistics Quarterly* **7** (1960), no. 2, 105–107.
78. E. Zermelo, On an application of set theory to the theory of the game of chess. *Proceedings of the 5th International Congress of Mathematicians* **2** (1913), 501–504 (in German).
    English translation: E. Rasmusen, *Readings in Games and Information*. Hoboken: Wiley-Blackwell, 79–82.

Part A

<hr>

APPENDIX

A

Open time-dependent scheduling problems

Open problems are important in every research domain, since as long as they exist the domain may develop. Time-dependent scheduling is no exception. Therefore, we complete the book with an appendix including a list of selected open time-dependent scheduling problems. Though our choice is inevitably subjective, problems discussed in the appendix seem to be important in the sense that establishing the status of any of them will clarify the status of some other time-dependent scheduling problems.

Appendix A is composed of three sections. In Sect. A.1, we discuss open time-dependent scheduling problems concerning a single machine. In Sects. A.2 and A.3, we formulate open time-dependent scheduling problems in parallel and dedicated machine environments, respectively. For each discussed open problem, we give a summary of current research on this problem. The appendix is completed with a list of references.

A.1 Open single machine scheduling problems

In this section, we discuss single machine open time-dependent scheduling problems with linear job processing times.

Problem $1|p_j = 1 + b_j t| \sum C_j$

If we consider single machine time-dependent scheduling with linearly deteriorating jobs and the $\sum C_j$ criterion, the situation is unclear. On the one hand, it seems that the problem is not difficult since by summing the right side of formula (7.11) for $1 \leqslant j \leqslant n$ we obtain the following formula for the total completion time:

$$\sum_{j=1}^{n} C_j(\sigma) = \sum_{i=1}^{n} \sum_{j=1}^{i} a_{\sigma_j} \prod_{k=j+1}^{i} (1 + b_{\sigma_k}) + t_0 \sum_{i=1}^{n} \prod_{j=1}^{i} (1 + b_{\sigma_j}), \qquad (A.1)$$

from which we can obtain formulae for special cases, e.g., when $a_{\sigma_j} = 0$ for $1 \leqslant j \leqslant n$ or $t_0 = 0$. Therefore, for a given schedule we can easily calculate job completion times.

On the other hand, the complexity of problem $1|p_j = a_j + b_j t| \sum C_j$ is still unknown, even if $a_j = 1$ for $1 \leqslant j \leqslant n$. The latter problem, $1|p_j = 1 + b_j t| \sum C_j$, formulated in 1991 by Mosheiov [10], is one of the main open problems in time-dependent scheduling. We suppose that the following conjecture holds.

Conjecture A.1. Problem $1|p_j = a_j + b_j t| \sum C_j$ is $\mathcal{NP}$-hard in the ordinary sense.

Some authors indicated polynomially solvable special cases of this problem. For example, Kubale and Ocetkiewicz [7] proved that if all jobs have distinct deterioration rates and if for any $1 \leqslant i \neq j \leqslant n$ inequality $b_i > b_j$ implies inequality $b_i \geqslant \frac{b_{\min}+1}{b_{\min}} b_j + \frac{1}{b_{\min}}$, where $b_{\min}$ is defined as in inequality (6.39), then problem $1|p_j = 1 + b_j t| \sum C_j$ is solvable in $O(n \log n)$ time.

Despite unknown complexity of problem $1|p_j = 1 + b_j t| \sum C_j$, some of its properties are known.

Property A.2. (Mosheiov [10]) If $k = \arg \max_{1 \leqslant j \leqslant n} \{b_j\}$, then in an optimal schedule for problem $1|p_j = 1 + b_j t| \sum C_j$ job J_k is scheduled as the first one.

Proof. Consider job J_k with the greatest deterioration rate, $k = \arg \max\{b_j\}$. (If there are several jobs that have the same rate, choose any of them.) Since the completion time of job J_k is $C_k = 1 + (1 + b_k)S_k$, this completion time (and thus the total completion time) will be the smallest if $S_k = 0$.   ∎

Notice that by Property A.2 we may reduce the original problem to the problem with only $n - 1$ jobs, since the greatest job (i.e., the one with the greatest deterioration rate) has to be scheduled as the first one. Moreover, since in this case this job contributes only one unit of time to the total completion time, we can also reformulate the criterion function. Namely, for any schedule σ for problem $1|p_j = 1 + b_j t| \sum C_j$ the equality $\sum C_j(\sigma) = g(\sigma) + n$ holds. Function g is defined in the next property.

Property A.3. (Mosheiov [10]) Let $g(\sigma) := \sum_{i=1}^{n} \sum_{k=1}^{i} \prod_{j=k}^{i}(1+b_{[j]})$ and let σ and $\bar{\sigma}$ be a schedule for problem $1|p_j = 1 + b_j t| \sum C_j$ and the reverse schedule to σ, respectively. Then $g(\sigma) = g(\bar{\sigma})$.

Proof. Since $\sum_{i=1}^{n} \sum_{j=1}^{i} \prod_{k=j}^{i}(1 + b_k) = \sum_{i=1}^{n} \sum_{j=1}^{i} \prod_{k=j}^{i}(1 + b_{n-k+1})$, the result follows.   ∎

By Property A.3 we obtain the following *symmetry property* of problem $1|p_j = 1 + b_j t| \sum C_j$.

Property A.4. (Mosheiov [10]) If σ and $\bar{\sigma}$ are defined as in Property A.3, then $\sum C_j(\sigma) = \sum C_j(\bar{\sigma})$.

Proof. We have $\sum C_j(\sigma) = g(\sigma) + n$. But, by Property A.3, $g(\sigma) = g(\bar{\sigma})$. Hence $\sum C_j(\sigma) = g(\sigma) + n = g(\bar{\sigma}) + n = \sum C_j(\bar{\sigma})$. ∎

From Property A.3 follows the next property.

Property A.5. (Mosheiov [10]) Let $k = \arg\min_{1 \leqslant j \leqslant n}\{b_j\}$. Then in the optimal schedule for problem $1|p_j = 1 + b_j t|\sum C_j$ job J_k is scheduled neither as the first nor as the last one.

Proof. Let $\sigma = (k, \sigma_2, \ldots, \sigma_n)$ be any schedule in which job J_k is scheduled as the first one. Consider schedule $\sigma' = (\sigma_2, k, \ldots, \sigma_n)$. Then we have $g(\sigma') - g(\sigma) = (b_k - b_2)\sum_{i=3}^{n}\prod_{j=3}^{i}(1 + b_j) \leqslant 0$, since $b_k - b_2 \leqslant 0$ by assumption. Hence schedule σ' is better than schedule σ.

By Property A.3, job J_k cannot be scheduled as the last one either. ☐

There is known a counterpart of Property A.5 for a special case of problem $1|p_j = 1 + b_j t|\sum C_j$, when in the set $\{b_1, b_2, \ldots, b_n\}$ there exist only k different values, i.e.,

$$b_j \in \{B_1, B_2, \ldots, B_k\}, \tag{A.2}$$

where $1 \leqslant j \leqslant n$ and $k < n$ is fixed. Then, without loss of generality, we can assume that

$$B_1 \geqslant B_2 \geqslant \ldots B_{k-1} \geqslant B_k.$$

Lemma A.6. (Gawiejnowicz et al. [6]) *If condition (A.2) is satisfied, then in the optimal schedule for problem $1|p_j = 1 + b_j t|\sum C_j$ the jobs with the smallest deterioration rate b_k are scheduled as one group without inserting between them jobs with deterioration rates greater than b_k.*

Proof. The result is a corollary from Theorem A.8. ☐

Property A.5 allows us to prove the following result.

Property A.7. (Mosheiov [10]) Let J_{i-1}, J_i and J_{i+1} be three consecutive jobs in a schedule for problem $1|p_j = 1 + b_j t|\sum C_j$. If $b_i > b_{i-1}$ and $b_i > b_{i+1}$, then this schedule cannot be optimal.

Proof. The main idea of the proof is to show that the exchange of jobs J_{i-1} and J_i or of jobs J_i and J_{i+1} leads to a better schedule.

Let σ, σ' and σ'' denote schedules in which jobs J_{i-1}, J_i, J_{i+1} are scheduled in the order $(\sigma^{(a)}, i-1, i, i+1, \sigma^{(b)})$, $(\sigma^{(a)}, i, i-1, i+1, \sigma^{(b)})$ and $(\sigma^{(a)}, i-1, i+1, i, \sigma^{(b)})$, respectively, where $\sigma^{(a)}$ ($\sigma^{(b)}$) denotes the part of schedule σ before (after) the jobs J_{i-1}, J_i, J_{i+1}.

Since

$$g(\sigma') - g(\sigma) = (b_i - b_{i-1})\sum_{k=1}^{i-2}\prod_{j=k}^{i-2}(1 + b_j) + (b_{i-1} - b_i)\sum_{k=i+1}^{n}\prod_{j=i+1}^{k}(1 + b_j)$$

and

$$g(\sigma'') - g(\sigma) = (b_{i+1} - b_i) \sum_{k=1}^{i-1} \prod_{j=k}^{i-1} (1 + b_j) + (b_i - b_{i+1}) \sum_{k=i+2}^{n} \prod_{j=i+2}^{k} (1 + b_j),$$

it can be shown (see [10, Lemma 2]) that the two differences cannot be both positive. Hence, either σ' or σ'' are better schedules than schedule σ. □

Properties A.5 and A.7 imply the following result which describes the so-called *V-shape property* for problem $1|p_j = 1 + b_j t| \sum C_j$.

Theorem A.8. (Mosheiov [10]) *The optimal schedule for problem $1|p_j = 1 + b_j t| \sum C_j$ is V-shaped with respect to the deterioration rates b_j.*

Proof. The result is a consequence of Property A.5 and Property A.7. □

Remark A.9. V-shaped sequences were introduced in Definition 1.3; see also Definitions 19.11, 19.16, 19.18 and 20.13.

There is known a counterpart of Theorem A.8. Before we state the result, we introduce a new notion.

Definition A.10. *A deterioration rate b is said to be of type b_k, if $b = b_k$.*

Theorem A.11. (Gawiejnowicz et al. [6]) *If condition (A.2) is satisfied, then the optimal schedule for problem $1|p_j = 1 + b_j t| \sum C_j$ is V-shaped with respect to the types of jobs.*

Proof. By mathematical induction with respect to the number of types, k, and by Lemma A.6, the result follows. □

Theorem A.8 allows us to decrease the number of possible optimal schedules for problem $1|p_j = 1 + b_j t$ from $n!$ to $2^{n-3} - 1 = O(2^n)$. In some cases we can obtain a V-shaped sequence that is optimal.

Definition A.12. (Mosheiov [10]) *A sequence (x_j), $1 \leqslant j \leqslant n$, is said to be perfectly symmetric V-shaped (to have a perfect V-shape), if it is V-shaped and $x_i = x_{n-i+2}$, $2 \leqslant i \leqslant n$.*

The following result shows the importance of perfectly symmetric V-shaped sequences for problem $1|p_j = 1 + b_j t| \sum C_j$.

Theorem A.13. (Mosheiov [10]) *If a perfectly symmetric V-shaped sequence can be constructed from the sequence of deterioration rates of an instance of problem $1|p_j = 1 + b_j t| \sum C_j$, then the sequence is optimal.*

Proof. See Mosheiov [10, Proposition 3]. ◇

Though problem $1|p_j = 1 + b_j t| \sum C_j$ has been studied for more than twenty five years, during all those years the bound $O(2^n)$ was considered as the best possible one. Gawiejnowicz and Kurc [2] improved this bound. Let $\beta_j = 1 + b_j$ for all j, and let

$$\Delta_k(r, q) = \sum_{i=1}^{q-k-1} \prod_{j=i}^{q-k-1} \beta_j - \sum_{i=q-k+1}^{q-1} \prod_{j=q-k+1}^{i} \beta_j - \frac{1}{a_q} \sum_{i=q+1}^{n} \prod_{j=q-k+1}^{i} \beta_j$$

and

$$\nabla_k(r, q) = \frac{1}{a_r} \sum_{i=1}^{r-1} \prod_{j=i}^{r+k-1} \beta_j + \sum_{i=r+1}^{r+k-1} \prod_{j=i}^{r+k-1} \beta_j - \sum_{i=r+k+1}^{n} \prod_{j=r+k+1}^{i} \beta_j,$$

where $1 \leqslant r < q \leqslant n$ and $k = 1, 2, \ldots, q - r$.

Theorem A.14. (Gawiejnowicz and Kurc [2]) *Let $\beta = (\beta_1, \beta_2, \ldots, \beta_n)$ be an optimal schedule for problem $1|p_j = 1 + b_j t| \sum C_j$. Then (i) β is V-shaped, the minimal element in β is β_m, where $1 < m < n$, and (ii) the inequalities*

$$\Delta_1(m - 1, m + 1) = \sum_{j=1}^{m-1} \prod_{k=j}^{m-1} \beta_k - \sum_{i=m+2}^{n} \prod_{k=m+2}^{i} \beta_k \geqslant 0$$

and

$$\nabla_1(m - 1, m + 1) = \sum_{j=1}^{m-2} \prod_{k=j}^{m-2} \beta_k - \sum_{i=m+1}^{n} \prod_{k=m+1}^{i} \beta_k \leqslant 0$$

hold.

Proof. By direct computation. □

Theorem A.14 implies an improved bound on the number of possible optimal schedules for problem $1|p_j = 1 + b_j t| \sum C_j$. Given an instance of this problem, let $V_I(\beta)$ and $V_{II}(\beta)$ denote the sets of all schedules which satisfy the V-shape property and Theorem A.14, respectively. Let $1 < \beta_{\min} < \beta_{\max}$, where

$$\beta_{\min} := \min_{1 \leqslant j \leqslant n} \{\beta_j\} \tag{A.3}$$

and

$$\beta_{\max} := \max_{1 \leqslant j \leqslant n} \{\beta_j\}. \tag{A.4}$$

Let us denote

$$d_n := n \times \frac{\log \beta_{\min}}{\log \beta_{\min} + \log \beta_{\max}} \tag{A.5}$$

and

$$g_n := 1 + n \times \frac{\log \beta_{\max}}{\log \beta_{\min} + \log \beta_{\max}}. \tag{A.6}$$

The following result estimates the position m of the minimal element β_m in a V-shaped schedule β.

Theorem A.15. *Let* $\beta = (\beta_1, \ldots, \beta_m, \ldots, \beta_n)$ *be a V-shaped schedule for problem* $1|p_j = 1 + b_j t| \sum C_j$ *such that* β_m *is the minimal element in* β. *Then* (i) *if* $1 < \beta_{\min} < \beta_{\max}$ *are arbitrary, then* $d_n < m < g_n$; (ii) *if* $\beta_{\min} \to \beta_{\max}$, *then* $\frac{n}{2} \leqslant m \leqslant \frac{n}{2} + 1$; (iii) *if* $\beta_{\max} \to +\infty$, *then* $1 < m < n$.

Proof. By applying to definitions (A.5) and (A.6) the values of $\beta_{\min}$ and $\beta_{\max}$ defined by (A.3) and (A.4) and direct computation. □

Theorem A.15 implies that the set of possible values of m is contained in the open interval (d_n, g_n), where $d_n < g_n$. Let

$$D := \{k \in \mathbb{N} : d_n < k < g_n, 1 < k < n\}, \tag{A.7}$$

be the set of indices between d_n and g_n and let $V_D(\beta^\circ)$ denote the set of all V-shaped sequences $\beta = (\beta_1, \ldots, \beta_m, \ldots, a_n)$, where $m \in D$, which can be generated from a given initial sequence β°.

It is clear that

$$V_{II}(\beta^\circ) \subseteq V_D(\beta^\circ) \subseteq V_I(\beta^\circ)$$

and that

$$|V_{II}(\beta^\circ)| \leqslant |V_D(\beta^\circ)| \leqslant |V_I(\beta^\circ)|.$$

Hence, applying definitions (A.5), (A.6), (A.7) and an asymptotic formula for binomial coefficients (cf. Odlyzko [13]), we can estimate the cardinality of set $V_{II}(\beta^\circ)$ by the cardinality of set $V_D(\beta^\circ)$ which, in view of Theorem A.8, is estimated by $|V_I(\beta^\circ)| = 2^{n-1}$.

Theorem A.16. (Gawiejnowicz and Kurc [2]) *Let*

$$c(n) = \sqrt{\frac{2}{\pi n}} \, 2^n \left(1 + O\left(\frac{1}{n}\right)\right).$$

Then

$$|V_{II}(\beta)| \leqslant |V_D(\beta)| \leqslant \left(1 + \frac{\log \beta_{\max} - \log \beta_{\min}}{\log \beta_{\max} + \log \beta_{\min}} n\right) \times c(n)$$

and, if $\beta_{\max}$ *is sufficiently close to* $\beta_{\min}$,

$$|V_D(\beta)| \geqslant c(n).$$

Theorem A.16 decreases the bound $O(2^n)$ on the number of possible optimal schedules for problem $1|p_j = 1 + b_j t| \sum C_j$ by the factor $O(n^{-\frac{1}{2}})$.

A few suboptimal algorithms are known for problem $1|p_j = 1 + b_j t| \sum C_j$. In Chap. 15, we considered *greedy* Algorithms 15.1 and 15.2 for this problem. Both are based on the properties of functions $S^-(\beta)$ and $S^+(\beta)$ called *signatures* (cf. Gawiejnowicz et al. [4, 5]).

In Sect. 16.3.2, we discussed for this problem heuristic Algorithms 16.15 and 16.16 (cf. Mosheiov [10]), which iteratively try to generate a suboptimal V-shaped schedule with equally balanced sides.

Numerical experiments show that Algorithms 15.1 and 15.2 generate better near-optimal schedules than those generated by Algorithms 16.15 and 16.16.

We also suppose that Algorithm 15.2 generates optimal schedules for regular sequences, i.e., that the following conjecture holds.

Conjecture A.17. If $b_j = j + 1$, then problem $1|p_j = 1 + b_jt| \sum C_j$ is solvable in polynomial time.

Ocetkiewicz [12] proposed an exact algorithm for problem $1|p_j = a_j + b_jt| \sum C_j$, based on elimination of non-dominated partial schedules.

Matheuristics combine local search (cf. Chap. 17) and mathematical programming techniques (see Maniezzo et al. [9] for a review). Gawiejnowicz and Kurc [3] proposed two matheuristics for problem $1|p_j = 1 + b_jt| \sum C_j$. Introductory results of numerical experiments show that these algorithms generate very good suboptimal schedules for instances with $n \leqslant 12$ jobs.

Some authors proposed approximation schemes for special cases of problem $1|p_j = 1 + b_jt| \sum C_j$. For example, Ocetkiewicz [11] proposed an FPTAS for this problem when all job deterioration rates are not smaller than a constant $u > 0$. This approximation scheme runs in $O(n^{1+6\log_{1+u} 2}(\frac{1}{\epsilon})^{2\log_{1+u} 2})$ time, where $\epsilon > 0$ is a given accuracy.

Problem $1|p_j = b_jt| \sum T_j$

Another open single machine time-dependent scheduling problem is problem $1|p_j = b_jt| \sum T_j$. The time complexity of this problem is unknown, though its counterpart with fixed job processing times is $\mathcal{NP}$-hard in the ordinary sense (see, e.g., Koulamas [8] for a recent review). Because time-dependent scheduling problems with proportional jobs have similar properties as their counterparts with fixed job processing times (cf. Gawiejnowicz and Kononov [1], see also Sect. 19.6), we suppose that the following conjecture holds.

Conjecture A.18. Problem $1|p_j = b_jt| \sum T_j$ is $\mathcal{NP}$-hard in the strong sense.

Problem $1|p_j = b_jt| \sum T_j$ is a special case of problem $1|p_j = b_j(a+bt)| \sum T_j$ for $a = 0$ and $b = 1$. Hence, knowing the time complexity status of the former problem we would establish the status of the latter problem, too.

Problems $1|p_j = b_jt, prec| \sum w_jC_j$ and $1|p_j = a_j + b_jt, prec|C_{\max}$

In Chap. 18, we considered single machine time-dependent scheduling problems with precedence constraints of special forms such as chains, trees or serial-parallel graphs. All these problems are solvable in polynomial time. As opposed to this, we do not know the time complexity of single machine time-dependent scheduling problems with arbitrary precedence constraints such as

- $1|p_j = b_jt, prec| \sum w_jC_j$,

- $1|p_j = b_j(a + bt), prec| \sum w_j C_j$ and
- $1|p_j = a_j + b_j t, prec|C_{\max}$.

Because the first of the above problems is a special case of the second one, establishing the time complexity status of the former will clarify the status of the latter. Moreover, in view of results of Sect. 19.4 (see, e.g., Corollary 19.38), it seems that problems $1|p_j = b_j t, prec| \sum w_j C_j$ and $1|p_j = a_j + b_j t, prec|C_{\max}$ may be related. This means, again, that in order to establish the status of both these problems it may be sufficient to establish the status of only one of them.

We suppose that the following conjecture holds.

Conjecture A.19. (a) Problem $1|p_j = a_j + b_j t, prec|C_{\max}$ is $\mathcal{NP}$-hard in the strong sense. (b) Problem $1|p_j = b_j t, prec| \sum w_j C_j$ is $\mathcal{NP}$-hard in the strong sense.

A.2 Open parallel machine scheduling problems

In this section, we formulate open parallel machine time-dependent scheduling problems with non-linear job processing times.

Problems $Pm|p_j = a_j + f(t)|C_{\max}$ and $Pm|p_j = b_j f(t)| \sum C_j$

In Chap. 8, we considered polynomially solvable parallel machine time-dependent scheduling problems. An interesting open question is whether there exist polynomially solvable special cases of parallel machine time-dependent scheduling problems $Pm|p_j = a_j + f(t)|C_{\max}$ and $Pm|p_j = b_j f(t)| \sum C_j$, with not very strong assumptions on function $f(t)$.

There are some arguments which support the conjecture that such polynomially solvable special cases may exist. First, problem $Pm|p_j = a_j + f(t)|C_{\max}$ is a generalization of problem $1|p_j = a_j + f(t)|C_{\max}$ (cf. Melnikov and Shafransky [15], cf. Theorem 7.25), while problem $Pm|p_j = b_j f(t)|C_{\max}$ is a generalization of polynomially solvable problem $1|p_j = b_j f(t)|C_{\max}$ (Kononov [14], cf. Theorem 7.28), so the parallel machine problems may have similar properties (cf. Theorem 8.2). Second, the replacement of addition by multiplication in job processing times does not change the status of the time complexity of the single machine problem $1|p_j = a_j + f(t)|C_{\max}$, so maybe a similar situation takes place in case of multiple machines. Third, the function appearing in job processing times of polynomially solvable problems is either monotonic, convex or concave, hence it is possible that a function of a similar form may lead to polynomially solvable problems, too. Therefore, we suppose that the following conjecture holds.

Conjecture A.20. There exist polynomially solvable cases of problems $Pm|p_j = a_j + f(t)|C_{\max}$ and $Pm|p_j = b_j f(t)| \sum C_j$, provided that function $f(t)$ satisfies some conditions.

A.3 Open dedicated machine scheduling problems

In this section, we formulate open dedicated machine time-dependent scheduling problems with proportional and proportionally-linear job processing times.

Problems $F2|p_{ij} = b_{ij}t| \sum C_j$ and $O2|p_{ij} = b_{ij}t| \sum C_j$

There exist at least four dedicated machine time-dependent scheduling problems which are open:

- $F2|p_{ij} = b_{ij}t| \sum C_j$,
- $O2|p_{ij} = b_{ij}t| \sum C_j$,
- $F2|p_{ij} = b_{ij}(a + bt)| \sum C_j$ and
- $F2|p_{ij} = b_{ij}(a + bt)| \sum C_j$.

Results concerning isomorphic scheduling problems (cf. Gawiejnowicz and Kononov [1], see Sect. 19.6) show that time-dependent scheduling problems with proportional (or proportional-linear) job processing times and the $C_{\max}$ criterion have similar properties as their counterparts with fixed job processing times. Moreover, in case of problems mentioned above we deal with the $\sum C_j$ criterion, which leads to $\mathcal{NP}$-hard problems, even if the counterpart of a given time-dependent scheduling problem with fixed job processing times is polynomially solvable: problem $P2|p_j = b_j t| \sum C_j$ is $\mathcal{NP}$-hard (cf. Kononov [19], Mosheiov [20], though its counterpart with fixed job processing times, $P2|| \sum C_j$, is solvable in $O(n \log n)$ time (cf. Conway et al. [17]; see Theorem 11.1). Finally, the counterparts of the first two of these problems with fixed job processing times, $F2|| \sum C_j$ and $O2|| \sum C_j$, are both $\mathcal{NP}$-hard (cf. Garey et al. [18] and Achugbue and Chin [16], respectively). Hence, we suppose that the following conjecture holds.

Conjecture A.21. (a) Problem $F2|p_{ij} = b_{ij}t| \sum C_j$ is $\mathcal{NP}$-hard in the strong sense. (b) Problem $O2|p_{ij} = b_{ij}t| \sum C_j$ is $\mathcal{NP}$-hard in the strong sense.

Because the first two of the problems listed above are special cases of the next two problems, establishing the time complexity status of the former will clarify the status of the latter ones.

This section completes the book. The author hopes that it will contribute to increased interest in time-dependent scheduling, a very attractive and dynamically developing research domain of modern scheduling theory. If this hope becomes true, the main aim of this book will have been achieved.

References

Open single machine scheduling problems

1. S. Gawiejnowicz and A. Kononov, Isomorphic scheduling problems. *Annals of Operations Research* **213** (2014), no. 1, 131–145.

2. S. Gawiejnowicz and W. Kurc, A new necessary condition of optimality for a single machine scheduling problem with deteriorating jobs. *The 13th Workshop on Models and Algorithms for Planning and Scheduling Problems*, 2017, 177–179.

3. S. Gawiejnowicz and W. Kurc, Two matheuristics for problem $1|p_j = 1 + b_j t| \sum C_j$. *The 2nd International Workshop on Dynamic Scheduling Problems*, 2018, 45–50.

4. S. Gawiejnowicz, W. Kurc and L. Pankowska, A greedy approach for a time-dependent scheduling problem. *Lecture Notes in Computer Science* **2328** (2002), 79–86.

5. S. Gawiejnowicz, W. Kurc and L. Pankowska, Analysis of a time-dependent scheduling problem by signatures of deterioration rate sequences. *Discrete Applied Mathematics* **154** (2006), no. 15, 2150–2166.

6. S. Gawiejnowicz, T-C. Lai and M-H. Chiang, Polynomially solvable cases of scheduling deteriorating jobs to minimize total completion time. In: P. Brucker et al. (eds.), *Extended Abstracts of the 7th Workshop on Project Management and Scheduling*, 2000, 131-134.

7. M. Kubale and K. Ocetkiewicz, A new optimal algorithm for a time-dependent scheduling problem. *Control & Cybernetics* **38** (2009), no. 3, 713–721.

8. C. Koulamas, The single-machine total tardiness scheduling problem: review and extensions. *European Journal of Operational Research* **202** (2010), no. 1, 1–7.

9. V. Maniezzo, T. Stützle and S. Voß (eds.), *Matheuristics: Hybrydizing Metaheuristics and Mathematical Programming*. Berlin-Heidelberg: Springer 2010.

10. G. Mosheiov, V-shaped policies for scheduling deteriorating jobs. *Operations Research* **39** (1991), no. 6, 979–991.

11. K. M. Ocetkiewicz, A FPTAS for minimizing total completion time in a single machine time-dependent scheduling problem. *European Journal of Operational Research* **203** (2010), no. 2, 316–320.

12. K. M. Ocetkiewicz, Partial dominated schedules and minimizing the total completion time of deteriorating jobs. *Optimization* **62** (2013), no. 10, 1341–1356.

13. A. M. Odlyzko, Asymptotic enumeration methods. In: R. L. Graham, M. Groetschel and L. Lovasz (eds.), *Handbook of Combinatorics*, Vol. 2. Amsterdam: Elsevier 1995, pp. 1063–1229.

Open parallel machine scheduling problems

14. A. Kononov, Single machine scheduling problems with processing times proportional to an arbitrary function. *Discrete Analysis and Operations Research* **5** (1998), no. 3, 17–37 (in Russian).

15. O. I. Melnikov and Y. M. Shafransky, Parametric problem of scheduling theory. *Kibernetika* **6** (1979), no. 3, 53–57 (in Russian). English translation: *Cybernetics and System Analysis* **15** (1980), no. 3, 352–357.

Open dedicated machine scheduling problems

16. J. O. Achugbue and F. Y. Chin, Scheduling the open shop to minimize mean flow time. *SIAM Journal on Computing* **11** (1982), no. 4, 709–720.

17. R. W. Conway, W. L. Maxwell and L. W. Miller, *Theory of Scheduling*. Reading: Addison-Wesley 1967.

18. M. R. Garey, D. S. Johnson and R. Sethi, The complexity of flowshop and jobshop scheduling. *Mathematics of OR* **1** (1976), no. 2, 117–129.

19. A. Kononov, Scheduling problems with linear increasing processing times. In: U. Zimmermann et al. (eds.), *Operations Research 1996*. Berlin-Heidelberg: Springer 1997, pp. 208–212.

20. G. Mosheiov, Multi-machine scheduling with linear deterioration. *Infor* **36** (1998), no. 4, 205–214.

List of Algorithms

© Springer-Verlag GmbH Germany, part of Springer Nature 2020   517
S. Gawiejnowicz, *Models and Algorithms of Time-Dependent Scheduling*,
Monographs in Theoretical Computer Science. An EATCS Series,
https://doi.org/10.1007/978-3-662-59362-2

List of Figures

© Springer-Verlag GmbH Germany, part of Springer Nature 2020
S. Gawiejnowicz, *Models and Algorithms of Time-Dependent Scheduling*,
Monographs in Theoretical Computer Science. An EATCS Series,
https://doi.org/10.1007/978-3-662-59362-2

List of Tables

© Springer-Verlag GmbH Germany, part of Springer Nature 2020
S. Gawiejnowicz, *Models and Algorithms of Time-Dependent Scheduling*,
Monographs in Theoretical Computer Science. An EATCS Series,
https://doi.org/10.1007/978-3-662-59362-2

Symbol Index

© Springer-Verlag GmbH Germany, part of Springer Nature 2020
S. Gawiejnowicz, *Models and Algorithms of Time-Dependent Scheduling*,
Monographs in Theoretical Computer Science. An EATCS Series,
https://doi.org/10.1007/978-3-662-59362-2

Subject Index

ageing
  effect, 69
    log-linear, 69
  index, 69
agents
  compatible, 477
  incompatible, 477
agreeable
  basic job processing times and due
    dates, 71
  basic job processing times and
    weights, 72, 73
  parameters, 159, 160, 246
algorithm, 20
  c-competitive, 24
  r-approximation, 26
  ABC, 349
  ACO, 338, 348, 473
  AMP, 173
  ant colony optimization, 338
  approximation, 26, 29
  asymptotically optimal, 267
  branch-and-bound, 25, 29, 239,
    242–246, 248–250, 253–257, 470,
    473, 479, 481, 482
    axioms, 29
    bounding, 239, 240
    branching, 239
    child solution, 239
    dominance rules, 240
    dominated nodes, 240
    lower bound, 240
    non-dominated nodes, 240

competitive ratio, 24
dynamic programming, 25, 242, 244,
  252, 460, 461, 463, 477
EA, 337, 346, 347
efficient, 22
enumeration, 25, 140, 160, 175, 247,
  251
Euclid's, 21, 29
evolutionary, 28, 30
  evaluation, 337
  initialization, 337
  offspring population, 337
  postselection, 337
  preselection, 337
  simple, 337
exact, 24, 140, 242
  exponential, 29
exponential, 22, 23
exponential-time, 23
GA, 336, 337, 346
genetic, 472
GLS, 330, 331
Gonzalez–Sahni's, 180
greedy, 27, 29, 285, 510
HC, 332
heuristic, 27, 30, 301
II, 332, 333
IIRS, 332
input, 20
iterative improvement, 332, 340
Johnson's, 178, 427
Lawler's, 150, 322, 359, 361, 494
LDR, 265

© Springer-Verlag GmbH Germany, part of Springer Nature 2020
S. Gawiejnowicz, *Models and Algorithms of Time-Dependent Scheduling*,
Monographs in Theoretical Computer Science. An EATCS Series,
https://doi.org/10.1007/978-3-662-59362-2

Printed in the United States
by Baker & Taylor Publisher Services

Printed in the United States
by Baker & Taylor Publisher Services